AF412228

Control of Coupled Partial Differential Equations

Karl Kunisch
Günter Leugering
Jürgen Sprekels
Fredi Tröltzsch
Editors

Birkhäuser

Basel · Boston · Berlin

Editors:

Karl Kunisch
Institute for Mathematics
University of Graz
Heinrichstraße 36
A-8010 Graz
Austria

karl.kunisch@uni-graz.at

Jürgen Sprekels
Institute for Mathematics
Humboldt University Berlin
Unter den Linden 6
D-10099 Berlin
Germany

sprekels@wias-berlin.de

Günter Leugering
Institute for Applied Mathematics
University of Erlangen-Nürnberg
Martensstraße 3
D-91058 Erlangen
Gemany

leugering@am.uni-erlangen.de

Fredi Tröltzsch
Institute for Mathematics
Technical University Berlin
Straße des 17 Juni 136
D-10623 Berlin
Germany

troeltzsch@math.tu-berlin.de

2000 Mathematics Subject Classification: Primary 49-XX, 65K10, 93B40; Secondary 00B25, 26E25, 34H05, 35B40, 35D05, 35K45, 35L45, 35Q30, 35R30, 93B05.

Library of Congress Control Number: 2007922150

Bibliographic information published by Die Deutsche Bibliothek. Die Deutsche Bibliothek lists this publication in the Deutsche Nationalbibliografie; detailed bibliographic data is available in the Internet at http://dnb.ddb.de

ISBN 978-3-7643-7720-5 Birkhäuser Verlag AG, Basel - Boston - Berlin

© 2007 Birkhäuser Verlag AG
Basel · Boston · Berlin
P.O. Box 133, CH-4010 Basel, Switzerland
Part of Springer Science+Business Media
Printed on acid-free paper produced from chlorine-free pulp. TCF ∞
Printed in Germany

ISBN-10: 3-7643-7720-8
ISBN-13: 978-3-7643-7720-5
9 8 7 6 5 4 3 2 1

e-ISBN-10: 3-7643-7721-6
e-ISBN-13: 978-3-7643-7721-2
www.birkhauser.ch

Contents

Preface vii

K. Ammari, M. Tucsnak and G. Tenenbaum
   A Sharp Geometric Condition for the Boundary Exponential
   Stabilizability of a Square Plate by Moment Feedbacks only 1

V. Barbu, I. Lasiecka and R. Triggiani
   Local Exponential Stabilization Strategies of the
   Navier-Stokes Equations, $d = 2, 3$, via Feedback
   Stabilization of its Linearization 13

A. Gaevskaya, R.H.W. Hoppe, Y. Iliash and M. Kieweg
   Convergence Analysis of an Adaptive Finite Element Method
   for Distributed Control Problems with Control Constraints 47

M. Gugat
   Optimal Boundary Control in Flood Management 69

V. Heuveline and H. Nam-Dung
   On Two Numerical Approaches for the Boundary Control
   Stabilization of Semi-linear Parabolic Systems: A Comparison 95

M. Hintermüller, S. Volkwein and F. Diwoky
   Fast Solution Techniques in Constrained Optimal Boundary
   Control of the Semilinear Heat Equation 119

M. Hinze and U. Matthes
   Optimal and Model Predictive Control of the
   Boussinesq Approximation 149

K. Ito and K. Kunisch
   Applications of Semi-smooth Newton Methods to
   Variational Inequalities 175

B. Kaltenbacher
Identification of Nonlinear Coefficients in Hyperbolic PDEs,
with Application to Piezoelectricity 193

C. Meyer
An SQP Active Set Method for a Semilinear Optimal Control
Problem with Nonlocal Radiation Interface Conditions 217

P.I. Plotnikov and J. Sokolowski
Shape Optimization for Navier-Stokes Equations 249

J.-P. Raymond
A Family of Stabilization Problems for the Oseen Equations 269

G. Turinici
Beyond Bilinear Controllability: Applications to
Quantum Control 293

D. Wachsmuth
Optimal Control Problems with Convex Control Constraints 311

J.-P. Zolésio
Control of Moving Domains, Shape Stabilization and
Variational Tube Formulations 329

Preface

The international **Conference on Optimal Control of Coupled Systems of Partial Differential Equations** was held at the **Mathematisches Forschungsinstitut Oberwolfach** (www.mfo.de) from April, 17 to 23, 2005. The scientific program included 30 talks covering various topics as controllability, feedback-control, optimality systems, model-reduction techniques, analysis and optimal control of flow problems and fluid-structure interactions, as well as problems of shape and topology optimization. The applications discussed during the conference range from the optimization and control of quantum mechanical systems, the design of piezo-electric acoustic micro-mechanical devices, optimal control of crystal growth, the control of bodies immersed into a fluid to airfoil design and much more. Thus the applications are across all time and length scales.

Optimization and control of systems governed by partial differential equations and more recently by variational inequalities is a very active field of research in Applied Mathematics, in particular in numerical analysis, scientific computing and optimization. In order to able to handle real-world applications, scalable and parallelizable algorithms have to be designed, implemented and validated. This requires an in-depth understanding of both the theoretical properties and the numerical realization of such structural insights. Therefore, a 'core' development within the field of optimization with PDE-constraints such as the analysis of control-and-state-constrained problems, the role of obstacles, multi-phases etc. and an interdisciplinary 'diagonal' bridging regarding applications and numerical simulation are most important.

The aim of the conference, therefore, was to bring together applied mathematicians and also engineers in order to provide a state-of-the-art and to establish new standards in the field. It became apparent that the analysis of state-constrained nonlinear optimal control problems, of such problems governed by variational inequalities and the analysis of free boundary value problems are a key issues. Moreover, shape and topology optimization becomes critical in material sciences, light-weight materials, complex chambers and flexible structures. Shape-calculus in combination with top-level optimization algorithms and in particular the combination of topological and shape gradients are subject to analysis and simulation.

The editors express their gratitude to the contributors of this volume, the Oberwolfach Institute, and the Birkhäuser-Verlag for publishing this volume.

K. Kunisch, G. Leugering,
J. Sprekels and F. Tröltzsch

International Series of Numerical Mathematics, Vol. 155, 1–11

A Sharp Geometric Condition for the Boundary Exponential Stabilizability of a Square Plate by Moment Feedbacks only

K. Ammari, M. Tucsnak and G. Tenenbaum

Abstract. We consider a boundary stabilization problem for the plate equation in a square. The feedback law gives the bending moment on a part of the boundary as function of the velocity field of the plate. The main result of the paper asserts that the obtained closed loop system is exponentially stable if and only if the controlled part of the boundary contains a vertical and a horizontal part of non-zero length (the geometric optics condition introduced by Bardos, Lebeau and Rauch in [2] for the wave equation is thus not necessary in this case). The proof of the main result uses the methodology introduced in Ammari and Tucsnak [1], where the exponential stability for the closed loop problem is reduced to an observability estimate for the corresponding uncontrolled system combined to a boundedness property of the transfer function of the associated open loop system. The second essential ingredient of the proof is an observability inequality recently proved by Ramdani, Takahashi, Tenenbaum and Tucsnak [7]

Keywords. Boundary stabilization, Dirichlet type boundary feedback, plate equation.

1. Introduction and main results

In this work we study the boundary stabilization of a square Euler-Bernoulli plate by means of a feedback acting on the bending moment on a part of the boundary. Let us first describe the open loop control problem. Let $\Omega \subset \mathbb{R}^2$ be an open bounded set representing the domain occupied by the plate. We denote by $\partial\Omega$ the boundary of Ω and we assume that $\partial\Omega = \overline{\Gamma}_0 \cup \overline{\Gamma}_1$, where Γ_0, Γ_1 are open subsets of $\partial\Omega$ with $\Gamma_0 \cap \Gamma_1 = \emptyset$. The transverse displacement of the plate at the point x and at the moment t will be denoted by $w(x,t)$. We assume that $\partial\Omega$ is fixed, that the plate is simply supported on Γ_0 and that a bending moment (the control) is acting on Γ_1.

With the above notation, the system modelling the vibrations of the plate with boundary control acting on the moment can be written as

$$\ddot{w} + \Delta^2 w = 0, \qquad\qquad x \in \Omega,\ t > 0, \qquad (1.1)$$

$$w(x,t) = 0, \qquad\qquad x \in \partial\Omega,\ t > 0, \qquad (1.2)$$

$$\Delta w(x,t) = 0, \qquad\qquad x \in \Gamma_0,\ t > 0 \qquad (1.3)$$

$$\Delta w(x,t) = u(x,t), \qquad\qquad x \in \Gamma_1,\ t > 0 \qquad (1.4)$$

$$w(x,0) = w_0(x), \quad \dot{w}(x,0) = w_1(x), \qquad\qquad x \in \Omega, \qquad (1.5)$$

where we have denoted by a dot differentiation with respect to the time t and ν stands for the unit normal vector of $\partial\Omega$ pointing towards the exterior of Ω. It is known (see, for instance, Lasiecka and Triggiani [4]) that for any input function $u \in L^2_{\mathrm{loc}}(0,\infty; L^2(\Gamma_1))$ the system (1.1)–(1.5) admits a unique solution $w \in C([0,\infty); H^1_0(\Omega)) \cap C^1([0,\infty); H^{-1}(\Omega))$ (this result has been proved for any smooth domain Ω). The controllability of the dynamical system determined by (1.1)–(1.5) has been investigated in several works such as Krabs, Leugering and Seidman [3], Leugering [6], [4], Lebeau [5] and in [7]. In [4] the exact controllability has been established in the case when the control is active on the whole boundary whereas in [5] the controlled part of the boundary was supposed to satisfy the geometric optics condition of Bardos, Lebeau and Rauch. In [7] the exact controllability of the system (1.1)–(1.5) has been established under the assumption that Ω is a square and under a much weaker assumption on the controlled part of the boundary (Γ_1 is only supposed to contain non-empty vertical and horizontal subsets).

The main result of the paper concerns a system obtained by giving the input u in (1.9) as function of $\dot{w}$. More precisely, we consider the equations

$$\ddot{w} + \Delta^2 w = 0, \qquad\qquad x \in \Omega,\ t > 0, \qquad (1.6)$$

$$w(x,t) = 0, \qquad\qquad x \in \partial\Omega,\ t > 0, \qquad (1.7)$$

$$\Delta w(x,t) = 0, \qquad\qquad x \in \Gamma_0,\ t > 0 \qquad (1.8)$$

$$\Delta w(x,t) = -\frac{\partial}{\partial\nu}(G\dot{w}), \qquad\qquad x \in \Gamma_1,\ t > 0 \qquad (1.9)$$

$$w(x,0) = w_0(x), \quad \dot{w}(x,0) = w_1(x), \qquad\qquad x \in \Omega. \qquad (1.10)$$

The operator G in (1.9) is defined as A_0^{-1}, where $A_0 : H^1_0(\Omega) \rightarrow H^{-1}(\Omega)$ is defined by $A_0\varphi = -\Delta\varphi$ for all $\varphi \in H^1_0(\Omega)$. The system (1.6)–(1.10) is obtained from (1.1)–(1.5) by giving the control u in the feedback form

$$u(x,t) = -\frac{\partial}{\partial\nu}(G\dot{w}), \quad x \in \Gamma_1,\ t > 0.$$

This choice of the feedback law is the simplest one which makes the mapping $t \mapsto \|(w(\cdot,t),\dot{w}(\cdot,t))\|^2_{H^1_0 \times H^{-1}}$ decreasing. The concept of solution of (1.6)–(1.10) will be made precise in Section 3. In the same section we also give a proof of the following result.

Proposition 1.1. *Assume that $\Omega \subset \mathbb{R}^2$ is an open set with smooth boundary or that Ω is a square. Moreover, suppose that $w_0 \in H_0^1(\Omega)$ and that $w_1 \in H_0^{-1}(\Omega)$. Then the initial and boundary value problem (1.6)–(1.10) admits a unique solution. In other words (1.6)–(1.10) determine a well-posed linear dynamical system with state space $H_0^1(\Omega) \times H^{-1}(\Omega)$.*

The above result has been first proved in [4]. However, for the sake of completeness we give here the proof. Moreover, the notation introduced for this proof will be useful in the proof of our main result.

The exponential stability of (1.6)–(1.10) has been studied in [4] where it has been shown that the system is exponentially stable if $\Gamma_1 = \partial\Omega$. In this paper we show that if Ω is a square we only need a much smaller control region. More precisely, the main result of this paper is the following theorem:

Theorem 1.2. *Assume that Ω is a square in $\mathbb{R}^2$. Then the following assertions are equivalent:*

1. *The linear dynamical system determined by (1.6)–(1.10) is exponentially stable in $H_0^1(\Omega) \times H^{-1}(\Omega)$.*
2. *Γ_1 contains both a horizontal and a vertical segment of non-zero length.*

The paper is organized as follows. In Section 2 we recall some results on a class of dynamical systems. In Section 3 we prove Proposition 1.1 asserting that (1.6)–(1.10) determine a dynamical system and we show that this system fits into the framework introduced in Section 2. The last section is devoted to the proof of Theorem 1.2.

2. Some background on a class of dynamical systems

In this section we recall, following Ammari and Tucsnak [1], some results on a class of systems which appears naturally in mathematical models of vibrating systems with damping. Let H be a Hilbert space, and let $A_1 : \mathcal{D}(A_1) \to H$ be a self-adjoint, positive and boundedly invertible operator. We introduce the scale of Hilbert spaces H_α, $\alpha \in \mathbb{R}$, as follows: for every $\alpha \geq 0$, $H_\alpha = \mathcal{D}(A_1^\alpha)$, with the norm $\|z\|_\alpha = \|A_1^\alpha z\|_H$. The space $H_{-\alpha}$ is defined by duality with respect to the pivot space H as follows: $H_{-\alpha} = H_\alpha^*$ for $\alpha > 0$. The operator A_1 can be extended (or restricted) to each H_α, such that it becomes a bounded operator

$$A_1 : H_\alpha \to H_{\alpha-1} \qquad \forall\, \alpha \in \mathbb{R}. \tag{2.11}$$

The second ingredient needed for our construction is a bounded linear operator $B_1 : U \to H_{-\frac{1}{2}}$, where U is another Hilbert space which will be identified with its dual.

The systems we consider are described by

$$\ddot{w}(t) + A_1 w(t) + B_1 B_1^* \dot{w}(t) = 0, \tag{2.12}$$

$$w(0) = w_0, \quad \dot{w}(0) = w_1, \tag{2.13}$$

4   K. Ammari, M. Tucsnak and G. Tenenbaum

where $t \in [0, \infty)$ is the time. The equation (2.12) is understood as an equation in $H_{-\frac{1}{2}}$, i.e., all the terms are in $H_{-\frac{1}{2}}$. Most of the linear equations modelling the damped vibrations of elastic structures can be written in the form (2.12), where w stands for the displacement field and the term $B_1 B_1^* \dot{w}(t)$, represents a viscous feedback damping. The system (2.12)–(2.13) is well posed. More precisely, the following classical result (see, for instance, Weiss and Tucsnak [8]) holds.

Proposition 2.1. *Suppose that $(w_0, w_1) \in \mathcal{D}(A_1^{\frac{1}{2}}) \times H$. Then the problem (2.12)–(2.13) admits a unique solution*

$$w \in C([0, \infty); \mathcal{D}(A_1^{\frac{1}{2}})) \cap C^1([0, \infty); H)$$

such that $B_1^* w(\cdot) \in H^1(0, T; U)$. Moreover w satisfies, for all $t \geq 0$, the energy estimate

$$\|(w_0, w_1)\|^2_{\mathcal{D}(A_1^{\frac{1}{2}}) \times H} - \|(w(t), \dot{w}(t))\|^2_{\mathcal{D}(A_1^{\frac{1}{2}}) \times H} = \int_0^t \left\| \frac{\mathrm{d}}{\mathrm{d}t} B_1^* w(s) \right\|^2_U \mathrm{d}s. \quad (2.14)$$

From (2.14) it follows that the mapping $t \mapsto \|(w(t), \dot{w}(t))\|^2_{\mathcal{D}(A_1^{\frac{1}{2}}) \times H}$ is non-increasing. In many applications it is important to know if this mapping decays exponentially when $t \to \infty$, i.e., if the system (2.12)–(2.13) is exponentially stable. One of the methods currently used for proving such exponential stability results is based on an observability inequality for the undamped system associated to (2.12)–(2.13). More precisely, consider the initial value problem

$$\ddot{\phi}(t) + A_1 \phi(t) = 0, \quad (2.15)$$

$$\phi(0) = w_0, \ \dot{\phi}(0) = w_1. \quad (2.16)$$

It is well known that (2.15)–(2.16) is well posed in $\mathcal{D}(A_1) \times \mathcal{D}(A_1^{\frac{1}{2}})$ and in $\mathcal{D}(A_1^{\frac{1}{2}}) \times H$. The result below, proved in [1], shows that, under a certain regularity assumption, the exponential stability of (2.12)–(2.13) is equivalent to an observability inequality for (2.15)–(2.16). More precisely, we have:

Theorem 2.2. *Assume that for any $\gamma > 0$ we have*

$$\sup_{\mathrm{Re}\lambda = \gamma} \left\| \lambda B_1^* (\lambda^2 I + A_1)^{-1} B_1 \right\|_{\mathcal{L}(U)} < \infty. \quad (2.17)$$

Then the system described by (2.12)–(2.13) is exponentially stable in $\mathcal{D}(A_1^{\frac{1}{2}}) \times H$ if and only if there exists $T > 0, C > 0$ such that

$$\int_0^T \|B_1^* \dot{\phi}(t)\|^2_U \, \mathrm{d}t \geq C \, \|(w_0, w_1)\|^2_{\mathcal{D}(A_1^{\frac{1}{2}}) \times H} \qquad \forall \, (w_0, w_1) \in \mathcal{D}(A_1) \times \mathcal{D}(A_1^{\frac{1}{2}}).$$

$$(2.18)$$

3. Proof of Proposition 1.1

Notation. Throughout this section, Ω, Γ_0 and Γ_1 are as in the first section. We denote by H the Sobolev space $\mathcal{H}^{-1}(\Omega)$ and

$$\mathcal{D}(A_1) = \left\{ \varphi \in \mathcal{H}_0^1(\Omega) \mid \Delta\varphi \in \mathcal{H}_0^1(\Omega) \right\}.$$

The operator $A_1 : \mathcal{D}(A_1) \to H$ is defined by $A_1\varphi = \Delta^2\varphi$ for all $\varphi \in \mathcal{D}(A_1)$. This operator is clearly the square of the operator Λ_0 in the introduction, thus it is clearly self-adjoint, bounded by below and the space $H_{\frac{1}{2}} = \mathcal{D}(A_1^{\frac{1}{2}})$ is given by $H_{\frac{1}{2}} = \mathcal{H}_0^1(\Omega)$. The input space $L^2(\Gamma_1)$ is denoted by U.

In this section we show that the system (1.6)–(1.10) can be written in the form (2.12)–(2.13), by an appropriate choice of the spaces and operators. Then, by using Proposition 2.1 we obtain a wellposedness result for (1.6)–(1.10).

We first define the concept of a weak solution of (1.6)–(1.10).

Definition 3.1. For $w_0 \in H_0^1(\Omega)$, $w_1 \in H^{-1}(\Omega)$, a function

$$w \in C([0,\infty); H_0^1(\Omega) \cap C^1([0,\infty), H^{-1}(\Omega))$$

is called a *weak solution of* (1.6)–(1.10) if $\frac{\partial}{\partial\nu}(G\dot{w}) \in L^2_{\mathrm{loc}}(0,\infty; L^2(\Gamma_1))$ and for all $t \geq 0$ and for all $\eta \in \mathcal{D}(A_0)$ we have

$$\langle \dot{w}(\cdot,t), \eta \rangle_{H^{-1},H_0^1} - \langle w_1, \eta \rangle_{H^{-1},H_0^1}$$

$$+ \int_0^t \int_\Omega \Delta w(x,s)\Delta\overline{\eta(x)}\,\mathrm{d}x\,\mathrm{d}s + \int_0^t \int_{\Gamma_1} \frac{\partial}{\partial\nu}(G\dot{w})\frac{\partial\overline{\eta}}{\partial\nu}(x)\,\mathrm{d}\Gamma\,\mathrm{d}s = 0. \qquad (3.19)$$

$$w(x,0) = w_0(x) \qquad \forall\, x \in \Omega. \qquad (3.20)$$

In order to show the existence and uniqueness of the weak solutions of the system (1.6)–(1.10), we will need the following simple result, a direct consequence of the Riesz representation theorem in $L^2(\Omega)$:

Proposition 3.2. *For every* $\mathrm{v} \in L^2(\Gamma)$, *there exists a unique function* $R\mathrm{v} \in \mathcal{H}^2(\Omega) \cap \mathcal{H}_0^1(\Omega)$ *such that*

$$\int_\Omega \Delta(R\mathrm{v})(x)\overline{\psi(x)}\,\mathrm{d}x = -\int_\Gamma \mathrm{v}\,\overline{\frac{\partial(G\psi)}{\partial\nu}}\,\mathrm{d}\Gamma \qquad \forall\,\psi \in L^2(\Omega). \qquad (3.21)$$

Moreover, the operator R defined above is linear and bounded from $L^2(\Gamma)$ into $L^2(\Omega)$.

Remark 3.3. The above defined mapping R has the following property: if we assume that v is continuous on $\partial\Omega$ and $R\mathrm{v} \in C^4(\Omega) \cap C^2(\overline{\Omega})$ then $R\mathrm{v}$ satisfies the conditions

$$\Delta^2(R\mathrm{v}) = 0 \quad \text{in} \quad \Omega,$$

$$\mathrm{v} = 0 \quad \text{on} \quad \partial\Omega.$$

$$\Delta(R\mathrm{v}) = \mathrm{v} \quad \text{on} \quad \partial\Omega.$$

This fact can be easily checked by using Green's formula.

6   K. Ammari, M. Tucsnak and G. Tenenbaum

Let us introduce the operator $B_1 \in \mathcal{L}(U, H_{-\frac{1}{2}})$ defined by

$$B_1 v = A_1 R v \qquad \forall\, v \in U, \tag{3.22}$$

where A_1 is considered as an operator from $H_{\frac{1}{2}}$ to $H_{-\frac{1}{2}}$ and $R : U \to L^2(\Omega)$ is the mapping defined in Proposition 3.2.

Proposition 3.4. *The adjoint of the operator B_1 defined in (3.22) is the operator $C_1 \in \mathcal{L}(H_{\frac{1}{2}}, U)$ defined by*

$$C_1 \phi = B_1^* \phi = \frac{\partial (G\phi)}{\partial \nu}\Big|\Gamma_1 \qquad \forall\, \phi \in H_{\frac{1}{2}} = \mathcal{H}_0^1(\Omega). \tag{3.23}$$

Proof. Let $\phi \in \mathcal{H}^2(\Omega) \cap \mathcal{H}_0^1(\Omega) \subset H_1$ and $u \in L^2(\Gamma_1)$. Then we have

$$\langle B_1 u, \phi \rangle_{H_{-\frac{1}{2}}, H_{\frac{1}{2}}} = \langle Ru, A_1 \phi \rangle_H = \langle Ru, \phi \rangle_{H_{\frac{1}{2}}},$$

which, by using the fact that $H_{\frac{1}{2}} = \mathcal{H}_0^1(\Omega)$ and by applying Green's formula implies that

$$\langle B_1 u, \phi \rangle_{H_{-\frac{1}{2}}, H_{\frac{1}{2}}} = -\int_\Omega \Delta(Ru)\phi\, \mathrm{d}x.$$

By using the density of $\mathcal{H}^2(\Omega) \cap \mathcal{H}_0^1(\Omega)$ in $\mathcal{H}_0^1(\Omega)$ and (3.21), it follows that, for every $\phi \in \mathcal{H}_0^1(\Omega)$ and for every $u \in L^2(\Gamma_1)$,

$$\langle B_1 u, \phi \rangle_{H_{-\frac{1}{2}}, H_{\frac{1}{2}}} = \int_{\Gamma_1} u\, \overline{\frac{\partial (G\phi)}{\partial \nu}}\, \mathrm{d}\Gamma.$$

We conclude that the adjoint of B_1 (using the pivot space $H = \mathcal{H}^{-1}(\Omega)$) is given by (3.23). $\qquad \square$

Proof of Proposition 1.1. The essential part of the proof consists in showing that a function $w : \Omega \times [0, \infty) \to \mathbb{C}$ is a weak solution of (1.6)–(1.10) if and only if the mapping $t \to w(\cdot, t)$ is a solution in $H_{-\frac{1}{2}}$ of

$$\ddot{w}(t) + A_1 w(t) + B_1 B_1^* \dot{w}(t) = 0, \tag{3.24}$$

$$w(0) = w_0, \quad \dot{w}(0) = w_1. \tag{3.25}$$

Since G maps $H_{\frac{1}{2}}$ onto H_1, for $\eta \in H_1$ there exists a unique $\varphi \in H_{\frac{1}{2}} = \mathcal{H}_0^1(\Omega)$ such that $\eta = G\varphi$. Thus relation (3.19) is equivalent to the fact that, for all $\varphi \in \mathcal{H}_0^1(\Omega)$, we have

$$\int_\Omega \dot{w}(x, t)\overline{G\varphi(x)}\, \mathrm{d}x - \int_\Omega w_1(x)\overline{G\varphi(x)}\, \mathrm{d}x$$

$$-\int_0^t \int_\Omega \Delta w(x, s)\overline{\varphi(x)}\, \mathrm{d}x\, \mathrm{d}s + \int_{\Gamma_1} \frac{\partial (G\dot{w})}{\partial \nu} \overline{\frac{\partial (G\varphi)}{\partial \nu}}\, \mathrm{d}\Gamma = 0. \tag{3.26}$$

Since $\dot{w} = -\Delta(G\dot{w})$, a simple application of Green's formula yields that

$$\int_\Omega \dot{w}(x, t)\overline{G\varphi(x)}\, \mathrm{d}x = \int_\Omega \nabla(G\dot{w}) \cdot \nabla(\overline{G\varphi})\, \mathrm{d}x.$$

The above relation and the definition of the inner product in H imply that

$$\int_\Omega \dot{w}(x,t)\overline{G\varphi(x)}\,\mathrm{d}x = \langle \dot{w},\varphi \rangle. \tag{3.27}$$

Similarly we can show that

$$\int_\Omega w_1(x)\overline{G\varphi(x)}\,\mathrm{d}x = \langle w_1,\varphi \rangle. \tag{3.28}$$

On the other hand

$$\int_\Omega \Delta w\overline{\varphi}\,\mathrm{d}x = \int_\Omega \nabla(Gw)\cdot\nabla(\Delta\overline{\varphi})\,\mathrm{d}x = \int_\Omega \nabla(Gw)\cdot\nabla(GA_1\overline{\varphi})\,\mathrm{d}x. \tag{3.29}$$

Consequently, we have

$$\int_\Omega w\overline{\varphi}\,\mathrm{d}x = \langle w, A_1\varphi \rangle \qquad \forall\,\varphi \in \mathcal{H}_0^1(\Omega).$$

By using (3.26)–(3.29), combined to Proposition 3.4 we obtain that w is a weak solution of (1.6)–(1.10) iff w satisfies (3.20) and, for all $\varphi \in H_1$ we have:

$$\langle \dot{w}(t) - v_0, \varphi \rangle + \int_0^t \langle w(\sigma), A_1\varphi \rangle\,\mathrm{d}\sigma = \int_0^t \langle B_1 B_1^* \dot{w}(\sigma), \varphi \rangle\,\mathrm{d}\sigma \qquad \forall\,t \in [0,\infty).$$

The above relation, combined to the fact that A_1 is self-adjoint, clearly imply that w is a weak solution of (1.6)–(1.10) if and only if the mapping $t \to w(\cdot,t)$ is a solution of (3.24)–(3.25) in $H_{-\frac{1}{2}}$.

The existence and uniqueness of a weak solution of (1.6)–(1.10) follows now from Proposition 2.1. $\qquad\square$

4. Proof of the main result

An important ingredient of the proof of Theorem 1.2 is the following technical result:

Lemma 4.1. *For every $\gamma > 0$ we have*

$$\sup_{\mathrm{Re}\lambda=\gamma,\,m\in\mathbb{N}} \sum_{n\geq 1} \left| \frac{\lambda}{\lambda^2 + (n^2+m^2)^2} \right| < \infty. \tag{4.1}$$

Proof. Let λ be a complex number with real part equal to $\gamma > 0$ and denote by β the imaginary part of λ. We set

$$S_m = \sum_{n\geq 1} \left| \frac{\lambda}{\lambda^2 + (n^2+m^2)^2} \right| \qquad \forall\,m \in \mathbb{N}.$$

We have

$$S_m \leq \sqrt{2} \sum_{n\geq 1} \frac{|\lambda|}{|\beta| + |(n^2+m^2)^2 + \gamma^2 - \beta^2|} \qquad \forall\,m \in \mathbb{N}. \tag{4.2}$$

By symmetry, it suffices to find an upper bound for the right-hand side of (4.2) which is valid for every $\beta \geq 0$.

We distinguish three cases.

Case 1. Assume that $0 \leq \beta \leq \frac{\gamma}{2}$. Then (4.2) implies that:

$$S_m \leq \frac{\gamma\sqrt{10}}{2} \sum_{n\geq 1} \frac{1}{\beta + \left|(n^2+m^2)^2 + \frac{3\gamma^2}{4}\right|} \qquad \forall\, m \in \mathbb{N}.$$

Consequently, for all $m \in \mathbb{N}$ we have

$$S_m \leq C_{1\gamma},$$

where

$$C_{1\gamma} = \frac{\gamma\sqrt{10}}{2} \sum_{n\geq 1} \frac{1}{\frac{3\gamma^2}{4} + n^4}\,.$$

Case 2. Assume that $\frac{\gamma}{2} < \beta \leq 2\gamma$. In this case (4.2) implies that:

$$S_n m \leq \gamma\sqrt{20} \sum_{1\leq n \leq 2\gamma} \frac{1}{\gamma} + \gamma\sqrt{10} \sum_{n>2\gamma} \frac{1}{\beta + |n^4 - 3\gamma^2|} \qquad \forall\, m \in \mathbb{N}.$$

Thus, in this case, for all $m \in \mathbb{N}$ we have

$$S_m \leq C_{2\gamma},$$

where

$$C_{2\gamma} = 2\gamma\sqrt{20} + \gamma\sqrt{10} \sum_{n>2\gamma} \frac{1}{n^4 - 3\gamma^2}\,.$$

Case 3. Assume that $\beta > 2\gamma$ and denote $\beta_\gamma := \sqrt{\beta^2 - \gamma^2}$. Then clearly $\frac{1}{2}\beta \leq \beta_\gamma \leq \beta$ and (4.2) implies that for all $m \in \mathbb{N}$ we have

$$S_m \leq 2 \sum_{n\geq 1} \frac{\beta}{\beta + (m^2 + n^2 + \beta)|m^2 + n^2 - \beta_\gamma|} \leq 2 \sum_{n\geq 1} \frac{1}{1 + |m^2 + n^2 - \beta_\gamma|}\,. \quad (4.3)$$

For $k, m \geq 1$ we next estimate the number of integers n, such that the inequality

$$2^{k-1} < m^2 + n^2 - \beta_\gamma \leq 2^k, \qquad (4.4)$$

holds true. Assume that

$$2^{k-1} + \beta_\gamma - m^2 > 0. \qquad (4.5)$$

In this case (4.4) implies that

$$\sqrt{2^{k-1} + \beta_\gamma - m^2} < n \leq \sqrt{2^k + \beta_\gamma - m^2}.$$

Consequently, (4.4) is satisfied for at most $\sqrt{2^k + \beta_\gamma - m^2} - \sqrt{2^{k-1} + \beta_\gamma - m^2} + 1$ integers n. Since

$$\sqrt{2^k + \beta_\gamma - m^2} - \sqrt{2^{k-1} + \beta_\gamma - m^2}$$
$$= \frac{2^{k-1}}{\sqrt{2^k + \beta_\gamma - m^2} + \sqrt{2^{k-1} + \beta_\gamma - m^2}} \leq 2^{\frac{k-1}{2}},$$

we have that, under the assumption (4.5), the inequality (4.4) is satisfied for at most $2^{\frac{k-1}{2}} + 1$ integers m. If we assume that

$$2^{k-1} + \beta_\gamma - m^2 \leq 0,$$

then (4.4) implies that

$$1 \leq n \leq 2^{\frac{k}{2}}.$$

We have thus shown that the inequality (4.4) is satisfied for at most $2^{\frac{k+2}{2}} + 1$ integers n. It can be shown in a completely similar way that for any $k, m \geq 1$ there exist at most $2^{\frac{k+2}{2}} + 1$ integers n such that

$$2^{k-1} < \beta_\gamma - m^2 - n^2 \leq 2^k.$$

Consequently, the inequality $2^{k-1} < |m^2 + n^2 - \beta_\gamma| \leq 2^k$ holds for at most $2^{\frac{k+4}{2}} + 2$ integers n. Moreover, the inequality $|m^2 + n^2 - \beta_\gamma| \leq 1$ holds for at most three integers n. Thus, we have

$$S_m \leq C_{3\gamma} \qquad \forall\, m \in \mathbb{N},$$

where

$$C_{3\gamma} = 6 + \sum_{k \geq 1} 2^{\frac{6-k}{2}}.$$

Consequently, for all $m \in \mathbb{N}$ and for all $\lambda \in \mathbb{C}$ with $\mathrm{Re}\,\lambda = \gamma > 0$, we have

$$S_m \leq \max\{C_{1\gamma}, C_{2\gamma}, C_{3\gamma}\},$$

which implies the conclusion (4.1). $\qquad\square$

Proof of Theorem 1.2. We have seen in the proof of Proposition 1.1 that (1.6)–(1.10) can be written in the form (3.24), (3.25) with A_1 and B_1 defined in Section 3. In order to apply Proposition 2.2 we need to estimate the norm in $\mathcal{L}(U)$ of the operator

$$H(\lambda) = \lambda B_1^*(\lambda^2 I + A_1)^{-1} B_1,$$

for $\lambda \in \mathbb{C}$ with $\mathrm{Re}\,\lambda = \gamma > 0$. This will be done by using the eigenvalues and the eigenvectors of A_1. It is easy to see that the eigenvalues of A_1 are

$$\mu_{m,n} = (m^2 + n^2)^2, \qquad \forall\, m, n \in \mathbb{N}^*.$$

A corresponding family of normalized (in $H = H^{-1}(\Omega)$), eigenfunctions of A_1 are

$$\Phi_{m,n}(x_1, x_2) = \frac{2\sqrt{m^2 + n^2}}{\pi} \sin(mx_1)\sin(nx_2) \qquad \forall\, m, n \in \mathbb{N}^*.$$

For $v \in U$ denote $\psi_\lambda = (\lambda^2 I + A_1)^{-1} B_1 v$. If we consider the decomposition

$$\psi_\lambda = \sum_{m,n \geq 1} a_{m,n}(\lambda) \Phi_{m,n}, \tag{4.6}$$

then we see that

$$a_{m,n}(\lambda) = \frac{\langle B_1 v, \Phi_{m,n}\rangle_{H_{-\frac{1}{2}}, H_{\frac{1}{2}}}}{\lambda^2 + (m^2 + n^2)^2} = \frac{\langle v, B_1^* \Phi_{m,n}\rangle_{L^2(\Gamma_1)}}{\lambda^2 + (m^2 + n^2)^2} \qquad \forall\, m,\, n \geq 1. \tag{4.7}$$

Notice that, by symmetry considerations, it suffices to consider the case when Γ_1 is one of the sides of the square. Therefore, in the remaining part of the proof we assume that $\Gamma_1 = (0, \pi) \times \{0\}$, which implies that $U = L^2(0, \pi)$. In this case, by using Proposition 3.4 we see that

$$B_1^* \Phi_{m,n}(x_1) = \frac{2n}{\pi \sqrt{m^2 + n^2}} \sin\left(m x_1\right) \qquad \forall\, m, n \geq 1. \tag{4.8}$$

If we consider the Fourier series of v, denoted by

$$v(x_1) = \sum_{k \geq 1} b_k \sin\left(k x_1\right),$$

then (4.7) and (4.8) imply that

$$a_{m,n} = \frac{n b_m}{\sqrt{m^2 + n^2}[\lambda^2 + (m^2 + n^2)^2]} \qquad \forall\, m,\, n \geq 1.$$

The above relation, combined to (4.6) and to (4.8) imply that, for almost all $x_1 \in (0, \pi)$, we have

$$[H(\lambda)v](x_1) = [\lambda B_1^* \psi_\lambda](x_1) = \lambda \sum_{m,n \geq 1} \frac{n^2 b_m}{(m^2 + n^2)[\lambda^2 + (m^2 + n^2)^2]} \sin\left(m x_1\right). \tag{4.9}$$

On the other hand, if $\mathrm{Re}\lambda = \gamma > 0$ then Lemma 4.1 implies the existence of a constant $K_\gamma > 0$ such that

$$\left| \lambda \sum_{n \geq 1} \frac{n^2 b_m}{(m^2 + n^2)[\lambda^2 + (m^2 + n^2)^2]} \right| \leq K_\gamma |b_m| \qquad \forall\, m \geq 1.$$

The above relation and (4.9) imply that

$$\|H(\lambda)\|_{\mathcal{L}(U)} \leq \sqrt{\frac{\pi}{2}} K_\gamma,$$

for all $\lambda \in \mathbb{C}$ with $\mathrm{Re}\lambda = \gamma$. Consequently, the pair (A_1, B_1) satisfies the assumption (2.17) in Theorem 2.2. This fact, combined with the result in [7], saying that (A_1, B_1) satisfies condition (2.18) in Theorem 2.2 if and only if Γ_1 contains both a horizontal and a vertical segment of non-zero length, yields the conclusion of the theorem.

References

[1] K. AMMARI AND M. TUCSNAK, *Stabilization of second order evolution equations by a class of unbounded feedbacks*, ESAIM COCV, **6** (2001), 361–386.

[2] C. BARDOS, G. LEBEAU AND J. RAUCH, *Sharp sufficient conditions for the observation, control and stabilization of waves from the boundary*, SIAM J. Control. Optim., **30** (1992), 1024–1065.

[3] W. KRABS, G. LEUGERING AND T. SEIDMAN, *On boundary controllability of a vibrating plate*, Appl. Math. Optim., **13** (1985), 205–229.

[4] I. LASIECKA AND R. TRIGGIANI, *Exact controllability and uniform stabilization of Euler-Bernoulli equations with boundary control only in* $\Delta w|_\Sigma$, Boll. Un. Mat. Ital. B, **7** (1991), 665–702.

[5] G. LEBEAU, *Contrôle de l'équation de Schrödinger*, Journal de Mathématiques Pures et Appliquées, **71**(1992), 267–291.

[6] G. LEUGERING, *Boundary control of a vibrating plate*, in Optimal control of partial differential equations (Oberwolfach, 1982), vol. 68 of Internat. Schriftenreihe Numer. Math., Birkhäuser, Basel, 1984, 167–172.

[7] K. RAMDANI, T. TAKAHASHI, G. TENENBAUM AND M. TUCSNAK, *A spectral approach for the exact observability of infinite dimensional systems with skew-adjoint generator*, J. Funct. Anal., **226** (2005), 193–229.

[8] G. WEISS AND M. TUCSNAK, *How to get a conservative well-posed linear system out of thin air. I. Well-posedness and energy balance*, ESAIM Control Optim. Calc. Var., 9 (2003), pp. 247–274.

K. Ammari
Département de Mathématiques
Faculté des Sciences de Monastir
5019 Monastir, Tunisie
e-mail: `kais.ammari@fsm.rnu.tn`

M. Tucsnak and G. Tenenbaum
Institut Élie Cartan
Département de Mathématiques
Université de Nancy I
F-54506 Vandoeuvre lès Nancy Cedex
e-mail: `tucsnak@iecn.u-nancy.fr`
e-mail: `Gerald.Tenenbaum@iecn.u-nancy.fr`

International Series of Numerical Mathematics, Vol. 155, 13–46
© 2007 Birkhäuser Verlag Basel/Switzerland

Local Exponential Stabilization Strategies of the Navier-Stokes Equations, $d = 2, 3$, via Feedback Stabilization of its Linearization

Viorel Barbu, Irena Lasiecka and Roberto Triggiani

Abstract. We review recent results on the boundary and interior feedback stabilization of Navier-Stokes equations, $d = 2, 3$, and provide new ones.

Mathematics Subject Classification (2000). Primary 35B40, 35Q30; Secondary 76D05, 7655.

Keywords. Navier-Stokes equations, boundary and interior feedback stabilization, optimal control, Riccati equation, steady-state solution.

0. Orientation

Opening question: What axiomatic properties, and in what topological setting, depending on the dimension $d = 2, 3$, need to be satisfied by the *linearized* Navier-Stokes equations, in order to guarantee local exponential stabilization of the (*full nonlinear*) Navier-Stokes equations, in the vicinity of a steady-state solution, by means of the same boundary feedback mechanism, which has proved successful in the linearized dynamics?

The core of this question was the motivating strategy behind the approaches successfully pursued in achieving local exponential stabilization, near an equilibrium solution, of the Navier-Stokes equations in the authors' recent work: first, by means of a *localized interior feedback control, $d = 2, 3$,* [B.1], [B-T.1]; and next, more challengingly, by means of a *tangential boundary feedback control, $d = 2, 3$,* [B-L-T.1], [B-L-T.2]. In these works, implementation of this strategy was based on a basic well-known fact in optimal control theory: that the linear quadratic control problem not only identifies, in feedback form, the optimal solution of the corresponding minimization problem; but, moreover, it yields that such optimal

Research partially supported by the National Science Foundation under Grant DMS-0104305, and by the Army Research Office under Grant DAAD19-02-1-0179.

feedback solution has the additional attractive bonus of being uniformly, exponentially stable. That is, the optimization problem forces dissipation. This tenet, therefore, prompts the following guiding directive, first implemented in [La.3] (in the case of nonlinear hyperbolic problems): in order to inject dissipation as to force local exponential stabilization of the steady-state solutions, an Optimal Control Problem (OCP) with a quadratic cost functional over an infinite time horizon is introduced for the linearized N-S equations. As a result, the same Riccati-based, optimal boundary feedback controller which is obtained in the *linearized* OCP is then selected and implemented also in the full N-S system. Implementation of this known idea depends on how 'rough' the nonlinear term in the full nonlinear dynamics is; and, consequently, what is the topological setting that the nonlinear term imposes and dictates on the problem.

In the case of Navier-Stokes equations, this strategy has been successfully implemented first to obtain stabilizing, Riccati-based, *localized interior* feedback controls, $d = 2, 3$ [B.1], [B-T.1]; and, subsequently, with the critical use of [B-T.1], to obtain stabilizing, Riccati-based, tangential boundary controls, $d = 2, 3$ [B-L-T.1], [B-L-T.2]. In all cases, the class of initial conditions, topologically depending on the dimension $d = 2, 3$, is tangential on the boundary. Paper [B-L-T.1] was mainly devoted to the more challenging case $d = 3$, as well as its supporting, preliminary background: the establishment of an optimal regularity theory of the mixed (initial, boundary value) problem for the linearized N-S equations and corresponding adjoint problem; the derivation of the abstract model for the mixed linearized N-S equations, in particular of the control operator "B" and its corresponding (trace operator) adjoint "B^*"; open-loop well-posedness of the optimization (linear quadratic) problem for the linearized N-S equations, at the high topological level that is forced and imposed by the nonlinear term for $d = 3$; corresponding linear quadratic and algebraic Riccati theory, etc. The more amenable case $d = 2$ may be made to fall into standard OCP and Riccati theory as in [La.1], [La.2], [L-T.1]–[L-T.4], and is therefore relegated to [B-L-T.1, Appendix B].

Having at hand a definite, Riccati-based solution of the local exponential stabilization problem for Navier-Stokes equations with either localized interior [B-T.1] or else tangential boundary feedback controls [B-L-T.1], the next phase of investigation was to address the opening question in its full force and generality. Are different feedback controls, other than Riccati-based, also possible to likewise obtain local exponential feedback stabilization of the full nonlinear N-S equations in a vicinity of an equilibrium solution? Attractive alternatives include: spectral-based feedback controls; or feedback controls induced by approximating numerical schemes. In this broader setting, the issue of local feedback stabilization of N-S equations for $d = 2, 3$, was then revisited in [B-L-T.2]. Here, guided by the concrete solutions in [B-T.1] and [B-L-T.1] for $d = 2, 3$, and [R.1] for $d = 2$, the key essential ingredients occurring in the treatment of the special Riccati-based solution presented in these references were singled out. They were then elevated to two abstract settings – each involving *exclusively* well-posedness and stability properties of the *linearized* N-S problem – having, as a desired distinguishing feature, the capability to imply and guarantee local, tangential feedback stabilization of the equilibrium solution for the *full N-S problem*. They are labeled Setting #1 and Setting #2, the second being contained in the first. In short: this is an illustration of the principle of "local feedback stabilization of the nonlinear dynamics by virtue of global feedback stabilization of its linearized part." The correct topological setting is critical, however.

In this paper, we present the topic of N-S feedback stabilization by following the more general setting of [B-L-T.2]. This way, we shall present not only the original, Riccati-based, local feedback stabilization results of Navier-Stokes equations of the aforementioned references, but also the additional, new, spectral-based counterparts, as cast within the aforementioned abstract setting, by verifying the corresponding abstract assumptions. To be sure, the abstract assumptions will not define the problem away; rather, it is plain that actual verification of these abstract assumptions relies critically on the technical analysis and results of the original work [B-L-T.1]. Within the limited space, our goal is therefore to espouse the guiding ideas behind the principle of "local feedback stabilization of the nonlinear system by global feedback stabilization of its linearized part," while deferring most of the proofs to the original sources. There will be two exceptions: Section 4 and Section 5. In Section 4, we shall re-obtain the main local exponential feedback stabilization result of [B-T.1] by means of a localized interior *Riccati-based* feedback controls, as an illustration of both the abstract Setting #1 and the abstract Setting #2, by verifying the corresponding abstract assumptions. This phase will rest critically on the technical analysis and result of [B-T.1], of course. Finally, in Section 5, we shall go one step further and present a new result as an application, again, of abstract Setting #1 and abstract Setting #2: a corresponding local exponential feedback stabilization of N-S equations, $d = 2, 3$, by means of, this time, *spectral-based* localized interior feedback controls.

High-gain versus low-gain feedback stabilizing controllers. Cumulatively, [B-L-T.1] and [B-L-T.2] yield local feedback tangential boundary feedback stabilization results for $d = 2, 3$, either with high-gain, Riccati-based feedback controllers, or else with low-gain, Riccati-based or spectral-based feedback controllers. More precisely: for $d = 3$, high-gain, Riccati-based feedback controllers ([B-L-T.1] and [B-L-T.2, Section 2.1]), while for $d = 2$, either high-gain, Riccati-based feedback controllers ([B-L-T.1] and [B-L-T.2, Section 2.2]), or else low-gain, Riccati-based ([B-L-T.1, Appendix B], [B-L-T.2, Section 3.1]), or spectral-based ([B-L-T.2, Section 3.2]) feedback controllers. Paper [R.1] deals with low-gain, Riccati-based feedback controllers for $d = 2$: it served as a further motivation for us to revisit the problem. What are then the advantages and disadvantages of high-gain versus low-gain implementation of feedback controllers?

The main advantage of the low-gain, Riccati-based feedback controller (case $d = 2$) is the validity of Theorem 3.1.1 below, which is taken from [B-L-T.1, Propositio B.4.1], [B-L-T.2]. It contains, in particular, the following features: (i) the feedback operator is bounded; (ii) the s.c. feedback semigroup is analytic;

(iii) the Algebraic Riccati Equation (ARE) is simple and transparent on H: see Eqn. (3.1.3) below, which is valid for all testing functions $y, z \in H$; hence, as an operator equation. In contrast, the high-gain, Riccati-based feedback controller (case $d = 3$) offers a mixed balance. To begin with, it has an unbounded feedback. On the one hand, it yields the distinct advantage that the feedback generator A_R is *dissipative* with respect to the inner product $(\,\cdot\,, R\,\cdot\,)_H$, R the Riccati operator unbounded on H [B-L-T.1, Proposition 4.5.1, Eqn. (4.5.1)], which is *equivalent* to the inner product of the state space $W \subset H$ [the $|x|_W$-norm is equivalent to the $|R^{\frac{1}{2}}x|_H$-norm, $x \in W$, see below: (2.1.4) for $d = 3$ and (2.2.11) for $d = 2$]. The nonlinear dynamics is accordingly *dissipative* on a desirable high-topological level state space W with high coercivity ("$\frac{3}{2} + \epsilon$" derivative in space and L_2 in time). As a consequence, feedback controllers with high gain are expected to be "robust," with respect to a larger class of perturbations. This feature is, in particular, expected to play an important role in the corresponding numerical construction of feedback stabilizers. (We intend to provide details thereof in a subsequent analysis elsewhere.)

On the other hand, this approach (for $d = 3$) inherits the disadvantage that the ARE is valid for all testing function in the domain $\mathcal{D}(A_R^2)$ [B-L-T.1, Eqn. (4.5.1)].

We further remark that the low-gain setting and high-gain setting both contribute extra regularity of the linearized dynamics, but for different reasons. In the low-gain setting, extra regularity of the linearized dynamics is obtained due to the analyticity of the feedback semigroup; while in the high-gain setting, it is the unbounded observation that contributes extra regularity of the linearized dynamics.

It may not always be possible to have both options available, e.g., $d = 3$ in the boundary case.

The case $d = 2$. A main claim that we wish to make at the outset is this: that, for $d = 2$, the optimal control theory and related Riccati theory of the optimal control problem in Section 3.1, Eqn. (3.1.1), at the low- (that is, $H-$, hence $(L_2(\Omega))^2$-) topological level, with identity on H as observation, can be essentially borrowed from long-established (late 70's–mid 80's [La.1–2], [L-T.1–2]) parabolic optimal control theory, as reported in book-form in [L-T.4, Chapter 2], once the following two main ingredients are in place:

(i) Optimal regularity theory on H-based spaces of the linearized N-S (tangential) *mixed* problem (1.1.20), or its abstract version (1.1.21); that is, under the action of a non-homogeneous tangential boundary term (control) in the Dirichlet (no-slip) boundary conditions. This theory, which generalizes to N-S problems the mixed theory for *classical* parabolic problems as in [L-M.1, Vol. 2] (to cover even a critical range of the interpolating parameter not included in [L-M.1, Vol. 2] as pointed out in [B-L-T.1, Remark 3.1.1]) is given in [B-L-T.1, Section 3.1, Theorems 3.1.4 through 3.1.8].

(ii) The abstract models (1.1.21) (linearized N-S problem) and (1.2.8) (full, nonlinear N-S problem), established in [B-L-T.1, Section 3.1, Section 5] for the corresponding PDE-versions (1.2.20) and (1.1.12) with tangential controls. These are then complemented by the critical trace result (1.1.19): $D_k^* A_k^* = -\nu_0 \frac{\partial}{\partial \nu}$ on $\mathcal{D}(A^*)$, which is established in [B-L-T.1, Proposition 3.2.1].

The validity of this latter paragraph is verified in the statements of Theorem 3.1.1 and Proposition 3.1.2: see, in particular, the preliminaries, as well as the actual proof of Proposition 3.1.2 given in [B-L-T.2, Section 6].

1. Introduction; setting; main results

1.1. Setting of the problem; goal

Boundary controlled Navier-Stokes equations. We consider the controlled Navier-Stokes equations (see [C-F.1, p. 45], [Te.1, p. 253] for the uncontrolled case $u \equiv 0$) with boundary control u in the Dirichlet (no-slip) B.C.:

$$\begin{cases} y_t(x,t) - \nu_0 \Delta y(x,t) + (y \cdot \nabla)y(x,t) &= f_e(x) + \nabla p_1(x,t) & \text{in } Q; & (1.1.1\text{a}) \\ \nabla \cdot y &= 0 & \text{in } Q; & (1.1.1\text{b}) \\ y &= u & \text{on } \Sigma; & (1.1.1\text{c}) \\ y(x,0) &= y_0(x) & \text{in } \Omega. & (1.1.1\text{d}) \end{cases}$$

Here, $Q \equiv \Omega \times (0,\infty)$; $\Sigma = \partial\Omega \times (0,\infty)$ and Ω is an open smooth bounded domain of R^d, $d = 2,3$; $u \in L^2(0,T;(L^2(\partial\Omega))^d)$ is the boundary control input; and $y = (y_1, y_2, \ldots, y_d)$ is the state (velocity) of the system. The constant $\nu_0 > 0$ is the viscosity coefficient. The functions $y_0, f_e \in (L^2(\Omega))^d$ are given, the latter being a body force, while $-p_1$ is the unknown pressure. The boundary $\partial\Omega$ is assumed to be of class C^2.

Steady-state solutions. Let $(y_e, p_e) \in ((H^2(\Omega))^d \cap V) \times H^1(\Omega)$ be a steady-state (equilibrium) solution to equations (1.1.1), i.e.,

$$\begin{cases} -\nu_0 \Delta y_e + (y_e \cdot \nabla)y_e &= f_e + \nabla p_e & \text{in } \Omega; & (1.1.2\text{a}) \\ \nabla \cdot y_e &= 0 & \text{in } \Omega; & (1.1.2\text{b}) \\ y_e &= 0 & \text{on } \partial\Omega. & (1.1.2\text{c}) \end{cases}$$

A steady-state solution is known to exist for $d = 2,3$ [C-F.1, Theorem 7.3, p. 59]. Here, [C-F.1, p. 9], [Te.1, p. 18],

$$V = \{y \in (H_0^1(\Omega))^d; \ \nabla \cdot y = 0\} \text{ with norm } \|y\|_V \equiv \|y\| = \left\{ \int_\Omega |\nabla y(x)|^2 d\Omega \right\}^{\frac{1}{2}}.$$

$$(1.1.3)$$

For large Reynolds number $\frac{1}{\nu_0}$, the steady-state (stationary) solutions y_e are unstable and cause turbulence in their surroundings. Let y_e be one such unstable steady-state solution.

Ambient state space; boundary control space. Throughout the paper, we shall make reference to the classical space [C-F.1, p. 7], [Te.1, p. 15]:

$$H \equiv \{y \in (L^2(\Omega))^d; \ \nabla \cdot y = 0; \ y \cdot \nu = 0 \text{ on } \partial\Omega\}, \qquad (1.1.4)$$

so that the following orthogonal decomposition holds true:

$$(L^2(\Omega))^d = H + H^\perp, \ H^\perp \equiv \{y \in (L^2(\Omega))^d : \ y = \text{grad } p, \ p \in H^1(\Omega)\}. \quad (1.1.5)$$

Here, ν is the unit outward normal to the boundary $\partial\Omega$ of Ω. Because of the divergence theorem, we have: $\int_\Omega \nabla \cdot y \, d\Omega = \int_\Gamma y \cdot \nu \, d\Gamma$, $\Gamma = \partial\Omega$. Thus, by (1.1.1b–c), we must require, at least, the integral boundary compatibility condition $\int_\Gamma u \cdot \nu \, d\Gamma = 0$ on the boundary control function. Actually, in order for the velocity y to fall into the space H in (1.1.4), a more stringent condition will be imposed on u. Indeed, throughout this paper, we choose the control space $\mathcal{U}$ defined by

$$\mathcal{U} \equiv L^2(0, \infty; U); \quad U = \{\mu \in (L^2(\Gamma))^d : \ \mu \cdot \nu \equiv 0 \text{ a.e.}\}. \qquad (1.1.6)$$

Thus, in line with [B-L-T.1], we restrict to the class of boundary controls with purely *tangential* boundary action, at each point of $\partial\Omega \equiv \Gamma$.

Goal. Our goal is to identify a more specific and natural space $W \subset H$ (possibly depending on the dimension d), and a linear feedback operator

$$F : \ W \supset \mathcal{D}(F) \to U \subset (L^2(\Gamma))^d, \qquad (1.1.7)$$

densely defined in W, such that: If the I.C. $y_0 \in \mathcal{V}_\rho$ for $\rho > 0$ sufficiently small, where

$$\mathcal{V}_\rho = \{y_0 \in W : \ |y_0 - y_e|_W < \rho\}, \qquad (1.1.8)$$

and if the control u is given by the following feedback tangential control law

$$u = F(y - y_e), \quad \text{ on } \Sigma, \qquad (1.1.9)$$

then the resulting N-S problem (1.1.1), obtained from inserting (1.1.9) into (1.1.1c) [i.e., with (1.1.1c) replaced by $y = F(y - y_e)$ on Σ], possesses the following two features:

  (i) it is semigroup well-posed on W;

 (ii) it exponentially stabilizes the flow of (1.1.1), (1.1.9) in W: There exist constants $M \geq 1$ and $\omega > 0$, independent of $\rho > 0$ such that its solution y satisfies

$$|y(t) - y_e|_W \leq Me^{-\omega t}|y_0 - y_e|_W, \ t \geq 0. \qquad (1.1.10)$$

Translated nonlinear N-S problem: PDE and abstract versions. It is natural, as in [B-L-T.1], to consider the nonlinear problem (1.1.1) around the equilibrium solution y_e. Thus, we introduce the new variables

$$\eta = y - y_e; \qquad p \equiv p_1 - p_e \qquad (1.1.11)$$

for velocity and pressure. Then, by substitution of (1.1.11) in (1.1.1), we obtain the PDE version of the *translated nonlinear N-S problem* in η and p:

$$\begin{cases} \eta_t - \nu_0 \Delta\eta + (\eta \cdot \nabla)\eta + (y_e \cdot \nabla)\eta + (\eta \cdot \nabla)y_e = \nabla p & \text{in } Q; & (1.1.12\text{a}) \\ \nabla \cdot \eta \equiv 0 & \text{in } Q; & (1.1.12\text{b}) \\ \eta \equiv u & \text{on } \Sigma; & (1.1.12\text{c}) \\ \eta_0(x) = y_0(x) - y_e(x) & \text{in } \Omega. & (1.1.12\text{d}) \end{cases}$$

It is shown in [B-L-T.1, Section 3.1 and Appendix A.2] that the *abstract model of the N-S problem* (1.1.12) *projected on the space H* defined by (1.1.4) is given by

$$\eta_t - \mathcal{A}\eta + B\eta = -\mathcal{A}Du \in [\mathcal{D}(\mathcal{A}^*)]', \ \eta_0 \in H, \ u \cdot \nu \equiv 0 \text{ on } \Sigma. \qquad (1.1.13)$$

Here: (i) the operator $\mathcal{A}$ in (1.1.13) is the *extension by transposition* $\mathcal{A} : H \to [\mathcal{D}(\mathcal{A}^*)]'$, duality with respect to H as a pivot space, of the original (differential) Oseen operator

$$\mathcal{A} = -(\nu_0 A + A_0), \ \mathcal{D}(\mathcal{A}) = \mathcal{D}(A) = (H^2(\Omega))^d \cap V \to H; \qquad (1.1.14)$$

$$Av = -P\Delta v, \ \forall \, v \in \mathcal{D}(A); \ A_0 v = P((y_e \cdot \nabla)v + (v \cdot \nabla)y_e); \ \mathcal{D}(A_0) = V \equiv \mathcal{D}(A^{\frac{1}{2}}), \qquad (1.1.15)$$

where $P : (L^2(\Omega))^d \to H$ is the Leray projector [C-F.1, p. 9], which is orthogonal on $(L^2(\Omega))^d$. Thus, applying P to Eqn. (1.1.12a) eliminates the pressure term on H by virtue of (1.1.5); moreover, $P\eta_t = \eta_t$, as $\eta \cdot \nu \equiv u \cdot \nu \equiv 0$ on Σ, as imposed in (1.1.6). See [B-L-T.1, Appendix A.1] for the required extension of P outside the space $(L^2(\Omega))^d$. The following results are well known. The operator $-\nu_0 A$ ($\nu_0 > 0$, the viscosity coefficient) is negative self-adjoint and has compact resolvent on H. Thus, $-\nu_0 A$ generates a s.c. analytic (self-adjoint) semigroup on H. Moreover, the perturbed operator $\mathcal{A}$ in (1.1.14) likewise has *compact resolvent* and *generates a s.c. analytic semigroup on H*. Finally, the operator $\mathcal{A}$ has a finite number N of eigenvalues λ_j with Re $\lambda_j \geq 0$ (the unstable eigenvalues):

$$\text{Re } \lambda_{N+1} < 0 \leq \text{Re } \lambda_N \leq \cdots \leq \text{Re } \lambda_1. \qquad (1.1.16)$$

The eigenvalues are repeated according to their algebraic multiplicity ℓ_j. Finally, $\ell_1 + \ell_2 + \cdots + \ell_M = N$, where M denotes the number of *distinct* unstable eigenvalues of $\mathcal{A}$.

(ii) The operator $B : V \to V'$ in (1.1.13) is defined by [C-F.1, p. 54], [Te.1, p. 162]:

$$Bv = P[(v \cdot \nabla)v], \quad B \in \mathcal{L}(V; V'). \qquad (1.1.17)$$

(iii) The definition of the 'Dirichlet map' D in (1.1.13) is more delicate and is given as follows [B-L-T.1, Section 3.1, complemented by Appendix A.2]. Essentially, we may begin by introducing the Dirichlet map D_k which solves the static Oseen problem translated by a suitable constant k, due to a tangential boundary datum g on Γ, $g \cdot \nu = 0$ on Γ. There is no essential loss of generality to take $k = 0$, and write D for $D_{k=0}$, as in (1.1.13).

20   V. Barbu, I. Lasiecka and R. Triggiani

Two features are important:

(i) [B-L-T.1, Eqn. (3.1.3)]

$$D : \text{continuous } (H^s(\Gamma))^d \to (H^{s+\frac{1}{2}}(\Omega))^d \cap H, \quad s \geq 0; \tag{1.1.18}$$

(ii) [B-L-T.1, Lemma 3.3.1]

$$D^* \mathcal{A}^* \varphi = -\nu_0 \frac{\partial \varphi}{\partial \nu}, \qquad \varphi \in \mathcal{D}(\mathcal{A}^*) = \mathcal{D}(\mathcal{A}). \tag{1.1.19}$$

Linearization of translated N-S problem: PDE and abstract versions. The linearized version of the translated nonlinear PDE η-problem (1.1.12) is

$$\begin{cases} v_t - \nu_0 \Delta v + (y_e \cdot \nabla)v + (v \cdot \nabla)y_e = \nabla p & \text{in } Q; & (1.1.20a) \\ \nabla \cdot v \equiv 0 & \text{in } Q; & (1.1.20b) \\ v = u & \text{in } \Sigma; & (1.1.20c) \\ v_0(x) = y_0(x) - y_e(x) & \text{in } \Omega. & (1.1.20d) \end{cases}$$

Its abstract version on H is then, from (1.1.13),

$$v_t = \mathcal{A}(v - Du), \qquad v(0) = v_0. \tag{1.1.21}$$

1.2. Two abstract settings and statement of corresponding main, local, exponential, feedback stabilization results of N-S systems [B-L-T.2]

In this section, we shall introduce two abstract settings (two abstract sets of assumptions) on the *linearization*

$$v_t(t) = \mathcal{A}(I - DF)v(t), \qquad v(0) = v_0 \in W \tag{1.2.1}$$

in feedback form of the translated N-S problem. Eqn. (1.2.1) is obtained from (1.1.21) by using the feedback control $u = Fv$, $v = y - y_e$, see (1.1.9), where $W \subset H$ is a natural, specialized state space, which along with the corresponding output space Z will be specified below. In line with [B-L-T.1], throughout this paper we shall take two sets of (state, output)-spaces, depending on the dimension. More precisely, let $\epsilon_0 > 0$ be arbitrarily small and fixed once and for all; throughout this paper, we shall consider the following two selections [B-L-T.1], where $\hat{\mathcal{A}} \equiv \omega I - \mathcal{A}$, for a sufficiently large $\omega > 0$, fixed once and for all, so that the s.c. analytic semigroup $e^{-\hat{\mathcal{A}}t}$ on H is, moreover, exponentially stable and the fractional powers $\hat{\mathcal{A}}^\theta$, $0 < \theta < 1$, of $\hat{\mathcal{A}}$ are well defined. They are:

$$\text{for } d = 2: W \equiv (H^{\frac{1}{2}-\epsilon_0}(\Omega))^d \cap H \equiv \mathcal{D}(\hat{\mathcal{A}}^{\frac{1}{4}-\frac{\epsilon_0}{2}}); \quad Z \equiv (H^{\frac{3}{2}-\epsilon_0}(\Omega))^d \cap H; \tag{1.2.1a}$$

$$-\hat{\mathcal{A}} \equiv \mathcal{A} - \omega I = -\nu_0 A - (A_0 + \omega I); \tag{1.2.1b}$$

$$\text{for } d = 3: W \equiv (H^{\frac{1}{2}+\epsilon_0}(\Omega))^d \cap H; \ Z \equiv (H^{\frac{3}{2}+\epsilon_0}(\Omega))^d \cap H. \tag{1.2.2}$$

See [B-L-T.1, Eqns. (1.13)–(1.17)] and [W.1] for coincidence between Sobolev spaces and domain of fractional powers. In addition, prompted and inspired by [B-L-T.1], we shall introduce two distinct sets of abstract settings:

Setting #1 (Unbounded feedback): Assumption (H.1). Here, we assume the existence of a linear (feedback) operator F

$$F: W \supset \mathcal{D}(F) \to U, \tag{1.2.3}$$

densely defined on W (see (1.2.2), (1.2.3) for W; (1.1.6) for U), such that the following three properties hold true:

(H.1i) the feedback operator of (1.2.1)

$$A_F \equiv \mathcal{A}(I - DF): W \supset \mathcal{D}(A_F) \to W; \tag{1.2.4a}$$

$$\mathcal{D}(A_F) \equiv \{w \in W: \mathcal{A}(w - DFw) \in W\}, \tag{1.2.4b}$$

generates a s.c. semigroup $e^{A_F t}$ on W: $v(t; v_0) = e^{A_F t} v_0$, $v_0 \in W$.;

(H.1ii) the semigroup $e^{A_F t}$ is uniformly (exponentially) stable on W: there exist constants $C \geq 1$ and $\delta > 0$ such that

$$|e^{A_F t}|_{\mathcal{L}(W)} \leq C e^{-\delta t}, \quad t \geq 0; \tag{1.2.5}$$

(H.1iii) for each $w \in W$, we have $e^{A_F t} w \in L^2(0, \infty; Z)$; thus for some positive constant c

$$\int_0^\infty |e^{A_F t} w|_Z^2 \, dt \leq c |w|_W^2, \quad \forall \, w \in W. \tag{1.2.6}$$

Main result for Setting #1. Under Assumption (H.1) for the linearized v-problem (1.2.1) in feedback form, the main result of the present paper is the following local exponential stabilization result for the nonlinear problem (1.1.1).

Theorem 1.2.1. *Assume hypothesis (H.1(i), (ii), (iii)) on the linearized v-problem (1.2.1). Then, the following local exponential stabilization on W holds true for the nonlinear translated N-S η-problem (1.1.12) or (1.1.13) in feedback form $u = F\eta$: for each $\eta_0 \in W$ with $|\eta_0|_W \leq \rho$ for ρ sufficiently small, the corresponding N-S problem (1.1.13), which is obtained with u given in feedback form as $u = F\eta$ (as in (1.1.9)), thus resulting via (1.2.5) in*

$$\eta_t + B\eta = \mathcal{A}(\eta - DF\eta) = A_F \eta, \quad \eta(0) = \eta_0 \in W, \tag{1.2.7}$$

is well posed on W and satisfies the following regularity properties:

$$\eta \in C([0, \infty]; W) \cap L^2(0, \infty; Z). \tag{1.2.8a}$$

Moreover, $\eta(t)$ satisfies the following local exponential decay

$$|\eta(t)|_W \leq M e^{-\omega t} |\eta_0|_W, \quad t \geq 0; \quad |\eta_0|_W \leq \rho, \tag{1.2.8b}$$

for constants $M \geq 1$, $\omega > 0$, independent of $\rho > 0$.

Accordingly, the original nonlinear N-S problem (1.1.1) with u given in feedback form as $u = F(y - y_e)$, as in (1.1.9), satisfies the statement in the aforementioned goal: For all $y_0 \in \mathcal{V}_\rho$ in (1.1.8), with $\rho > 0$ sufficiently small, the corresponding problem (1.1.1), (1.1.9) is well posed on W, and satisfies (1.2.9a–b) with $\eta(t) = y(t) - y_e$, in particular the exponential decay (1.1.10).

The proof of Theorem 1.2.1 is given in [B-L-T.2, Section 5].

Comments on Theorem 1.2.1. Theorem 1.2.1 allows for the construction of a stabilizing feedback for 2- and 3-d Navier-Stokes flows, provided that a corresponding feedback is available for the linearization. It can indeed be applied to both cases $d = 2$ and $d = 3$, see subsequent sections. However, the main issue that in the application of Theorem 1.2.1 differentiates between the two- and three-dimensional cases is the construction of the feedback linear operator F with the properties postulated in assumption (H.1). The level of difficulty in the construction of such operator F greatly depends on the dimension d. This is clearly expected, if one refers to the pair of (state, output)-spaces W, Z defined in (1.2.2) for $d = 2$, and in (1.2.3) for $d = 3$. Indeed, for $d = 3$, the spaces W and Z in (1.2.3) recognize, respectively, Dirichlet and Dirichlet and Neumann B.C., unlike the case $d = 2$ in (1.2.2), where W does not recognize any B.C., while Z does not recognize the Neumann B.C. When B.C. are recognized, compatibility conditions of the data on the boundary must be accounted for.

As a consequence, the theory is no longer "perturbation-based" and instead requires more challenging means to achieve stabilization of the linearized v-dynamics (1.2.1), of which optimization is a critical one. The corresponding results are given in [B-L-T.1].

We also note that [B-L-T.1, Appendix B] provides an open-loop infinite-dimensional (or even finite-dimensional, under the Finite-Dimensional Spectral Assumption [FDSA] recalled in Section 1.3) tangential boundary controller $u \in L^2(0, \infty; (L^2(\Gamma_1))^d)$, Γ_1 an arbitrary subset of $\Gamma = \partial\Omega$, meas $\Gamma_1 > 0$, yielding $\eta \in L^2(0, \infty; (H^{\frac{3}{2}-\epsilon_0}(\Omega))^d \cap H)$, for $\eta_0 \in (H^{\frac{1}{2}-\epsilon_0}(\Omega))^d \cap H$, in both cases $d = 2, 3$. Thus, the resulting Riccati theory (or the explicit finite-dimensional construction in [B-L-T.1, Appendix B2] under the FDSA) do provide a feedback operator A_F satisfying properties (H.1(i), (ii), (iii)) with $W \equiv H^{\frac{1}{2}-\epsilon_0}(\Omega) \cap H$, $Z \equiv (H^{\frac{3}{2}-\epsilon_0}(\Omega))^d \cap H$. However, then, only the case $d = 2$ can be carried over to obtain, in Theorem 1.2.1, the local exponential stabilization, by the same feedback operator, of the full nonlinear N-S η-problem (1.1.12), or the original y-problem (1.1.1) in the vicinity of a stationary solution. This is so since a key point in the proof in [B-L-T.2, Section 5] of the nonlinear stabilization result, Theorem 1.2.1, is estimate [B-L-T.2, Eqn. (5.21)] on the nonlinear term $|Bz|_W \leq k|z|_Z^2$. And this is established precisely for the pair of {state space, output space} $= \{W, Z\}$ with W and Z defined by (1.2.2a) for $d = 2$; and (1.2.3) for $d = 3$.

Setting #2 (Bounded feedback): Assumption (H.2). Here, we assume at the outset that the feedback operator F is *bounded*, as an operator from H to U (recall (1.1.4) and (1.1.6)):

(H.2i) $$F \in \mathcal{L}(H; U), \tag{1.2.9}$$

so that the s.c. *analytic* semigroup $e^{A_F t}$ on H, $t > 0$, generated by

$$\begin{cases} A_F &= \mathcal{A}(I - DF) : H \supset \mathcal{D}(A_F) \to H; & (1.2.10a) \\ \mathcal{D}(A_F) &= \{h \in H : h - DFh \in \mathcal{D}(\mathcal{A})\}, & (1.2.10b) \end{cases}$$

satisfies the following three additional properties:

(H.2ii) $e^{A_F t}$ is uniformly (exponentially) stable on H: there exist constants C, $\delta > 0$, such that

$$|e^{A_F t}|_{\mathcal{L}(H)} \leq Ce^{-\delta t}, \quad t > 0; \tag{1.2.11}$$

(H.2iii) for each $w \in W$, we have $e^{A_F t}w \in L^2(0,1;Z)$ and then for some positive constant $c > 0$,

$$\int_0^1 |e^{A_F t}w|_Z^2\, dt \leq c|w|_W^2, \quad \forall\, w \in W; \tag{1.2.12}$$

(H.2iv) moreover, there is a positive number $r > 0$ such that

$$\mathcal{D}((-A_F)^r) \subset Z. \tag{1.2.13}$$

Comments on Setting #2: Assumption (H.2(i)–(iv)).

(a) Setting #2 will imply Setting #1 for $d = 2$; in other words, the implication

$$\text{assumption (H.2)} \Rightarrow \text{assumption (H.1)} \tag{1.2.14}$$

holds true with $W \equiv (H^{\frac{1}{2}-\epsilon_0}(\Omega))^d \cap H$, and Z the space postulated in (1.2.13), (1.2.14). Thus, this result applies to the case $d = 2$ in (1.2.2). This is the content of Theorem 1.2.2 and is shown in [B-L-T.2, Section 4].

(b) Assumption (H.2) (involving only finite time regularity) may be easier to verify. However, its applicability is, so far, restricted to the 2-dimensional case; either with Riccati-based or with spectral-based feedback operators, $d = 2$. See Section 3.

(c) It is shown in [B-L-T.2, Lemma 4.1 of Section 4] that the s.c. analytic exponentially stable semigroup $e^{A_F t}$ on H – generated by the operator A_F in (1.2.11), satisfying (H.2)(i), (ii) – preserves the properties of being a s.c. analytic exponentially stable semigroup also on the space W defined in (1.2.2) in the 2-d case.

(d) The $L^2(0,\infty;Z)$-stability property (1.2.7) under Setting #1 is now replaced, under Setting #2, by the finite time $L^2(0,1;Z)$-condition in (1.2.13), along with the additional regularity requirement (H.2iv) = (1.2.14) on $\mathcal{D}(A_F)$. The technical justification is given in [B-L-T.2, Lemma 4.2 in Section 4].

Main result for Setting #2. Under Assumption (H.2) for the linearized problem (1.2.1), local exponential stabilization for the nonlinear N-S problem (1.1.1) follows, at least for $d = 2$. The following result is proved in [B-L-T.2, Section 4].

Theorem 1.2.2. *For $d = 2$, assumption (H.2) of Setting #2 implies Assumption (H.1) of Setting #1. As a consequence, all the statements of Theorem 1.2.1 hold true under Assumption (H.2).*

Comments on Theorem 1.2.2 vis-a-vis Theorem 1.2.1. Both Theorems 1.2.1 and 1.2.2 allow for the construction of a stabilizing feedback for 2-d and 3-d Navier-Stokes flows, provided that a corresponding feedback is available for the linearization.

Theorem 1.1.1 provides an abstract framework within which (i) linearized stability in W as in (1.2.6) and $L_2(0, \infty; Z)$-regularity as in (1.2.7) imply local exponential stability of a Navier-Stokes flow on W. Of course, the key difficulty rests at the level of verifying these two assumptions imposed on the linearized semigroup. However, this setting is not empty: in fact, verification of hypothesis (H.1) – which includes these two properties (1.2.6) and (1.2.7) – has already been performed, for $d = 3$ as well, in reference [B-L-T.1] in the case where the stabilizing feedback operator F is a 'high gain' Riccati operator $F = R$; that is, R is unbounded on H, and satisfies an Algebraic Riccati Equation on the domain $\mathcal{D}(A_R^2)$ of the square of the feedback operator A_R, which is generally not explicit [B-L-T.1, Proposition 4.5.1]. On the other hand, A_R is dissipative on $\mathcal{D}(A_R)$ in the inner product $(\,\cdot\,, R\,\cdot\,)_H$ [B-L-T.1, Proposition 4.5.1], which is equivalent to the W inner product [see (2.1.7) for $d = 3$, and (2.2.11) for $d = 2$]. Thus, our Setting #1 (Assumption (H.1)) represents an abstraction of 'concrete' results of [B-L-T.1] for $d = 3$ and $d = 2$ over the spaces $\{W, Z\}$ in (1.2.2), (1.2.3). This is verified in the forthcoming Section 2.

In contrast, the framework of Theorem 1.2.2 is more limited, as it requires that the feedback operator F be *bounded*, as in (1.2.10), so that the linearized dynamics defines an analytic semigroup. This requirement is in fact possible for $d = 2$; indeed, for two choices of the feedback operator F: (1) a first choice, where F is a 'low gain' Riccati operator arising from an optimal control problem with $L_2(0, \infty; H)$-observation as verified in the forthcoming Section 3.1; (2) a second choice, where F is a finite-dimensional operator, at least when a Finite-Dimensional Spectral Assumption (FDSA), stated at the outset of Section 1.3, is assumed to hold true [B-L-T.1] as verified in the forthcoming Section 3.2. This property FDSA is believed to be generically true, modulo a small perturbation of the domain Ω [H.1].

Remark 1.2.1. In the bounded feedback, low-gain Setting #2, it is the *analyticity* of the semigroup $e^{A_F t}$ (see below (1.2.10)) that is responsible for extra regularity of the linearized dynamics. By contrast in the unbounded feedback, high-gain Setting #1, extra regularity of the linearized dynamics is obtained due to the *unbounded observation* (see (1.2.7)). $\qquad\qquad\square$

1.3. The finite-dimensional spectral assumption and its consequences

Following [B-T.1], [B-L-T.1, Section 3.6], we introduce the following Finite-Dimensional Spectral Assumption.

FDSA: We assume that for each of the distinct unstable eigenvalues $\lambda_1, \ldots, \lambda_M$ of the Oseen operator $\mathcal{A}$, see (1.1.16), algebraic and geometric multiplicity coincide.

Denote by the same symbol H the complexification of the original space H. Let

$$P_N = -\frac{1}{2\pi i}\int_C (\lambda I - \mathcal{A})^{-1}d\lambda; \qquad P_N^* = -\frac{1}{2\pi i}\int_{\bar{C}} (\lambda I - \mathcal{A}^*)^{-1}d\lambda;$$

$$: H \xrightarrow{\text{onto}} Z_N^u \qquad\qquad\qquad\qquad : H \xrightarrow{\text{onto}} (Z_N^u)^* \qquad (1.3.1)$$

where $\mathcal{C}$ (respect. $\overline{\mathcal{C}}$) is a simple, closed curve surrounding $\{\lambda_i\}_{i=1}^{M}$ (respect. $\{\overline{\lambda}_i\}_{i=1}^{M}$). The complexified space H can be decomposed in complementary, non-necessarily orthogonal subspaces as in [K.1, p. 178]

$$H = Z_N^u \oplus Z_N^s; \quad Z_N^u = P_N H; \quad Z_N^s = (I - P_N)H; \quad \dim Z_N^u = N, \qquad (1.3.2)$$

where each of the subspaces Z_N^u and Z_N^s is invariant under $\mathcal{A}$. We set

$$\mathcal{A}_N^u = P_N \mathcal{A} = \mathcal{A}|_{Z_N^u}; \quad \mathcal{A}_N^s - (I - P_N)\mathcal{A} - \mathcal{A}|_{Z_N^s} \qquad (1.3.3)$$

for the restrictions of $\mathcal{A}$ to Z_N^u and Z_N^s, respectively. We then have that the spectra of $\mathcal{A}$ on Z_N^u and Z_N^s coincide with $\{\lambda_j\}_{j=1}^{N}$ and $\{\lambda_j\}_{j=N+1}^{\infty}$, respectively.

We denote by $\{\varphi_{ij}\}_{j=1}^{\ell_i}$, $\{\varphi_{ij}^*\}_{j=1}^{\ell_i}$, the (normalized) linearly independent eigenfunctions corresponding to each unstable distinct eigenvalue λ_i of $\mathcal{A}$ and $\overline{\lambda}_i$ of $\mathcal{A}^*$, respectively:

$$\mathcal{A}\varphi_{ij} = \lambda_i \varphi_{ij}; \quad \mathcal{A}^* \varphi_{ij}^* = \overline{\lambda}_i \varphi_{ij}^*. \qquad (1.3.4)$$

Under the FDSA, we have

$$Z_N^u = P_N H = \mathrm{span}\{\varphi_{ij}\}_{i=1 \;\; j=1}^{M \;\; \ell_i}; \quad (Z_N^u)^* = P_N^* H = \mathrm{span}\{\varphi_{ij}^*\}_{i=1 \;\; j=1}^{M \;\; \ell_i} \qquad (1.3.5)$$

(without the FDSA, Z_N^u is the span of the *generalized* eigenfunctions of $\mathcal{A}$ corresponding to its unstable eigenvalues and similarly for $(Z_N^u)^*$). In other words, the FDSA says that the restriction $\mathcal{A}_N^u = \mathcal{A}|_{Z_N^u}$ of $\mathcal{A}$ on Z_N^u is *diagonalizable* or that the operator $\mathcal{A}_N^u$ is a normal operator on Z_N^u. In the terminology of [K.1], $\mathcal{A}_N^u$ is *semi-simple*. It is believed that the FDSA is generically true, as is the case for the Laplacian on a bounded domain [H.1].

Complexified dynamics and its decomposition. [B-L-T.1, Section 3.4]. The complexified version on $H \oplus iH$ of the linearized dynamics (1.1.21) is then

$$\frac{dz}{dt} - \mathcal{A}z = -\mathcal{A}Du \in [\mathcal{D}(\mathcal{A}^*)]', \quad z(0) = z_0, \; u \cdot \nu = 0 \text{ on } \Sigma. \qquad (1.3.6)$$

Then the z-system can accordingly be decomposed as

$$z = z_N + \zeta_N, \quad z_N = P_N z, \quad \zeta_N = (I - P_N)z_N, \qquad (1.3.7)$$

where applying P_N and $(I - P_N)$ (which commute with $\mathcal{A}$) on (1.3.6), we obtain via (1.3.3),

$$\text{on } Z_N^u : \quad z_N' - \mathcal{A}_N^u z_N = -P_N(\mathcal{A}Du) = -\mathcal{A}_N^u P_N Du, \; z_N(0) = P_N z_0; \qquad (1.3.8)$$

$$\text{on } Z_N^s : \quad \zeta_N' - \mathcal{A}_N^s \zeta_N = -(I - P_N)(\mathcal{A}Du) = -\mathcal{A}_N^s (I - P_N)Du,$$

$$\zeta_N(0) = (I - P_N)z_0, \qquad (1.3.9)$$

respectively.

2. Application of Theorem 1.2.1 with a topologically high-gain, Riccati-based feedback operator

In the present section, we illustrate the application of Theorem 1.2.1 to two topological settings: the first refers to the case $d = 3$, as given in (1.2.3) (Subsection 2.1); while the second refers to the case $d = 2$, as given in (1.2.2) (Subsection 2.2). They both rest on 'concrete' results already established in [B-L-T.1]. Both involve high-gain, Riccati-based feedback operators.

2.1. Case $d = 3$: $W = (H^{\frac{1}{2}+\epsilon_0}(\Omega))^3 \cap H$; $Z \equiv (H^{\frac{3}{2}+\epsilon_0}(\Omega))^3 \cap H$

As a first illustration on the applicability of Theorem 1.2.1, in the case $d = 3$, we can take the stabilizing feedback operators that were constructed in [B-L-T.1]. This construction rests on an optimization problem with (topologically) 'high observation,' hence via a corresponding 'high-gain' Riccati operator, which arises in an optimal control problem with topologically high cost functional. See (2.1.2), (2.1.3) below. The analysis of the resulting non-standard Riccati equation is complicated and is given in [B-L-T.1, Section 4 along with Appendix C], to which we refer for details. The key point, however, is that the Riccati-based stabilizing feedback operators constructed in [B-L-T.1] for $d = 3$ satisfy in full Assumption (H.1), parts (i), (ii), (iii), on the required feedback operator F. Below we shall extract only the main features of the procedure and refer to [B-L-T.1] for details. We shall use the notation of [B-L-T.1] when applicable. In line with (1.2.3), the present setting is as follows:

$$\text{Case } d = 3. \quad W = (H^{\frac{1}{2}+\epsilon_0}(\Omega))^3 \cap H; \ Z \equiv (H^{\frac{3}{2}+\epsilon_0}(\Omega))^3 \cap H. \tag{2.1.1}$$

Let $R \in \mathcal{L}(W; W')$, W' being the dual of W with respect to H as a pivot space, be the Riccati operator – which is positive self-adjoint on H – which defines the value function

$$(Rv_0, v_0)_W = \min_{u \in L^2(0,\infty;U)} J(v, u, v_0) = J^*(v_0); \tag{2.1.2}$$

$$J(v; u, v_0) \equiv \int_0^\infty [|v(t; v_0)|_Z^2 + |u(t)|_U^2]dt, \quad v_0 \in W \tag{2.1.3}$$

of the optimal control problem with state space W and output space Z, defined in (2.1.1). Here $v(t; v_0)$ is the solution of the *linearized* problem (1.1.20) in PDE-form, or else of (1.1.21) in abstract form, due to the I.C. $v_0 \in W$ and the boundary control $u \in L^2(0, \infty; U)$. The space U is defined in (1.1.6): it requires purely *tangential* boundary controls. In addition to the stringent requirement that the class of Dirichlet boundary controls be tangential at each point of the boundary $\partial\Omega$, the OCP (2.1.2), (2.1.3) faces two additional difficulties that set it apart and definitely outside the setting of established optimal control theory for parabolic systems with boundary controls: (1) the high degree of unboundedness of the *boundary* control operator, of order $(\frac{3}{4} + \epsilon)$ as expressed in terms of fractional powers of the basic free-dynamic generator $\mathcal{A}$ (or A); and (2) the high degree of unboundedness of the 'penalization' or 'observation' operator of order also $(\frac{3}{4} + \epsilon)$, as expressed in terms of fractional powers of the basic free-dynamics generator. This yields a 'combined index' of unboundedness *strictly greater* than $\frac{3}{2}$. In contrast, the established (and rich) optimal control theory of boundary control parabolic problems and corresponding algebraic Riccati theory requires a 'combined index' of unboundedness *strictly less* than 1 [L-T.4, vol. 1, in particular, pp. 501–503], [B-D-D-M.1], which is the maximum limit handled by perturbation theory of analytic semigroups. Implementation of this program is carried out in [B-L-T.2] by critically relying on the analysis and results of [B-L-T.1]. It then leads to the result described below.

The analysis of [B-L-T.1, Section 4] also yields that the Riccati operator R in (2.1.2) defining the optimal value, $R \in \mathcal{L}(W; W')$, is an isomorphism of W onto W'. One then obtains [B-L-T.1, Eqn. (4.1.13a)]

$$c|x|_W^2 \leq (Rx, x)_H \leq C|x|_W^2, \quad \forall\, x \in W, \tag{2.1.4}$$

for some constants $0 < c < C < \infty$, so that $|R^{\frac{1}{2}} x|_H$-norm is equivalent to the $|x|_W$-norm. In summary: The main result for the present case $d = 3$ is the statement that the feedback operator

$$u = F(y - y_e) = \nu_0 \frac{\partial R}{\partial \nu}(y - y_e), \tag{2.1.5}$$

– which turns out to be pointwise tangential on $\partial\Omega$ [B-L-T.1, Proposition D.1] – once inserted in the RHS of Eqn. (1.1.1c), exponentially stabilizes in W the N-S flow (1.1.1), in the W-vicinity of its equilibrium solution y_e. To measure the W-vicinity of y_e, we introduce the set

$$\mathcal{V}_\rho \equiv \{y_0 \in W \equiv (H^{\frac{1}{2}+\epsilon_0}(\Omega))^3 \cap H : |y_0 - y_e|_W < \rho\} \tag{2.1.6}$$

of initial conditions of (1.1.1) whose W-distance from y_e is less than $\rho > 0$.

Theorem 2.1.1. [B-L-T.1, Theorem 2.3] *Let $\rho > 0$ in (2.1.6) be sufficiently small. Then: for each $y_0 \in \mathcal{V}_\rho$, there exists a unique mild solution y (obtained by fixed-point (contraction mapping) in [B-L-T.1, Theorem 5.1]) of the following closed-loop problem:*

$$\left\{ \begin{array}{rclr} y_t(x,t) - \nu_0 \Delta y(x,t) + (y \cdot \nabla)y(x,t) &=& f_e(x) + \nabla p(x,t) & \text{in } Q; \quad (2.1.7a) \\ \nabla \cdot y &=& 0 & \text{in } Q; \quad (2.1.7b) \\ y &=& \nu_0 \dfrac{\partial}{\partial \nu} R(y - y_0) & \text{on } \Sigma; \quad (2.1.7c) \\ y(x,0) &=& y_0(x) & \text{in } \Omega \quad (2.1.7d) \end{array} \right.$$

obtained from (1.1.1) by replacing u with the boundary feedback control in (2.1.5) tangential to $\partial\Omega$, having the following regularity and asymptotic properties where W and Z are defined in (2.1.1):

(i) $$(y - y_e) \in C([0, \infty); W) \cap L^2(0, \infty; Z), \tag{2.1.8}$$

continuously in $y_0 \in W \equiv (H^{\frac{1}{2}+\epsilon_0}(\Omega))^3 \cap H$, where $Z = (H^{\frac{3}{2}+\epsilon_0}(\Omega))^3 \cap H$, that is,

$$|y(t) - y_e|_W^2 + \int_0^\infty |y(t) - y_e|_Z^2 dt \leq C|y_0 - y_e|_W^2, \quad t \geq 0; \tag{2.1.9}$$

(ii) *there exist constants $M \geq 1$, $\omega > 0$ (independent of $\rho > 0$) such that such solution $y(t)$ satisfies*

$$|y(t) - y_e|_W \leq Me^{-\omega t}|y_0 - y_e|_W, \quad t \geq 0. \tag{2.1.10}$$

Here R is the Riccati operator described at the outset of this subsection, defined by (2.1.2) and satisfying (in particular) (2.1.4).

2.2. Case $d = 2$: $\tilde{W} \equiv (H^{\frac{1}{2}-\epsilon_0}(\Omega))^2 \cap H$; $\tilde{Z} \equiv (H^{\frac{3}{2}-\epsilon_0}(\Omega))^2 \cap H$

The two-dimensional case, $d = 2$, is, as expected, more regular. Here, various functional settings are possible. In the present subsection, we focus on the functional setting for W and Z given in (1.2.2), repeated here (with a superscript $\sim$ for clarity)

$$\text{Case } d = 2 : \quad \tilde{W} \equiv (H^{\frac{1}{2}-\epsilon_0}(\Omega))^2 \cap H; \quad \tilde{Z} \equiv (H^{\frac{3}{2}-\epsilon_0}(\Omega))^2 \cap H, \tag{2.2.1}$$

$\epsilon_0 > 0$ arbitrary and fixed, to which we apply Theorem 1.2.1. The control space is the one given by (1.1.6). The corresponding Riccati-based feedback operator will accordingly be 'high-gain' (topologically). A topologically lower level treatment with, accordingly, a low-gain Riccati-based feedback operator, will be presented in the subsequent Subsection 3, as an illustration of Theorem 1.2.2 this time. The counterpart of the 3-dimensional case of Subsection 2.1 is as follows: We let $\tilde{R} \in \mathcal{L}(\tilde{W}; \tilde{W}')$, $\tilde{W}$ as in (2.2.1), $\tilde{W}'$ its dual with respect to H as the pivot space, be the Riccati operator (which is unbounded, positive, self-adjoint on H) defining the value function

$$(\tilde{R}v_0, v_0)_{\tilde{W}} = \min_{u \in L^2(0,\infty;U)} \tilde{J}(v, u; v_0) \equiv \tilde{J}^*(v_0); \tag{2.2.2}$$

$$\tilde{J}(u, v; v_0) \equiv \int_0^\infty \left[|v(t; v_0)|_{\tilde{Z}}^2 + |u(t)|_U^2 \right] dt, \quad v_0 \in \tilde{W} \tag{2.2.3}$$

of the optimal control problem with state space $\tilde{W}$ and output space $\tilde{Z}$, defined in (2.2.1). Here, $v(t; v_0)$ is the solution of the *linearized* problem (1.1.20) in PDE-form, or else of (1.1.21) in abstract form, due to the I.C. $v_0 \in \tilde{W}$ and the boundary control $u \in L^2(0, \infty; U)$, U in (1.1.6), i.e., with pointwise tangential control u. *A fortiori* from the case $d = 3$ reviewed in Section 2.1, it is established in [B-L-T.1, Appendix B] that the optimal control problem (2.2.2), (2.2.3) satisfies the Finite Cost Condition within the class of $L^2(0, \infty; U)$-controls – i.e., the functional $\tilde{J}(u, v; v_0)$ is proper for each $v_0 \in \tilde{W}$ in the language of optimization theory – indeed, even within the class of $L^2(0, \infty; U_1)$-controls, where

$$\left\{ \begin{array}{l} U_1 = \{\mu \in (L^2(\Gamma_1))^d, \ \mu \equiv 0 \text{ on } \Gamma \setminus \Gamma_1; \ \mu \cdot \nu \equiv 0 \text{ a.e. on } \Gamma_1\} \text{ where} \\ \Gamma_1 \text{ is an arbitrarily preassigned portion of positive measure of the} \\ \text{boundary } \Gamma \equiv \partial\Omega. \end{array} \right. \tag{2.2.4}$$

The main result for the present case $d = 2$ is the statement that the feedback operator

$$u = \begin{cases} F(y - y_e) = \nu_0 \dfrac{\partial \tilde{R}}{\partial \nu}(y - y_e) & \text{on } \Sigma_1 = \Gamma_1 \times (0, \infty) \\ 0 & \text{on } \Sigma/\Sigma_1 \equiv \Sigma_2 = \Gamma_2 \times (0, \infty), \end{cases} \tag{2.2.5}$$

acting on an arbitrarily small portion Γ_1 of the boundary – which turns out to be pointwise tangential on $\partial\Omega$ [B-L-T.1, Proposition D.1] – once inserted in the RHS of Eqn. (1.1.1c), exponentially stabilizes in $\tilde{W}$ in (2.2.1) the N-S flow (1.1.1), in the $\tilde{W}$-vicinity of its equilibrium solution y_e. To measure the $\tilde{W}$-vicinity of y_e, we introduce the set

$$\tilde{\mathcal{V}}_\rho \equiv \{y_0 \in \tilde{W} \equiv (H^{\frac{1}{2}-\epsilon_0}(\Omega))^2 \cap H : |y_0 - y_e|_{\tilde{W}} < \rho\} \tag{2.2.6}$$

of initial conditions of (1.1.1) whose $\tilde{W}$-distance from y_e is less than $\rho > 0$.

Theorem 2.2.2. [B-L-T.1, Theorem 2.5] *Let $d = 2$. Let $\rho > 0$ in (2.2.6) be sufficiently small. Then, for each $y_0 \in \tilde{\mathcal{V}}_\rho$, there exists a unique mild solution y (obtained by fixed-point (contraction mapping)) of the following closed-loop problem:*

$$\begin{cases} y_t(x,t) - \nu_0 \Delta y(x,t) + (y \cdot \nabla)y(x,t) = f_e(x) + \nabla p(x,t) & \text{in } Q; & (2.2.7a) \\[2mm] \nabla \cdot y \equiv 0 & \text{in } Q; & (2.2.7b) \\[2mm] y = \begin{cases} \nu_0 \dfrac{\partial \tilde{R}}{\partial \nu}(y - y_0) & \text{on } \Sigma_1; \\ 0 & \text{on } \Sigma_2; \end{cases} & & (2.2.7c) \\[2mm] y(x,0) = y_0(x) & \text{in } \Omega & (2.2.7d) \end{cases}$$

obtained from (1.1.1) by replacing u with the boundary feedback control in (2.2.5), having the following regularity and asymptotic properties, where $\tilde{W}$ and $\tilde{Z}$ are defined in (2.2.1):

(i)
$$(y - y_e) \in C([0, \infty); \tilde{W}) \cap L^2(0, \infty; \tilde{Z}) \tag{2.2.8}$$

continuously in $y_0 \in \tilde{W} \equiv (H^{\frac{1}{2}-\epsilon_0}(\Omega))^2 \cap H$, where $\tilde{Z} = (H^{\frac{3}{2}-\epsilon_0}(\Omega))^2 \cap H$, that is,

$$|y(t) - y_e|_{\tilde{W}}^2 + \int_0^\infty |y(t) - y_e|_{\tilde{Z}}^2 dt \le C|y_0 - y_e|_{\tilde{W}}^2, \quad t \ge 0; \tag{2.2.9}$$

(ii) *there exist constants $M \ge 1$, $\omega > 0$ (independent of $\rho > 0$), such that said solution $y(t)$ satisfies*

$$|y(t) - y_e|_{\tilde{W}} \le Me^{-\omega t}|y_0 - y_e|_{\tilde{W}}, \quad t \ge 0. \tag{2.2.10}$$

Here, $\tilde{R}$ is the Riccati operator described at the outset of this subsection, defined by (2.2.2), satisfying (in particular)

$$c|x|_{\tilde{W}}^2 \leq (\tilde{R}x, x)_H \leq C|x|_{\tilde{W}}^2, \quad \forall\, x \in \tilde{W}, \tag{2.2.11}$$

for constants $0 < c < C < \infty$, so that the $|\tilde{R}^{\frac{1}{2}}x|_H$-norm is equivalent to the $|x|_{\tilde{W}}$-norm.

3. Applications of bounded, low-level gain Theorem 1.2.2, Case $d = 2$: (i) Riccati-based feedback and (ii) Spectral feedback

In the present section we shall illustrate the applicability of the more amenable Theorem 1.2.2. The treatment is restricted to $d = 2$. We shall consider two types of feedback operators: (i) Riccati-based feedbacks, and (ii) spectral feedbacks.

3.1. Theorem 1.2.2 with low-level Riccati-Based feedback operators, $d = 2$

In the present subsection, we let $R_0 \in \mathcal{L}(H)$ be the bounded positive self-adjoint operator, which defines the value function

$$(R_0 v_0, v_0)_H \equiv \inf_{u \in L^2(0,\infty;U_1)} J_0(u, v; v_0) = J_0^*(v_0); \tag{3.1.1}$$

$$J_0(u, v; v_0) \equiv \int_0^\infty [|v(t; v_0)|_H^2 + |u(t)|_{U_1}^2]dt, \quad v_0 \in H. \tag{3.1.2}$$

U_1 defined in (2.2.4), of the optimal control problem defined in (3.1.1), with state and output space equal to H (the observation operator is the identity on H). Here, $v(t; v_0)$ is the solution of the *linearized* problem (1.1.20) in PDE-form, or else of (1.1.21) in abstract form, due to the I.C. $v_0 \in H$ and the boundary control $u \in L^2(0, \infty; U_1)$, U_1 as in (2.2.4). By virtue of [B-L-T.1] – *a fortiori* from Section 2 – the optimal control problem (3.1.1), (3.1.2) satisfies the Finite Cost Condition, i.e., the functional $J_0(u, v; v_0)$ is proper for each $v_0 \in H$, in the language of optimization theory. Thus, the optimal control problem (3.1.1) for the dynamics (1.1.20) in PDE-form, or else (1.1.21) in abstract form has a unique optimal pair $\{u^0(t; v_0), v^0(t; v_0)\}$. Accordingly, established Optimal Control Theory (since the 80's [L-T.2]) for parabolic problems yields parts (i), almost part (ii), through (iv) of the following

Theorem 3.1.1. ([L-T.4, Theorem 2.2.1, p. 125]) *With reference to the OCP for the parabolic dynamics (1.1.20) in PDE-form, or its abstract version (1.1.21), the following results hold true:*

 (i) *the operator $R_0 \in \mathcal{L}(H)$ defined by (3.1.1) is the unique positive self-adjoint solution of the following Algebraic Riccati Equation*

$$(R_0 y, \mathcal{A}^* z)_H + (\mathcal{A}y, R_0 z)_H + (y, z)_H = \nu_0^2 \left(\frac{\partial}{\partial\nu} R_0 y, \frac{\partial}{\partial\nu} R_0 z\right)_{(L^2(\Gamma_1))^2}, \tag{3.1.3}$$

$$\forall\, y, z \in H,$$

$\Gamma_1 =$ arbitrarily small portion of $\Gamma = \partial\Omega$, meas $\Gamma_1 > 0$. [The general theory gives the ARE satisfied only for $y, z \in \mathcal{D}(\hat{\mathcal{A}}^\epsilon)$, $\epsilon > 0$ arbitrary, $\hat{\mathcal{A}}$ as in (1.2.2).] For the present Oseen operator in (1.1.14) we can improve the general statement to all $y, z \in H$, as in (3.1.3). This is so by property (ii) below, in line with the special cases noted in [L-T.4, Theorem 2.2.1(a_3), p. 126, and Chapter 2, Appendix 2B, p. 168];

(ii) *for the Oseen operator in (1.1.14), we further have*

$$\mathcal{A}^* R_0 \in \mathcal{L}(H); \quad A R_0 \in \mathcal{L}(H); \ or$$
$$R_0 y \in \mathcal{D}(A) \subset (H(\Omega))^2 \cap H \ for \ y \in H; \tag{3.1.4a}$$

hence,

$$R_0 A_F \in \mathcal{L}(H), \tag{3.1.4b}$$

and

$$F \equiv \nu_0 \frac{\partial}{\partial\nu} R_0 \in \mathcal{L}(H; (H^{\frac{1}{2}}(\Gamma_1))^2) \subset \mathcal{L}(H; U_1); \tag{3.1.5}$$

(iii) *the feedback operator $A_F \equiv \mathcal{A}(I - DF)$ defined in (1.2.11) generates a s.c. analytic semigroup $e^{A_F t}$ on H, $t \geq 0$, so that the optimal pair $\{u^0, v^0\}$ of problem (3.1.1) for the dynamics (1.1.20) (or (1.1.21)) is*

$$v^0(t; v_0) = e^{A_F t} v_0 = optimal \ solution; \tag{3.1.6a}$$

$$u^0(t; v_0) = -D^* \mathcal{A}^* R_0 v^0(t; v_0) = \nu_0 \frac{\partial}{\partial\nu} R_0 v^0(t; v_0) = \nu_0 \frac{\partial}{\partial\nu} R_0 e^{A_F t} v_0$$
$$= v^0(t; v_0)|_{\Gamma_1} = optimal \ control, \tag{3.1.6b}$$

while $u^0(t; v_0) \equiv 0$ on $\Gamma_2 = \Gamma \setminus \Gamma_1$.

(iv) *$e^{A_F t}$ is uniformly (exponentially) stable on H: there exist constants $M \geq 1$, $\delta > 0$ such that*

$$|e^{A_F t}|_{\mathcal{L}(H)} \leq M e^{-\delta t}, \ t \geq 0. \tag{3.1.7}$$

(v) *$e^{A_F t}$ in (iii)–(iv) continues to be a s.c. analytic semigroup also on the space $W = (H^{\frac{1}{2} - \epsilon_0}(\Omega))^2 \cap H$ in (1.2.2), which, moreover, remains uniformly (exponentially) stable on W with the same bound: $|e^{A_F t}|_{\mathcal{L}(W)} \leq M e^{-\delta t}, \quad t \geq 0$.*

Proof. Parts (i), (iii), (iv) are standard results from [L-T.2], [L-T.4], except that trace theory is used for F in (3.1.5). Part (v) is a consequence of Parts (iii), (iv), as established in [B-L-T.2, Section 4]. Part (ii) is an "ϵ-improvement" of established theory, see Remark 3.1.1. It is established in [B-L-T.1, Appendix B.4]. $\qquad\square$

Remark 3.1.1. We note that Optimal Control Theory [L-T.2], [L-T.4, Theorem 2.2.1(a_3), p. 126] gives, in general, $\hat{\mathcal{A}}^{*1-\epsilon} R_0 = (\omega I - \mathcal{A}^*)^{1-\epsilon} R_0 \in \mathcal{L}(H), \ \forall \ \epsilon > 0$, equivalently $\hat{\mathcal{A}}^{1-\epsilon} R_0 \in \mathcal{L}(H)$, by interpolation, since $\mathcal{D}(\mathcal{A}) = \mathcal{D}(\mathcal{A}^*)$ in the present case of the Oseen operator. The regularity noted in (3.1.4a) shows that we can presently take $\epsilon = 0$. This is due to two features of our present optimal control problem:

(a) the observation operator in (3.1.2) is the identity on H;

(b) the Oseen operator $\mathcal{A} = -(\nu_0 A + A_0)$ in (1.1.14) is a lower-order perturbation of the self-adjoint operator $-\nu_0 A$, A the Stokes operator, $\mathcal{D}(A_0) = \mathcal{D}(A^{\frac{1}{2}})$, see (1.1.15). Thus, a suitable more technical modification of the proof in [L-T.4, Appendix 2B, p. 168], given in [B-L-T.1, Appendix B.4], shows that we can take $\epsilon = 0$.

Remark 3.1.2. Theorem 3.1.1 was actually obtained in the '80s, see [L-T.2], following references [La.1–2], [L-T.1] which referred to the same Optimal Control Problem, however, over a finite horizon. These results were first collected in [L-T.3] and then, in book form, in [L-T.4].

We next provide regularity properties of the optimal pair $\{u^0, v^0\}$. These results are the counterpart of regularity properties of the optimal pair for *classical* parabolic boundary control problems given in [La.1], [La.2, Section 7], [L-T.2], later collected in [L-T.3], [L-T.4, Section 3.2, p. 187]. The translation to the present Navier-Stokes problem, $d = 2$, is obtained via the critical optimal regularity theory of non-homogeneous Navier-Stokes equations (with tangential Dirichlet boundary controls), given in [B-L-T.1, Section 3.1, Theorem 3.1.4 through Theorem 3.1.8], as well as (1.1.19): $D^*\mathcal{A}^* = -\nu_0 \frac{\partial}{\partial \nu}$. In effect, one could essentially obtain the next Proposition 3.1.2 by quoting [La.1], [La.2, Thm. 7.1, p. 320], [L-T.4, Section 3.2, p. 187, also p. 116], by use this time of the aforementioned non-homogeneous regularity results [B-L-T.1, Section 3.1] and of (1.1.19). For clarity, a complete proof is provided in [B-L-T.2, Section 6]. As in [B-L-T.1, Section 3.1, Eqns. (3.1.28), (3.1.29)], motivated by classical parabolic theory [L-M.1, Vol. 2], we introduce the following Sobolev spaces, for $r, s \geq 0$:

$$H^{r,s}(Q_T) \equiv L^2(0, T; (H^r(\Omega))^2 \cap H) \cap H^s(0, T; H); \tag{3.1.8}$$

$$H^{r,s}(\Sigma_T) \equiv L^2(0, T; H^r(\Gamma_1)^2) \cap H^s(0, T; (L^2(\Gamma_1))^2), \quad u \cdot \nu \equiv 0, \tag{3.1.9}$$

subject further to the tangential condition $u \cdot \nu \equiv 0$ on Γ. The following result gives regularity properties of the optimal pair $\{u^0, v^0\}$ in terms of the above spaces. It will be critical in verifying the validity of property (H.2iii) = (1.2.13) in [B-L-T.2, Section 6].

Proposition 3.1.2. (*Compare with* [La.2, Thm. 7.1], [L-T.2], [L-T.4, p. 116 and 187].) *With reference to the optimal control problem* (3.1.1) *for the linearized dynamics* (1.1.20), *or* (1.1.21), *the following regularity properties of the optimal pair* $\{u^0(\,\cdot\,; v_0), v^0(\,\cdot\,; v_0)\}$ *hold true:*

$$v_0 \in \tilde{W} = \mathcal{D}(\hat{\mathcal{A}}^{\frac{1}{4}-\frac{\epsilon_0}{2}}) \equiv \mathcal{D}((\omega I - \mathcal{A})^{\frac{1}{4}-\frac{\epsilon_0}{2}}) = (H^{\frac{1}{2}-\epsilon_0}(\Omega))^2 \cap H \tag{3.1.10}$$

$$\Rightarrow \begin{cases} \hat{v}^0(t; v_0) \;\equiv\; e^{-\omega t} v^0(t; v_0) \in H^{\frac{3}{2}-\epsilon_0, \frac{3}{4}-\frac{\epsilon_0}{2}}(Q_\infty) & (3.1.11a) \\[2mm] \qquad\qquad \subset\; C([0, \infty]; (H^{\frac{1}{2}-\epsilon_0}(\Omega))^2 \cap H)) & (3.1.11b) \\[2mm] \hat{u}^0(t; v_0) \;\equiv\; e^{-\omega t} u^0(t; v_0) \in H^{2-\epsilon_0, 1-\frac{\epsilon_0}{2}}(\Sigma_\infty), & (3.1.12) \end{cases}$$

continuously. Indeed, a slightly more precise result for $\hat{v}^0$ *is shown in* [B-L-T.2, Eqn. (6.21)].

Theorem 3.1.1 and Proposition 3.1.2 on the linearized problem (1.2.21) permit one in [B-L-T.2, Section 6] to verify assumption (H.2) in full and thus to apply Theorem 1.2.2 and obtain the following main local exponential decay in $\tilde{W}$ for the original N-S flow (1.1.1) on the $\tilde{W}$-vicinity of the equilibrium solution y_e, where $\tilde{W} \equiv (H^{\frac{1}{2}-\epsilon_0}(\Omega))^2 \cap H$, as defined in (2.2.1), or (1.2.2).

Theorem 3.1.3. *Let $d = 2$. Recall that $y_e \in (H^2(\Omega))^2 \cap V$ (see below (1.1.1d)). Let $\rho > 0$ in (2.2.6) be sufficiently small. Then:*

(a) *For each $y_0 \in \tilde{V}_\rho$ (defined in (2.2.6)), there exists a unique solution y (obtained by a fixed point/contraction mapping) of the following closed loop problem*

$$\begin{cases} y_t(x,t) - \nu_0\Delta y(x,t) + (y \cdot \nabla)y(x,t) = f_e(x) + \nabla p(x,t) & in \ Q; & (3.1.13a) \\ \nabla \cdot y \equiv 0 & in \ Q; & (3.1.13b) \\ y = \begin{cases} \nu_0 \dfrac{\partial R_0}{\partial \nu}(y - y_e) & on \ \Sigma_1; \\ 0 & on \ \Sigma/\Sigma_1; & (3.1.13c) \end{cases} \\ y(x,0) = y_0(x) & in \ \Omega; & (3.1.13d) \end{cases}$$

obtained from (1.1.1) by replacing u with the boundary feedback control

$$u \equiv \begin{cases} \nu_0 \dfrac{\partial R_0}{\partial \nu}(y - y_e) & in \ \Sigma_1; \\ 0 & in \ \Sigma_2, \end{cases} \qquad (3.1.14)$$

having all the regularity and asymptotic properties in (2.2.8), (2.2.9), (2.2.10), with $\tilde{W} \equiv (H^{\frac{1}{2}-\epsilon_0}(\Omega))^2 \cap H$. Here R_0 is the Riccati operator of Theorem 3.1.1.

(b) *Assume, in addition, the FDSA, stated at the outset of Section 1.3, that the Oseen operator $\mathcal{A}$, restricted over the unstable subspace associated with the unstable eigenvalues $\{\lambda_1, \ldots, \lambda_N\}$ in (1.1.16) be diagonalizable on such finite-dimensional unstable subspace. Then one may replace the feedback control u in (3.1.14) with a finite-dimensional feedback control of the form*

$$u \equiv \begin{cases} \nu_0 \displaystyle\sum_{i=1}^{K}\left(\dfrac{\partial R_0}{\partial \nu}(y - y_e), w_i\right)_{L_2(\Gamma_1))^2} w_i & on \ \Sigma_1; \\ 0 & on \ \Sigma_2, \end{cases} \qquad (3.1.15)$$

for suitable vectors $w_i \in (H^{\frac{1}{2}}(\Gamma_1))^2$, Γ_1 arbitrarily preassigned portion of $\Gamma \equiv \partial\Omega$, meas $\Gamma_1 > 0$, and then all the conclusions of part (a) hold true with (3.1.13c) replaced by (3.1.15).

Remark 3.1.3. Theorem 3.1.3 is more general than the result of [R.1] obtained for a 2-d N-S flow. Indeed, while [R.1], [B-L-T.2], as well as the prior work [B-L-T.1] all use the topological level $(H^{\frac{1}{2}-\epsilon_0}(\Omega))^2$ for the claimed local stabilization result, and

34   V. Barbu, I. Lasiecka and R. Triggiani

while [R.1], [B-L-T.1, Appendix B4] and [B-L-T.2] use the Riccati operator corresponding to the penalization of $v(t; v_0)$ in H, as in (3.1.2), the major differences between [R.1] and [B-L-T.2] in the present 2-d case are three:

(i) Our Theorem 3.1.3(a) provides a feedback control with purely tangential action (as in (3.1.9)): there is no need for controlling the normal component of the velocity vector, as done in [R.1];

(ii) Our Theorem 3.1.3(b) yields a (tangential) feedback control acting only on the arbitrary portion Γ_1 which, moreover, is finite-dimensional as in (3.1.15), under the FDSA (believed to be generically true [H.1]). There is no comparable result in [R.1];

(iii) The treatment of [R.1] relies, ultimately, on Carleman estimates, while ours does not.

Remark 3.1.4. References [B.2], [Tr.3] study the problem of stabilization of a 2-d linearized Navier-Stokes channel by purely *wall-normal* controllers. In particular, [Tr.3] employs a finite-dimensional, wall-normal, boundary controller with arbitrarily small support on the top wall of the 2-d linearized N-S channel flow, which is periodic in the stream-wise direction.

3.2. Theorem 1.2.2 with low-level spectral-based feedback operators, $d = 2$

We return to the FDSA introduced in Section 1.3, believed to be generically true [H.1]. Recall the eigenfunctions φ_{ij}^* of $\mathcal{A}^*$ in (1.3.4) and, following [B-L-T.1, Section 3.6], introduce the following space of traces

$$\mathcal{F} \equiv \operatorname{span}\{\partial_\nu \varphi_{ij}^*\}_{i=1}^{M} \, _{j=1}^{\ell_i} \subset (H^{\frac{1}{2}}(\Gamma))^d, \tag{3.2.1}$$

where containment follows by trace theory on $\varphi_{ij}^* \in (H^2(\Omega))^d$. For any $f \in \mathcal{F}$, we have $f \cdot \nu \equiv 0$ on Γ [B-L-T.1, Lemma 3.6.1, Lemma 3.3.1]. Under the FDSA, it is proved in [B-L-T.1, Eqn. (3.6.20b)] that, if Γ_1 is any portion of $\Gamma = \partial\Omega$ of positive surface measure, then, for each $i = 1, \ldots, M$,

$$\text{the system } \{\partial_\nu \varphi_{ij}^*\}_{j=1}^{\ell_i} \text{ is linearly independent in } (L^2(\Gamma_1))^d. \tag{3.2.2}$$

Further, if $w_1, w_2, \ldots, w_N \in (L^2(\Gamma))^d$, we introduce the $\ell_i \times N$ matrix W_i, for $i = 1, \ldots, M$ [B-L-T.1, Eqn. (3.6.10)],

$$W_i = \begin{bmatrix} (w_1, \partial_\nu \varphi_{i1}^*)_{\Gamma_1} & \cdots & (w_N, \partial_\nu \varphi_{i1}^*)_{\Gamma_1} \\ (w_1, \partial_\nu \varphi_{i2}^*)_{\Gamma_1} & \cdots & (w_N, \partial_\nu \varphi_{i2}^*)_{\Gamma} \\ \vdots & & \\ (w_1, \partial_\nu \varphi_{i\ell_i}^*)_{\Gamma_1} & \cdots & (w_N, \partial_\nu \varphi_{i\ell_i}^*)_{\Gamma_1} \end{bmatrix} ; \ (\ , \ )_{\Gamma_1} = (\ , \ )_{(L^2(\Gamma_1))^d}. \tag{3.2.3}$$

In view of property (3.2.2), the following assumption can be fulfilled, indeed, by infinitely many choices of the vectors $w_1, \ldots, w_N \in \mathcal{F} \subset (H^{\frac{1}{2}}(\Gamma_1))^d$,

$$\operatorname{rank} W_i = \ell_i, \quad i = 1, 2, \ldots, M, \text{ where } w_i \cdot \nu = 0 \text{ on } \Gamma. \tag{3.2.4}$$

Lemma 3.2.1. [B-L-T.1, Lemma 3.6.1]. *Assume the FDSA stated at the outset of Section 1.3. Given $\gamma_1 > 0$ arbitrarily large, there is a controller $u = u_N(t) \equiv \sum_{j=1}^{J} u_N^j(t) w_j$, $J \leq N$ (conservatively) for infinitely many choices of suitable vectors $w_j \in \mathcal{F}$ in (3.2.1), $w_j \cdot \nu = 0$ on Γ, which will have to satisfy the rank conditions (3.2.4), such that, once inserted in (1.3.8), yields the estimate*

$$|z_N(t)|_H + |u_N(t)|_{(H^{\frac{1}{2}}(\Gamma))^d} + |\dot{u}_N(t)|_{(H^{\frac{1}{2}}(\Gamma))^d} \leq C_{\gamma_1} e^{-\gamma_1 t} |P_N z_0|_H, \quad t \geq 0. \quad (3.2.5)$$

Moreover, the vectors w_j can be supported on an arbitrary portion Γ_1 of Γ. Here, z_N is the solution to (1.3.8) corresponding to such u. Moreover, the controller $u = u_N$ may be chosen in feedback form; i.e., in the form $u_N^j(t) = (z_N(t), p_j)_H$, with suitable vectors $p_j \in Z_N^u$, depending on γ_1. In conclusion, z_N in (3.2.5) is the solution of the following equation on Z_N^u

$$z_N' - \mathcal{A}_N^u z_N = -\mathcal{A}_N^u P_N D \left(\sum_{j=1}^{J} (z_N(t), p_j)_H w_j \right), \quad (3.2.6)$$

rewritten as $z_N' = \bar{A}^u z_N$, $z_N(0) = P_N z_0$, $z_N(t) = e^{\bar{A}^u t} P_N z_0$.

We now return from the complexified version to the corresponding real version of the above problem. To this end, we notice that the proof of Lemma 3.6.1 given in [B-L-T.1] shows that, under the FDSA, the finite-dimensional z_N-dynamics in (1.3.8) is controllable within the class of N open-loop, complex-valued controls of the form $u = u_N(t) = \sum_{j=1}^{N} u_N^j(t) w_j$, provided that the vectors w_j's satisfy the rank conditions (3.2.4). Then, since

$$\mathrm{Re} \left(\sum_{j=1}^{N} u_N^j(t) w_j \right) = \sum_{j=1}^{N} (\mathrm{Re}\ u_N^j(t))(\mathrm{Re}\ w_j) - (\mathrm{Im}\ u_N^j(t))(\mathrm{Im}\ w_j) \quad (3.2.7a)$$

$$= \sum_{j=1}^{N} (\mathrm{Re}\ (z_N(t), p_j)_H)(\mathrm{Re}\ w_j)$$

$$- \sum_{j=1}^{N} (\mathrm{Im}\ (z_N(t), p_j)_H (\mathrm{Im}\ w_j), \quad (3.2.7b)$$

we deduce that *a-fortiori* the real value $(\mathrm{Re}\ z_N)$-dynamics is likewise controllable within the class of $K = 2N$ open loop, real-valued controls of the form (3.2.7b). [However, because of the presence of $\mathrm{Im}\ z_N$, such controls in (3.2.7b) are only open-loop and not closed-loop.] Accordingly, by standard, finite-dimensional control theory, the linear dynamics in $(\mathrm{Re}\ z_N)$ can be stabilized by a $K = 2N$-dimensional real-valued, closed-loop feedback control of the same feedback law as its complex-valued counterpart. Thus, henceforth in this section, we set $K = 2N$ ($K = N$ if all unstable eigenvalues λ_j, $j = 1, \ldots, N$, in (1.1.16) are real). Let

$$\begin{cases} \tilde{w}_j \equiv \mathrm{Re}\ w_j, \text{ for } j = 1, \ldots, N; \quad \tilde{w}_{j+N} \equiv -\mathrm{Im}\ w_j, \quad \text{for } j = 1, \ldots, N; & (3.2.8a) \\ \tilde{Z}_N^u = \mathrm{span}\{\mathrm{Re}\ \varphi_{ij}, \mathrm{Im}\ \varphi_{ij}\}_{i=1}^{M}\ {}_{j=1}^{\ell_i}, \quad \varphi_{ij} \text{ in (1.3.4).} & (3.2.8b) \end{cases}$$

We thus obtain the following result for $d = 2$ which corresponds to [B-L-T.1, Theorem 2.6].

Theorem 3.2.2. *Let $d = 2$ and assume the FDSA. Let Γ_1 be any portion of the boundary $\Gamma = \partial\Omega$, meas $\Gamma_1 > 0$. Recall that $y_e \in (H^2(\Omega))^2$. Let $\rho > 0$ be sufficiently small. Then: (a) for each $y_0 \in \tilde{\mathcal{V}}_\rho$ (defined by (2.2.6)), given vectors $\{\tilde{w}_1, \dots, \tilde{w}_K\} \in (H^{\frac{1}{2}}(\Gamma))^2$ satisfying the rank conditions (3.2.4), there exist suitable vectors $\{p_1, \dots, p_K\} \in \tilde{Z}_N^u$ (see (3.2.8b)), such that there exists a unique solution y (obtained by a fixed point/contraction mapping) of the following closed loop problem:*

$$\begin{cases} y_t(x,t) - \nu_0\Delta y(x,t) + (y \cdot \nabla)y(x,t) = f_e(x) + \nabla p(x,t) & \text{in } Q; & (3.2.9\text{a}) \\[2mm] \nabla \cdot y \equiv 0 & \text{in } Q; & (3.2.9\text{b}) \\[2mm] y = \displaystyle\sum_{i=1}^{K} (P_N(y - y_e), p_i)_H \; \tilde{w}_i & \text{in } \Sigma_1 & (3.2.9\text{c}) \\[2mm] y = 0 & \text{in } \Sigma/\Sigma_1 & (3.2.9\text{d}) \\[2mm] y(x,0) = y_0(x) & \text{in } \Omega, & (3.2.9\text{e}) \end{cases}$$

obtained from (1.1.1) by replacing u with the boundary feedback control

$$u = \begin{cases} \displaystyle\sum_{i=1}^{K} (P_N(y - y_e), p_i)_H \; \tilde{w}_i & \text{in } \Sigma_1 & (3.2.10\text{a}) \\[2mm] 0 & \text{in } \Sigma_2 = \Sigma/\Sigma_1, & (3.2.10\text{b}) \end{cases}$$

tangential on Γ, having all the regularity and asymptotic properties in (2.2.8), (2.2.9), (2.2.10) with $\tilde{W} \equiv (H^{\frac{1}{2}-\epsilon_0}(\Omega))^2 \cap H$, $\tilde{Z} = (H^{\frac{3}{2}-\epsilon_0}(\Omega))^2 \cap H$.

4. The case of exponential stabilization of N-S equations, $d = 2, 3$, by means of localized interior, Riccati-based feedback controls

Consider the following controlled Navier-Stokes equations, $d = 2, 3$ [B-T.1]:

$$\begin{cases} y_t(x,t) - \nu_0\Delta y(x,t) + (y \cdot \nabla)y(x,t) \\[2mm] \qquad\qquad = m(x)u(x,t) + f_e(x) - \nabla p(x,t) & \text{in } Q \equiv \Omega \times (0, \infty); & (4.1\text{a}) \\[2mm] \qquad \nabla \cdot y \;\; \equiv \;\; 0 & \text{in } Q; & (4.1\text{b}) \\[2mm] \qquad\quad\; y \;\; \equiv \;\; 0 & \text{in } \Sigma \equiv \partial\Omega \times (0, \infty); & (4.1\text{c}) \\[2mm] \quad y(x,0) \;\; = \;\; y_0(x) & \text{in } \Omega, & (4.1\text{d}) \end{cases}$$

with the same symbols as in (1.1.1), except that now m is the characteristic function of an arbitrary open smooth (C^2-) subset, $\omega \subset \Omega$ of positive measure; and u is the interior control, therefore (mu) is the *localized interior* control supported on $\omega \times (0, \infty)$.

High-gain stabilizing feedback controllers. In this section we briefly return to the problem of exponential feedback stabilization of N-S equations, $d = 2, 3$, in a vicinity of an equilibrium solution, by means – this time – of arbitrarily localized interior feedback control (mu) in (4.1a). This problem was first solved in [B.1], by an infinite-dimensional feedback controller, and later in [B-T.1]. This latter reference yields (in a constructive way) a class of finite-dimensional feedback stabilizing controllers, of minimal dimension $2K$ which, moreover, are Riccati-based. They have the following form:

$$u = -\sum_{i=1}^{2K}(R_N(y - y_e), \psi_i)_{L^2(\omega)}\psi_i = -B^*R_N(y - y_e), \tag{4.2}$$

where $R_N \in \mathcal{L}(\mathcal{D}(A^{\frac{1}{4}})) \cap \mathcal{L}(\mathcal{D}(A^{\frac{1}{2}}); H)$ is a Riccati operator (positive, self-adjoint on $\mathcal{D}(A^{\frac{1}{4}})$ and solution of the algebraic Riccati equation

$$-(\mathcal{A}y, R_N y)_H + \frac{1}{2}\sum_{i=1}^{2K}(\psi_i, R_N y)^2_{L^2(\omega)} = \frac{1}{2}|A^{\frac{3}{4}}y|^2_H, \quad \forall\, y \in \mathcal{D}(A), \tag{4.3}$$

$\mathcal{A}$ in (1.1.14), while the vectors $\psi_i \in L^2(\omega)$ are explicitly *constructed* in [B-T.1, Eqn. (3.52) of Lemma 3.8]. In order to inject 'dissipation' into the N-S equation (4.1), [B-T.1] introduces the optimal control problem (OCP):

$$(R_N v_0, v_0)_{\mathcal{D}(A^{\frac{1}{4}})} = \min_{u \in L^2(0,\infty;\mathbb{R}^{2K})} J(v, u; v_0) = J^*(v_0), \quad v_0 \in \mathcal{D}(A^{\frac{1}{4}}); \tag{4.4a}$$

$$J(v, u; v_0) = \int_0^\infty \left[|A^{\frac{3}{4}}v(t; v_0)|^2_H + |u(t)|^2_{\mathbb{R}^{2K}}\right] dt, \quad v_0 \in \mathcal{D}(A^{\frac{1}{4}}), \tag{4.4b}$$

which penalizes the $L_2(0,\infty;\mathbb{R}^{2K})$-norm of the (finite-dimensional) controls as well as the $\mathcal{D}(A^{\frac{3}{4}})$-norm (i.e., $H^{\frac{3}{2}}(\Omega))^d \cap H$-norm) of the solution v of the *linearized* (and translated, by subtracting the equilibrium solution) N-S equations [B-T.1, Eqn. (2.5)], given by (see (1.1.14), (1.1.15)):

$$v_t = -(\nu_0 A + A_0)v + Pm\sum_{i=1}^{2K}u_i\psi_i = \mathcal{A}v + Pmu, \quad v(0) = v_0, \tag{4.5}$$

P being the Leray projector. The optimal solution of the OCP is given in feedback form via the Riccati operator R_N. Using this same feedback in the full nonlinear N-S equations, [B-T.1, Theorem 2.2] provides the following exponential, local, feedback stabilization result, in the vicinity $\mathcal{V}_\rho$:

$$\mathcal{V}_\rho \equiv \{y_0 \in \mathcal{D}(A^{\frac{1}{4}}) : |A^{\frac{1}{4}}(y_0 - y_e)|_H < \rho\} \tag{4.6}$$

of the equilibrium solution y_e, defined by (1.1.2).

Theorem 4.1. [B-T.1, Theorem 2.2] *Let ω be an arbitrary subdomain ω of Ω of class C^2. Let $\rho > 0$ be sufficiently small, and let the I.C. $y_0 \in \mathcal{V}_\rho$ defined by (4.6). Then:*

(a) *there exists a Riccati linear operator*

$$R_N \in \mathcal{L}(\mathcal{D}(A^{\frac{1}{4}}); [\mathcal{D}(A^{\frac{1}{4}})]') \cap \mathcal{L}(\mathcal{D}(A^{\frac{1}{2}}); H),$$

positive, self-adjoint on $\mathcal{D}(A^{\frac{1}{4}})$, *and satisfying* (4.3) *and* (4.4); *moreover,* $|R_N y|_H$ *is norm equivalent to* $|A^{\frac{1}{4}} y|_H$, $y \in \mathcal{D}(A^{\frac{1}{4}})$;

(b) *there exist (constructively) suitable vector* $\psi_i \in L^2(\omega)$, $i = 1, \ldots, 2K$ *identified in* [B-T.1, Eqn. (3.52)], *such that the closed loop N-S system*

$$
\begin{cases}
y_t(x,t) - \nu_0 \Delta y(x,t) + (y \cdot \nabla) y(x,t) + m\left\{ \displaystyle\sum_{i=1}^{2K} (R_N(y - y_e), \psi_i)_{L^2(\omega)} \psi_i \right\} \\
\qquad\qquad = \quad f_e(x) + \nabla p(x,t) \quad in \ Q \equiv \Omega \times (0, \infty); \qquad\qquad (4.7a) \\
\qquad\quad \nabla \cdot y \quad \equiv \quad 0 \qquad\qquad\qquad\qquad in \ Q; \qquad\qquad\qquad\qquad (4.7b) \\
\qquad\qquad\; y \quad \equiv \quad 0 \qquad\qquad\qquad\qquad on \ \Sigma \equiv \partial\Omega \times (0, \infty); \qquad (4.7c) \\
\qquad y(x,0) \quad = \quad y_0(x) \qquad\qquad\qquad in \ \Omega, \qquad\qquad\qquad\qquad (4.7d)
\end{cases}
$$

– which is obtained by replacing u *in* (4.1a) *in the feedback form given by* (4.2) *– [or, correspondingly, its abstract version, as projected onto* H:

$$\frac{dy}{dt} + \nu_0 A y + B y + P m \left\{ \sum_{i=1}^{2K} (R_N(y - y_e), \psi_i)_{L^2(\omega)} \psi_i \right\} = P f_e, \quad t \geq 0, \ y(0) = y_0,$$

$$(4.8)$$

$P = $ *Leray projector] possesses a weak solution*

$$
\begin{cases}
\quad y \quad \in \quad L^\infty(0, T; H) \cap L^2(0, T; V); & (4.9a) \\[2mm]
\dfrac{dy}{dt} \quad \in \quad L^{\frac{4}{3}}(0, T; V'), \ d = 3, \qquad \forall\, T > 0, & (4.9b) \\[2mm]
\dfrac{dy}{dt} \quad \in \quad L^2(0, T; V'), \ d = 2; & (4.9c)
\end{cases}
$$

such that the following asymptotic properties hold true:

$$\text{(i)} \qquad \int_0^\infty |A^{\frac{3}{4}}(y(t) - y_e)|_H^2 \, dt \leq C |A^{\frac{1}{4}}(y_0 - y_e)|_H; \qquad ((4.10))$$

$$\text{(ii)} \qquad |A^{\frac{1}{4}}(y(t) - y_e)|_H \leq M e^{-at} |A^{\frac{1}{4}}(y_0 - y_e)|_H, \quad t \geq 0, \quad ((4.11))$$

for some $M \geq 1$, $a > 0$. [*For* $d = 2$, *the solution of* (4.7) *is strong and unique.*]

Goal of section. The purpose of this section is to re-examine the above localized interior feedback stabilization problem within the context of both the abstract Setting #1 and the abstract Setting #2, introduced in Section 1.2 in connection with the *boundary* stabilization problem of the N-S system (1.1.1). Thus, in the expressions (1.2.5a) and (1.2.11a) of the generator A_F, one should take $D =$ Identity. Moreover, in (1.2.4) and (1.2.10), U is replaced by W and H, respectively.

Within the context of Setting #1, we shall reprove Theorem 4.1 (in a slightly more precise version, claiming a unique fixed point solution), with high-gain feedback controller, by verifying the validity of assumptions (H.1(i), (ii), (iii)) in (1.2.5)–(1.2.7), and then appealing to Theorem 1.2.1. Of course, verification of (H.1) rests heavily on [B-T.1].

Within the context of the simplified or specialized version of the abstract Setting #2, we shall instead obtain a new result, labeled Theorem 4.2 below, which is a variation of Theorem 4.1, this time, however, with a low (that is, $H-$) topological level of a new OCP given by (4.19). Theorem 4.2 will be obtained by verifying the validity of assumptions (H.2(i), (ii), (iii), (iv)) – a task that again heavily relies on [B-T.1], of course – and then invoking Theorem 1.2.2.

Low-gain stabilizing feedback controller. As announced, we shall complement Theorem 4.1 with the following new result, a variation of Theorem 4.1 at the low-gain level.

Theorem 4.2. *Assume the hypotheses of Theorem 4.1. Thus, ω is an arbitrary subdomain of Ω of class C^2 and the I.C. $y_0 \in \mathcal{V}_\rho$ in (4.6), with $\rho > 0$ sufficiently small. Then:*

(a) *there exists a Riccati operator $R_N^0 \in \mathcal{L}(H)$ (positive self-adjoint on H), identified in (4.14) and also (4.19a) below;*

(b) *there exist (constructively) suitable vectors $\chi_i \in L^2(\omega)$, $i = 1, \ldots, 2K$, such that the closed loop N-S system obtained from (4.7a–d) [or (4.8)] by replacing*

$$R_N \text{ with } R_N^0; \quad \psi_i \text{ with } \chi_i, \tag{4.12}$$

possesses a unique fixed point solution y satisfying

$$(y - y_e) \in L^2(0, \infty; \mathcal{D}(A^{\frac{3}{4}})) \cap C([0, \infty]; \mathcal{D}(A^{\frac{1}{4}})) \tag{4.13}$$

continuously in $(y_0 - y_e) \in \mathcal{D}(A^{\frac{1}{4}})$. Thus, the counterpart of the asymptotic property (4.10) holds true. Moreover, the counterpart of the uniform exponential decay in (4.11) holds true as well. The operator R_N^0 is defined by (4.19a) below and is the unique, positive, self-adjoint solution in $\mathcal{L}(H)$, satisfying the following ARE, where $\mathcal{A}$ is the Oseen operator in (1.1.14):

$$(\mathcal{A}z, R_N^0 z)_H + \frac{1}{2}|z|_H^2 = \frac{1}{2}\sum_{i=1}^{2K}(R_N^0 z, \chi_i)_{L^2(\omega)}^2, \quad \forall\, z \in H, \tag{4.14}$$

since now $\mathcal{A}^* R_N^0 \in \mathcal{L}(H)$ (refer to Theorem 3.1.1(ii) and Remark 3.1.1 referring to [B-L-T.1, Appendix B.4], ultimately to [L-T.4, Appendix 2B, p. 168], since:

(i) *$\mathcal{A}$ is a lower-order perturbation of the self-adjoint operator $-\nu_0 A$ on H, by (1.1.14), (1.1.15) and $R_N^0 \in \mathcal{L}(H)$;*

(ii) *the observation operator in the cost functional (4.19b) is the identity on H).*

Proof of Theorem 4.1 via Theorem 1.2.1. We need to verify the validity of assumption (H.1(i), (ii), (iii)) of Setting #1. This will be done by invoking results of [B-T.1]. First, by [B-T.1, Lemma 3.8], the OCP (4.4) satisfies the Finite Cost condition [L-T.4] (the functional J in (4.4b) is proper for any $v_0 \in \mathcal{D}(A^{\frac{1}{4}})$) within the class of $L^2(0,\infty;\mathbb{R}^{2K})$-controllers $u_N(x,t) = \sum_{i=1}^{2K} u_N^i(t)\psi_i(x)$, for suitable vectors $\{\psi_i\}_{i=1}^{2K}$ identified there in [B-T.1, Eqn. (3.52)]. Accordingly, the optional solution $v^*(t;v_0)$ of the OCP (4.4) is given by the s.c. feedback semigroup

$$
v^*(t;v_0) \equiv e^{A_F t}v_0, \quad A_F = \mathcal{A} + P\left(m \sum_{i=1}^{2K} (R_N \cdot, \psi_i)_{L^2(\omega)}\psi_i \right) \tag{4.15}
$$

[B-T.1, Eqn. (3.81)], where the s.c. feedback semigroup $e^{A_F t}$ is analytic on $\mathcal{D}(A^{\frac{1}{4}})$ and, moreover, is (by Datko's theorem) exponentially stable on $\mathcal{D}(A^{\frac{1}{4}})$ [B-T.1, Eqn. (3.82)]

$$
|e^{A_F t}|_{\mathcal{L}(\mathcal{D}(A^{\frac{1}{4}}))} \leq C_\gamma e^{-\gamma t}, \quad t \geq 0, \tag{4.16}
$$

for some constant $\gamma > 0$. Of course, we have

$$
\int_0^\infty |e^{A_F t}v_0|_Z^2 = \int_0^\infty |A^{\frac{3}{4}}v^*(t;v_0)|_H^2 dt \leq C|v_0|_{\mathcal{D}(A^{\frac{1}{4}})}^2. \tag{4.17}
$$

Accordingly, if we set, for $d = 2, 3$:

$$
W \equiv \mathcal{D}(A^{\frac{1}{4}}); \quad Z \equiv \mathcal{D}(A^{\frac{3}{4}}); \quad F \equiv P\left(m \sum_{i=1}^{2K} (R_N \cdot; \psi_i)_{L^2(\omega)}\psi_i \right), \tag{4.18}
$$

with ψ_i identified by [B-T.1, Eqn. (3.52)], we see that *properties* (4.15), (4.16), (4.17) *verify*, respectively, *properties* (H.1i), (H.1ii), (H.1iii), *in* (1.2.5), (1.2.6), (1.2.7). We then invoke Theorem 1.2.1 and obtain Theorem 4.1, in the more refined version that the local solution of (4.7a–d), or of (4.8), is a unique fixed point solution (see [B-L-T.2, Section 5]).

Proof of Theorem 4.2 via Theorem 1.2.2. We shall use Setting #2. We need to verify assumptions (H.2(i), (ii), (iii), (iv)) in (1.2.10)–(1.2.14). To this end, instead of the OCP (4.4), we consider the new OCP

$$
(R_N^0 v_0, v_0)_H = \inf_{u \in L^2(0,\infty;\mathbb{R}^{2K})} J_0(u, v; v_0) = J_0^*(v_0), \quad v_0 \in H; \tag{4.19a}
$$

$$
J_0(u, v; v_0) = \int_0^\infty \left[|v(t;v_0)|_H^2 + |u(t)|_{\mathbb{R}^{2K}}^2 \right] dt, \quad v_0 \in H, \tag{4.19b}
$$

$v(\cdot;v_0)$ the solution of the linearized equation (4.5) where $\psi_i = \tilde{\psi}_i$. This is a standard OCP where the control operator (Pm) is *bounded* on H. Thus, *a fortiori*, amply contained in [L-T.4, Theorem 2.2.1], see also [Bal.1].

In particular, the following results hold true:

(i) the operator

$$A_F = \mathcal{A} + Pm\left(\sum_{i=1}^{2K}(R_N^0\,\cdot\,,\tilde{\psi}_i\right)_{L^2(\omega)}\tilde{\psi}_i \tag{4.20}$$

generates a s.c. analytic semigroup $e^{A_F t}$ on H, for all $\tilde{\psi}_i \in L^2(\omega)$.

(ii) As in [B-T.1, Section 3] (Lemma 3.8 in [B-T.1] is a *finite*-dimensional result), we can select (in infinitely many ways) suitable vectors $\tilde{\psi}_i = \chi_i \in L^2(\omega)$ such that $e^{A_F t}$ is, moreover, uniformly (exponentially) stable on H: $|e^{A_F t}|_{\mathcal{L}(H)} \le M_1 e^{-\delta_1 t}$, $t \ge 0$, $\delta_1 > 0$. Henceforth, we shall restrict to such vectors $\tilde{\psi}_i = \chi_i$.

(iii) R_N^0 is a positive, self-adjoint operator in $\mathcal{L}(H)$, the unique such solution of an ARE (4.14). We henceforth select the following spaces

$$W \equiv \mathcal{D}(A^{\frac{1}{4}}); \qquad Z \equiv \mathcal{D}(A^{\frac{3}{4}}), \tag{4.21}$$

in both cases $d = 2, 3$. Since

$$\mathcal{D}(A_F) = \mathcal{D}(\mathcal{A}) = \mathcal{D}(A), \text{ hence } \mathcal{D}((-A_F)^\theta) = \mathcal{D}(A^\theta), \quad 0 \le \theta \le 1, \tag{4.22}$$

by interpolation, we have that:

(1) $e^{A_F t}$ continues to be a s.c. analytic semigroup also on the space $W = \mathcal{D}(A^{\frac{1}{4}})$.

(2) $e^{A_F t}$ remains uniformly (exponentially) stable on W: $|e^{A_F t}|_{\mathcal{L}(W)} \le M_1 e^{-\delta_1 t}$, $t \ge 0$, $\delta_1 > 0$.

In addition, we have

$$(3) \qquad \int_0^\infty |e^{A_F t}v_0|_Z^2 dt \le \text{const}|v_0|_W^2. \tag{4.23}$$

Proof of (4.23). Since $Z = \mathcal{D}(A^{\frac{3}{4}}) = \mathcal{D}((-A_F)^{\frac{3}{4}})$ and $W = \mathcal{D}(A^{\frac{1}{4}}) = \mathcal{D}((-A_F)^{\frac{1}{4}})$ by (4.21), we have for $v_0 \in W$:

$$\int_0^\infty |e^{A_F t}v_0|_Z^2 dt$$

$$= \int_0^\infty |e^{A_F t}v_0|_{\mathcal{D}((-A_F)^{\frac{3}{4}})}^2 dt = \int_0^\infty |(-A_F)^{\frac{3}{4}}e^{A_F t}v_0|_H^2 dt \tag{4.24}$$

$$= \int_0^\infty |(-A_F)^{\frac{1}{2}}e^{A_F t}(-A_F)^{\frac{1}{4}}v_0|_H^2 dt \le C|(-A_F)^{\frac{1}{4}}v_0|_H^2 = c|v_0|_W^2, \tag{4.25}$$

and (4.23) is established. All the steps are justified since $e^{A_F t}$ is a s.c. analytic semigroup on H, generated by $A_F = -\nu_0 A + (-A_0 + Pm)$ which is a lower-order perturbation $\mathcal{D}(-A_0 + Pm) = \mathcal{D}(A_0) = \mathcal{D}(A^{\frac{1}{2}})$ of the negative, self-adjoint operator $-\nu_0 A$. [A more complicated way to show (4.23) is to recall that $v^0(\,\cdot\,;v_0) = e^{A_F t}v_0 = $ optimal solution of the OCP (4.19) and invoke regularity properties of the optimal pair in a boot-strap argument as in Proposition 3.1.2 proved in [B-L-T.2, Section 6], ultimately [L-T.4].] Finally, we have

$$(4) \qquad\qquad \mathcal{D}(A_F) = \mathcal{D}(A) \subset \mathcal{D}(A^{\frac{3}{4}}) = Z. \qquad\qquad (4.26)$$

Thus, properties (1), (2), (3), (4) are the counterpart of the properties (H.1(i), (ii), (iii), (iv)) of the abstract Setting #2. Thus, the counterpart of Theorem 1.2.2 holds true and – via the proof of [B-L-T.2, Section 5] – produces a unique fixed point solution of problem (4.7) with the following Riccati-based feedback control

$$u = -\sum_{i=1}^{2K} (R_N^0(y - y_e), \tilde{\psi}_i)_{L^2(\omega)} \tilde{\psi}_i \equiv -(B^0)^* R_N^0(y - y_e).$$

Theorem 4.2 is proved. $\qquad\qquad\qquad\qquad\qquad\qquad\qquad\qquad\qquad\qquad$ $\square$

5. The case of exponential stabilization of N-S equations, $d = 2, 3$, by means of localized interior, spectral-based feedback controls

In this section we shall appeal to results of [B-T.1, Section 3] in order to verify the validity of the assumptions (H.2(i), (ii), (iii), (iv)) in (1.2.10)–(1.2.14). This way, we will be able to invoke Theorem 1.2.2 and establish the following local exponential stabilization of the N-S equations, $d = 2, 3$, with a localized interior, finite-dimensional, *spectral-based* feedback controllers of the form

$$u = -\sum_{i=1}^{2K} (y - y_e, \tilde{p}_i)_{L^2(\omega)} \tilde{\psi}_i, \qquad\qquad (5.1)$$

for suitably constructed (via [B-T.1, Lemma 3.8]) vectors $\tilde{p}_i$, $\tilde{\psi}_i \in L^2(\omega)$.

Theorem 5.1. *Let ω be an arbitrary subdomain of Ω of class C^2. Let $\rho > 0$ be sufficiently small, and let the I.C. $y_0 \in V_\rho$ defined by (4.6). Then there exist (constructively) suitable vectors $\tilde{\psi}_i, \tilde{p}_i \in L^2(\omega)$, $i = 1, \ldots, 2K$ identified in [B-T.1, Eqn. (3.52)], such that the closed loop N-S system*

$$\left\{ \begin{aligned} y_t(x,t) - \nu_0 \Delta y(x,t) &+ (y \cdot \nabla) y(x,t) + m\left\{ \sum_{i=1}^{2K} ((y - y_e), \tilde{p}_i)_{L^2(\omega)} \tilde{\psi}_i \right\} \\ &= f_e(x) + \nabla p(x,t) \quad in \ Q \equiv \Omega \times (0, \infty); \qquad\qquad (5.2\text{a}) \\ \nabla \cdot y &\equiv 0 \qquad\qquad\qquad in \ Q; \qquad\qquad\qquad\qquad (5.2\text{b}) \\ y &\equiv 0 \qquad\qquad\qquad on \ \Sigma \equiv \partial\Omega \times (0, \infty); \qquad (5.2\text{c}) \\ y(x,0) &= y_0(x) \qquad\qquad in \ \Omega, \qquad\qquad\qquad\qquad (5.2\text{d}) \end{aligned} \right.$$

– which is obtained by replacing u in (4.1a) in the feedback form given by (5.1) – [or, correspondingly, its abstract version], as projected onto H:

$$\frac{dy}{dt} + \nu_0 Ay + By + Pm\left\{\sum_{i=1}^{2K}((y-y_e),\tilde{p}_i)_{L^2(\omega)}\tilde{\psi}_i\right\} = Pf_e, \quad t \geq 0, \; y(0) = y_0, \quad (5.3)$$

P = Leray projector] possesses a unique fixed point solution such that the following asymptotic properties hold true:

(i)
$$\int_0^\infty |A^{\frac{3}{4}}(y(t)-y_e)|_H^2 dt \leq C|A^{\frac{1}{4}}(y_0-y_e)|_H; \quad (5.4)$$

(ii)
$$|A^{\frac{1}{4}}(y(t)-y_e)|_H \leq Me^{-at}|A^{\frac{1}{4}}(y_0-y_e)|_H, \quad t \geq 0, \quad (5.5)$$

for some $M \geq 1$, $a > 0$.

Proof of Theorem 5.1. The linearized semigroup is analytic on H. One begins with he complexified finite-dimensional projection [B-T.1, Eqn. (3.5)] on the unstable subspace $Z_N^u = P_N H$ in (1.3.5), (1.3.1):

$$\text{on } Z_N^u: \quad z_N' = \mathcal{A}_N^u z_N + P_N(Pmv), \quad z_N(0) = P_N z_0, \quad (5.6)$$

with finite-dimensional controllers $v = v_N = \sum_{i=1}^K v_N^i(t)w_i$, $w_i \in Z_N^u \subset H$, so that $mw_i \in H$ and $P_N mw_i = P_N Pmw_i$, $K = \max\{\ell_i, i = 1, \ldots, M\}$. We shall eventually seek controllers in feedback form $v_N^i(t) = (z_N(t), p_i)$. Let the vectors $w_i \in Z_N^u$ satisfy the algebraic rank conditions in [B-T.1, Eqn. (3.15)]. Then, the dynamics (5.6) is controllable on Z_N^u [B-T.1, Proof of Lemma 3.2]. Accordingly, (5.6) satisfies Popov's pole assignment property. In particular, given any $\gamma_1 > 0$, there exist (constructively) suitable vectors $\{p_1, \ldots, p_K\} \in Z_N^u$ such that the corresponding feedback problem

$$z_N' - \mathcal{A}_N^u z_N = \sum_{i=1}^K P_N(mw_i)(z_N, p_i), \quad z_N(0) = P_N z_0 \quad (5.7)$$

is exponentially stable with rate $\gamma_1 > 0$:

$$|z_N(t)| + |v_N(t)| \leq C_{\gamma_1} e^{-\gamma_1 t}|P_N z_0|, \quad t > 0 \quad (5.8)$$

[B-T.1, Lemma 3.1]. Next, we return from the complexified version to the corresponding real version of the problem. Since

$$u_N \equiv \text{Re } v_N = \text{Re}\sum_{i=1}^K v_N^i(t)w_i = \sum_{i=1}^K(\text{Re } v_N^i(t))(\text{Re } w_i) - \sum_{i-1}^K(\text{Im } v_N^i(t))(\text{Im } w_i) \quad (5.9)$$

$$u_N \equiv \text{Re } v_N = \sum_{i=1}^K(\text{Re}(z_N(t),p_i))(\text{Re } w_i) - \sum_{i=1}^K(\text{Im}(z_N(t),p_i))(\text{Im } w_i), \quad (5.10)$$

and with $w_i \in Z_N^u \in H$,

$$\text{Re } P_N Pmv_N = \text{Re } P_N mv = P_N m\text{Re } v_N = P_N m u_N, \quad (5.11)$$

we deduce that *a fortiori*, the real value (Re z_N)-dynamics is likewise controllable within the class of $2K$ open-loop, real-valued controls of the form (5.10), (5.11). [However, because of the presence of Im z_N, such controls in (5.10), (5.11) are only open-loop, and not closed loop. Accordingly, by standard, finite-dimensional control theory, the linear dynamics in (Rez_N) can be stabilized by a 2K-dimensional, real-valued, closed-loop feedback control of the same feedback low as its complex-valued counterpart. Hence, there exist real vectors $\tilde{p}_i \in \tilde{Z}_N^u$, $i = 1, \ldots, 2K$, such that the real feedback dynamics

$$(\mathrm{Re}\ z_N)_t = \mathcal{A}_N^u(\mathrm{Re}\ z_N) + \sum_{i=1}^{2K}(\mathrm{Re}\ z_N(t), \tilde{p}_i)\psi_i, \quad \mathrm{Re}\ z_N(0) = \mathrm{Re}\ z_0 \quad (5.12)$$

satisfies

$$|(\mathrm{Re}\ z_N)(t)|_H \leq C_{\gamma_1} e^{-\gamma_1 t}|\mathrm{Re}\ z_0|_H, \quad t \geq 0, \quad (5.13)$$

where we have set

$$\left\{ \begin{array}{rcl} \psi_i & = & \mathrm{Re}\ w_i, \ i = 1, \ldots, K; \ \psi_{i+K} = -\mathrm{Im}\ w_i, \ i = 1, \ldots, K; \quad (5.14a) \\ \tilde{Z}_N^u & = & \mathrm{span}\{\mathrm{Re}\ \varphi_i, \ \mathrm{Im}\ \varphi_{ij}\}_{i=1}^M {}_{j=1}^{\ell_i}, \ \varphi_{ij}\ \mathrm{in}\ (1.3.4). \quad (5.14b) \end{array} \right.$$

Finally, the argument of [B-T.1, Lemma 3.8] shows that the real linearized dynamics

$$v_t = -(\nu_0 A + A_0)v + Pmu_N = \mathcal{A}v + Pm\left\{\sum_{i=1}^{2K}(v(t), \tilde{p}_i)\psi_i\right\}, \quad (5.15)$$

that is,

$$v_t = A_F v, \ v(t) = e^{A_F t}v_0; \ A_F = \mathcal{A} + Pm\sum_{i=1}^{2K}(\cdot, \tilde{p}_i)\psi_i \quad (5.16)$$

satisfies the asymptotic pointwise decays

$$|v(t)|_H = |e^{A_F t}v_0|_H \leq C_{\gamma_0}e^{-\gamma_0 t}|v_0|_H, \quad t \geq 0, \quad (5.17)$$

$$|A^{\frac{1}{4}}v(t)|_H = |A^{\frac{1}{4}}e^{A_F t}v_0|_H \leq C_{\gamma_0}e^{-\gamma_0 t}|A^{\frac{1}{4}}v_0|_H, \quad t \geq 0, \quad (5.18)$$

with γ_0 any constant such that $0 < \gamma_0 < \mathrm{Re}\ \lambda_{N+1}$, see [B-T.1, (3.5.1a–b)]. Moreover, the argument of [B-T.1, Section 3.4, Step 1, p. 1475] then shows that

$$\int_0^\infty |A^{\frac{3}{4}}e^{A_F t}v_0|_H dt = \int_0^\infty |A^{\frac{3}{4}}v(t)|_H dt \leq C|A^{\frac{1}{4}}v_0|_H, \quad v_0 \in \mathcal{D}(A^{\frac{1}{4}}). \quad (5.19)$$

Finally, we set

$$W \equiv \mathcal{D}(A^{\frac{1}{4}}); \quad Z \equiv \mathcal{D}(A^{\frac{3}{4}}); \quad F \equiv Pm\sum_{i=1}^{2K}(\cdot, \tilde{p}_i)\psi_i. \quad (5.20)$$

We then have that (5.15), (5.17), (5.19) satisfy assumptions (H.2i), (H.2ii), (H.2iii). Moreover,

$$\mathcal{D}(A_F) = \mathcal{D}(A) \subset \mathcal{D}(A^{\frac{3}{4}}) \equiv Z, \quad (5.21)$$

and (H.2iv) is likewise satisfied. Having verified assumption (H.2), we invoke Theorem 1.2.2 and, this way, obtain Theorem 5.2. $\square$

Remark 5.1. Using (5.18), one could also verify (H.1) and then invoke Theorem 1.2.1 to obtain Theorem 5.2. $\square$

References

[Bal.1]   A.V. Balakrishnan, *Applied Functional Analysis*, Springer-Verlag, 1981.

[B-D-D-M.1] A. Bensoussan, G. Da Prato, M. Delfour, S. Mitter, *Representation and Control of Infinite-Dimensional Systems*, Vol. 1, Birkhäuser, 1992.

[B.1]   V. Barbu, Feedback stabilization of Navier-Stokes equations, *ESAIM Control Optim. Calc. Var.* 9 (2003), 197–206 (electronic).

[B.2]   V. Barbu, Stabilization of a plane channel flow by wall-normal controllers, to appear.

[B-L-T.1] V. Barbu, I. Lasiecka, and R. Triggiani, Tangential boundary stabilization of Navier-Stokes equations, *Memoir AMS*, Number 852, Volume 181, ISSN 0065-9266, 128 pp., May 2006.

[B-L-T.2] V. Barbu, I. Lasiecka, and R. Triggiani, Abstract settings for tangential boundary stabilization of Navier-Stokes equations by high- and low-gain feedback controllers, *J. Nonlinear Anal.*, to appear.

[B-T.1]   V. Barbu and R. Triggiani, Internal stabilization of Navier-Stokes equations with finite-dimensional controllers, *Indiana University Mathematics Journal* 53(5) (2004), 1443–1494.

[C-F.1]   P. Constantin and C. Foias, *Navier-Stokes Equations*, University of Chicago Press, Chicago-London, 1989.

[F-L-T.1] F. Flandoli, I. Lasiecka, and R. Triggiani, Algebraic Riccati equations with non-smoothing observation arising in hyperbolic and Euler-Bernoulli equations, *Ann. di Matematica Pura e Appl.* CLII (1988), 307–382.

[Fu.1]   A.V. Fursikov, Real processes of the 3D Navier-Stokes systems and its feedback stabilization from the boundary, in *AMS Translations*, Series 2, Vol. 206, *Partial Differential Equations*, Mark Vishik's Seminar; M.S. Agranovich, M.A. Shubin, editors (2002), 95–123.

[Fu.2]   A.V. Fursikov, Feedback stabilization for the 2D Oseen equations: Additional remarks, Proc. 8th Conference on Control of Distributed Parameter Systems, *Int. Ser. Numerical Math.*, Birkhäuser-Verlag 143 (2002), 169–187.

[Fu.3]   A.V. Fursikov, Stabilization for the 3D Navier-Stokes system of feedback boundary control, *Disc. Cont. Dynam. Systems* 10 (182) (2004)m 289–314.

[H.1]   D. Henry, *Perturbation of the Boundary in Boundary Value Problems*, London Mathematical Society Lecture Notes 318, Cambridge University Press, 2005.

[K.1]   T. Kato, *Perturbation Theory of Linear Operators*, Springer-Verlag, New York-Berlin, 1966.

[La.1]   I. Lasiecka, Boundary control of parabolic systems: Regularity of optimal solutions, *Appl. Math. & Optimiz.* 4 (1978), 301–327.

[La.2]   I. Lasiecka, Unified theory for abstract parabolic boundary problems: A semigroup approach, *Appl. Math. & Optimiz.* 6 (1980), 287–333.

[La.3]   I. Lasiecka, Exponential stabilization of hyperbolic systems with nonlinear, unbounded perturbations: A Riccati operator approach, *Appl. Anal.* 42 (1991), 2434–261.

[L-T.1]   I. Lasiecka an R. Triggiani, Dirichlet boundary control problems for parabolic equations with quadratic cost: Analyticity and Riccati feedback synthesis, *SIAM J. Control & Optimiz.* 21(1) (1983), 41–67.

[L-T.2] I. Lasiecka an R. Triggiani, The regulator problem for parabolic equations with Dirichlet boundary control. Part I: Riccati's feedback synthesis and regularity of optimal solution, *Appl. Math. & Optimiz.* 16 (1986), 147–168.

[L-T.3] I. Lasiecka an R. Triggiani, *Differential and Algebraic Riccati Equations with Applications to Boundary/Point Control Problems: Continuous Theory and Approximation Theory*, Springer Verlag Lecture Notes, LNCIS, Vol. 164 (1991), 160 pp.

[L-T.4] I. Lasiecka an R. Triggiani, *Control Theory for Partial Differential Equations: Continuous and Approximation Theories I: Abstract Parabolic Systems*, Encyclopedia of Mathematics and its Applications 74, Cambridge University Press (2002), 648 pp.

[L-M.1] J.L. Lions and E. Magenes, *Nonhomogeneous Boundary Value Problems and Applications*, Vol. 1 (1970); Vol. 2 (1972), Springer-Verlag.

[R.1] J.P. Raymond, Feedback boundary stabilization of the two-dimensional Navier-Stokes equations, preprint, 2005.

[Te.1] R. Temam, *Navier-Stokes Equations*, Studies in Mathematics and its Applications, Vol. 1, North Holland Publishing Co., Amsterdam, 1979.

[Tr.1] R. Triggiani, On the stabilizability problem in Banach spaces, *J. Math. Anal. & Appl.* (1975), 383–403.

[Tr.2] R. Triggiani, Boundary feedback stabilizability of parabolic equations, *Appl. Math. Optimiz.* 6 (1980), 201–220.

[Tr.3] R. Triggiani, Exponential feedback stabilization of a 2-D linearized Navier-Stokes channel flow by finite-dimensional wall normal boundary controllers, with arbitrarily small support, *Discrete and Continuous Dynamical Systems, B*, to appear.

[W.1] W. von Wahl, *The Equations of Navier-Stokes and Abstract Parabolic Equations*, Vieweg & Sohn, Braunschweig, 1985.

Viorel Barbu
Dept. of Mathematics
University "Al. I. Cuza"
RO-6600 Iasi, Romania
e-mail: vb41@uaic.ro

Irena Lasiecka and Roberto Triggiani
Department of Mathematics
University of Virginia
Charlottesville, VA 22903, USA
e-mail: il2v@virginia.edu
e-mail: rt7u@virginia.edu

International Series of Numerical Mathematics, Vol. 155, 47–68
© 2007 Birkhäuser Verlag Basel/Switzerland

Convergence Analysis of an Adaptive Finite Element Method for Distributed Control Problems with Control Constraints

A. Gaevskaya, R.H.W. Hoppe, Y. Iliash and M. Kieweg

Abstract. We develop an adaptive finite element method for a class of distributed optimal control problems with control constraints. The method is based on a residual-type a posteriori error estimator and incorporates data oscillations. The analysis is carried out for conforming P1 approximations of the state and the co-state and elementwise constant approximations of the control and the co-control. We prove convergence of the error in the state, the co-state, the control, and the co-control. Under some additional non-degeneracy assumptions on the continuous and the discrete problems, we then show that an error reduction property holds true at least asymptotically. The analysis uses the reliability and the discrete local efficiency of the a posteriori estimator as well as quasi-orthogonality properties as essential tools. Numerical results illustrate the performance of the adaptive algorithm.

Mathematics Subject Classification (2000). Primary 65K10; Secondary 49M15.

Keywords. Distributed optimal control, control constraints, adaptive finite elements, residual-type a posteriori error estimators, convergence analysis.

1. Introduction

We present a convergence analysis of adaptive finite element approximations of a distributed optimal control problem with control constraints. In particular, assuming $\Omega \subset \mathbb{R}^2$ to be a bounded, polygonal domain with boundary $\Gamma := \partial\Omega$ and given data $y^d \in L^2(\Omega)$ and $f \in L^2(\Omega), \psi \in H^1(\Omega) \cap L^\infty(\Omega)$ as well as a parameter $0 < \alpha \le 1$, we consider the following distributed optimal control problems with

The second author has been partially supported by the NSF under Grant No. DMS-0411403 and Grant No. DMS-0511611. The fourth author acknowledges the support by the elite graduate school TopMath.

bound constrained controls

$$\text{minimize } J(y, u) \; := \; \frac{1}{2} \, \|y - y^d\|_{0,\Omega}^2 \; + \; \frac{\alpha}{2} \, \|u\|_{0,\Omega}^2 \tag{1.1a}$$

over $(y, u) \in H_0^1(\Omega) \times L^2(\Omega)$,

$$\text{subject to } - \Delta\, y \; = \; f \; + \; u \;, \tag{1.1b}$$

$$u \in K \; := \; \{v \in L^2(\Omega) \mid v \leq \psi \text{ a.e. in } \Omega\} \;. \tag{1.1c}$$

It is well known (cf., e.g., [15, 20, 21]) that (1.1a)–(1.1c) admits a unique solution $(y, u) \in H_0^1(\Omega) \times L^2(\Omega)$. The optimality conditions involve the existence of a co-state $p \in H_0^1(\Omega)$ and a co-control $\sigma \in L_+^2(\Omega)$ such that y, p, u, σ satisfy

$$a(y, v) \; = \; (f + u, v)_{0,\Omega} \quad, \quad v \in H_0^1(\Omega) \;, \tag{1.2a}$$

$$a(p, v) \; = \; -\, (y - y^d, v)_{0,\Omega} \quad, \quad v \in H_0^1(\Omega) \;, \tag{1.2b}$$

$$u \; = \; \frac{1}{\alpha} \, (p - \sigma) \in K \;, \tag{1.2c}$$

$$(\sigma, u - v)_{0,\Omega} \; \geq \; 0 \quad, \quad v \in K \;. \tag{1.2d}$$

Here, $(\cdot, \cdot)_{0,\Omega}$ refers to the standard L^2 inner product and $a(\cdot, \cdot)$ stands for the bilinear form

$$a(w, z) \; := \; \int_\Omega \nabla w \cdot \nabla z \; dx \quad, \quad w, z \in H_0^1(\Omega) \;.$$

We note that the variational inequality (1.2d) can be equivalently stated as the complementarity condition

$$\sigma \in L_+^2(\Omega) \;, \; \psi - u \in L_+^2(\Omega) \;, \tag{1.3}$$
$$(\sigma, \psi - u)_{0,\Omega} \; = \; 0 \;.$$

We define the active control set $\mathcal{A}(u)$ as the maximal open set $A \subset \Omega$ such that $u(x) = \psi(x)$ f.a.a. $x \in A$ and the inactive control set $\mathcal{I}(u)$ according to $\mathcal{I}(u) := \bigcup_{\varepsilon > 0} B_\varepsilon$, where B_ε is the maximal open set $B \subset \Omega$ such that $u(x) \leq \psi(x) - \varepsilon$ for almost all $x \in B$. Further, we refer to $\mathcal{F}(u) := \partial \mathcal{A}(u)$ as the free boundary between the active and inactive sets.

The control problem (1.1a)–(1.1c) will be approximated by Lagrangian type finite elements with respect to an adaptively generated hierarchy of simplicial triangulations of the computational domain. We note that adaptive finite element methods (AFEM) are efficient and reliable algorithmic tools in the numerical solution of partial differential equations. AFEMs typically consist of successive loops of the sequence

$$\text{SOLVE} \to \text{ESTIMATE} \to \text{MARK} \to \text{REFINE} \;. \tag{1.4}$$

Here, SOLVE stands for the numerical solution of the finite element discretized problem, ESTIMATE requires the a posteriori estimation of the global discretization error in some appropriate norm or with respect to a goal oriented error functional. The step MARK is devoted to the selection of elements and edges for refinement, and the final step REFINE takes care of the technical realization of the refinement process.

The development, analysis and implementation of efficient and reliable a posteriori error estimators has been the subject of intensive research in the past two decades and has actually reached some level of maturity (see, e.g., the monographs [1, 3, 4, 14, 25, 26] and the references therein). On the other hand, a rigorous convergence analysis of (1.4) relying on appropriate error reduction properties has so far only been done for conforming AFEMs [8, 13, 24] and, very recently, by Carstensen and the second author for mixed and nonconforming finite element methods in [10, 11] as well as for edge element methods for eddy current equations in [12].

As far as the a posteriori error analysis of adaptive finite element schemes for optimal control problems is concerned, the unconstrained case has been considered in [4, 6], whereas residual-type a posteriori error estimators in the control constrained case have been derived and analyzed in [17, 19, 21, 22]. No convergence analysis has been addressed so far.

This contribution aims to provide a convergence analysis of AFEM for (1.1a)–(1.1c). The paper is organized as follows:

In Section 2, we consider the finite element approximation of (1.1a)–(1.1c) and present details of the adaptive loop (1.4) focusing on a residual-type a posteriori error estimator in the step ESTIMATE and a bulk criterion for the selection of edges and elements for refinement in the step MARK. The reliability of the estimator and its discrete local efficiency are shown in Section 3. In Section 4, we prove convergence of the discrete states and co-states in $H_0^1(\Omega)$ and of the discrete controls and co-controls in $L^2(\Omega)$. Under the assumptions of strict complementarity and non-degeneracy, in Section 5 we show that an error reduction property holds true at least asymptotically. Finally, Section 6 illustrates the performance of the estimator by an illustrative numerical example.

2. Finite element discretization

For the finite element discretization of (1.1a)–(1.1c) we assume that $\mathcal{T}_\ell$ is a shape-regular simplicial triangulation of Ω. We refer to $\mathcal{N}_\ell(D)$, $\mathcal{E}_\ell(D)$, and $\mathcal{T}_\ell(D)$, $D \subseteq \overline{\Omega}$, as the sets of vertices, edges and elements of $\mathcal{T}_\ell$ in $D \subseteq \overline{\Omega}$. We set $h_\ell := \max\{h_T | T \in \mathcal{T}_\ell\}$ where h_T stands for the diameter of an element $T \in \mathcal{T}_\ell$ and we denote by h_E the length of an edge $E \in \mathcal{E}_\ell$. Further, we refer to g_T as the integral mean of $g \in L^2(\Omega)$ on $T \in \mathcal{T}_\ell$, i.e., $g_T = |T|^{-1} \int_T g \, dx$. We denote by

$$V_\ell := \{ v_\ell \in C_0(\Omega) \mid v_\ell|_T \in P_1(T) , \ T \in \mathcal{T}_\ell \} ,$$

the standard conforming P1 finite element space and by

$$W_\ell := \{ w_\ell \in L^2(\Omega) \mid w_\ell|_T \in P_0(T) , \ T \in \mathcal{T}_\ell \}$$

the linear space of elementwise constant functions on Ω. We refer to $y_\ell \in V_\ell$ and $u_\ell \in W_\ell$ as finite element approximations of the state y and the control u, respectively. The upper obstacle ψ is approximated by the elementwise constant function $\psi_\ell \in W_\ell$ with $\psi_\ell|_T := \psi_T, T \in \mathcal{T}_\ell$.

Then, the finite element approximation of the distributed optimal control problem (1.1a)–(1.1c) reads as follows:

$$\text{minimize } J_\ell(y_\ell, u_\ell) := \frac{1}{2} \, \|y_\ell - y^d\|_{0,\Omega}^2 + \frac{\alpha}{2} \|u_\ell\|_{0,\Omega}^2 \ , \tag{2.1a}$$

$$\text{over } (y_\ell, u_\ell) \in V_\ell \times W_\ell \ ,$$

$$\text{subject to } a(y_\ell, v_\ell) = (f + u_\ell, v_\ell)_{0,\Omega} \ , \quad v_\ell \in V_\ell \ , \tag{2.1b}$$

$$u_\ell \in K_\ell := \{ w_\ell \in W_\ell \mid w_\ell|_T \le \psi_T \ , \ T \in \mathcal{T}_\ell \} \ . \tag{2.1c}$$

The optimality conditions for (2.1a)–(2.1c) again give rise to the existence of a co-state $p_\ell \in V_\ell$ and a co-control $\sigma_\ell \in W_\ell$ such that

$$a(y_\ell, v_\ell) \ = \ (f + u_\ell, v_\ell)_{0,\Omega} \ , \quad v_\ell \in V_\ell \ , \tag{2.2a}$$

$$a(p_\ell, v_\ell) \ = \ - \, (y_\ell - y^d, v_\ell)_{0,\Omega} \ , \quad v_\ell \in V_\ell \ , \tag{2.2b}$$

$$u_\ell \ = \ \frac{1}{\alpha} \, (M_\ell p_\ell - \sigma_\ell) \in K_\ell \ , \tag{2.2c}$$

$$(\sigma_\ell, v_\ell - u_\ell)_{0,\Omega} \ = \ 0 \ v_\ell \in K_\ell \ . \tag{2.2d}$$

Here, $M_\ell : H_0^1(\Omega) \to W_\ell$ stands for the operator given by

$$(M_\ell v)_T \ := \ v_T \ = \ |T|^{-1} \int_T v(x) \, dx \ , \ T \in \mathcal{T}_\ell \ . \tag{2.3}$$

As in the continuous case, (2.2d) can be stated as the complementarity condition

$$\sigma_\ell \ge 0 \quad , \quad \psi_\ell - u_\ell \ge 0 \ , \tag{2.4}$$

$$(\sigma_\ell, \psi_\ell - u_\ell)_{0,\Omega} \ = \ 0 \ .$$

We define $\mathcal{A}(u_\ell)$ and $\mathcal{I}(u_\ell)$ as the discrete active and inactive control sets according to

$$\mathcal{A}(u_\ell) \ := \ \bigcup \{ \, T \in \mathcal{T}_\ell \mid u_\ell|_T = \psi_\ell|_T \, \} \ , \tag{2.5a}$$

$$\mathcal{I}(u_\ell) \ := \ \bigcup \{ \, T \in \mathcal{T}_\ell \mid u_\ell|_T < \psi_\ell|_T \, \} \tag{2.5b}$$

and refer to $\mathcal{F}(u_\ell) := \partial \mathcal{A}(u_\ell)$ as the discrete free boundary between the discrete active and inactive sets.

We note that the discrete state and co-state $y_\ell, p_\ell \in V_\ell$ may also be considered as finite element approximations of an auxiliary state $y(u_\ell) \in H_0^1(\Omega)$ and an auxiliary co-state $p(u_\ell) \in H_0^1(\Omega)$ as given by the coupled elliptic system

$$a(y(u_\ell), v) \ = \ (f + u_\ell, v)_{0,\Omega} \ , \quad v \in H_0^1(\Omega) \ , \tag{2.6a}$$

$$a(p(u_\ell), v) \ = \ - \, (y(u_\ell) - y^d, v)_{0,\Omega} \ , \quad v \in H_0^1(\Omega) \ . \tag{2.6b}$$

Obviously, we have the Galerkin orthogonality

$$a(y_\ell - y(u_\ell), v_\ell) \;=\; 0 \quad , \quad v_\ell \in V_\ell \;. \tag{2.7}$$

Furthermore, there holds

$$|y(u_\ell) - y|_{1,\Omega} \;\leq\; c_F(\Omega)\, \|u - u_\ell\|_{0,\Omega} \;, \tag{2.8a}$$

$$|p(u_\ell) - p|_{1,\Omega} \;\leq\; c_F(\Omega)\, \|y - y(u_\ell)\|_{0,\Omega} \;, \tag{2.8b}$$

where $c_F(\Omega) > 0$ is the constant in the Poincaré-Friedrichs inequality

$$\|v\|_{0,\Omega} \;\leq\; c_F(\Omega)\, |v|_{1,\Omega} \;, \; v \in H_0^1(\Omega) \;. \tag{2.9}$$

Throughout the rest of this paper, we assume that the coupled system (2.6a), (2.6b) is $H^{1+\gamma}$-regular for some $\gamma > 0$ which implies the existence of a constant $C_r > 0$, depending only on the shape regularity of the triangulations, such that

$$\|y_\ell - y(u_\ell)\|_{0,\Omega} \;\leq\; C_r h_\ell^\gamma \, |y_\ell - y(u_\ell)|_{1,\Omega} \;. \tag{2.10}$$

3. The adaptive loop

In the step SOLVE of the adaptive loop, for the computation of the solution of (2.1a)–(2.1c) we use the primal-dual active set strategy as described in [7]. In the step ESTIMATE, we use the residual type error estimator

$$\eta \;:=\; \left(\eta_y^2 \,+\, \eta_p^2 \right)^{1/2} \;, \tag{3.1}$$

$$\eta_y \;:=\; \left(\sum_{T \in \mathcal{T}_\ell} \eta_{y,T}^2 \,+\, \sum_{E \in \mathcal{E}_\ell} \eta_{y,E}^2 \right)^{1/2} \;, \tag{3.2}$$

$$\eta_p \;:=\; \left(\sum_{T \in \mathcal{T}_\ell} \sum_{i=1}^{2} (\eta_{p,T}^{(i)})^2 \,+\, \sum_{E \in \mathcal{E}_\ell} \eta_{p,E}^2 \right)^{1/2} \;. \tag{3.3}$$

The estimator consists of easily computable element residuals and edge residuals. In particular, for $T \in \mathcal{T}_\ell$ the element residuals $\eta_{y,T}$ and $\eta_{p,T}^{(i)}, 1 \leq i \leq 2$, are as follows

$$\eta_{y,T} \;:=\; h_T\, \|f + u_\ell\|_{0,T} \;, \tag{3.4}$$

$$\eta_{p,T}^{(1)} \;:=\; h_T\, \|y^d - y_\ell\|_{0,T} \;, \tag{3.5}$$

$$\eta_{p,T}^{(2)} \;:=\; \|M_\ell p_\ell - p_\ell\|_{0,T} \;, \tag{3.6}$$

whereas for $E \in \mathcal{E}_\ell$ the edge residuals $\eta_{y,E}, \eta_{p,E}$ are given by

$$\eta_{y,E} \;:=\; h_E^{1/2}\, \|\boldsymbol{\nu}_E \cdot [\nabla y_\ell]\|_{0,E} \;, \tag{3.7}$$

$$\eta_{p,E} \;:=\; h_E^{1/2}\, \|\boldsymbol{\nu}_E \cdot [\nabla p_\ell]\|_{0,E} \;. \tag{3.8}$$

Here, $E = T_1 \cap T_2, T_\nu \in \mathcal{T}_\ell, 1 \leq \nu \leq 2$, and $\boldsymbol{\nu}_E$ is the exterior unit normal vector on E directed towards T_2, whereas $[\nabla y_\ell]$ and $[\nabla p_\ell]$ denote the jumps of $\nabla y_\ell, \nabla p_\ell$ across E.

Moreover, the convergence analysis invokes data oscillations in y^d, f and ψ

$$\mathrm{osc}_\ell(y^d) := \Big(\sum_{T \in \mathcal{T}_\ell} \mathrm{osc}_T(y^d)^2 \Big)^{1/2} , \mathrm{osc}_T(y^d) := h_T \|y^d - y_\ell^d\|_{0,T} , \tag{3.9a}$$

$$\mathrm{osc}_\ell(f) := \Big(\sum_{T \in \mathcal{T}_\ell} \mathrm{osc}_T(f)^2 \Big)^{1/2} , \mathrm{osc}_T(f) := h_T \|f - f_\ell\|_{0,T} , \tag{3.9b}$$

$$\mathrm{osc}_\ell(\psi) := \Big(\sum_{T \in \mathcal{T}_\ell} \mathrm{osc}_T(\psi)^2 \Big)^{1/2} , \ \mathrm{osc}_T(\psi) := h_T \|\nabla\psi\|_{0,T} . \tag{3.9c}$$

where $y_\ell^d \in W_\ell$ and $f_\ell \in W_\ell$ with $y_\ell^d|_T := y_T^d, f_\ell|_T := f_T, T \in \mathcal{T}_\ell$.
Given universal constants $\Theta_i, 1 \le i \le 2$ with $0 < \Theta_i < 1$, in the bulk criterion of step MARK we select a set of edges $\mathcal{M}_\ell^E \subset \mathcal{E}_\ell$ and a set of elements $\mathcal{M}_\ell^T := \mathcal{M}_\ell^{\eta,T} \cup \mathcal{M}^{\mathrm{osc},T} \subset \mathcal{T}_\ell$ such that

$$\Theta_1 \sum_{E \in \mathcal{E}_\ell} (\eta_{y,E}^2 + \eta_{p,E}^2) \le \sum_{E \in \mathcal{M}^E} (\eta_{y,E}^2 + \eta_{p,E}^2) , \tag{3.10}$$

$$\Theta_2 \Big(\sum_{T \in \mathcal{T}_\ell} (\eta_{y,T}^2 + (\eta_{p,T}^{(1)})^2 + (\eta_{p,T}^{(2)})^2) \Big)$$
$$\le \sum_{T \in \mathcal{M}^{\eta,T}} (\eta_{y,T}^2 + (\eta_{p,T}^{(1)})^2 + (\eta_{p,T}^{(2)})^2) . \tag{3.11}$$

The bulk criteria are realized by a greedy algorithm (cf., e.g., [17]).

In the final step REFINE, an element T selected in the bulk criterion is refined by successive bisection such that at least one interior nodal point is generated ('interior node property'). If two or three edges of an element have been marked for refinement, the triangle is subdivided into four subtriangles by joining the midpoints of the edges, whereas simple bisection is used, if only one edge has been selected in the bulk criterion. Bisection is also used in case of newly created nodes at midpoint of edges not contained in $\mathcal{M}^E$ in order to provide a geometrically conforming new triangulation $\mathcal{T}_{\ell+1}$. Setting

$$\mathrm{osc}_\ell := \big(\mathrm{osc}_\ell^2(y^d) + \mathrm{osc}_\ell^2(f) + \mathrm{osc}_\ell^2(\psi) \big)^{1/2} , \tag{3.12}$$

we assume that $\mathcal{T}_{\ell+1}$ is such that there exists $0 \le \rho_2 < 1$ satisfying

$$\mathrm{osc}_{\ell+1}^2 \le \rho_2 \, \mathrm{osc}_\ell^2 . \tag{3.13}$$

In practice, the oscillation term osc_ℓ is included in the bulk criteria of step MARK (cf., e.g., [24] for a thorough discussion of the oscillation term and see [17] for details of the algorithmic realization).

4. Reliability and discrete local efficiency

The reliability of the estimator has been established in [17].

Theorem 4.1. *Let* (y, p, u, σ) *and* $(y_\ell, p_\ell, u_\ell, \sigma_\ell)$ *be the solutions of* (1.2a)–(1.2d) *and* (2.2a)–(2.2d), *and let* η *and* $\mu_\ell(\psi)$ *be the residual error estimator and the data oscillations as given by* (3.1) *and* (3.9c), *respectively. Then, there exists a constant* $C_1 > 1$, *depending only on* α *and on the shape regularity of the triangulations, such that*

$$|y - y_\ell|^2_{1,\Omega} + |p - p_\ell|^2_{1,\Omega} + \|u - u_\ell\|^2_{0,\Omega} \tag{4.1}$$
$$+ \|\sigma - \sigma_\ell\|^2_{0,\Omega} \leq C_1 \left(\eta^2 + \mathrm{osc}^2_\ell(\psi) \right) .$$

For the discrete local efficiency of the estimator we have to show that for refined elements T and edges E the local components of the estimator can be bounded from above by the norms of the differences of the fine and coarse mesh approximations on T and the patches ω_E, respectively.

Lemma 4.2. *For a refined element* $T \in \mathcal{T}_\ell$ *there holds*

$$\eta^2_{y,T} \;\lesssim\; |y_\ell - y_{\ell+1}|^2_{1,T} + h^2_T \|u_\ell - u_{\ell+1}\|^2_{0,T} + \mathrm{osc}^2_T(f) , \tag{4.2}$$

$$(\eta^{(1)}_{p,T})^2 \;\lesssim\; |p_\ell - p_{\ell+1}|^2_{1,T} + |y_\ell - y_{\ell+1}|^2_{1,T} + \mathrm{osc}^2_T(y^d) , \tag{4.3}$$

$$(\eta^{(2)}_{p,T})^2 \;\lesssim\; |p_\ell - p_{\ell+1}|^2_{0,T} + \alpha^2 \|u_\ell - u_{\ell+1}\|^2_{0,T} + \|\sigma_\ell - \sigma_{\ell+1}\|^2_{0,T} . \tag{4.4}$$

Proof. Let $\varphi^a_{\ell+1} \in V_{\ell+1}$ be a nodal basis function associated with an interior point $a \in \mathcal{N}_{\ell+1}(T)$ and $D_a := \mathrm{supp}(\varphi^a_{\ell+1})$. Then, the function $z_{\ell+1} := (f_T + u_\ell)\varphi^a_{\ell+1}$ satisfies

$$\|f_T + u_\ell\|^2_{0,T} \;\lesssim\; (f_T + u_\ell, z_{\ell+1})_{0,T} , \tag{4.5}$$

$$\|z_{\ell+1}\|_{0,T} \;\lesssim\; \|f_T + u_\ell\|_{0,T} \quad , \quad |z_{\ell+1}|_{1,T} \;\lesssim\; h^{-1}_T \|f_T + u_\ell\|_{0,T} . \tag{4.6}$$

Using (4.5) and (4.6) we find

$$\eta^2_{y,T} = h^2_T \|f_T + u_\ell\|^2_{0,T} \;\lesssim\; h^2_T (f + u_{\ell+1}, z_{\ell+1})_{0,T} \tag{4.7}$$

$$+ h^2_T \left(\|u_\ell - u_{\ell+1}\|_{0,T}) + \|f - f_T\|_{0,T} \right) \|z_{\ell+1}\|_{0,T} .$$

Since $z_{\ell+1}$ is an admissible test function in (2.5a) (with ℓ replaced by $\ell + 1$), we have

$$a(y_{\ell+1}, z_{\ell+1}) = (f + u_{\ell+1}, z_{\ell+1})_{0,T} .$$

Observing $\Delta y_\ell = 0$ on T and $z_{\ell+1}|_{\partial D_a} = 0$, a simple integration by parts shows

$$a|_T(y_\ell, z_{\ell+1}) = \sum_{T' \in \mathcal{T}_{\ell+1}(D_a)} \int_{T'} \nabla y_\ell \cdot \nabla z_{\ell+1} \, dx$$

$$= - \sum_{T' \in \mathcal{T}_{\ell+1}(D_a)} \int_{T'} \Delta y_\ell z_{\ell+1} \, dx + \int_{\partial D_a} \nu_{\partial D_a} \cdot \nabla y_\ell z_{\ell+1} \, ds = 0 .$$

Consequently, we obtain

$$h_T^2(f + u_{\ell+1}, z_h)_{0,T} \tag{4.8}$$
$$= h_T^2 a(y_{\ell+1} - y_\ell, z_{\ell+1}) \leq h_T^2 |y_{\ell+1} - y_\ell|_{1,T} |z_{\ell+1}|_{1,T} .$$

Inserting (4.8) into (4.7) and using (4.6) as well as Young's inequality gives the assertion.

The proof of (4.3) follows by similar arguments, this time choosing $z_{\ell+1} = (y_\ell^d - \hat{y}_\ell)\varphi_{\ell+1}^a$, where $\hat{y}_\ell$ is the integral mean of y_ℓ on T.

For the proof of (4.4), the triangle inequality readily gives

$$\eta_{p,T}^{(2)} \leq \|M_\ell p_\ell - M_{\ell+1} p_{\ell+1}|_{0,T} + \|p_\ell - p_{\ell+1}\|_{0,T} + \|M_{\ell+1} p_\ell - p_\ell\|_{0,T} . \tag{4.9}$$

Using the relationship (2.2c) both for the coarse and the fine mesh, for the first term on the right-hand side in (4.9) we obtain

$$\|M_\ell p_\ell - M_{\ell+1} p_{\ell+1}|_{0,T} \leq \alpha \|u_\ell - u_{\ell+1}\|_{0,T} + \|\sigma_\ell - \sigma_{\ell+1}\|_{0,T} . \tag{4.10}$$

For the third term on the right-hand side in (4.9), there exists $0 \leq q < 1$ such that

$$\|M_{\ell+1} p_\ell - p_\ell\|_{0,T} \leq q \|M_\ell p_\ell - p_\ell\|_{0,T} . \tag{4.11}$$

Taking advantage of (4.10), (4.11) in (4.9) yields

$$\eta_{p,T}^{(2)} \leq \frac{1}{1-q} \left(\|p_\ell - p_{\ell+1}\|_{0,\Omega} + \alpha \|u_\ell - u_{\ell+1}\|_{0,\Omega} + \|\sigma_\ell - \sigma_{\ell+1}\|_{0,\Omega} \right) .$$

$\square$

Lemma 4.3. *For a refined edge $E \in \mathcal{E}_\ell$ there holds*

$$\eta_{y,E}^2 \lesssim |y_\ell - y_{\ell+1}|_{1,\omega_E}^2 + h_T^2 \|u_\ell - u_{\ell+1}\|_{0,\omega_E}^2 + \eta_{y,\omega_E}^2 , \tag{4.12}$$
$$\eta_{p,E}^2 \lesssim |p_\ell - p_{\ell+1}|_{1,\omega_E}^2 + |y_\ell - y_{\ell+1}|_{1,\omega_E}^2 + \eta_{p,\omega_E}^2 . \tag{4.13}$$

Proof. Let $\varphi_{\ell+1}^{mid_E} \in V_{\ell+1}$ be the nodal basis function associated with $\mathrm{mid}(E) \in \mathcal{N}_{\ell+1}(\Omega)$. Then, the function $z_{\ell+1} := [\boldsymbol{\nu}_E \cdot \nabla y_\ell]\varphi_{\ell+1}^{mid_E}$ satisfies

$$\|[\boldsymbol{\nu}_E \cdot \nabla y_\ell]\|_{0,E}^2 \lesssim ([\boldsymbol{\nu}_E \cdot \nabla y_\ell], z_{\ell+1})_{0,E} , \tag{4.14}$$
$$\|z_{\ell+1}\|_{0,\omega_E} \lesssim h_E^{1/2} \|[\boldsymbol{\nu}_E \cdot \nabla y_\ell]\|_{0,E} , \tag{4.15}$$
$$|z_{\ell+1}|_{1,\omega_E} \lesssim h_E^{-1/2} \|[\boldsymbol{\nu}_E \cdot \nabla y_\ell]\|_{0,E} . \tag{4.16}$$

Using (4.14)–(4.16) and the fact that $z_{\ell+1}$ is an admissible test function in (2.5a) (with ℓ replaced by $\ell + 1$), we find

$$\eta_{y,E}^2 = h_E \|[\boldsymbol{\nu}_E \cdot \nabla y_\ell]\|_{0,E}^2 \lesssim h_E([\boldsymbol{\nu}_E \cdot \nabla y_\ell], z_{\ell+1})_{0,E} \tag{4.17}$$

$$= h_E \left(a|_{\omega_E}(y_\ell - y_{\ell+1}, z_{\ell+1}) + (u_\ell - u_{\ell+1}, z_{\ell+1})_{0,\Omega_E} - (f + u_\ell, z_{\ell+1})_{0,\Omega_E} \right)$$

$$\lesssim h_E^{1/2} \|[\boldsymbol{\nu}_E \cdot \nabla y_\ell]\|_{0,E} \left(|y_\ell - y_{\ell+1}|_{1,\omega_E} + h_T \|u_\ell - u_{\ell+1}\|_{0,\omega_E} + \eta_{y,\omega_E} \right) ,$$

which immediately leads to (4.12). The estimate (4.13) is shown in exactly the same way.

$\square$

Summarizing the results of Lemma 4.2 and Lemma 4.3 and taking account that the union of the patches ω_E has a finite overlap, we obtain

Theorem 4.4. *Let* (y, p, u, σ) *and* $(y_k, p_k, u_k, \sigma_k), k \in \{\ell, \ell+1\}$, *be the solutions of* (1.2a)–(1.2d) *and* (2.2a)–(2.2d), *and let* η *and* $\mathrm{osc}_\ell(y^d), \mathrm{osc}_\ell(f)$ *be the residual error estimator and the data oscillations as given by* (3.1) *and* (3.9a), (3.9b), *respectively. Then, there exists a constant* $C_2 > 1$, *depending only on the constants* $O_i, 1 \leq i \leq 2$, *in the bulk criteria* (3.10), (3.11), *and on the shape regularity of the triangulations, such that*

$$\sum_{T \in \mathcal{M}_\ell^T} \left(\eta_{y,T}^2 + (\eta_{p,T}^{(1)})^2 + (\eta_{p,T}^{(2)})^2 \right) + \sum_{E \in \mathcal{M}_\ell^E} \left(\eta_{y,E}^2 + \eta_{p,E}^2 \right)$$

$$\leq C_2 \left(|y_\ell - y_{\ell+1}|_{1,\Omega}^2 + |p_\ell - p_{\ell+1}|_{1,\Omega}^2 + \alpha^2 \|u_\ell - u_{\ell+1}\|_{0,\Omega}^2 \right.$$

$$\left. + \|\sigma_\ell - \sigma_{\ell+1}\|_{0,\Omega}^2 + \mathrm{osc}_\ell^2(y^d) + \mathrm{osc}_\ell^2(f) \right) . \tag{4.18}$$

5. Convergence result

In this section, we will prove convergence of the discrete states, co-states, controls, and co-controls to its continuous counterparts.

The reliability (4.1), the bulk criteria (3.10), (3.11), and the discrete local efficiency (4.18) imply

$$|y - y_\ell|_{1,\Omega}^2 + |p - p_\ell|_{1,\Omega}^2 + \|u - u_\ell\|_{0,\Omega}^2 + \|\sigma - \sigma_\ell\|_{0,\Omega}^2$$

$$\leq C_3 \left(\alpha |y_\ell - y_{\ell+1}|_{1,\Omega}^2 + |p_\ell - p_{\ell+1}|_{1,\Omega}^2 \right) + C_4 \left(\alpha^2 \|u_\ell - u_{\ell+1}\|_{0,\Omega}^2 \right.$$

$$\left. + \|\sigma_\ell - \sigma_{\ell+1}\|_{0,\Omega}^2 \right) + C_5 \left(\mathrm{osc}_\ell^2(y^d) + \mathrm{osc}_\ell^2(f) + \mathrm{osc}_\ell^2(\psi) \right) , \tag{5.1}$$

where $C_3 := C_1 C_2 \alpha^{-1}, C_4 := C_1 C_2$ and $C_5 := \max(C_1, C_2)$.

Now, using the fundamental relationships

$$|s_\ell - s_{\ell+1}|_{1,\Omega}^2 \tag{5.2}$$

$$= |s - s_\ell|_{1,\Omega}^2 - |s - s_{\ell+1}|_{1,\Omega}^2 + 2a(s - s_{\ell+1}, s_\ell - s_{\ell+1}) ,$$

$$\|w_\ell - w_{\ell+1}\|_{0,\Omega}^2 \tag{5.3}$$

$$= \|w - w_\ell\|_{0,\Omega}^2 - \|w - w_{\ell+1}\|_{0,\Omega}^2 + 2(w - w_{\ell+1}, w_\ell - w_{\ell+1})_{0,\Omega}$$

for $s = y, s_k = y_k$ and $s = p, s_k = p_k$ and for $w = u, w_k = u_k$ and $w = \sigma, w_k = \sigma_k, k \in \{\ell, \ell+1\}$, we would be able to deduce not only convergence, but even an error reduction property, if we had Galerkin orthogonality of the AFEM. However, Galerkin orthogonality does not apply here. Instead, we will establish some quasi-orthogonality properties which allow to prove convergence.

Lemma 5.1. *Let* (y, p, u, σ) *and* $(y_k, p_k, u_k, \sigma_k), k \in \{\ell, \ell+1\}$, *be the solutions of* (1.2a)–(1.2d) *and* (2.2a)–(2.2d), *and let* $y(u_{\ell+1}) \in V$ *be the auxiliary state as given* *by* (2.6). *Then, there holds*

$$\alpha\, a(y - y_{\ell+1}, y_\ell - y_{\ell+1}) + a(p - p_{\ell+1}, p_\ell - p_{\ell+1}) \tag{5.4}$$
$$= (y_{\ell+1} - y(u_{\ell+1}), p_\ell - p_{\ell+1})_{0,\Omega} + (y(u_{\ell+1}) - y, \sigma_\ell - \sigma_{\ell+1})_{0,\Omega}$$
$$+\ (y(u_{\ell+1}) - y, (I - M_\ell)p_\ell - (I - M_{\ell+1})p_{\ell+1})_{0,\Omega}$$
$$+\ \alpha\ (u_{\ell+1} - u, (y(u_\ell) - y_\ell) + (y_{\ell+1} - y(u_{\ell+1})))_{0,\Omega}\ ,$$

$$\alpha^2\ (u - u_{\ell+1}, u_\ell - u_{\ell+1})_{0,\Omega} + (\sigma - \sigma_{\ell+1}, \sigma_\ell - \sigma_{\ell+1}) \tag{5.5}$$
$$\le \alpha\ (\sigma, u_{\ell+1} - u_\ell)_{0,\Omega} + \alpha\ (\sigma_{\ell+1} - \sigma, \psi_\ell - \psi_{\ell+1})_{0,\Omega}$$
$$+\ \alpha\ (\sigma, \psi_\ell - \psi_{\ell+1})_{0,\Omega} + \alpha\ (\sigma_{\ell+1} - \sigma_\ell, u - \psi)_{0,\Omega}$$
$$+\ \alpha\ (\sigma_\ell - \sigma_{\ell+1}, \psi_{\ell+1} - \psi)_{0,\Omega} + (p - M_{\ell+1}p_{\ell+1}, M_\ell p_\ell - M_{\ell+1}p_{\ell+1})_{0,\Omega}.$$

Proof. In view of (1.2a), (2.2a) and (2.6), (2.7), we readily get

$$\alpha a(y - y_{\ell+1}, y_\ell - y_{\ell+1}) \tag{5.6}$$
$$=\ \alpha a(y - y(u_{\ell+1}), y_\ell - y_{\ell+1})\ =\ \alpha(u - u_{\ell+1}, y_\ell - y_{\ell+1})_{0,\Omega}\ .$$

On the other hand, observing (1.2b), (1.2c) and (2.2b), (2.2c) as well as (2.8), we find

$$a(p - p_{\ell+1}, p_\ell - p_{\ell+1})\ =\ (y_{\ell+1} - y(u_{\ell+1}), p_\ell - p_{\ell+1})_{0,\Omega} \tag{5.7}$$
$$+\ (y(u_{\ell+1}) - y, \sigma_\ell - \sigma_{\ell+1} + (I - M_\ell)p_\ell - (I - M_{\ell+1})p_{\ell+1})_{0,\Omega}$$
$$+\ \alpha(y(u_{\ell+1}) - y, u_\ell - u_{\ell+1})_{0,\Omega}\ .$$

Moreover, since $y(u_{\ell+1}) - y = (-\Delta)^{-1}(u_{\ell+1} - u)$ and $(-\Delta)^{-1}(u_\ell - u_{\ell+1}) = y(u_\ell) - y(u_{\ell+1})$, we obtain

$$\alpha(y(u_{\ell+1}) - y, u_\ell - u_{\ell+1})_{0,\Omega} \tag{5.8}$$
$$= \alpha(u_{\ell+1} - u, y(u_\ell) - y(u_{\ell+1}))_{0,\Omega} = \alpha(u_{\ell+1} - u, y_\ell - y_{\ell+1})_{0,\Omega}$$
$$+\ \alpha(u_{\ell+1} - u, (y(u_\ell) - y_\ell) + (y_{\ell+1} - y(u_{\ell+1})))_{0,\Omega}\ .$$

Using (5.8) in (5.7) and combining (5.6) and (5.7) results in (5.4).

As far as the proof of (5.5) is concerned, using again (1.2c), (2.2c), we find

$$\alpha^2(u - u_{\ell+1}, u_\ell - u_{\ell+1})_{0,\Omega} + (\sigma - \sigma_{\ell+1}, \sigma_\ell - \sigma_{\ell+1}) \tag{5.9}$$
$$=\ \alpha(\sigma_{\ell+1} - \sigma, u_\ell - u_{\ell+1})_{0,\Omega} + \alpha(u - u_{\ell+1}, \sigma_{\ell+1} - \sigma_\ell)_{0,\Omega}$$
$$+\ (p - M_{\ell+1}p_{\ell+1}, M_\ell p_\ell - M_{\ell+1}p_{\ell+1})_{0,\Omega}\ .$$

For the first term on the right-hand side in (5.9), we obtain

$$\alpha(\sigma_{\ell+1} - \sigma, u_\ell - u_{\ell+1})_{0,\Omega} \tag{5.10}$$
$$=\ \alpha(\sigma_{\ell+1}, u_\ell - \psi_\ell)_{0,\Omega} + \alpha(\sigma_{\ell+1}, \psi_\ell - \psi_{\ell+1})_{0,\Omega}$$
$$+\ \alpha(\sigma_{\ell+1}, \psi_{\ell+1} - u_{\ell+1})_{0,\Omega} + \alpha(\sigma, u_{\ell+1} - u_\ell)_{0,\Omega}$$
$$\le\ \alpha(\sigma_{\ell+1}, \psi_\ell - \psi_{\ell+1})_{0,\Omega} + \alpha(\sigma, u_{\ell+1} - u_\ell)_{0,\Omega}\ ,$$

where we have used that due to (2.4)

$$(\sigma_{\ell+1}, u_\ell - \psi_\ell)_{0,\Omega} \le 0 \quad \text{and} \quad (\sigma_{\ell+1}, \psi_{\ell+1} - u_{\ell+1})_{0,\Omega} = 0 \, .$$

For the second term, similar arguments yield

$$\alpha(\sigma_{\ell+1} - \sigma_\ell, u - u_{\ell+1})_{0,\Omega} \tag{5.11}$$
$$= \alpha(\sigma_{\ell+1} - \sigma_\ell, u - \psi)_{0,\Omega} + \alpha(\sigma_{\ell+1} - \sigma_\ell, \psi_{\ell+1} - \psi)_{0,\Omega} \, .$$

Using (5.10), (5.11) in (5.9) gives the assertion. $\qquad\square$

For the terms in (5.4), (5.5) involving the averaging operators M_ℓ and $M_{\ell+1}$ we provide the following result.

Lemma 5.2. *Under the assumptions of Lemma 5.1 there holds*

$$(y_{\ell+1} - y, (I - M_\ell)p_\ell - (I - M_{\ell+1})p_{\ell+1})_{0,\Omega} \tag{5.12}$$
$$\lesssim h_\ell^2 \left(|y - y_{\ell+1}|_{1,\Omega}^2 + |p - p_\ell|_{1,\Omega}^2 + |p - p_{\ell+1}|_{1,\Omega}^2\right) + \mu_\ell^2(p) \, ,$$
$$(p - M_{\ell+1}p_{\ell+1}, M_\ell p_\ell - M_{\ell+1}p_{\ell+1})_{0,\Omega} \tag{5.13}$$
$$\lesssim h_\ell^2 \left(|p - p_\ell|_{1,\Omega}^2 + |p - p_{\ell+1}|_{1,\Omega}^2\right) + \mu_\ell^2(p) \, ,$$

where

$$\mu_\ell(p) := \Big(\sum_{T \in \mathcal{T}_\ell} \mu_T^2(p) \Big)^{1/2} \, , \quad \mu_T(p) := h_T \, |p|_{1,T} \, . \tag{5.14}$$

Proof. We split the left-hand side in (5.9) according to

$$(y_{\ell+1} - y, (I - M_\ell)p_\ell - (I - M_{\ell+1})p_{\ell+1})_{0,\Omega}$$
$$= (y_{\ell+1} - y, (I - M_{\ell+1})(p_\ell - p_{\ell+1}))_{0,\Omega}$$
$$+ (y_{\ell+1} - y, (M_{\ell+1} - M_\ell)p_\ell)_{0,\Omega} \, .$$

For $T \in \mathcal{T}_{\ell+1}$ we set $\hat{y}_{\ell+1} := |T|^{-1} \int_T (y_{\ell+1} - y) dx$. Since $M_{\ell+1}(p_\ell - p_{\ell+1})$ has zero integral mean on $T \in \mathcal{T}_{\ell+1}$, an elementwise application of Poincaré's inequality and of Young's inequality gives

$$(y_{\ell+1} - y, (I - M_{\ell+1})(p_\ell - p_{\ell+1}))_{0,T}$$
$$= (y_{\ell+1} - y - \hat{y}_{\ell+1}, (I - M_{\ell+1})(p_\ell - p_{\ell+1}))_{0,T}$$
$$\lesssim h_T \, |y - y_{\ell+1}|_{1,T} \, h_T \, |p_\ell - p_{\ell+1}|_{1,T}$$
$$\lesssim h_T^2 \, |y - y_{\ell+1}|_{1,T}^2 + h_T^2 \, |p_\ell - p_{\ell+1}|_{1,T}^2 \, .$$

Summing over all $T \in \mathcal{T}_{\ell+1}$, we obtain

$$(y_{\ell+1} - y, (I - M_\ell)p_\ell - (I - M_{\ell+1})p_{\ell+1})_{0,\Omega}$$
$$\lesssim h_\ell^2 \left(|y - y_{\ell+1}|_{1,\Omega}^2 + |p_\ell - p_{\ell+1}|_{1,\Omega}^2\right).$$

Moreover, using similar arguments

$$(y_{\ell+1} - y, (M_{\ell+1} - M_\ell)p_\ell)_{0,\Omega}$$
$$\lesssim h_\ell^2 \left(|y - y_{\ell+1}|_{1,\Omega}^2 + |p - p_\ell|_{1,\Omega}^2\right) + \mu_\ell^2(p) \, .$$

Combing both inequalities proves (5.12). The proof of (5.13) is along the same lines. $\qquad\square$

Since the $|\cdot|_{1,\Omega}$-norm of the co-state p can be bounded from above by means of the given data of the problem (cf., e.g., [20]), we may interpret $\mu_\ell(p)$ as a data term. As far as the reduction of that data term is concerned, we may assume the existence of $0 \leq \rho_3 < 1$ such that

$$\mu_{\ell+1}^2(p) \;\leq\; \rho_3\,\mu_\ell^2(p) \;. \tag{5.15}$$

For the convergence proof, we set $z := (y, p, u, \sigma)$, $z_\ell := (y_\ell, p_\ell, u_\ell, \sigma_\ell)$, $\ell \in \mathbb{N}_0$, and introduce the norm

$$|||z - z_\ell||| := \left(|y - y_\ell|_{1,\Omega}^2 + |p - p_\ell|_{1,\Omega}^2 + \|u - u_\ell\|_{0,\Omega}^2 + \|\sigma - \sigma_\ell\|_{0,\Omega}^2 \right)^{1/2}. \tag{5.16}$$

We establish convergence with respect to $||| \cdot |||$ in the sense that the sequence $\{|||z - z_\ell|||\}_{\mathbb{N}_0}$ belongs to ℓ^2.

Theorem 5.3. *Let (y, p, u, σ) and $(y_\ell, p_\ell, u_\ell, \sigma_\ell)$ be the solutions of (1.2a)–(1.2d) and (2.2a)–(2.2d) and let $\mathrm{osc}_\ell(y^d), \mathrm{osc}_\ell(f), \mathrm{osc}_\ell(\psi), \mu_\ell(p)$ be the data oscillations and data terms given by (3.9a)–(3.9c) and (5.14). Assume that (3.13) and (5.15) are satisfied. Then, there exists a constant $\Lambda > 0$, depending on the data of the problem, the constants $\Theta_i, 1 \leq i \leq 2$, in the bulk criteria (3.10), (3.11) and on the shape regularity of the triangulations, such that*

$$\sum_{\ell=0}^{\infty} |||z - z_\ell|||^2 \;\leq\; \Lambda \;. \tag{5.17}$$

Proof. In addition to Lemma (5.2), we provide further estimates for the remaining terms on the right-hand side in (5.4). In particular, by means of (2.8a), (2.9) and (2.10), setting $c_\Omega := \max(1, c_F(\Omega))$ we obtain

$$
\begin{aligned}
(y_{\ell+1} - y(u_{\ell+1}), p_\ell - p_{\ell+1})_{0,\Omega} \;&\leq\; C_r c_\Omega h_\ell^\gamma \left(|y - y_{\ell+1}|_{1,\Omega} \right. \\
&\quad + |y - y(u_{\ell+1})|_{1,\Omega} \Big) \left(|p - p_\ell|_{1,\Omega} + |p - p_{\ell+1}|_{1,\Omega} \right) \\
&\leq\; \frac{1}{2} C_r c_\Omega h_\ell^\gamma \left(|y - y_{\ell+1}|_{1,\Omega}^2 + 4|p - p_\ell|_{1,\Omega}^2 + 4|p - p_{\ell+1}|_{1,\Omega}^2 \right) \\
&\quad + \frac{1}{2} C_r c_\Omega^2 h_\ell^\gamma \|u - u_{\ell+1}\|_{0,\Omega}^2 \;,
\end{aligned}
\tag{5.18}
$$

$$
\begin{aligned}
\alpha(u_{\ell+1} - u, (y(u_\ell) - y_\ell) + (y_{\ell+1} - y(u_{\ell+1}))_{0,\Omega} \;&\leq\; \alpha C_r h_\ell^\gamma \|u - u_{\ell+1}\|_{0,\Omega} \left(|y - y_\ell|_{1,\Omega} + |y - y_{\ell+1}|_{1,\Omega} \right. \\
&\quad + c_\Omega \left(\|u - u_\ell\|_{0,\Omega} + \|u - u_{\ell+1}\|_{0,\Omega} \right) \Big) \\
&\leq\; \alpha C_r h_\ell^\gamma \left(|y - y_\ell|_{1,\Omega}^2 + |y - y_{\ell+1}|_{1,\Omega}^2 \right. \\
&\quad + c_\Omega^2 \|u - u_\ell\|_{0,\Omega}^2 + (1 + c_\Omega^2) \|u - u_{\ell+1}\|_{0,\Omega}^2 \Big) \;.
\end{aligned}
\tag{5.19}
$$

Moreover, in view of (2.8a), (2.9) and Young's inequality

$$
\begin{aligned}
(y(u_{\ell+1}) - y, \sigma_\ell - \sigma_{\ell+1})_{0,\Omega} \;&\leq\; c_\Omega^2 \|u - u_{\ell+1}\|_{0,\Omega} \|\sigma_\ell - \sigma_{\ell+1}\|_{0,\Omega} \\
&\leq\; \varepsilon_1 \|u - u_{\ell+1}\|_{0,\Omega}^2 + \frac{c_\Omega^4}{4\varepsilon_1} \|\sigma_\ell - \sigma_{\ell+1}\|_{0,\Omega}^2 \;,
\end{aligned}
\tag{5.20}
$$

where $\varepsilon_1 > 0$ can be arbitrarily chosen. Likewise, for some arbitrary $\varepsilon_2 > 0$ we get

$$\alpha(\sigma_{\ell+1} - \sigma, \psi_\ell - \psi_{\ell+1})_{0,\Omega}$$

$$\leq \alpha \|\sigma - \sigma_{\ell+1}\|_{0,\Omega} \left(\|\psi - \psi_\ell\|_{0,\Omega} + \|\psi - \psi_\ell\|_{0,\Omega} \right) \tag{5.21}$$

$$\leq \varepsilon_2 \, \|\sigma - \sigma_{\ell+1}\|_{0,\Omega}^2 + \frac{\alpha^2}{4\varepsilon_2} (1 + \rho_2) \, \mathrm{osc}_\ell^2(\psi) \, ,$$

$$\alpha(\sigma_\ell - \sigma_{\ell+1}, \psi_{\ell+1} - \psi)_{0,\Omega} \tag{5.22}$$

$$\leq \varepsilon_2 \left(\|\sigma - \sigma_\ell\|_{0,\Omega}^2 + \|\sigma - \sigma_{\ell+1}\|_{0,\Omega}^2 \right) + \frac{\alpha^2}{2\varepsilon_2} \, \rho_2 \, \mathrm{osc}_\ell^2(\psi) \, ,$$

where we have used (3.13) in both estimates.

Now, we choose $\varepsilon_1 > 0, \varepsilon_2 > 0$ according to

$$\varepsilon_1 \; := \; \alpha/(16C_4) \, , \quad \varepsilon_2 \; := \; \alpha/(16C_4(\alpha + 4c_\Omega^4))$$

and $h^* \in \mathbb{R}_+$ by means of

$$h^* \; := \; (\alpha/(120C_4 \mathrm{max}(C_6, C_7, C_r)c_\Omega^4))^{1/\gamma} \, .$$

Then, there exists $\ell^* \in \mathbb{N}$ such that $h_\ell \leq h^*$ for $\ell \geq \ell^*$. If we take advantage of (5.18)–(5.22) as well as (5.12), (5.13) from Lemma 5.2 in (5.4), (5.5) and use the result in (5.1), setting $C_8 := 4C_3 c_\Omega^4 + C_4 \alpha^2$ and $C_9 := 4C_3 c_\Omega^4 + C_4$, for $\ell \geq \ell^*$ we get

$$|||z - z_\ell|||^2 \leq (C_4 + \frac{1}{12})|y - y_\ell|_{1,\Omega}^2 \; - \; (C_4 - \frac{1}{4})|y - y_{\ell+1}|_{1,\Omega}^2 \tag{5.23}$$

$$+ \; (C_3 + \frac{1}{4}) \, |p - p_\ell|_{1,\Omega}^2 \; - \; (C_3 - \frac{1}{4}) \, |p - p_{\ell+1}|_{1,\Omega}^2$$

$$+ \; (C_8 + \frac{1}{4}) \, \|u - u_\ell\|_{0,\Omega}^2 \; - \; (C_8 - \frac{1}{4}) \, \|u - u_{\ell+1}\|_{0,\Omega}^2$$

$$+ \; (C_9 + \frac{1}{2}) \, \|\sigma - \sigma_\ell\|_{0,\Omega}^2 \; - \; (C_9 - \frac{1}{4}) \, \|\sigma - \sigma_{\ell+1}\|_{0,\Omega}^2$$

$$+ \; C_{10} \, (\mathrm{osc}_\ell^2 + \mu_\ell^2(p)) \; + \; 2\alpha(\sigma, u_{\ell+1} - u_\ell)_{0,\Omega}$$

$$+ \; 2\alpha(\sigma, \psi_\ell - \psi_{\ell+1})_{0,\Omega} \; + \; 2\alpha(\sigma_{\ell+1} - \sigma_\ell, u - \psi)_{0,\Omega} \, ,$$

where $C_{10} := C_5 + 2C_3 C_6 + 2C_9(C_7 + 4(1 + \rho_2))$. We define constants $0 < \kappa_i \leq 1, 1 \leq i \leq 3$, and $0 < \rho_1 < 1$ according to

$$\kappa_1 \; := \; \frac{C_4 - 1/4}{C_9 - 1/4} \quad , \quad \kappa_2 \; := \; \frac{C_3 - 1/4}{C_9 - 1/4} \, , \tag{5.24}$$

$$\kappa_3 \; := \; \frac{C_8 - 1/4}{C_9 - 1/4} \quad , \quad \rho_1 \; := \; \frac{C_9 - 1/2}{C_9 - 1/4}$$

(observe $C_4 < C_9, C_3 < C_9$, and $C_8 \leq C_9$). We further introduce the weighted norm

$$|||z - z_\ell|||_\kappa \; := \; \left(\kappa_1 |y - y_\ell|_{1,\Omega}^2 + \kappa_2 |p - p_\ell|_{1,\Omega}^2 + \kappa_3 \|u - u_\ell\|_{0,\Omega}^2 + \|\sigma - \sigma_\ell\|_{0,\Omega}^2 \right)^{1/2} \, . \tag{5.25}$$

Then, from (5.23) we deduce that

$$|||z - z_{\ell+1}|||^2_\kappa \tag{5.26}$$
$$\leq \rho|||z - z_\ell|||^2_\kappa + 2a(\sigma, u_{\ell+1} - u_\ell)_{0,\Omega} + 2a(\sigma, \psi_\ell - \psi_{\ell+1})_{0,\Omega}$$
$$+ 2a(\sigma_{\ell+1} - \sigma_\ell, u - \psi)_{0,\Omega} + C_{11}\left(\operatorname{osc}^2_l + \mu^2_\ell(p)\right),$$

where $C_{11} := C_{10}/(C_9 - 1/4)$. Summing in (5.26) over ℓ from $\ell = \ell^*$ to $\ell = n > \ell^*$ results in

$$(1 - \rho)\sum_{\ell=\ell^*}^{n-1}|||z - z_{\ell+1}|||^2_\kappa + |||z - z_{n+1}|||^2_\kappa \leq \rho|||z - z_{\ell^*}|||^2_\kappa \tag{5.27}$$
$$+ 2a(\sigma, u_{n+1} - u_{\ell^*})_{0,\Omega} + 2a(\sigma, \psi_{\ell^*} - \psi_{n+1})_{0,\Omega}$$
$$+ 2a(\sigma_{n+1} - \sigma_{\ell^*}, u - \psi)_{0,\Omega} + C_7\sum_{\ell=\ell^*}^{n}\left(\operatorname{osc}^2_\ell + \mu^2_\ell(p)\right).$$

Now, taking (1.3) and (2.4) into account, we have

$$(\sigma, u_{n+1} - u_{\ell^*})_{0,\Omega} + (\sigma, \psi_{\ell^*} - \psi_{n+1})_{0,\Omega} \leq (\sigma, \psi_{\ell^*} - u_{\ell^*})_{0,\Omega},$$
$$(\sigma_{n+1} - \sigma_{\ell^*}, u - \psi)_{0,\Omega} \leq (\sigma_{\ell^*}, \psi - u)_{0,\Omega}.$$

Moreover, due to (3.13) and (5.15)

$$\sum_{\ell=\ell^*}^{\infty}\left(\operatorname{osc}^2_\ell + \mu^2_\ell(p)\right) \leq \frac{C_7}{1 - \rho_2}\operatorname{osc}^2_{\ell^*} + \frac{C_7}{1 - \rho_3}\mu^2_{\ell^*}(p).$$

Using the preceding estimates in (5.27) implies the existence of a constant ϑ such that

$$\min_{1 \leq i \leq 3} \kappa_i \sum_{\ell=0}^{\infty}|||z - z_\ell|||^2 \leq \vartheta,$$

which gives the assertion. $\square$

Corollary 5.4. *Under the assumptions of Theorem 5.3 there holds*

$$|y - y_\ell|_{1,\Omega}, |p - p_\ell|_{1,\Omega}, \|u - u_\ell\|_{0,\Omega}, \|\sigma - \sigma_\ell\|_{0,\Omega} \to 0 \quad as \quad \ell \to \infty. \tag{5.28}$$

6. Error reduction property

An error reduction property of the adaptive finite element approximation of the obstacle problem in the weighted norm $||| \cdot |||_\kappa$ can be established under some additional assumptions. In particular, we suppose that the sequence $\{W_\ell\}_{\mathbb{N}_0}$ of spaces of elementwise constants is limit dense in $L^2(\Omega)$ in the sense

(L)   For each $w \in L^2(\Omega)$, there is a sequence $\{w_\ell\}_{\mathbb{N}}, w_\ell \in W_\ell, \ell \in \mathbb{N}$, such that $w_\ell \to w$ in $L^2(\Omega)$ as $\ell \to \infty$.

We further assume strict complementarity of the continuous problem

(C) $$\sigma|_{\mathcal{I}(u)} > 0 \,,$$

as well as the following non-degeneracy properties of the discrete control problems:

(N$_1$) There exist $\varepsilon_1^* > 0$ and $C_1 > 0$ such that for all $0 < \varepsilon < \varepsilon_1^*$ and for all sufficiently large $\ell \in \mathbb{N}$

$$\text{meas}(\{x \in \mathcal{I}(u_\ell) \mid 0 < \psi_\ell(x) - u_\ell(x) < \varepsilon^2\}) \leq C_1 \, \varepsilon \,.$$

(N$_2$) There exist $\varepsilon_2^* > 0$ and $C_2 > 0$ such that for all $0 < \varepsilon < \varepsilon_2^*$ and for all sufficiently large $\ell \in \mathbb{N}$

$$\{x \in \mathcal{I}(u_\ell) \mid 0 < \psi_\ell(x) - u_\ell(x) < \varepsilon^2\} \subseteq \{x \in \mathcal{I}(u_\ell) \mid \text{dist}(x, \mathcal{F}_\ell) < C_2\varepsilon\} \,.$$

(N$_3$) There exist $\varepsilon_3^* > 0$ and $C_3 > 0$ such that for all $0 < \varepsilon < \varepsilon_3^*$ and for all sufficiently large $\ell \in \mathbb{N}$

$$\{x \in \mathcal{I}(u_\ell) \mid \text{dist}(x, \mathcal{F}_\ell) < \varepsilon\} \subseteq \{x \in \mathcal{I}(u_\ell) \mid 0 < \psi_\ell(x) - u_\ell(x) < C_3\varepsilon^2\}.$$

The error reduction property holds asymptotically, i.e., once the continuous free boundary has been sufficiently resolved by its discrete counterpart. We enhance the resolution of the free boundary by an extension of the bulk criteria. To this end, we define the sets

$$\hat{\mathcal{A}}(u_\ell) := \text{int}\Big(\bigcup\{T \in \mathcal{T}_\ell \mid u_\ell|_{T'} = \psi_\ell|_{T'} \,, \ T' \in \mathcal{T}_\ell, T' \cap T \neq \emptyset\}\Big) \,, \tag{6.1}$$

$$\hat{\mathcal{I}}(u_\ell) := \text{int}\Big(\bigcup\{T \in \mathcal{T}_\ell \mid \psi_\ell|_T - u_\ell|_T \geq \hat{\varepsilon} > 0\}\Big) \,, \tag{6.2}$$

$$\hat{\mathcal{F}}(u_\ell) := \Omega \setminus \big(\hat{\mathcal{A}}(u_\ell) \cup \hat{\mathcal{I}}(u_\ell)\big) \tag{6.3}$$

for some $\hat{\varepsilon} > 0$ in (6.2). Then, the extension of the bulk criteria (3.10)–(3.11) is as follows:

(E) In the step 'MARK' of the adaptive loop, all edges $E \in \mathcal{E}_\ell(\hat{\mathcal{F}}_\ell)$ are marked for refinement.

Proposition 6.1. *Assume that the discrete problem* (2.2a)–(2.2d) *satisfies* **(N1)**, **(N2)** *and that the refinement is done based on the bulk criteria* (3.10), (3.11) *and its extension* **(E)**. *Then, there exists a subsequence* $\mathbb{N}^* \subset \mathbb{N}$ *such that for all* $\ell \in \mathbb{N}^*$

$$\hat{\mathcal{I}}(u_\ell) \subseteq \hat{\mathcal{I}}(u_{\ell+1}) \quad , \quad \hat{\mathcal{A}}(u_\ell) \subseteq \hat{\mathcal{A}}(u_{\ell+1}) \,, \tag{6.4}$$

$$\hat{\mathcal{F}}(u_{\ell+1}) \subset \hat{\mathcal{F}}(u_\ell) \,. \tag{6.5}$$

Proof. If the assertion does not hold true, we have $\hat{\mathcal{I}}(u_{m+1}) \subset \hat{\mathcal{I}}(u_m)$ and $\hat{\mathcal{A}}(u_{m+1}) \subset \hat{\mathcal{A}}(u_m)$ for $m > \ell$ which implies $\hat{\mathcal{F}}(u_m) \subseteq \hat{\mathcal{F}}(u_{m+1})$. Hence, in view of **(N1)** and **(E)**, there exists $T \in \hat{\mathcal{F}}(u_m)$ such that $\text{dist}(x, \mathcal{F}(u_m)) > \tau, x \in T$, where $\tau > C_2 h_m$ with C_2 from **(N2)**, and $0 = \psi_m(a_\nu) - u_m(a_\nu) < \psi_m(a_\mu) - u_m(a_\mu)$ for some vertices $a_\nu, a_\mu, \nu \neq \mu$, of T. Then, we find $\mathcal{U}(a_\nu) := \{x \in T' | 0 < \psi_m(x) - u_m(x) < h_m^2\}$ and **(N2)** implies $\text{dist}(x, \mathcal{F}(u_m)) < C_2 h_m < \tau$ contradicting $\text{dist}(x, \mathcal{F}(u_m)) > \tau, x \in \mathcal{U}(a_\nu) \subset T$. Note that (6.5) is a direct consequence of (6.4). $\qquad\square$

Proposition 6.2. *Assume that* (1.2a)–(1.2d) *and* (2.2a)–(2.2d) *satisfy* **(S)** *and* **(N$_1$)**, **(N$_2$)**, **(N$_3$)**, *respectively, and that* **(L)** *holds true. Then, there exists $\ell^* \in N$ such that*

$$\hat{\mathcal{I}}(u_\ell) \subseteq \mathcal{I}(u) \quad \text{for all } \ell \geq \ell^* . \tag{6.6}$$

Proof. If (6.6) does not hold true, there is a subsequence $\mathbb{N}' \subset \mathbb{N}$ such that $\hat{\mathcal{I}}(u_\ell) \cap \mathcal{A}(u) \neq \emptyset$ for all $\ell \in \mathbb{N}'$. Hence, we find $D_\ell \subset \mathcal{A}, \operatorname{meas}(D_\ell) \neq 0$ such that $D_\ell \subset \hat{\mathcal{I}}(u_\ell), \ell \in \mathbb{N}'$. If $D := \cap_{\ell \in \mathbb{N}'} D_\ell$ is such that $\operatorname{meas}(D) \neq 0$, for $w \in L^2(D)_+, w \not\equiv 0$, there is $\{w_\ell\}_{\mathbb{N}'}$ with $w_\ell \in L^2(\hat{\mathcal{I}}(u_\ell))W_\ell \cap L^2(\hat{\mathcal{I}}(u_\ell)), \ell \in \mathbb{N}'$, such that $w_\ell \to w$ in $L^2(\Omega)$ as $\ell \to \infty$ whence

$$\langle \sigma_\ell, w_\ell \rangle_{*, \hat{\mathcal{I}}(u_\ell)} \to \langle \sigma, w \rangle_{*, D} \quad \text{as } \ell \to \infty .$$

But $\langle \sigma_\ell, w_\ell \rangle_{*, \hat{\mathcal{I}}(u_\ell)} = 0, \ell \in \mathbb{N}'$, due to (2.4), and hence, $\langle \sigma, w \rangle_{*, D} = 0$ contradicting $\sigma|_D > 0, D \subset \mathcal{A}$ (cf. **(C)**). If $\operatorname{meas}(D) = 0$, for the Hausdorff distance $d_H(\mathcal{F}, \partial \hat{\mathcal{N}}_\ell)$ we must have $d_H(\mathcal{F}, \partial \hat{\mathcal{N}}_\ell) \to 0$ as $\ell \to \infty$ whence $d_H(\mathcal{F}, \mathcal{F}_\ell) \to 0$ as $\ell \to \infty$, since otherwise we arrive at a contradiction to (6.5). Consequently, there exist $x_\ell \in \hat{\mathcal{N}}_\ell, \ell \in \mathbb{N}'$, such that $u_\ell(x_\ell) - \chi(x_\ell) \geq \varepsilon > 0$ and $\operatorname{dist}(x_\ell, \mathcal{F}_\ell) \to 0$ as $\ell \to \infty$ contradicting **(N3)**. $\qquad\square$

We are now in a position to prove an error reduction property. The essential ingredient is a refined quasi-orthogonality property that can be derived by a more subtle treatment of the terms $a(\sigma_\ell - \sigma, u_\ell - u_{\ell+1})_{0,\Omega}$ and $a(\sigma_{\ell+1} - \sigma, u - u_{\ell+1})_{0,\Omega}$ in (5.6) of the proof of Lemma 5.1.

Lemma 6.3. *Under the same assumptions as in Proposition* 6.1, *for any $\varepsilon > 0$ and $\ell \geq \ell^*$ there holds*

$$a(\sigma_\ell - \sigma, u_\ell - u_{\ell+1})_{0,\Omega} \leq a\varepsilon \left(\|u - u_\ell\|_{0,\Omega}^2 + \|\sigma - \sigma_{\ell+1}\|_{0,\Omega}^2 \right) \tag{6.7}$$

$$+ a(\varepsilon + \frac{1}{4\varepsilon})\|\sigma\|_{0,\hat{\mathcal{F}}(u_\ell)}^2 + \frac{1}{4\varepsilon} \left(\operatorname{osc}_\ell^2(\psi) + \operatorname{osc}_{\ell+1}^2(\psi) \right) ,$$

$$a(\sigma_{\ell+1} - \sigma, u - u_{\ell+1})_{0,\Omega} \tag{6.8}$$

$$\leq a\varepsilon \left(\|\sigma - \sigma_\ell\|_{0,\Omega}^2 + \|\sigma - \sigma_{\ell+1}\|_{0,\Omega}^2 \right) + \frac{1}{4\varepsilon} \left(\|\psi - u\|_{0,\hat{\mathcal{F}}(u_\ell)}^2 \right.$$

$$\left. + \operatorname{osc}_{\ell+1}^2(\psi) \right) .$$

Proof. Taking advantage of the complementarity conditions (1.3), (2.4), we obtain

$$\alpha \left(\sigma_{\ell+1} - \sigma, u_\ell - u_{\ell+1} \right)_{0,\Omega} = \alpha \left(\sigma_{\ell+1}, \psi_{\ell+1} - u_{\ell+1} \right)_{0,\Omega} \tag{6.9}$$

$$+ \alpha \left(\sigma_{\ell+1}, u_\ell - \psi_\ell \right)_{0,\Omega} + \alpha \left(\sigma, u_{\ell+1} - \psi_{\ell+1} \right)_{0,\Omega}$$

$$+ \alpha \left(\sigma_{\ell+1} - \sigma, \psi_\ell - \psi_{\ell+1} \right)_{0,\Omega} + \alpha \left(\sigma, \psi_\ell - u_\ell \right)_{0,\Omega}$$

$$\leq \alpha \left(\sigma, \psi_\ell - u_\ell \right)_{0,\Omega} + \alpha \left(\sigma_{\ell+1} - \sigma, \psi_\ell - \psi_{\ell+1} \right)_{0,\Omega} ,$$

where we have used that the first term after the equality sign is zero, whereas the second and the term is non-positive.

The first term on the right-hand side is obviously zero in $\mathcal{I}(u)$ and in $\hat{\mathcal{A}}(u_\ell)$. Due to Proposition 6.1 we thus have

$$(\sigma, \psi_\ell - u_\ell)_{0,\Omega} \;=\; (\sigma, \psi_\ell - u_\ell)_{0,\hat{\mathcal{F}}(u_\ell)} \;.$$

We further get

$$(\sigma, \psi_H - u_H)_{0,\hat{\mathcal{F}}(u_\ell)} \;=\; (\sigma, u - u_H)_{0,\hat{\mathcal{F}}(u_\ell)} + (\sigma, \psi_H - \psi)_{0,\hat{\mathcal{F}}(u_\ell)} \;, \tag{6.10}$$

where we have used that $(\sigma, \psi - u)_{0,\hat{\mathcal{F}}(u_\ell)} = 0$ due to (1.3). Using (6.10) in (6.9) and applying Cauchy's and Young's inequality results in (6.7).

The proof of (6.8) is done by similar arguments. $\qquad\square$

In view of Lemma 6.3, we define

$$\mu_\ell(u,\sigma) \;:=\; \Big(\sum_{T \in \hat{\mathcal{F}}(u_\ell)} \big(\|\sigma\|_T^2 + \|\psi - u\|_T^2 \big) \Big)^{1/2} \;. \tag{6.11}$$

We assume that the sequence $\{\mathcal{T}_\ell\}_{\mathbb{N}}$ of triangulations, generated by (3.10), (3.11) and **(E)**, is such that there exists $0 \le \rho_4 < 1$ satisfying

$$\mu_{\ell+1}^2(u,\sigma) \;\le\; \rho_4\, \mu_\ell^2(u,\sigma) \quad, \quad \ell \in \mathbb{N}^* \;. \tag{6.12}$$

Theorem 6.4. *Let* (y,p,u,σ) *and* $(y_\ell, p_\ell, u_\ell, \sigma_\ell)$ *be the solutions of* (1.2a)–(1.2d) *and* (2.2a)–(2.2d) *and let* $\mathrm{osc}_\ell, \mu_\ell(p), \mu_\ell(u,\sigma)$ *be the data oscillations and data terms given by* (3.12), (5.14) *and* (6.11). *Assume that* (2.10), **(L)**,**(C)**, **(N$_1$)** $-$ **(N$_3$)** *and* (3.13), (5.15), (6.12) *are satisfied. Then, there exist constants* $0 \le \rho_1 < 1$ *and* $\Lambda > 0$, *depending on the data of the problem, the constants* $\Theta_i, 1 \le i \le 2$, *in the bulk criteria* (3.10), (3.11) *and on the shape regularity of the triangulations, such that*

$$\begin{pmatrix} |\!|\!|z - z_{\ell+1}|\!|\!|_\kappa^2 \\ \mathrm{osc}_{\ell+1}^2 \\ \mu_{\ell+1}^2(p) \\ \mu_{\ell+1}^2(u,\sigma) \end{pmatrix} \;\le\; \begin{pmatrix} \rho_1 & \Lambda & \Lambda & \Lambda \\ 0 & \rho_2 & 0 & 0 \\ 0 & 0 & \rho_3 & 0 \\ 0 & 0 & 0 & \rho_4 \end{pmatrix} \begin{pmatrix} |\!|\!|z - z_\ell|\!|\!|_\kappa^2 \\ \mathrm{osc}_\ell^2 \\ \mu_\ell^2(p) \\ \mu_\ell^2(u,\sigma) \end{pmatrix} \;.$$

Proof. Using the results of Lemma 6.3, as in the proof of Theorem 5.3 we find constants $0 < \kappa_i < 1, 1 \le i \le 3$, and $0 \le \rho_1 < 1$ such that for some $\Lambda > 0$

$$|\!|\!|z - z_{\ell+1}|\!|\!|_\kappa^2 \;\le\; \rho_1\, |\!|\!|z - z_\ell|\!|\!|_\kappa^2 \;+\; \Lambda \left(\mathrm{osc}_\ell^2 + \mu_\ell^2(p) + \mu_\ell^2(u,\sigma) \right) \;. \qquad\square$$

7. Numerical results

We provide numerical results that illustrate the performance of the adaptive finite element approximation for a distributed optimal control problems where the data are given as follows

$$\Omega = (0,1)^2 \;,\; u^d = f = 0 \;,\; \psi = 1 \;,\; \alpha \;=\; 10^{-k} \;,\; 1 \le k \le 5 \;,$$

$$y^d := \begin{cases} 200 x_1 x_2 (x_1 - 1/2)^2 (1 - x_2) &,\; 0 \le x_1 \le 1/2 \\ 200(x_1 - 1)x_2(x_1 - 1/2)^2(1 - x_2) &,\; 1/2 < x_1 \le 1 \end{cases} \;.$$

Figure 1 shows a visualization of the optimal state and the optimal control for various values of α. The flat region in the visualization of the control corresponds to the active set.

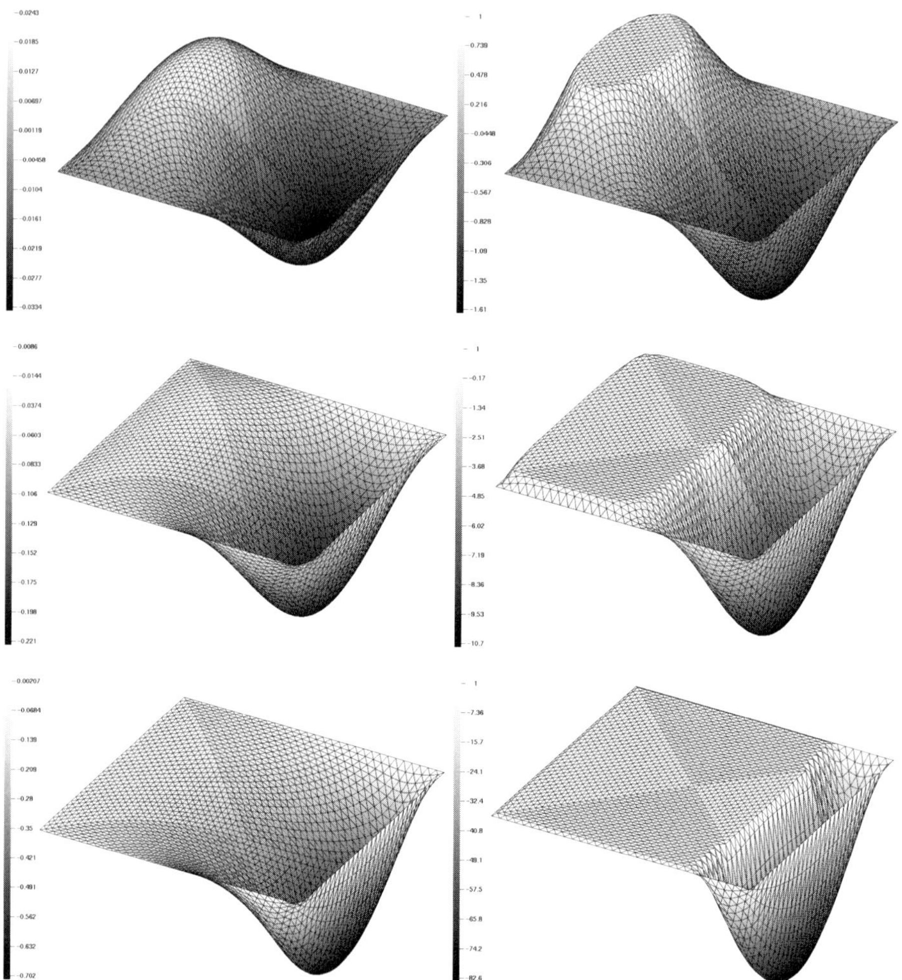

FIGURE 1. Visualization of the optimal state y and the optimal control u for $\alpha = 0.01$ (top), $\alpha = 0.001$ (middle) and $\alpha = 0.00001$ (bottom)

The initial simplicial triangulation $\mathcal{T}_0$ was chosen according to a subdivision of Ω by joining the four vertices resulting in one interior nodal point and four congruent triangles. Since u^d, f and ψ are constant, we have $\mathrm{osc}_\ell(u^d) = \mathrm{osc}_\ell(f) = \mathrm{osc}_\ell(\psi) = 0, \ell \in \mathbb{N}_0$.

For various values of α, Figure 2 displays the adaptively generated triangulations after six refinement steps with $\Theta_i = 0.6, 1 \leq i \leq 2$, in the bulk criteria. In

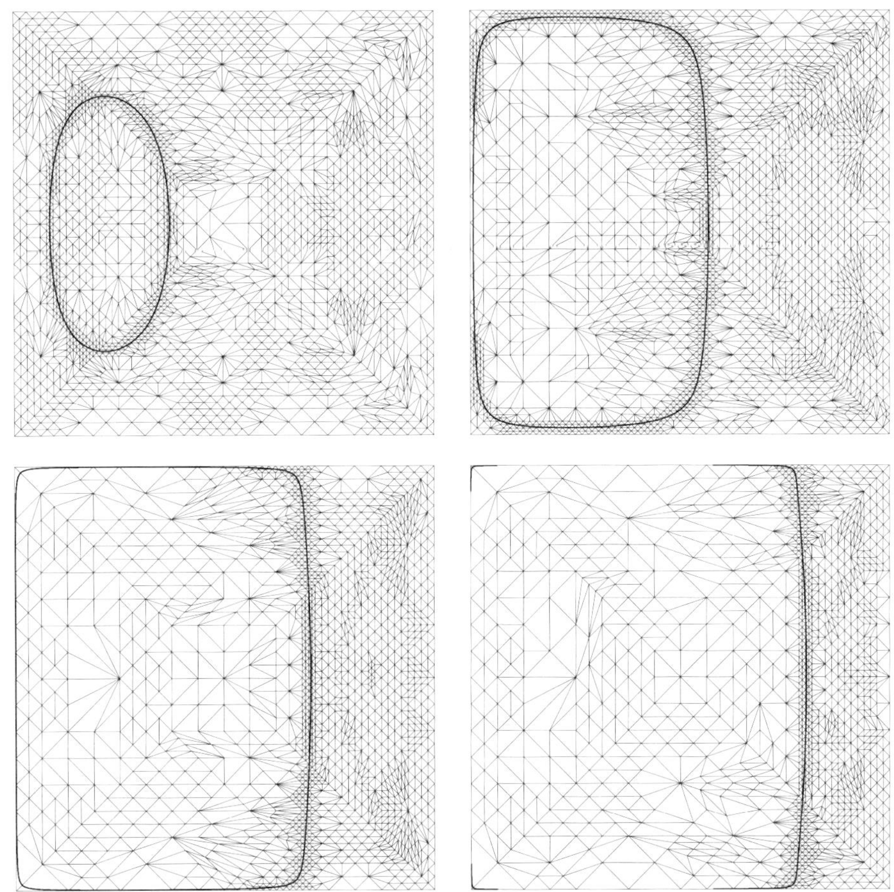

FIGURE 2. Adaptively generated grids after 6 refinement steps ($\alpha =$ 0.01 (top left), $\alpha = 0.001$ (top right), $\alpha = 0.0001$ (bottom left) and $\alpha = 0.00001$ (bottom right))

case $\alpha = 0.01$, the elliptically shaped area in the left part represents the active set. We observe that the active set is growing for decreasing α. The continuous free boundary between the active and inactive sets, displayed by a black curve, is well resolved by the adaptive refinement due to the extension (**E**) of the bulk criteria.

More detailed information is given in Table 1–Table 3. In particular, Table 1 displays the error reduction in the total error

$$|||z - z_\ell||| := (|y - y_\ell|_{1,\Omega}^2 + |p - p_\ell|_{1,\Omega}^2 \|u - u_\ell\|_{0,\Omega}^2 + \|\sigma - \sigma_\ell\|_{0,\Omega}^2)^{1/2}$$

and the errors in the state, the co-state, the control, and the co-control, whereas the actual element and edge related components of the residual type a posteriori

        A. Gaevskaya, R.H.W. Hoppe, Y. Iliash and M. Kieweg

TABLE 1. Total discretization error, discretization errors in the state, co-state, control, and co-control ($\alpha = 0.01$)

l	N_{dof}	$\|\|z - z_\ell\|\|$	$\|y - y_\ell\|_1$	$\|p - p_\ell\|_1$	$\|u - u_\ell\|_0$	$\|\sigma - \sigma_\ell\|_0$
1	13	9.38e-02	5.69e-02	3.18e-02	4.17e-01	1.37e-03
2	41	5.37e-02	3.35e-02	1.76e-02	2.07e-01	6.63e-04
3	134	3.02e-02	1.89e-02	9.67e-03	1.30e-01	3.24e-04
4	319	2.24e-02	1.39e-02	7.47e-03	8.07e-02	1.98e-04
5	795	1.47e-02	9.14e-03	4.84e-03	5.92e-02	1.10e-04
6	1998	1.02e-02	6.35e-03	3.33e-03	3.87e-02	9.16e-05
7	4373	7.16e-03	4.46e-03	2.37e-03	2.69e-02	6.70e-05
8	10612	4.93e-03	3.08e-03	1.61e-03	1.83e-02	4.60e-05
9	23019	3.44e-03	2.14e-03	1.13e-04	1.32e-02	3.24e-05

TABLE 2. Components of the error estimator and data oscillations ($\alpha = 0.01$)

l	N_{dof}	$\eta_{y,\text{T},\ell}$	$\eta_{p,\text{T},\ell}$	$\eta_{y,\text{E},\ell}$	$\eta_{p,\text{E},\ell}$	$\text{osc}_\ell(y^{\text{d}})$
1	13	2.54e-01	2.23e-01	1.56e-01	9.97e-02	9.76e-02
2	41	1.70e-01	1.10e-01	1.09e-01	6.50e-02	2.88e-02
3	134	1.03e-01	5.86e-02	6.63e-02	3.63e-02	1.03e-02
4	319	6.43e-02	3.83e-02	4.74e-02	2.63e-02	5.09e-03
5	795	4.18e-02	2.48e-02	3.25e-02	1.78e-02	2.21e-03
6	1998	2.80e-02	1.66e-02	2.30e-02	1.24e-02	1.02e-03
7	4373	1.90e-02	1.15e-02	1.64e-02	8.95e-03	5.01e-04
8	10612	1.28e-02	7.63e-03	1.15e-02	6.14e-03	2.53e-04
9	23019	8.75e-03	5.30e-03	8.35e-03	4.41e-03	1.30e-04

TABLE 3. Percentages of elements/edges selected for refinement by the bulk criteria and its extension ($\alpha = 0.01$)

l	N_{dof}	$M_{\eta,\text{T}}$	$M_{\eta,\text{E}}$	$M_{\text{osc},\text{E}}$	$M_{\text{fb},\text{E}}$
0	5	50.0	75.0	75.0	0.0
1	13	25.0	20.0	43.8	0.0
2	41	23.4	20.5	29.7	21.9
3	134	18.8	20.6	10.3	13.2
4	319	17.5	13.2	8.7	10.4
5	795	16.0	13.6	6.6	8.2
6	1998	15.4	11.8	5.8	6.4
7	4373	16.3	13.0	5.0	5.8
8	10612	15.7	12.5	2.6	4.7
9	23019	15.2	11.8	1.8	4.4

error estimator are given in Table 2. Table 3 contains the percentages of elements and edges that have been marked for refinement according to the bulk criteria and their extension (**E**). Here, $M_{\eta,T}$ stands for the level l elements marked for refinement due to the element residuals and the data oscillations. On the other hand, $M_{\eta,E}$, $M_{\text{osc},E}$ and $M_{fb,E}$ refer to the edges marked for refinement with regard to the edge residuals, data oscillations and the extension (**E**) of the bulk criteria (resolution of the free boundary). On the coarsest grid, the sum of the percentages exceeds 100 %, since an edge may satisfy more than one criterion in the adaptive refinement process. The refinement is initially dominated by the resolution of the free boundary and the data oscillations, whereas at a later stage edge and element residuals dominate.

References

[1] M. Ainsworth and J.T. Oden, *A Posteriori Error Estimation in Finite Element Analysis*. Wiley, Chichester, 2000.

[2] I. Babuska and W. Rheinboldt, *Error estimates for adaptive finite element computations*. SIAM J. Numer. Anal. **15** (1978), 736–754.

[3] I. Babuska and T. Strouboulis, *The Finite Element Method and its Reliability*. Clarendon Press, Oxford, 2001.

[4] W. Bangerth and R. Rannacher, *Adaptive Finite Element Methods for Differential Equations*. Lectures in Mathematics. ETH-Zürich. Birkhäuser, Basel, 2003.

[5] R.E. Bank and A. Weiser, *Some a posteriori error estimators for elliptic partial differential equations*. Math. Comput. **44** (1985), 283–301.

[6] R. Becker, H. Kapp, and R. Rannacher, *Adaptive finite element methods for optimal control of partial differential equations: Basic concepts*. SIAM J. Control Optim. **39** (2000), 113–132.

[7] M. Bergounioux, M. Haddou, M. Hintermüller, and K. Kunisch, *A comparison of a Moreau-Yosida based active set strategy and interior point methods for constrained optimal control problems*. SIAM J. Optim. **11** (2000), 495–521.

[8] P. Binev, W. Dahmen, and R. DeVore, *Adaptive Finite Element Methods with Convergence Rates*. Numer. Math. **97**, (2004), 219–268.

[9] C. Carstensen and S. Bartels, *Each averaging technique yields reliable a posteriori error control in FEM on unstructured grids. Part I: Low order conforming, nonconforming, and mixed FEM*. Math. Comput. **71** (2002), 945–969.

[10] C. Carstensen and R.H.W. Hoppe, *Error reduction and convergence for an adaptive mixed finite element method*. Math. Comp. (2006) (in press).

[11] C. Carstensen and R.H.W. Hoppe, *Convergence analysis of an adaptive nonconforming finite element method*. Numer. Math. (2006) (in press).

[12] C. Carstensen and R.H.W. Hoppe, *Convergence analysis of an adaptive edge finite element method for the 2d eddy current equations*. J. Numer. Math. **13** (2005), 19–32.

[13] W. Dörfler, *A convergent adaptive algorithm for Poisson's equation*. SIAM J. Numer. Anal. **33** (1996), 1106–1124.

[14] K. Eriksson, D. Estep, P. Hansbo, and C. Johnson, *Computational Differential Equations*. Cambridge University Press, Cambridge, 1995.

[15] H.O. Fattorini, *Infinite-Dimensional Optimization and Control Theory*. Cambridge University Press, Cambridge, 1999.

[16] M. Hintermüller, *A primal-dual active set algorithm for bilaterally control constrained optimal control problems*. Quarterly of Applied Mathematics **LXI** (2003), 131–161.

[17] M. Hintermüller, R.H.W. Hoppe, Y. Iliash, and M. Kieweg, *An a posteriori error analysis of adaptive finite element methods for distributed elliptic control problems with control constraints.* to appear in ESAIM (2006).

[18] J.B. Hiriart-Urruty and C. Lemaréchal, *Convex Analysis and Minimization Algorithms.* Springer, Berlin-Heidelberg-New York, 1993.

[19] R. Li, W. Liu, H. Ma, and T. Tang, *Adaptive finite element approximation for distributed elliptic optimal control problems.* SIAM J. Control Optim. **41** (2002), 1321–1349.

[20] J.L. Lions, *Optimal Control of Systems Governed by Partial Differential Equations.* Springer, Berlin-Heidelberg-New York, 1971.

[21] W. Liu and N. Yan, *A posteriori error estimates for distributed optimal control problems.* Adv. Comp. Math. **15** (2001), 285–309.

[22] W. Liu and N. Yan, *A posteriori error estimates for convex boundary control problems.* Preprint, Institute of Mathematics and Statistics, University of Kent, Canterbury (2003).

[23] X.J. Li and J. Yong, *Optimal Control Theory for Infinite-Dimensional Systems.* Birkhäuser, Boston-Basel-Berlin, 1995.

[24] P. Morin, R.H. Nochetto, and K.G. Siebert, *Data oscillation and convergence of adaptive FEM.* SIAM J. Numer. Anal. **38** (2000), 466–488.

[25] P. Neittaanmäki and S. Repin, *Reliable methods for mathematical modelling. Error control and a posteriori estimates.* Elsevier, New York, 2004.

[26] R. Verfürth, *A Review of A Posteriori Estimation and Adaptive Mesh-Refinement Techniques.* Wiley-Teubner, New York, Stuttgart, 1996.

A. Gaevskaya, Y. Iliash and M. Kieweg
Institute of Mathematics
Universität Augsburg
D-86159 Augsburg, Germany

e-mail: `gaevskaya@math.uni-augsburg.de`
e-mail: `iliash@math.uni-augsburg.de`
e-mail: `Michael.Kieweg@gmx.de`

R.H.W. Hoppe
Department of Mathematics
University of Houston
Houston, TX 77204-3008, USA
e-mail: `rohop@math.uh.edu`

International Series of Numerical Mathematics, Vol. 155, 69–94

Optimal Boundary Control in Flood Management

Martin Gugat

Abstract. In active flood hazard mitigation, lateral flow withdrawal is used to reduce the impact of flood waves in rivers. Through emergency side channels, lateral outflow is generated. The optimal outflow controls the flood in such a way that the cost of the created damage is minimized. The flow is governed by a networked system of nonlinear hyperbolic partial differential equations, coupled by algebraic node conditions. Two types of integrals appear in the objective function of the corresponding optimization problem: Boundary integrals (for example, to measure the amount of water that flows out of the system into the floodplain) and distributed integrals.

For the evaluation of the derivative of the objective function, we introduce an adjoint backwards system. For the numerical solution we consider a discretized system with a consistent discretization of the continuous adjoint system, in the sense that the discrete adjoint system yields the derivatives of the discretized objective function. Numerical examples are included.

Mathematics Subject Classification (2000). 35L45 35L50 35L65 93C20 .

Keywords. St. Venant equations, subcritical states, adjoint system, optimal boundary control, necessary optimality conditions, classical solutions.

1. Introduction

In flood management, the aim is to minimize the damage caused by a flood by active flow control (see [29]). In the present paper, we consider a problem where the aim is to find a compromise with minimal cost between flood in a city that occurs if the water level rises above a certain upper bound and the cheaper outflow through emergency side channels, for example to a floodplain. Often damage is caused if the water level in a certain area rises above a given upper bound. The control function in the corresponding problem of optimal boundary control is the

---

This work was supported by DFG-research cluster: real-time optimization of complex systems; grant number Le595/13-1.

discharge from the emergency side channel; in practice the control is realized by opening underflow gates.

The flow in the considered channel network is described by a networked system of conservation laws. On each edge of the corresponding graph, the flow is described by a hyperbolic quasilinear system of partial differential equations, namely de St. Venant's equations (see [11]), that model conservation of mass and the evolution of momentum. On the vertices, the flow variables are coupled by algebraic node conditions.

We work with continuously differentiable solutions of the system equations. Our strategy is to control the flow in such a way that no singularities are generated. This implies that also the boundary control functions have to be continuously differentiable. Questions of boundary controllability within the class of continuously differentiable solutions have been studied in [8], [9], [12], [16], [26]. The results in these papers show that a large class or states can be reached with continuously differentiable solutions of the system equations.

The nature of the boundary conditions changes with the speed of the flow. Here we consider subcritical flow, that is the flow velocity remains below the wave celerity. In this case, one scalar boundary condition appears at each boundary node. Our problem is related to studies of the optimization of the operation of hydro power stations, see for example [27].

This work is based upon the many studies of hyperbolic conservation laws, see for example [5], [10], [23], [25], and the references therein. Classical solutions are studied in [35]. In this paper, we want to study problems of optimal control where a hyperbolic networked system is controlled. Such a networked system has been considered for example in [17], [26] for networks of channels and in [6], [15], [22] for a model of traffic flow.

Our objective function is a sum of integrals of different types: First boundary integrals, where a function of the values of the state at a fixed boundary node is integrated over time and secondly distributed integrals, where the state in a whole rectangle in space-time contributes to the integral. The existence of directional derivatives of such objective functions has been proved in [14] for systems in diagonal form. In [13], this result is given for systems in conservative form.

To evaluate the derivatives of such objective functions, we present an adjoint sensitivity calculus, first for the infinite-dimensional problem, then also for a discretized problem that is obtained by upwind discretization that mimics the propagation of information along the characteristic curves in the sense that for the equation with a positive eigenvalue, information travels from left to right and for the equation with a negative eigenvalue, information travels from right to left. With our consistent discretization of the continuous adjoint problem we obtained gradient information that was used to approximate optimal controls. This is related to studies by Hager for ordinary differential equations (see [20], [21]). Hager characterizes time discretizations of the state equation such that the discrete adjoint equation leads to a highly accurate adjoint state.

Adjoint problems for single channels have been presented from the engineering point of view in [30], [31], where a detailed exposition of the application can be found. The adjoint sensitivity method is an important tool in optimal control, see for example [33] and for applications to flow problems [4], [18] and also [7], where the optimal boundary control af aeroaccoustic noise governed by the two-dimensional unsteady compressible Euler equations is considered. Recent mathematical studies of adjoint-based sensitivity calculations that emphasize the computation of the sensitivities of shock speeds are given in [36], [37] for scalar equations and in [2], [3]. The problem that we study here is of a different type, since the aim is to choose controls that generate classical solutions and do not cause shocks.

In contrast to what can be found in the literature our paper provides an adjoint sensitivity calculus for networked systems; in particular, we show how to derive adjoint interior node conditions. Moreover, we do not restrict our attention to distributed integrals but consider boundary integral as well (and combinations of the two) as objective functions. Finally, the quasilinear hyperbolic systems with source terms that is studied here is not scalar.

The numerical results illustrate that our sensitivity calculus allows the numerical solution of problems of optimal boundary control for networked systems that arise naturally in the application that we consider here.

The paper has the following structure. First we present the model for our networked system: The flow through each single channel is modelled by St. Venant's equation, that can be transformed to diagonal form. An appropriate initial condition has to be provided. The water flow through the nodes in our channel graph is governed by algebraic node conditions. The control acts on the system through the boundary conditions.

Then the cost functional is introduced, that is essential for the statement of our problem of optimal control. For the evaluation of the derivative of the cost functional, we use the results from [14]. The corresponding adjoint backwards problem is stated: The adjoint system equation, the end conditions, the adjoint node conditions and the adjoint boundary conditions.

Finally we present the discretized model for our system, the discretized objective function and a consisten discretization of the adjoint problem, which yields the exact derivatives for the discretized objective function.

2. The model for the network flow

Let a finite graph (V, E) be given, where V denotes the set of nodes and E is the set of edges. In each edge of the graph, the flow is governed by the equations of free surface flow. In the vertices, these equations are coupled by algebraic node conditions.

2.1. Shallow water equations

To model the flow in a single edge e of our network, we use de St. Venant's system (see [11], [34]) that models the conservation of mass and the evolution of

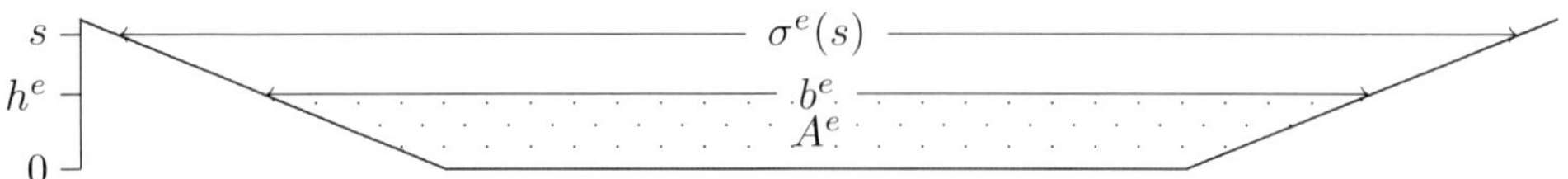

FIGURE 1. Cross section of the channel

momentum. Let an edge e of our network be given. This edge corresponds to an interval $[0, L_e]$, that is a one-dimensional model for a channel of length L_e. The initial state of the system is given by an initial condition of the form

$$v^e(x, 0) = v_0^e(x), \ x \in [0, L_e], \tag{1}$$

where for all edges $e \in E$, the function v_0^e is continuously differentiable.

We consider a prismatic channel with constant bottom slope γ^e. If γ^e is negative, the channel proceeds downward. Let A^e denote the area of the wetted cross section and V^e the average velocity in the channel. Let b^e denote the breadth of the water surface and let the width of the channel at elevation s be given by the value $\sigma^e(s)$ of the piecewise continuously differentiable function σ^e.

Then the wetted cross section corresponding to the water height h^e is

$$A^e(h^e) = \int_0^{h^e} \sigma^e(s) \, ds$$

and the water height corresponding to a wetted cross section A^e can be computed as a function $h^e(A^e)$. For example for a triangular channel we have $\sigma^e(s) = 2\tau s$ and $A^e(h^e) = \tau h^2 / 2$, hence $h^e(A^e) = \sqrt{(2A^e/\tau)}$.

The conservation of mass yields our first system equation

$$A_t^e + (V^e A^e)_x = 0. \tag{2}$$

The flux of momentum is modelled by the equation

$$V_t^e + (g h^e(A^e) + (V^e)^2/2)_x + S_2^e = 0.$$

The source term S_2^e has the form $S_2^e = g(\gamma^e + \eta^e(A^e, V^e))$, and the friction slope $\eta^e(A^e, V^e)$ can be modelled in the form of the Manning–Strickler relation (see for example [28])

$$\eta^e(A^e, V^e) = \frac{C_e^2 |V^e| \, V^e}{(A^e / P^e(A^e))^{4/3}}, \tag{3}$$

with a constant C^e and where $P^e(A^e)$ is the wetted perimeter corresponding to A^e, for example

$$P^e(A^e) = b^e + 2 \, h^e(A^e)$$

for a rectangular cross section. Note that wind stress can be modelled by adding another term to S_2^e. Our system of conservation laws can be written in the form

$$v_t^e + F_e(v^e)_x + S_e(v^e) = 0, \tag{4}$$

where $v^e = (A^e, V^e)^T$. We have

$$(F^e)'(v^e) = \begin{pmatrix} V^e & A^e \\ g(h^e)'(A^e) & V^e \end{pmatrix}.$$

A set of left eigenvectors for the system matrix $(F^e)'$ is

$$L_+^e(A^e, V^e) = (\sqrt{g(h^e)'(A^e)}/\sqrt{A^e}, 1), \ \ L_-^e(A^e, V^e) = (-\sqrt{g(h^e)'(A^e)}/\sqrt{A^e}, 1),$$

with the corresponding eigenvalues

$$\lambda_+^e(A^e, V^e) = V^e + \sqrt{gA^e(h^e)'(A^e)}, \ \ \lambda_-^e(A^e, V^e) = V^e - \sqrt{gA^e(h^e)'(A^e)}.$$

In the subcritical case that we consider in this paper, we have $\lambda_-^e < 0 < \lambda_+^e$.

Let $A_0^e > 0$ be given. Define $G^e(A^e) = \sqrt{g(h^e)'(A^e)}/\sqrt{A^e}$ and

$$\varphi^e(A^e) = \int_{A_0^e}^{A^e} G^e(a)\, da = \sqrt{g} \int_{h^e(A_0^e)}^{h^e(A^e)} \sqrt{\frac{d}{dh}\ln(A(h))}\, dh, \tag{5}$$

where the last equation follows by substitution. Note that $(A^e)'(h^e) = \sigma^e(h^e)$, hence $(h^e)'(A^e) = 1/(\sigma^e(h^e(A^e))$. Then we can work with the Riemann invariants (see [26]) $R_+^e(A^e, V^e) = V^e + \varphi^e(A^e)$, $R_-^e(A^e, V^e) = V^e - \varphi^e(A^e)$. In terms of the functions (R_+^e, R_-^e), our system equation (4) has the diagonal form

$$\begin{pmatrix} R_+^e \\ R_-^e \end{pmatrix}_t + D^e \begin{pmatrix} R_+^e \\ R_-^e \end{pmatrix}_x + \begin{pmatrix} S_2^e(R_+^e, R_-^e) \\ S_2^e(R_+^e, R_-^e) \end{pmatrix} = 0 \tag{6}$$

where D^e is a diagonal matrix that contains the eigenvalues λ_+^e, λ_-^e as functions of the Riemann invariants and the source term S^e is also written as a function of the Riemann invariants. We have $V^e = (R_+^e + R_-^e)/2$ and $A^e = (\varphi^e)^{-1}((R_+^e - R_-^e)/2)$.

2.2. Junctions: The interior node conditions

In [34], the conditions to be satisfied at a junction are chosen to be conservation of mass and continuity of the water surface. In general these geometric conditions do not guarantee the conservation of energy for the flow through the node. An alternative is to require the conservation of mass and the continuity of the specific energy, as proposed in [26]. In general this leads to a solution with a discontinuity in the water surface. In [32], an analysis of the balance of momentum along the streamlines leads to another more complicated set of node conditions that takes into account the boundary friction force. In these conditions two parameters appear that vary with the junction angle.

We consider here only subcritical flow, which means that our system matrix D^e has everywhere one positive and one negative eigenvalue, so at each end of each channel we have one characteristic curve that enters the channel and one that leaves the channel. This implies that in a node of our network with n adjacent channels we need a system of n equations as node conditions.

Let a vertex ω of our network be given. Let $E_0(\omega)$ denote the set of adjacent edges. Assume that the set $E_0(\omega)$ has more than one element, that is that ω is an interior node of the network.

For $e \in E_0(\omega)$, $x_e(\omega) \in \{0, L_e\}$ denotes the end of the interval $[0, L_e]$ that corresponds to the vertex. Define $\varepsilon_e(\omega) = 1$ if $x_e(\omega) = L_e$ and $\varepsilon_e(\omega) = -1$ if $x_e(\omega) = 0$. Think of $\varepsilon_e(\omega)$ as a one-dimensional outer normal vector.

Then the first node condition (*conservation of mass*) is

$$\sum_{e \in E_0(\omega)} A^e(x_e(\omega), t)\, V^e(x_e(\omega), t)\, \varepsilon_e(\omega) = 0 \text{ for all } t \geq 0. \tag{7}$$

Our second node condition (*continuity of water surface*) is

$$h^e(x_e(\omega), t) = h^f(x_f(\omega), t) \quad \text{for} \quad \text{all} \quad e, f \in E_0(\omega),\ t \geq 0. \tag{8}$$

Define $h^\omega(t)$ as the water height at the junction. Note that here we assume that the bed in all adjacent edges has the same height. Assume now that the beds of all adjacent channels have the same shape, that is there is a function σ^ω and for all $e \in E_0(\omega)$ there exist numbers $\beta^e > 0$ such that

$$\sigma^e(s) = \beta^e \sigma^\omega(s). \tag{9}$$

Assumption (9) is valid for nodes where all adjacent channels are rectangular and for nodes where all adjacent channels are triangular,

Due to (9) we have $A^e(h^e) = \beta^e \int_0^{h^e} \sigma^\omega(s)\, ds$, hence (7) and (8) yield

$$\sum_{e \in E_0(\omega)} \beta^e\, V^e(x_e(\omega), t)\, \varepsilon_e(\omega) = 0, \tag{10}$$

an equation where only the velocities V^e appear.

We choose the areas A_0^e in the definition (5) of φ in such a way that for all $e, f \in E_0(\omega)$ we have $h_e(A_0^e) = h_f(A_0^f)$. Then Equation (9) implies that

$$\varphi^e(A^e) = \int_{h^e(A_0^e)}^{h^e(A^e)} \sqrt{g\sigma^w(h)} \Big/ \sqrt{\int_0^h \sigma^\omega(s)\, ds}\ \ dh,$$

hence the node condition (8) can be replaced by the equation $\varphi^e(A^e) = \varphi^f(A^f)$ for all $e, f \in E_0(\omega)$. Therefore our node conditions can be transformed to a system of linear equations for the Riemann invariants, namely

$$\sum_{e \in E_0(\omega)} \beta^e(R_+^e + R_-^e)\, \varepsilon_e(\omega) \quad = \quad 0, \tag{11}$$

$$R_+^e - R_-^e = R_+^f - R_-^f \quad \text{for} \quad \text{all} \quad e, f \in E_0(\omega). \tag{12}$$

This fact is very useful, since we can solve this system analytically which simplifies the numerical implementation.

2.3. The boundary conditions

A boundary node ω is a node where the set $E_0(\omega)$ of adjacent edges has only one element. For a boundary node ω, let $e(\omega)$ denote this edge. Let V_B denote the set of all boundary nodes of our network. In our network, we have two types of boundary nodes: The controlled nodes and the uncontrolled nodes. At a controlled node ω with $x_e(\omega) = 0$ for the adjacent edge e, the boundary condition has the form

$$R_+^e(0, t) = u^\omega(t), \qquad (13)$$

where u^ω is the control function. The value of the other Riemann invariant R_-^e is transported to the boundary from the interior of the interval along the characteristic curves. In the applications, the boundary condition will usually be given in the implicit form $f(R_+^e(0,t), R_-^e(0,t), u_1^\omega(t)) = 0$ (for example with u_1^ω denoting the opening height of an underflow gate) but it can be transformed to the form (13) with u^ω given as a function of $R_-^e(0,t)$ and $u_1^\omega(t)$. Our policy is to solve the problem of optimal control in terms of the Riemann invariants as control functions; from the computed optimal control and the corresponding generated state in the system, we can then compute the controls in the form that is needed to steer the involved machinery. In this way we can work in our computations with the diagonal form of the system. At a controlled node ω with $x_e(\omega) = L^e$ for the adjacent edge e, the boundary condition is

$$R_-^e(L^e, t) = u^\omega(t). \qquad (14)$$

At an uncontrolled node with $x_e(\omega) = 0$ for the adjacent edge e, the boundary condition has the form

$$R_+^e(0, t) = w^\omega(t). \qquad (15)$$

Here the function w^ω contains the information that enters the system from outside, in our application through the expected hydrographs. Again this condition will usually be given implicitly, for example in the form

$$f(R_+^e(0,t), \ R_-^e(0,t), \ w_1^\omega(t)) = 0,$$

where w_1^ω denotes the expected flow rate. At an uncontrolled node with $x_e(\omega) = L^e$ for the adjacent edge e, the boundary condition has the form

$$R_-^e(L^e, t) = w^e(t). \qquad (16)$$

If the velocity is positive, the function w^e contains the information that enters the system from downstream. In our examples, we will prescribe a constant value for R_-^e at such a node where outflow occurs, which yields absorbing boundary conditions. Important work on boundary data for hyperbolic problems can be found for example in [1].

3. The objective function and a problem of optimal control

We consider a problem of control for our network, where a nonempty set V_B^c of controlled boundary nodes is given. For a controlled boundary node $\omega \in V_B^c$, the corresponding control costs are given in terms of boundary integrals of the form (17), (18) with integrands f^ω, that give the volume of the water that flows from the node ω into the floodplain during the time-interval $[0, T]$.

Let V^0 denote the set of boundary nodes that are at the end zero of the adjacent edge and V^L denote the set of boundary nodes that are at the end L_e of the adjacent edge e. We consider objective functions of the form

$$J(u) = \sum_{\omega \in V^0} \int_0^T f^\omega(R_+^{e(\omega)}(0,t), \, R_-^{e(\omega)}(0,t)) \, dt \tag{17}$$

$$+ \sum_{\omega \in V^L} \int_0^T f^\omega(R_+^{e(\omega)}(L_{e(\omega)},t), \, R_-^{e(\omega)}(L_{e(\omega)},t)) \, dt \tag{18}$$

$$+ \sum_{e \in E} \int_0^T \int_0^{L_e} f^e(R_+^e(x,t), \, R_-^e(x,t)) \, dx \, dt. \tag{19}$$

The functions f^ω, and f^e are assumed to be continuously differentiable, the function u is the control function and (R^+, R^-) is the solution of the networked system governed by the system equation (6), with the initial conditions (1), the interior node conditions (11), (12) and the boundary conditions (13), (14) at the controlled boundary nodes and (15), (16) at the uncontrolled boundary nodes that is generated by this control function u.

The cost of a flood in the floodplain will in general be much lower than the cost of a flood event in a city, which is given in terms of a distributed integral of the form (19), where the function f^e is for example of the form

$$f^e(x) = w^e(x)g^e(R_+^e(x,t), \, R_-^e(x,t)),$$

with a positive weight function w^e, whose mass is concentrated near the city center. This part of the objective function can for example penalize water heights that are greater than a given upper bound.

For a given time horizon $[0, T]$ and boundary data w^ω (for $\omega \in V_B$, $\omega \notin V_B^c$), we are looking for a continuously differentiable control function u that generates a continuously differentiable state with minimal cost. Then the set of feasible controls is

$$U = \{u: \quad for \;\; all \;\; \omega \in V_c^0, u_+^e \in C^1(0,T); \quad for \;\; all \;\; \omega \in V_c^L, u_-^e \in C^1(0,T);$$

the solution of the system equation (6) with the initial conditions (1) and the boundary conditions (13), (14), (15), (16) and the interior node conditions (25) ((27) respectively) is a continuously differentiable function of the time and space variables and a subcritical state.} Our optimization problem is

$$\min_{u \in U} J(u). \tag{20}$$

We do not discuss the question of the existence of a solution of the optimization problem here. This will be the subject of future studies.

4. The adjoint problem

The existence of the directional derivatives of J has been proved in [14]. In order to evaluate the derivatives of the objective function J with respect to the boundary controls, we will work with an adjoint backwards system. In this problem, end conditions for the adjoint state are prescribed and we proceed backwards in time, so the roles of ingoing and outgoing characteristics are exchanged. The characteristic curves are the same as for the forward problem, so our assumption that the state is continuously differentiable implies that no shocks are generated by the intersection of characteristic curves. Since the adjoint equation has also diagonal form, we can consider solutions of the adjoint problem that satisfy integral equations along the characteristic curves.

4.1. The adjoint system equations

Let $\mu^e = (\mu^e_+, \mu^e_-)$ denote the adjoint variables corresponding to $(R^e_+, R^e_-)^T$. The adjoint variables can be thought of as Lagrange multipliers for our pde-constrained optimization problem. The adjoint system equation is

$$(\mu^e_+, \mu^e_-)_t + \big((\mu^e_+, \mu^e_-)D^e\big)_x - (\mu^e_+, \mu^e_-)Z^e - (\mu^e_+, \mu^e_-)B^e = (\partial_+ f^e,\ \partial_- f^e)$$

or in short notation

$$\mu^e_t + (\mu^e D^e)_x - \mu^e Z^e - \mu^e B^e = (\nabla f^e)^T, \qquad (21)$$

where the 2×2 matrix

$$Z^e = \left(\begin{array}{cc} Z^e_{++} & Z^e_{+-} \\ Z^e_{-+} & Z^e_{--} \end{array} \right)$$

has the entries

$$Z^e_{++} = \partial_{R^e_+} \lambda^e_+(R^e_+, R^e_-)\, \partial_x R^e_+$$
$$Z^e_{+-} = \partial_{R^e_-} \lambda^e_+(R^e_+, R^e_-)\, \partial_x R^e_+$$
$$Z^e_{-+} = \partial_{R^e_+} \lambda^e_-(R^e_+, R^e_-)\, \partial_x R^e_-$$
$$Z^e_{--} = \partial_{R^e_-} \lambda^e_-(R^e_+, R^e_-)\, \partial_x R^e_-$$

and the 2×2 matrix B^e is given by the equation

$$B^e = \left(\begin{array}{cc} \partial_{R^e_+} S^e_2(R^e_+, R^e_-) & \partial_{R^e_-} S^e_2(R^e_+, R^e_-) \\ \partial_{R^e_+} S^e_2(R^e_+, R^e_-) & \partial_{R^e_-} S^e_2(R^e_+, R^e_-) \end{array} \right).$$

Since we work with continuously differentiable solutions of system (4), for each eigenvalue of the system matrix the characteristic curves, (i.e. the integral curves of this eigenvalue) generate a curvilinear system of coordinates, hence they do not intersect and there are no rarefaction fans. These curves are the same for the adjoint equation (21). Hence we can consider solutions of (21) that are defined by the evolution of the corresponding Riemann invariants along the characteristic curves (characteristic solutions). Note that this process does not necessarily produce continuously differentiable solutions, since singularities generated by the initial/boundary conditions are transported as contact discontinuities; however, this is the only type of singularities that occurs here.

The end conditions for the adjoint system are

$$\mu^e(x, T) = 0, \ x \in [0, L^e],$$ (22)

with the time T that is the endpoint of the time interval $[0, T]$ where we consider our system. If the friction slope η^e is given by (3), for the case of rectangular channels (i.e., $A^e = b^e h^e$) with the notation $\tau^e = 2/b^e + 1/h^e$ we have

$$\partial_{R^e_+} S^e_2 = (C^e)^2 (\tau^e)^{1/3} |V^e| \left[g\tau^e - V^e(R^e_+ - R^e_-)/(6(h^e)^2) \right],$$ (23)

$$\partial_{R^e_-} S^e_2 = (C^e)^2 (\tau^e)^{1/3} |V^e| \left[g\tau^e + V^e(R^e_+ - R^e_-)/(6(h^e)^2) \right].$$ (24)

4.2. The adjoint interior node conditions

Consider now a node δ with three adjacent edges e, f, g that model channels that satisfy (9). We consider two cases.

Case 1: Assume that $x_e(\delta) = L_e$, $x_f(\delta) = L_f$ and $x_g(\delta) = 0$ (see Figure 2). The

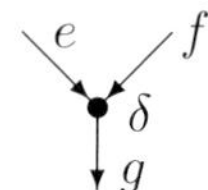

FIGURE 2. A junction of three channels

corresponding node conditions are

$$\beta^e(R^e_+ + R^e_-) + \beta^f(R^f_+ + R^f_-) - \beta^g(R^g_+ + R^g_-) = 0,$$

$$R^e_+ - R^e_- = R^f_+ - R^f_- = R^g_+ - R^g_-.$$

Since we consider subcritical flow, R^e_+, R^f_+ and R^g_- are determined from the interior of the channels and R^e_-, R^f_- and R^g_+ are determined from the node conditions. We can write the interior node conditions in the form

$$\begin{pmatrix} R^e_- \\ R^f_- \\ R^g_+ \end{pmatrix} = M^\omega \begin{pmatrix} R^e_+ \\ R^f_+ \\ R^g_- \end{pmatrix}$$ (25)

where the matrix M^ω is given by the equation

$$M^\omega = \frac{1}{\beta^e + \beta^f + \beta^g} \begin{pmatrix} -\beta^e + \beta^f + \beta^g & -2\beta^f & 2\beta^g \\ -2\beta^e & \beta^e - \beta^f + \beta^g & 2\beta^g \\ 2\beta^e & 2\beta^f & \beta^e + \beta^f - \beta^g \end{pmatrix}.$$

The eigenvalues of the matrix M^ω are -1 with a one-dimensional eigenspace and 1 with a two-dimensional eigenspace.

$$(\lambda^e_+ \mu^e_+, \lambda^f_+ \mu^f_+, -\lambda^g_- \mu^g_-) = (-\lambda^e_- \mu^e_-, -\lambda^f_- \mu^f_-, \lambda^g_+ \mu^g_+) \, M^\omega.$$ (26)

That these conditions are meaningful can be seen as follows. For each node, the ingoing characteristic curves for the forward problem where information flows into the node with time going forwards are for the adjoint problem with time going backwards the curves where information flows out of the node, hence the adjoint variables corresponding to these curves have to be prescribed in terms of the remaining variables.

Case 2: Assume that $x_e(\delta) = L_e$, $x_f(\delta) = 0$ and $x_g(\delta) = 0$ (see Figure 3). In this

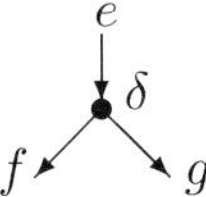

FIGURE 3. A junction of three channels

case, R_+^e, R_-^f and R_-^g are determined from the interior of the channels and R_-^e, R_+^f and R_+^g are determined from the node conditions that can be written in the form

$$\begin{pmatrix} R_-^e \\ R_+^f \\ R_+^g \end{pmatrix} = M^\omega \begin{pmatrix} R_+^e \\ R_-^f \\ R_-^g \end{pmatrix} \tag{27}$$

where the matrix M^ω is given by the equation

$$M^\omega = \frac{1}{\beta^e + \beta^f + \beta^g} \begin{pmatrix} -\beta^e + \beta^f + \beta^g & 2\beta^f & 2\beta^g \\ 2\beta^e & \beta^e - \beta^f + \beta^g & -2\beta^g \\ 2\beta^e & -2\beta^f & \beta^e + \beta^f - \beta^g \end{pmatrix}.$$

In this case, the adjoint interior node conditions are

$$(\lambda_+^e \mu_+^e, -\lambda_-^f \mu_-^f, -\lambda_-^g \mu_-^g) = (-\lambda_-^e \mu_-^e, \lambda_+^f \mu_+^f, \lambda_+^g \mu_+^g)\, M^\omega. \tag{28}$$

We have discussed in detail the adjoint interior node conditions for a node with three adjacent edges. In the sequel we assume that each node of our network is of the type considered in Case 1 or in Case 2.

4.3. The adjoint boundary conditions and derivatives

In this section we introduce the adjoint boundary conditions that allow the evaluation of derivatives of J. At a boundary node $\omega \in V^0$ the adjoint boundary condition has the form

$$\mu_-^{e(\omega)}(0,t) = \partial_- f^\omega(R_+^{e(\omega)}(0,t), R_-^{e(\omega)}(0,t))/\lambda_-^{e(\omega)}(0,t). \tag{29}$$

Here ∂_- denotes the partial derivative with respect to $R_-^{e(\omega)}$.

At a boundary node $\omega \in V^L$ the adjoint boundary condition has the form

$$\mu_+^{e(\omega)}(L_{e(\omega)},t) \;=\; -\partial_+ f^\omega(R_+^{e(\omega)}(L_{e(\omega)},t), R_-^{e(\omega)}(L_{e(\omega)},t))/\lambda_+^{e(\omega)}(L_{e(\omega)},t).$$

Here ∂_+ denotes the partial derivative with respect to $R_+^{e(\omega)}$.

We can evaluate the directional derivative

$$D_d J(u) = \lim_{h \to 0} [J(u + hd) - J(u)]/h$$

with respect to the boundary control function u using the adjoint solution μ that satisfies for all edges $e \in E$ the end condition (22), the adjoint system equation (21), in the interior nodes the adjoint interior node conditions (26), (28) respectively and at the boundary nodes the adjoint boundary conditions (29), (30).

Define the set $V_c^0 = \{\omega \in V_B^c : \text{for all } e \in E_0(\omega),\ x_e(\omega) = 0\}$, that is the set of all controlled boundary nodes with zero end for the adjacent edge, and analogously $V_c^L = \{\omega \in V_B^c : \text{for all } e \in E_0(\omega),\ x_e(\omega) = L_e\}$. Then we have

$$D_d J(u) = \sum_{\omega \in V_c^0} \int_0^T \left[\partial_+ f^\omega (R_+^{e(\omega)}(0,t),\, R_-^{e(\omega)}(0,t)) \right] d_+^\omega(t)\, dt \tag{30}$$

$$- \sum_{\omega \in V_c^0} \int_0^T \left[\mu_+^{e(\omega)}(0,t)\, \lambda_+^{e(\omega)}(0,t) \right] d_+^\omega(t)\, dt \tag{31}$$

$$+ \sum_{\omega \in V_c^L} \int_0^T \left[\partial_- f^\omega (R_+^{e(\omega)}(L_{e(\omega)},t),\, R_-^{e(\omega)}(L_{e(\omega)},t)) \right] d_-^\omega(t)\, dt \tag{32}$$

$$+ \sum_{\omega \in V_c^L} \int_0^T \left[\mu_-^{e(\omega)}(L_{e(\omega)},t)\, \lambda_-^{e(\omega)}(L_{e(\omega)},t) \right] d_-^\omega(t)\, dt. \tag{33}$$

Here for a controlled boundary node $\omega \in V_B^c$, d_+^ω (d_-^ω respectively) gives the direction in which the directional derivative is evaluated for the control function at ω. This representation of the directional derivative is valid for a finite network under the assumptions that the control u generates a continuously differentiable state and that the direction d satisfies the compatibility conditions $d(0) = d'(0) = 0$ (see [14], Theorem 3).

5. Necessary optimality conditions

We consider directions d_+^ω, for $\omega \in V_c^0$, d_-^ω for $\omega \in V_c^L$ that are compatible with continuously differentiable solutions, that is with zero value and zero derivative at zero. Assume that u^* is an optimal control function that generates a continuously differentiable state and solves (20). Then the directional derivative

$$J_d(u^*) = \lim_{h \to 0+} \frac{J(u^* + hd) - J(u^*)}{h}$$

in such a direction d is greater than or equal to zero, that is $J_d(u^*) \geq 0$ for all such directions. The representations of the directional derivatives given in (30)–(33) yields the following adjoint problem:

For all $e \in E$, the end condition is (22), that is $\mu^e(x,T) = 0$ for all $x \in [0, L_e]$, and the adjoint system equation is $\mu_t^e + (\mu^e\, D^e)_x = \mu^e Z^e + \mu^e B^e + (\nabla f^e)^T$. The interior node conditions are given by (26), (28) respectively. With the adjoint boundary conditions (29), (30) we have defined a backwards problem for a given control function u^*, that yields the necessary optimality condition

$$\partial_+ f^\omega - \mu_+^{e(\omega)} \lambda_+^{e(\omega)} = 0 \quad \text{for} \quad \text{all } \omega \in V_c^0 \tag{34}$$

$$\partial_- f^\omega + \mu_-^{e(\omega)} \lambda_-^{e(\omega)} = 0 \quad \text{for} \quad \text{all } \omega \in V_c^L, \tag{35}$$

on the time-interval $[0, T]$.

6. Stationary states on the network

It makes sense to choose a stationary state as an initial state for our simulation. Therefore in this section we study stationary states on our network. A stationary state (R_+^e, R_-^e) on the network satisfies the ordinary differential equation

$$D^e \begin{pmatrix} R_+^e \\ R_-^e \end{pmatrix}_x = \begin{pmatrix} S_2^e \\ S_2^e \end{pmatrix}$$

for all $e \in E$, and the interior node conditions (25), (27) respectively.

If such a state is constant, it satisfies for all $e \in E$ the equation

$$S_2^e(R_+^e, R_-^e) = 0. \tag{36}$$

Consider now the discharge $Q^e = A^e V^e$. The conservation of mass (2) implies that for every stationary state, Q^e is constant for all $e \in E$. In a channel with zero slope $\gamma^e = 0$, all states with $Q^e = 0$, that is with zero velocity are stationary.

It is interesting to study constant stationary states in the variables (Q^e, h^e). The second node condition (8) implies that the water height is constant on the whole network, say it has value $h_s > 0$. Equation (36) determines the corresponding values of the discharges Q^e, $e \in E$. Only if these values satisfy the first node condition (7), we have obtained a stationary solution. Whether this is the case depends on the geometry, the slopes and the roughness of the adjacent channels. In a network of horizontal channels with $\gamma^e = 0$ for all $e \in E$, we have $Q^e = 0$ for all $e \in E$ and this yields indeed a constant stationary state. Nonconstant stationary states have been studied for the supercritical case in [17].

Consider now a junction of three channels as in 4.2, Case 1, with the friction slope η^e as in (3). For a constant stationary state, (36) implies the equation $\eta^e = -\gamma^e$, hence we have $Q^e = -(\text{sign}\gamma^e) \sqrt{|\gamma^e|} A^e [A^e/(P^e(A^e))]^{2/3}/C^e$. At the node δ, we have $Q^e + Q^f = Q^g$ and $h^e = h^f = h^g$. If the data for the edges e and f are known and the geometry of the channel on edge e is also known, but the friction constant C^g and the slope γ^g are unknown, they must satisfy the equation

$$-(\text{sign}\gamma^g) \sqrt{|\gamma^g|}/C^g = (Q^e + Q^f)/[A^g(A^g/(P^g(A^g)))^{2/3}].$$

7. The discretized problem

Since we are unable to solve our infinite-dimensional optimization problem analytically, we use a discretized finite-dimensional problem that allows us to obtain numerical results.

7.1. The discretized system equations

For our numerical computations, we replace the original system equation (6) by a difference approximation obtained through upwind discretization and flux-vector splitting [38]. On each edge e of the graph, we have a grid with the gridpoints $x_j^e = jh^e$, $j \in \{0, 1, \ldots, N^e\}$ with the space step $h^e = L^e/N^e$. We introduce a time step $k = T/N^T > 0$ and replace the time-interval $[0, T]$ by the grid that consists of the points $t_0, t_1, \ldots, t_{N^T}$, with $t_n = nk$. So we have a time-space grid consisting for each $e \in E$ of the gridpoints (x_j^e, t_n), $j \in \{0, 1, \ldots, N^e\}$, $n \in \{0, 1, \ldots, N^T\}$. Gridfunctions will be denoted by

$$R_{+,n,j}^e = R_+^e(x_j^e, t_n), \ R_{-,n,j}^e = R_-^e(x_j^e, t_n), \ S_2^e(R_{+,n,j}^e, R_{-,n,j}^e) = S_{n,j}^e.$$

We use the following approximation based upon forward differences in time and backward differences in space for the positive eigenvalues and forward differences in space for the negative eigenvalues. This discretization mimics the evolution of the Riemann invariants along the characteristic curves.

$$R_{+,n+1,j}^e = R_{+,n,j}^e - (k/h)\lambda_{+,n,j-1}^e \left(R_{+,n,j}^e - R_{+,n,j-1}^e\right) - kS_{n,j-1}^e \quad (37)$$
$$\text{for } j \in \{1, \ldots, N^e\}, \ n \in \{0, \ldots, N^T - 1\};$$
$$R_{-,n+1,j}^e = R_{-,n,j}^e - (k/h)\lambda_{-,n,j+1}^e \left(R_{-,n,j+1}^e - R_{-,n,j}^e\right) - kS_{n,j+1}^e \quad (38)$$
$$\text{for } j \in \{0, \ldots, N^e - 1\}, \ n \in \{0, \ldots, N^T - 1\}.$$

It is well known that k/h should be chosen in such a way that the Courant–Friedrichs–Lewy condition (see for example [19]) $k/h \leq \min\{1/|\lambda_+|, \ 1/|\lambda_-|\}$ is valid. The values $R_{+,n+1,0}^e$ are determined from the boundary conditions for the controlled nodes (13) and (15) for the uncontrolled nodes and for the interior nodes with adjacent end 0 from (25), (27) respectively. Analogously the values of $R_{-,n+1,N^T}^e$ are obtained from (14), (16), (25), (27) respectively.

7.2. Boundary conditions for the discretized system

Let u denote the discretized control function; for a controlled node $\omega \in V_B$, the corresponding control vector has the components u_n^ω, $n = 0, \ldots, N^T$. The boundary conditions for the discretized problem are at the controlled nodes

$$R_{+,n,0}^{e(\omega)} = u_n^\omega \text{ for } n \in \{0, \ldots, N^T\}, \ \omega \in V_c^0, \quad (39)$$
$$R_{-,n,N^{e(\omega)}}^{e(\omega)} = u_n^\omega \text{ for } n \in \{0, \ldots, N^T\}, \omega \in V_c^L, \quad (40)$$

and at the uncontrolled nodes

$$R_{+,n,0}^{e(\omega)} = w_n^\omega \text{ for } n \in \{0, \ldots, N^T\}, \ \omega \in V^0, \omega \notin V_c^0, \quad (41)$$
$$R_{-,n,N^{e(\omega)}}^{e(\omega)} = w_n^\omega \text{ for } n \in \{0, \ldots, N^T\}, \omega \in V^L, \omega \notin V_c^L. \quad (42)$$

The interior node conditions (25), (27) remain unchanged, as well as the initial condition (1).

7.3. The discretized objective function

We consider objective functions K that are discretized integrals of the form

$$K(u) = \sum_{\omega \in V^0} \sum_{n=0}^{N^T-1} k f^\omega (R^{e(\omega)}_{+,n,0}, R^{e(\omega)}_{-,n,0})$$

$$+ \sum_{\omega \in V^L} \sum_{n=0}^{N^T-1} k f^\omega (R^{e(\omega)}_{+,n,N^{e(\omega)}}, R^{e(\omega)}_{-,n,N^{e(\omega)}})$$

$$+ \sum_{e \in E} \sum_{n=0}^{N^T-1} \sum_{j=1}^{N^e-1} k h f^e (R^e_{+,n,j}, R^e_{-,n,j}). \tag{43}$$

Here (R^+, R^-) is the solution of the discretized system governed by the system equations (37), (38) with initial values prescribed by the initial conditions (1), the interior node conditions (25), (27) and the boundary conditions (39), (40) at the controlled boundary nodes and (41), (42) at the uncontrolled boundary nodes that is generated by the discretized control function u.

For our numerical solution, we consider the discretized control problem

$$\min_u K(u). \tag{44}$$

For a numerical solution of this problem based upon gradient-based optimization methods it is useful to be able to evaluate the exact derivatives of K. On the other hand, an approximation for an optimal control should approximatively satisfy the necessary optimality conditions (34), (35). Therefore we consider a discretization of the adjoint problem that yields the exact derivatives of K.

7.4. The discretized adjoint equations

For the adjoint backwards equation (21) we use a discretization that allows the exact evaluation of the derivatives of the discretized objective function K. For optimal control problems with ordinary differential equations, this problem of consistency has been studied in the paper by Hager [20].

With the notation (see (23), (24))

$$S^e_{+,n,j} = \partial_{R^e_+} S^e_2(R^e_{+,n,j}, R^e_{-,n,j}), \qquad S^e_{-,n,j} = \partial_{R^e_-} S^e_2(R^e_{+,n,j}, R^e_{-,n,j}),$$
$$\Delta = k/h$$
$$\partial_+ \lambda^e_{+,n,j} = \partial_{R^e_+} \lambda^e_+(R^e_{+,n,j}, R^e_{-,n,j}), \quad \partial_- \lambda^e_{+,n,j} = \partial_{R^e_-} \lambda^e_+(R^e_{+,n,j}, R^e_{-,n,j}),$$
$$\partial_+ \lambda^e_{-,n,j} = \partial_{R^e_+} \lambda^e_-(R^e_{+,n,j}, R^e_{-,n,j}), \quad \partial_- \lambda^e_{-,n,j} = \partial_{R^e_-} \lambda^e_-(R^e_{+,n,j}, R^e_{-,n,j}),$$

we have: For $j \in \{0, \ldots, N^e - 2\}$, $n \in \{1, \ldots, N^T\}$:

$$
\begin{aligned}
\mu^e_{+,n-1,j} &= \\
\mu^e_{+,n,j} &+ \Delta\left(\lambda^e_{+,n,j+1}\mu^e_{+,n,j+1} - \lambda^e_{+,n,j}\mu^e_{+,n,j}\right) \\
&\quad -\Delta\partial_+\lambda^e_{+,n,j+1}(R^e_{+,n,j+2} - R^e_{+,n,j+1})\mu^e_{+,n,j+1} \\
&\quad -\Delta\partial_+\lambda^e_{-,n,j+1}(R^e_{-,n,j+1} - R^e_{-,n,j})\ \mu^e_{-,n,j+1} \\
&\quad -kS^e_{+,n,j+1}(\mu^e_{+,n,j+1} + \mu^e_{-,n,j+1}) - k\,\partial_+ f^e_{n,j+1};
\end{aligned}
\tag{45}
$$

For $j \in \{2, \ldots, N^e\}$, $n \in \{1, \ldots, N^T\}$:

$$
\begin{aligned}
\mu^e_{-,n-1,j} &= \\
\mu^e_{-,n,j} &+ \Delta\left(\lambda^e_{-,n,j}\mu^e_{-,n,j} - \lambda^e_{-,n,j-1}\mu^e_{-,n,j-1}\right) \\
&\quad -\Delta\partial_-\lambda^e_{+,n,j-1}(R^e_{+,n,j} - R^e_{+,n,j-1})\ \mu^e_{+,n,j-1} \\
&\quad -\Delta\partial_-\lambda^e_{-,n,j-1}(R^e_{-,n,j-1} - R^e_{-,n,j-2})\mu^e_{-,n,j-1} \\
&\quad -kS^e_{-,n,j-1}(\mu^e_{+,n,j-1} + \mu^e_{-,n,j-1}) - k\,\partial_- f^e_{n,j-1};
\end{aligned}
\tag{46}
$$

For $n \in \{1, \ldots, N^T\}$:

$$
\begin{aligned}
\mu^e_{+,n-1,N^e-1} &= \\
\mu^e_{+,n,N^e-1} &+ \Delta\left(\lambda^e_{+,n,N^e}\mu^e_{+,n,N^e} - \lambda^e_{+,n,N^e-1}\mu^e_{+,n,N^e-1}\right) \\
&\quad -\Delta\partial_+\lambda^e_{-,n,j+1}(R^e_{-,n,N^e} - R^e_{-,n,N^e-1})\mu^e_{-,n,N^e} \\
&\quad -kS^e_{+,n,N^e}\mu^e_{-,n,N^e}
\end{aligned}
$$

For $n \in \{1, \ldots, N^T\}$:

$$
\begin{aligned}
\mu^e_{-,n-1,1} &= \\
\mu^e_{-,n,1} &+ \Delta\left(\lambda^e_{-,n,1}\mu^e_{-,n,1} - \lambda^e_{-,n,0}\mu^e_{-,n,0}\right) \\
&\quad -\Delta\partial_-\lambda^e_{+,n,0}(R^e_{+,n,1} - R^e_{+,n,0})\ \mu^e_{+,n,0} \\
&\quad -kS^e_{-,n,0}\mu^e_{+,n,0}.
\end{aligned}
$$

The remaining values $\mu^e_{+,n-1,N^e}$ and $\mu^e_{-,n-1,0}$ come from the boundary conditions and the node conditions. Note that in (45) the role of the backward and forward differences in space is exchanged compared with (37), which is a consequence of the fact that the adjoint problem is a backwards equation; similarly the role of backward and forward differences is exchanged in (46) compared with (38). This corresponds to the fact that the adjoint solution is obtained by travelling backwards in time along the characteristic curves.

7.5. Boundary and node conditions for the adjoint discretized system

For boundary nodes $\omega \in V^0$, the adjoint boundary condition (29) for the continuous system is replaced by the following adjoint boundary condition for the discretized system

$$
\mu^{e(\omega)}_{-,n,0} = \partial_- f^\omega_n / \lambda^{e(\omega)}_{-,n,0}.
\tag{47}
$$

Instead of (30), the corresponding adjoint boundary condition for the discretized system is

$$\mu^{e(\omega)}_{+,n,N^{e(\omega)}} = -\partial_+ f^\omega_n / \lambda^{e(\omega)}_{+,n,N^{e(\omega)}}.$$
(48)

In the adjoint interior node conditions (26), λ^e_+, λ^f_+ and λ^g_- remain unchanged, λ^e_- (and analogously λ^f_-) is replaced by

$$\tilde\lambda^e_- = \lambda^e_{-,n,N^e} + \partial_- \lambda^e_{-,n,N^e}(R^e_{-,n,N^e} - R^e_{-,n,N^e-1}) + hS^e_{,n,N^e}$$
(49)

and λ^g_+ is replaced by

$$\tilde\lambda^g_+ = \lambda^g_{+,n,0} - \partial_+ \lambda^g_{+,n,0}(R^g_{+,n,1} - R^g_{+,n,0}) - hS^g_{+,n,0}.$$
(50)

In (28), λ^e_+, λ^f_- and λ^g_- remain unchanged, λ^e_- is replaced by (49) and λ^g_+ (and analogously λ^f_+) is replaced by (50).

7.6. Derivatives for the discretized objective function

With a solution of the discrete adjoint system with the end condition

$$\mu^e_{+,N^T,j} = 0 = \mu^e_{-,N^T,j}, \ j \in \{0,\dots,N^e\}, \ e \in E$$
(51)

(corresponding to (22)), the system equations (45), (46), the adjoint interior node conditions (with $\tilde\lambda$ as defined in (49), (50))

$$\left(\lambda^e_+ \mu^e_+, \lambda^f_+ \mu^f_+, -\lambda^g_- \mu^g_- \right)$$
(52)

$$= \left(-\tilde\lambda^e_- \mu^e_-, -\tilde\lambda^f_- \mu^f_-, \tilde\lambda^g_+ \mu^g_+ \right) M^\omega,$$
(53)

in Case 1 and

$$\left(\lambda^e_+ \mu^e_+, -\lambda^f_- \mu^f_-, -\lambda^g_- \mu^g_- \right)$$
(54)

$$= \left(-\tilde\lambda^e_- \mu^e_-, \tilde\lambda^f_+ \mu^f_+, \tilde\lambda^g_+ \mu^g_+ \right) M^\omega,$$
(55)

in Case 2 and the adjoint boundary conditions (47), (48), we have the following representation of the derivative of K:

$$D_d K(u) = \sum_{\omega \in V^0_c} \sum_{n=1}^{N^T-1} k\partial_+ f^\omega(R^{e(\omega)}_{+,n,0}, \ R^{e(\omega)}_{-,n,0}) d^\omega_n$$
(56)

$$+ \sum_{\omega \in V^0_c} \sum_{n=1}^{N^T-1} k\mu^{e(\omega)}_{+,n,0}$$
(57)

$$\left(-\lambda^{e(\omega)}_{+,n,0} + \partial_+ \lambda^{e(\omega)}_{+,n,0}(R^{e(\omega)}_{+,n,1} - R^{e(\omega)}_{+,n,0}) + hS^{e(\omega)}_{+,n,0} \right) d^\omega_n$$

$$+ \sum_{\omega \in V^L_c} \sum_{n=1}^{N^T-1} k\partial_- f^\omega(R^{e(\omega)}_{+,n,N^{e(\omega)}}, \ R^{e(\omega)}_{-,n,N^{e(\omega)}}) d^\omega_n$$
(58)

$$+ \sum_{\omega \in V_c^L} \sum_{n=1}^{N^T-1} k\mu_{-,n,N^{e(\omega)}}^{e(\omega)} \left(\lambda_{-,n,N^{e(\omega)}}^{e(\omega)} \right. \tag{59}$$

$$\left. + \partial_- \lambda_{-,n,N^e}^e (R_{-,n,N^{e(\omega)}}^{e(\omega)} - R_{-,n,N^{e(\omega)}-1}^{e(\omega)}) + hS_{-,n,N^{e(\omega)}}^{e(\omega)} \right) d_n^\omega.$$

Here the direction d has the components d_n^ω, $\omega \in V_c^0 \cup V_c^L$, $n = 1, \ldots, N^T$.

For the direction d with $d_n^\omega = 1$ if $\omega \in V_c^0$ and $n = m$, $d_n^\omega = 0$ else, this yields the partial derivative

$$\partial_{u_m^\omega} K(u) = k\partial_+ f^\omega(R_{+,m,0}^{e(\omega)}, R_{-,m,0}^{e(\omega)})$$

$$+ k\mu_{+,m,0}^{e(\omega)}(-\lambda_{+,m,0}^{e(\omega)} + \partial_+ \lambda_{+,m,0}^{e(\omega)}(R_{+,m,1}^{e(\omega)} - R_{+,m,0}^{e(\omega)}) + hS_{+,m,0}^{e(\omega)}).$$

If $d_n^\omega = 1$ if $\omega \in V_c^L$ and $n = m$, $d_n^\omega = 0$ else, this yields the partial derivative

$$\partial_{u_m^\omega} K(u) = k\partial_- f^\omega(R_{+,m,N^{e(\omega)}}^{e(\omega)}, R_{-,m,N^{e(\omega)}}^{e(\omega)}) + k\mu_{-,m,N^e(\omega)}^{e(\omega)} \left(\lambda_{-,m,N^{e(\omega)}}^{e(\omega)} \right.$$

$$\left. + \partial_- \lambda_{-,m,N^{e(\omega)}}^{e(\omega)}(R_{-,m,N^{e(\omega)}}^{e(\omega)} - R_{-,m,N^{e(\omega)}-1}^{e(\omega)}) + hS_{-,m,N^{e(\omega)}}^{e(\omega)} \right).$$

Note that (56) can also be written in terms of $\tilde{\lambda}$ defined in (49), (50).

If we compare the representation (56) of the derivative of the discretized objective function K with the representation (30) of the derivative of J we see that in (56) additional terms appear, that do not appear automatically in a discretization of (30) but are necessary to obtain the exact derivative of K. Since these additional terms are of order order h, their influence becomes smaller if the discretization is refined. The additional terms can also be considered as a perturbation of the eigenvalues λ^e to $\tilde{\lambda}^e$, and exactly the same perturbation appears in the discretization of the node conditions (53), (55).

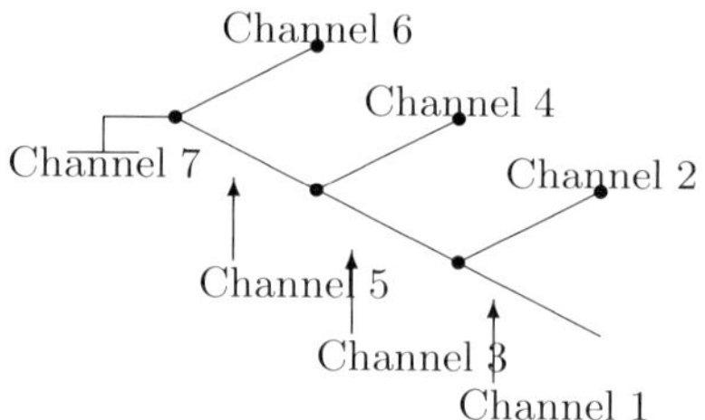

FIGURE 4. Example network

8. Example

As an example network, consider the graph depicted in Figure 4 that consists of seven channels. All the Channels e have the end L_e in the direction of the boundary node of Channel 7. The water enters the system at Channel 1. Channel 2, Channel 4 and Channel 6 are emergency side channels. The standard outflow of the system is at Channel 7. The objective is to keep the water height in Channel 3 and Channel 5 below a given upper bound $h_{\max} = 0.11$ by using outflow from the emergency side channels into the floodplain. Note that due to the orientation of the channel, emergency outflow means negative velocity in Channel 2, Channel 4 and Channel 6. The objective function is

$$J(u) = p \int_0^T \left(Q_2(0,t)\right)^2 + \left(Q_4(0,t)\right)^2 + \left(Q_6(0,t)\right)^2 \, dt$$

$$+ \int_0^T \int_0^{L_3} [(h_3(x,t) - h_{\max})_+]^2 dx \, dt + \int_0^T \int_0^{L_5} [(h_5(x,t) - h_{\max})_+]^2 dx \, dt$$

with $(-q)_+ = \max(-q, 0)$. We have considered the time $T = 10000$. The weight p equals 10^{-4}. The emergency side channels have slope zero and length $L_2 = L_4 = L_6 = 200$. The other channels have slope $\gamma = -10^{-4}$ and length $L_1 = 800$, $L_3 = 400$, $L_5 = 300$, $L_7 = 200$. All the channels are rectangular and have the same width $b = 1$.

We start with an initial state, where the water in the side channels is at rest, that is $V_2 = V_4 = V_6 = 0$. The water height throughout the system is $h_0 = 0.1$. We have $V_1 = V_3 = V_5 = V_7 = \sqrt{-\gamma} \, [bh_0/(b + 2h_0)]^{2/3}/C = \sqrt{-\gamma/\eta(A,1)}$, where $C = 0.02$ is the constant that appears in the Manning–Strickler relation (3).

Let $V^0 = \{\omega_1, \omega_2, , \omega_4, \omega_6\}$ where ω_1 is the vertex that is adjacent to Channel 1, etc. The boundary condition at ω_1 where the flood wave enters the system has the form $R_+^{e(\omega_1)}(0,t) = R_+^{e(\omega_1)}(0,0) + \sin^4(9\pi t/T)$ if $9t/T \le 1$ and $R_+^{e(\omega_1)}(0,t) = R_+^{e(\omega_1)}(0,0)$ if $9t/T > 1$.

At the node ω_7 at the end L_7 of channel 7 where outflow out of the system occurs we have the boundary conditions $R_-^{e(\omega_7)}(L_7,t) = R_-^{e(\omega_7)}(L_7,0)$.

At the controlled boundary nodes, that is for $\omega \in \{\omega_2, \omega_4, \omega_6\}$, the boundary conditions are $R_+^{e(\omega)}(0,t) = R_+^{e(\omega)}(0,0) + u^\omega(t)$. The discretized objective function is given by (43), with $f^\omega = 0$ if $\omega \in V^L = \{\omega_7\}$, and $f^\omega = p\,(g^\omega)^2$ if $\omega \in \{\omega_2, \omega_4, \omega_6\}$, where $g^\omega(R^+, R^-) = -b^{e(\omega)}(R^+ + R^-)(R^+ - R^-)^2/(32g)$ if $\omega \in \{\omega_2, \omega_4, \omega_6\}$. We have

$$f^e(R^+, R^-) = \left[\left((R^+ - R^-)^2/(16g) - h_{\max}\right)_+\right]^2$$

if the edge e corresponds to Channel 3 or Channel 5; otherwise we have $f^e = 0$. For the corresponding derivatives we have

$$\partial_+ g^\omega(R^+, R^-) \;=\; -b^{e(\omega)}(3R^+ + R^-)(R^+ - R^-)/(32g)$$

$$\partial_- g^\omega(R^+, R^-) \;=\; -b^{e(\omega)}(R^+ + 3R^-)(-R^+ + R^-)/(32g)$$

if $\omega \in \{\omega_2, \omega_4, \omega_6\}$. Moreover, if e is the edge corresponding to Channel 3 or Channel 5,

$$\partial_+ f^e(R^+, R^-) \;=\; \left((R^+ - R^-)^2/(16g) - h_{\max}\right)_+ (R^+ - R^-)/(4g)$$

$$\partial_- f^e(R^+, R^-) \;=\; \left((R^+ - R^-)^2/(16g) - h_{\max}\right)_+ (R^- - R^+)/(4g).$$

Let $u_n^2 = u_n^{\omega_2}$ denote a component of the control vector at the boundary node 0 of Channel 2. Then for the partial derivative with respect to this component, we have

$$\partial_{u_n^2} J(u) = k\partial_+ f^{\omega_2}(R_{+,n,0}^{e(\omega_2)}, \ R_{-,n,0}^{e(\omega_2)}) \tag{60}$$

$$+k\mu_{+,n,0}^{e(\omega_2)} \left[-\lambda_{+,n,0}^{e(\omega_2)} + 0.75(R_{+,n,1}^{e(\omega_2)} - R_{+,n,0}^{e(\omega_2)}) + hb_{+,n,0}^{e(\omega_2)} \right]$$

and the analogous results for ω_4 and ω_6 also hold.

We present numerical results for $N^T = 600$, $N^{e_1} = 33$, $N^{e_2} = N^{e_4} = N^{e_6} = N^{e_7} = 9$, $N^{e_3} = 17$, $N^{e_5} = 13$.

The iteration started with the zero control function $u^\omega = 0$ for all $\omega \in \{\omega_2, \omega_4, \omega_6\}$. The corresponding objective value is 1.28 with p times the sum of the pure time-integrals equal to $1.23 * 10^{-5}$.

We used the steepest descent method with the Armijo rule for step-length control (see for example [24]). The step-size in the line-search has to be chosen so small that the candidates for the control function all generate subcritical states. If the step-size is chosen too large, the corresponding control can generate a supercritical state that cannot be computed by our discretization of the state. In the algorithm, this was done by choosing the initial step-length in the line-search sufficiently small, the initial step was chosen ε times the gradient with $\varepsilon \in (0,1)$ sufficiently small.

With 80 gradient steps, we computed a solution with gradient norm $1.7 * 10^{-5}$ and value of the objective function $1.5 * 10^{-5}$. The value of the second part of the objective function (the space-time integrals where the water height appears) is much smaller, in fact $3.3 * 10^{-8}$, hence the value of the objective function is determined by the amount of water that flows out of the system, which is given by the pure time-integrals. During the iteration, in each step there was a decrease in the gradient norm. Of course the gradients obtained by the backwards problem can also be used for any other gradient-based optimization algorithm.

Figure 5 shows the computed control functions. The units at the x-axis denote the number of the points in the time grid. The control functions generate a depression wave before the flood wave arrives. We present several snapshots of the state. Figure 6 show the water height in the network. Channel 1 starts at the point $x = 0$ and Channel 7 ends at the point $x = 1700$. The line with '$\wedge$'s shows the upper bound for the water height. The stationary initial state is also plotted. The units of the x- and y-axis are m, and the unit of the z-axis is $10^{-4}m$. By far the most water flows into the first side channel at $x = 800$.

Figure 7 show the water height in Channel 3 generated by the computed control. The unit of the x- and z-axis is m, and the unit of the y-axis is seconds.

With zero control functions $u^\omega(t) = 0$ for $\omega \in \{\omega_2, \omega_4, \omega_6\}$ the objective value is 1.28, the gradient norm 3.206 and snapshots of the corresponding states are presented in Figure 9. The generated water height is shown in Figure 8.

For a **second example**, we changed the value of $h_{\max}$ to 0.102. Starting from the objective value 46.9 and the gradient norm 38.1, after 400 iterations we obtained a control with objective value 0.827, gradient norm 0.001 and value of p

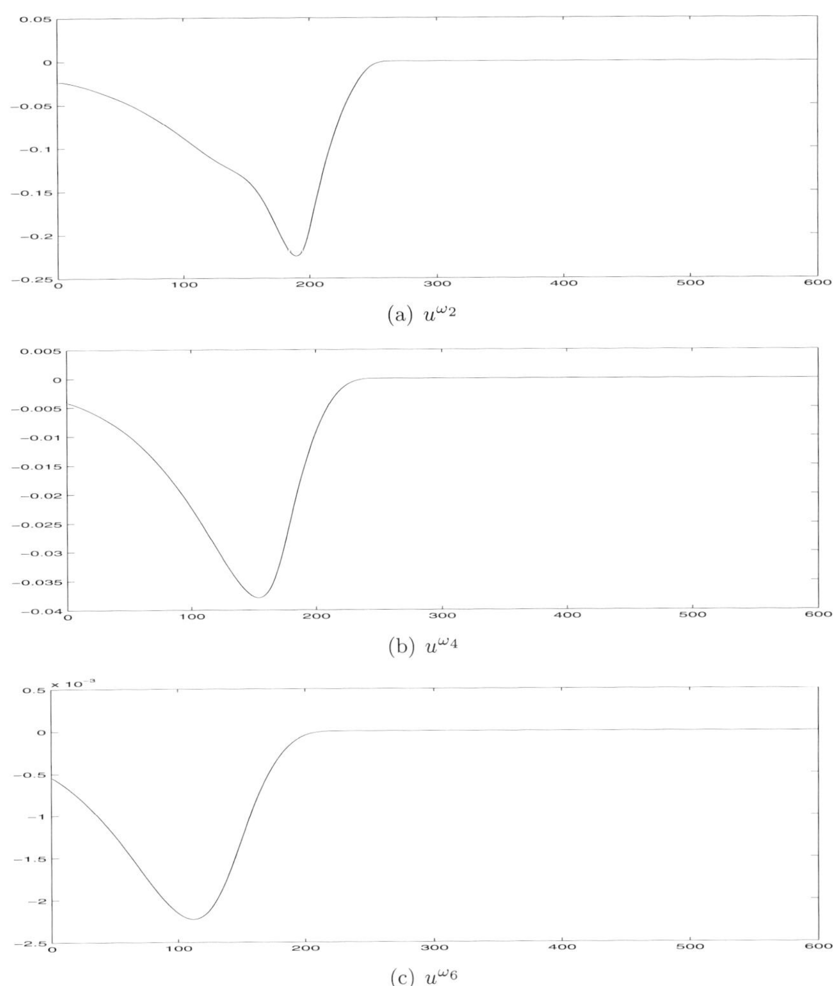

(a) u^{ω_2}

(b) u^{ω_4}

(c) u^{ω_6}

FIGURE 5. The computed approximations of the optimal controls

times the amount of water that flows out of the system equal to 10^{-4}. In this case, in the first twenty steps, the line-search was started with 0.1 times the gradient. After twenty steps, the first step in the line-search was 5 times the gradient.

Acknowledgements

I thank G. Leugering who proposed this topic of research for many encouraging and fruitful discussions.

M. Gugat

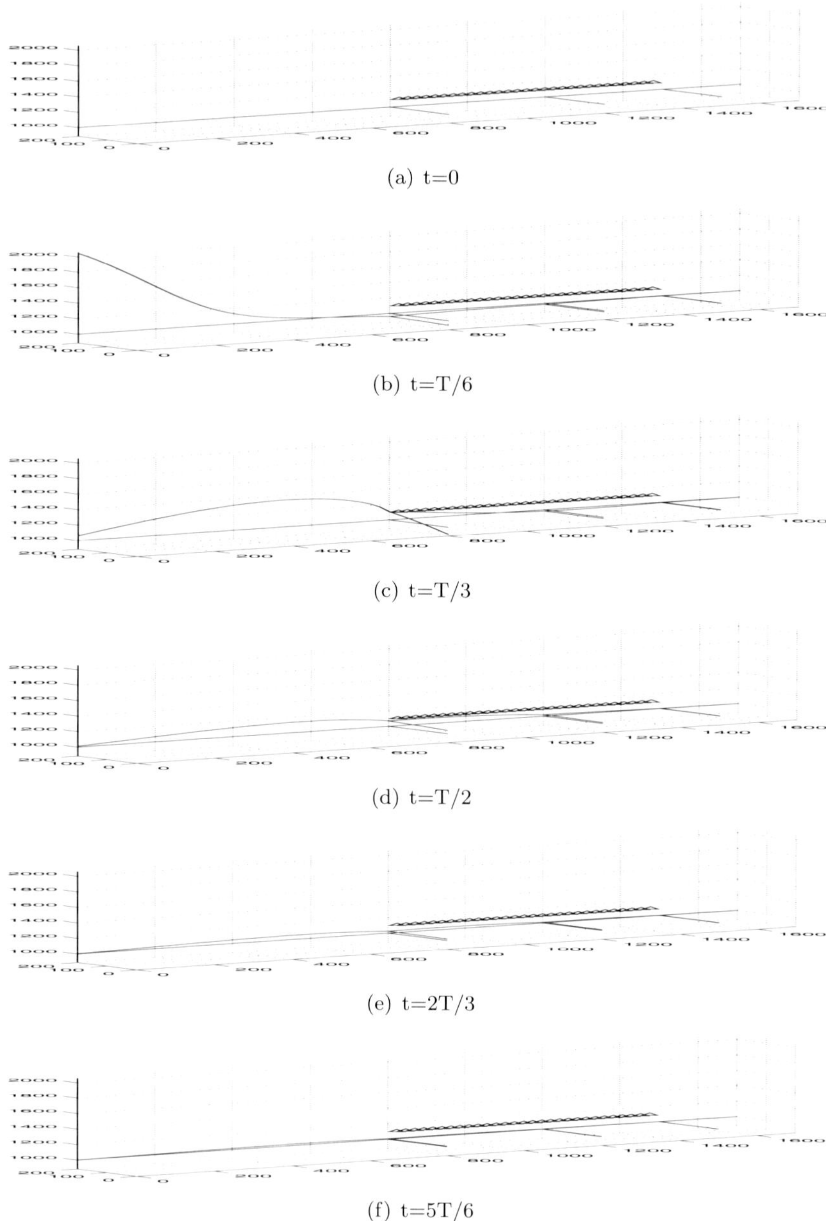

(a) t=0

(b) t=T/6

(c) t=T/3

(d) t=T/2

(e) t=2T/3

(f) t=5T/6

FIGURE 6. Water height with the computed control

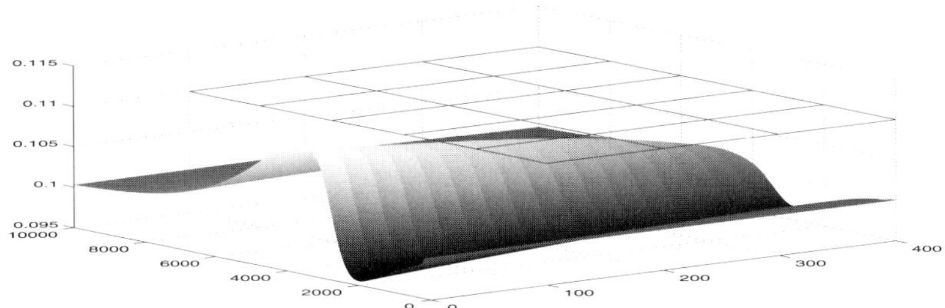

FIGURE 7. Water height in Channel 3 for the computed control

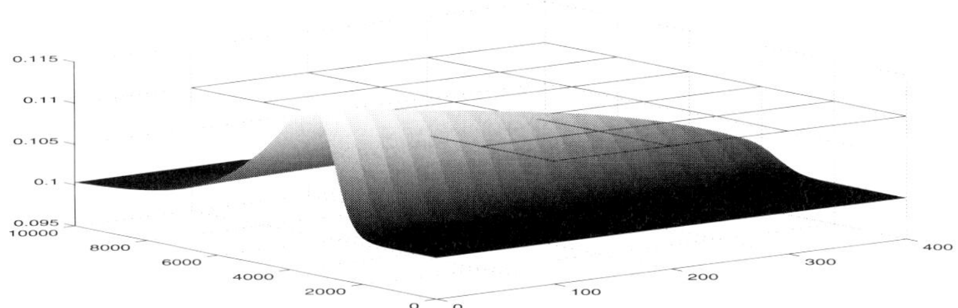

FIGURE 8. Water height in Channel 3 for zero control

References

[1] C. Bardos, A.Y. Leroux, and J.C. Nedelec. First order quasilinear equations with boundary conditions. *Comm. in Partial Differential Equations*, 4:1017–1034, 1979.

[2] C. Bardos and O. Pironneau. Sensitivities for Euler flows. *C. R. Acad. Sci., Paris, Ser. I*, 2002.

[3] C. Bardos and O. Pironneau. Derivatives and control in the presence of shocks. *Computational Fluid Dynamics Journal*, 12, 2003.

[4] J. Borggaard, J. Burns, E. Cliff, and M. Gunzburger. Sensitivity calculations for a 2d, inviscid, supersonic forebody problem. In *Identification and Control in Systems Governed by Partial Differential Equations*, pages 14–25. SIAM, Philadelphia, 1993.

[5] A. Bressan. *Hyperbolic Systems of conservation laws. The one-dimensional Cauchy Problem*. Oxford University Press, 2000.

[6] G.M. Coclite and B. Piccoli. Traffic flow on a road network. *S.I.S.S.A.*, 13:1–26, 2002.

[7] S. Scott Collis, Kaveh Ghayour, and Matthias Heinkenschloss. Optimal transpiration boundary control for aeroaccoustics. *Preprint*, pages 1–37, 2002.

[8] J.M. Coron. Local controllability of a 1-d tank containing a fluid modelled by the shallow water equations. *ESAIM:COCV*, 8:513–554, 2002.

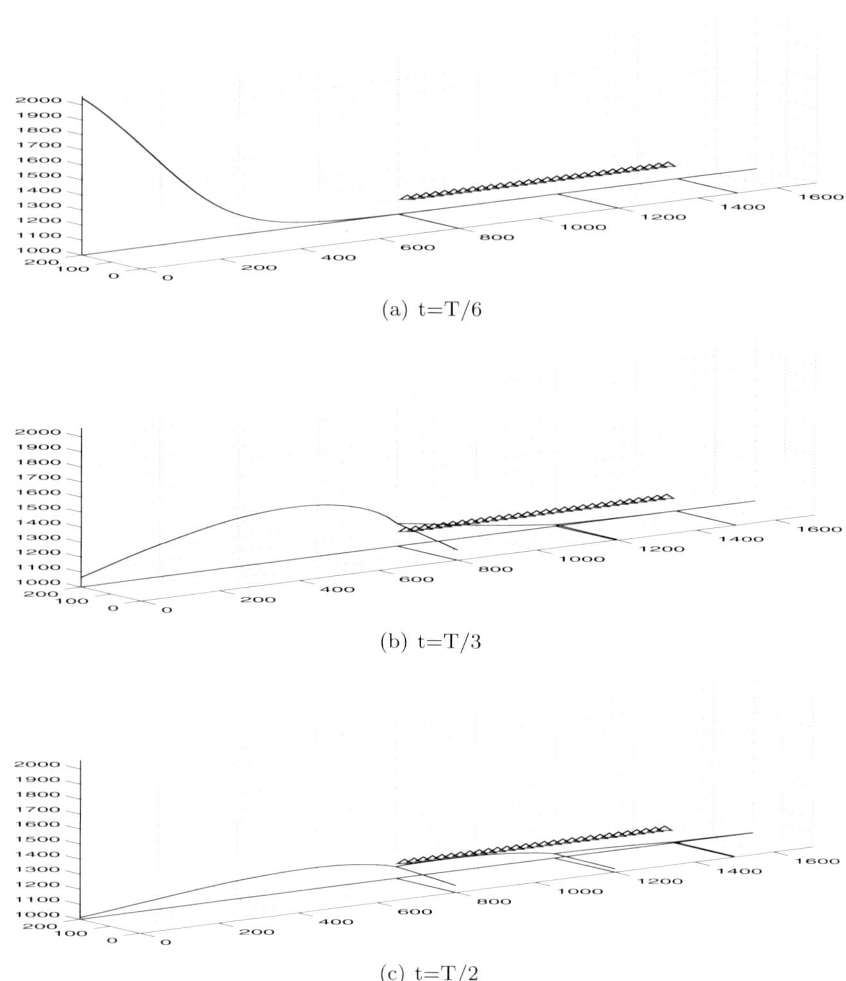

(a) t=T/6

(b) t=T/3

(c) t=T/2

FIGURE 9. Water height for zero control

[9] J.M. Coron, B. d'Andrea Novel, and G. Bastin. A Lyapunov approach to control irrigation canals modeled by Saint-Venant equations. In *ECC Karlsruhe*, 1999.

[10] C.M. Dafermos. *Hyperbolic Conservation Laws in Continuum Physics*. Springer, Berlin, 2000.

[11] B. de Saint-Venant. Théorie du mouvement non-permanent des eaux avec application aux crués des rivières et à l'introduction des marées dans leur lit. *Comptes Rendus Academie des Sciences*, 73:148–154,237–240, 1871.

[12] M. Gugat. Boundary controllability between sub- and supercritical flow. *SIAM Journal on Control and Optimization*, 42:1056–1070, 2003.

[13] M. Gugat. Nodal control of conservation laws on networks. In *Control and Boundary Analysis.* Cagnol, John (ed.) et al., Chapman & Hall/CRC, Boca Raton, FL. Lecture Notes in Pure and Applied Mathematics 240, 201–215, 2005.

[14] M. Gugat. Optimal nodal control of networked hyperbolic systems: Evaluation of derivatives. *Advanced Modeling and Optimization,* 7:9–37, 2005.

[15] M. Gugat, M. Herty, A. Klar, and G. Leugering. Optimal control for traffic flow networks. *J. Optimization Theory Appl.* 126:589–616, 2005.

[16] M. Gugat and G. Leugering. Global boundary controllability of the de Saint-Venant equations between steady states. *Annales de l'Institut Henri Poincaré, Nonlinear Analysis,* 20:1–11, 2003.

[17] M. Gugat, G. Leugering, and E.J.P.G. Schmidt. Global controllability between steady supercritical flows in channel networks. *Mathematical Methods in the Applied Sciences,* 27:781–802, 2004.

[18] Max D. Gunzburger. *Perspectives in Flow Control and Optimization.* SIAM, Philadelphia, 2002.

[19] B. Gustafsson, H.-O. Kreiss, and J. Oliger. *Time Dependent Problems and Difference Methods.* John Wiley, New York, 1995.

[20] W.W. Hager. Rates of convergence for discrete approximations to unconstrained optimal control problems. *SIAM J. Numer. Analysis,* 13:449–472, 1976.

[21] W.W. Hager. Runge–Kutta methods in optimal control and the transformed adjoint system. *Numer. Math.,* 87:247–282, 2000.

[22] H. Holden and N.H. Risebro. A mathematical model of traffic flow on a network of unidirectional roads. *SIAM J. Math. Analysis,* 26:999–1017, 1995.

[23] H. Holden and N.H. Risebro. *Front tracking for hyperbolic conservation laws.* Springer, Berlin, 2002.

[24] C.T. Kelley. *Iterative Methods for Optimization.* SIAM, Philadelphia, 1999.

[25] P.G. LeFloch. *Hyperbolic systems of conservation laws. The theory of classical and nonclassical shock waves.* Birkhäuser, Basel, 2002.

[26] G. Leugering and E.J.P. Georg Schmidt. On the modelling and stabilisation of flows in networks of open canals. *SIAM J. on Control and Optimization,* 41:164–180, 2002.

[27] P.O. Lindberg and A. Wolf. Optimization of the short term operation of a cascade of hydro power stations. In William H. Hager and P.M. Pardalos, editors, *Optimal control: theory, algorithms, and applications.,* 326–345. Kluwer Academic Publishers, Dordrecht, 1998.

[28] J.A. Roberson, J.J. Cassidy, and M.H. Chaudhry. *Hydraulic Engineering.* John Wiley, New York, 1995.

[29] B. F. Sanders and N.D. Katapodes. Active flood hazard mitigation. i: Bidirectional wave control. *Journal of Hydraulic Engineering,* 125:1057–1070, 1999.

[30] B.F. Sanders and N.D. Katapodes. Control of canal flow by adjoint sensitivity method. *Journal of Irrigation and Drainage Engineering,* 125:287–297, 1999.

[31] B.F. Sanders and N.D. Katapodes. Adjoint sensitivity analysis for shallow water wave control. *Journal of Irrigation and Drainage Engineering,* 126:909–919, 2000.

[32] S. Shabayek, P. Steffler, and F. Hicks. Dynamic model for subcritical combining flows in channel junctions. *Journal of Hydraulic Engineering,* 128(9):821–828, 2002.

[33] L.G. Stanley and D.L. Steward. *Design Sensitivity Analysis: Computational Issues of Sensitivity Equation Methods.* SIAM, Philadelphia, 2002.

[34] J.J. Stoker. *Water Waves.* Interscience, New York, 1957.

[35] Li Ta-tsien. *Global classical solutions for quasilinear hyperbolic systems.* Masson, Paris, 1994.

[36] S. Ulbrich. A sensitivity and adjoint calculus for discontinuous solutions of hyperbolic conservation laws with source terms. *SIAM J. Control Optim.*, 41:740–797, 2002.

[37] S. Ulbrich. Adjoint-based derivative computations for the optimal control of discontinuous solutions of hyperbolic conservation laws. *Systems and Control Letters*, 48:309–324, 2003.

[38] R. Le Veque. *Finite Volume Methods for Hyperbolic Problems.* Cambridge University Press, Cambridge, 2002.

Martin Gugat
Friedrich-Alexander-Universtität Erlangen-Nürnberg
Institut für Angewandte Mathematik
Lehrstuhl II für Angewandte Mathematik
Martensstr. 3
D-91058 Erlangen, Germany
e-mail: `gugat@am.uni-erlangen.de`

International Series of Numerical Mathematics, Vol. 155, 95–117

On Two Numerical Approaches for the Boundary Control Stabilization of Semi-linear Parabolic Systems: A Comparison

Vincent Heuveline and Hoang Nam-Dung

Abstract. The present article is concerned with boundary control stabilization of semi-linear parabolic systems which are unstable if uncontrolled. A particular emphasis is put on Dirichlet control in that context. We investigate two different numerical approaches to solve these problems. The first approach relies on the extension method proposed by A.V. Fursikov where the considered partial differential equations are first solved on an extended domain with suitable initial value leading a stable solution. The needed control is then defined as an appropriate trace of this solution. The second approach relies on the formulation of the stabilization problem as an optimization problem with constraints based on partial differential equations. We address the numerical issues related to both class of approaches toward a comparison of their specific stabilization properties. The considered methodology is applied to the solution of test parabolic problems assuming linear and nonlinear models.

Mathematics Subject Classification (2000). 35K45, 65N22, 93B40, 93B52, 93C20, 93D15, 93D21 .

Keywords. Dirichlet control, feedback control, stabilization, semi-linear parabolic systems, extension method, finite element method.

1. Introduction

In this article, we investigate two different numerical approaches for the boundary control stabilization of semi-linear parabolic systems which are unstable if uncontrolled. A particular emphasis is laid on Dirichlet control in that context. For the first class of methods, we consider the extension method proposed by A. Fursikov [12, 13, 14]. This approach relies on the solution of the considered partial differential equations in an extended domain. The initial value on the extended domain is chosen such that the extended solution is stable. The boundary control needed for the stabilization of the original problem is then defined to be an adequate trace of the extended solution. For the second class of methods, we consider an optimal control based solution process where the stabilization problem is formulated as an optimization problem with PDEs constraints. The considered approach is inspired from the method proposed by R. Glowinski et al. [19, 18]. In that context, we consider finite-difference methods for the time discretization, finite element methods for the space discretization, and the L-BFGS method for the solution of the resulting discrete control problem. The main goal of this article is to address some of the numerical issues associated to both class of approaches toward a comparison of their specific stabilization properties.

This article is organized as follows. In Section 2 we formulate the stabilization problem for semi-linear parabolic systems. In Section 3, we derive first the extension method for the stabilization problem and discuss then numerical aspects associated to the realization of the associated extension operator. The second approach relying on the formulation of the stabilization problem as an optimization problem with PDEs constraints is derived in Section 4. In Section 5, the proposed numerical schemes are validated and compared by means of numerical experiments assuming linear and nonlinear models.

2. Problem formulation

Let Ω be a bounded domain in $\mathbb{R}^d$ ($1 \leq d \leq 3$). We consider the following partial differential equation

$$
\begin{aligned}
\partial_t u - F(u) &= 0 && \text{in } \Omega \times (0, T), & (2.1)\\
u(x, 0) &= u_0(x) && \text{for } x \in \Omega, & (2.2)\\
u(x, t) &= 0 && \text{on } (\partial\Omega \backslash \Gamma) \times (0, T), & (2.3)\\
u(x, t) &= q(x, t) && \text{on } \Gamma \times (0, T), & (2.4)
\end{aligned}
$$

where $\Gamma \subset \partial\Omega$, u_0 is a given initial value, and F is a (semi-)elliptic operator. The goal of the stabilization problem of the above equation is to construct a control $q \in L^\infty(0, T; H_0^{1/2}(\Gamma))$ where

$$
L^\infty(0, T; H_0^{1/2}(\Gamma)) = \{q \mid q(\cdot, t) \in H_0^{1/2}(\Gamma) \ \forall t \in (0, T), \ \|q(\cdot, t)\|_{H^{1/2}(\Gamma)} \in L^\infty(0, T)\}
$$

such that the state solution u converges to a steady state solution u_s of the equation

$$
F(u) = 0, \quad u|_{\partial\Omega} = 0, \tag{2.5}
$$

as t tends to T. Note that the homogeneous Dirichlet boundary conditions (2.3) can be generalized to non homogeneous Dirichlet boundary conditions leading to non homogeneous Dirichlet boundary conditions in (2.5). This generalization would lead to minor additional technicalities in the derivation of the proposed stabilization methods and for simplicity of notation will not be considered in the sequel of this paper.

3. Extension method for the stabilization problem

3.1. Theoretical background

The basic idea of this method developed by Fursikov (see [12]) is to reformulate the stabilization problem associated to the partial differential equation (2.1)–(2.4) in an extended domain $G \supset \Omega$ where a stable solution can be comparatively more easily computed. The control q is then defined as an adequately chosen trace of the extended solution. Here the choice of G as well as of the extended initial value v_0 (i.e., $v_0|_\Omega = u_0$) are crucial.

First, we assume a linear parabolic problem (2.1)–(2.4) with

$$F(u) = A(x)u := -\sum_{i,j=1}^{d} a_{ij}(x)\partial_{ij}u + \sum_{i=1}^{d} b_i(x)\partial_i u + c(x)u,$$

$T = +\infty$ and $u_0 \in L^2(\Omega)$, where a_{ij}, b_i, and c are real-valued functions. We assume further $u_s = 0$ which can easily be obtained by substitution of u by $v = u - u_s$ in (2.1)–(2.4). Let $\sigma_0 > 0$ be a given number. Our goal is to construct a control $q \in L^\infty(0, T; H_0^{1/2}(\Gamma))$ such that the solution u of (2.1)–(2.4) satisfies

$$\|u(\cdot, t)\|_{L^2(\Omega)} \leq ce^{-\sigma_0 t} \quad \text{as } t \to \infty, \tag{3.1}$$

for some $c > 0$.

Let $\omega \subset \mathbb{R}^d$ be a bounded domain disjoint from Ω such that the intersection of the boundaries $\partial\omega$ and $\partial\Omega$ equals to $\overline{\Gamma}$, i.e.,

$$\omega \cap \Omega = \emptyset, \quad \text{Int}(\partial\omega \cap \partial\Omega) = \Gamma \neq \emptyset,$$

where Int $\mathcal{A}$ denotes the interior of set $\mathcal{A}$. Let

$$G = \text{Int}(\overline{\omega} \cup \overline{\Omega})$$

and assume that the domain ω guarantees that the boundary ∂G of G is a $(d-1)$-dimensional manifold which is of class C^2.

We extend the coefficients $a_{ij}(x)$, $b_i(x)$, $c(x)$ of the operator A from $\overline{\Omega}$ to real-valued functions $\widehat{a}_{ij}(x) = \widehat{a}_{ji}(x)$, $\widehat{b}_i(x)$, $\widehat{c}(x)$ of class $C^2(\overline{G})$ on $\overline{G}$ satisfying the ellipticity condition, i.e.,

$$\sum_{i,j=1}^{d} \widehat{a}_{ij}(x)\,\xi_i\,\xi_j \geq \beta\,|\xi|^2, \quad \forall \xi = (\xi_1, \ldots, \xi_d) \in \mathbb{R}^d,$$

where $\beta > 0$ does not depend on $x \in \overline{G}$. We consider the system

$$\partial_t v + \widehat{A}(x)\,v \;=\; 0 \quad \text{in } G \times \mathbb{R}_+, \tag{3.2}$$

$$v \;=\; 0 \quad \text{on } \partial G \times \mathbb{R}_+, \tag{3.3}$$

$$v(\cdot, t)\big|_{t=0} \;=\; v_0 \quad \text{in } G, \tag{3.4}$$

where

$$\widehat{A}(x)\,v := -\sum_{i,j=1}^{d} \widehat{a}_{ij}(x)\,\partial_{ij}v + \sum_{i=1}^{d} \widehat{b}_i(x)\,\partial_i v + \widehat{c}(x)\,v, \tag{3.5}$$

and v_0 is constructed in a special way from u_0.

We assume that σ_0 satisfies the condition

$$\sigma(-\widehat{A}) \cap \{\lambda \in \mathbb{C} \mid \operatorname{Re}\lambda = -\sigma_0\} = \emptyset,$$

where $\sigma(-\widehat{A})$ denotes the spectrum of the operator $-\widehat{A}$.

Let $\widehat{A}^*$ be the adjoint operator of $\widehat{A}$. There holds

$$\sigma(-\widehat{A}) = \sigma(-\widehat{A}^*).$$

In our case $\sigma(-\widehat{A})$ is a discrete set of points. For $\lambda \in \sigma(\widehat{A})$ we denote the canonical system of $-\widehat{A}^*$ associated to $-\lambda$ by

$$\varepsilon_0^{(k)}(-\lambda),\ \varepsilon_1^{(k)}(-\lambda),\ldots,\ \varepsilon_{m_k(-\lambda)}^{(k)}(-\lambda), \quad k = 1, 2, \ldots, N(-\lambda).$$

We number all the non-zero real and imaginary parts of the functions

$$\left\{\varepsilon_i^{(k)}(-\lambda) \ \middle|\ \lambda \in \sigma(\widehat{A});\ \operatorname{Re}(-\lambda) > -\sigma_0;\ k = 1,\ldots, N(-\lambda);\ i = 0,\ldots, m_k(-\lambda)\right\}$$

by $\varepsilon_1(x),\ \varepsilon_2(x),\ldots,\ \varepsilon_K(x)$, and denote

$$X_K(G) = \left\{v \in L^2(G) \ \middle|\ \int_G v(x)\,\varepsilon_j(x)\,dx = 0,\ j = 1, 2, \ldots, K\right\}.$$

Theorem 3.1 ([12], p. 611). *The system of the restrictions to ω of the functions $\varepsilon_1(x),\ \varepsilon_2(x),\ldots,\ \varepsilon_K(x)$ are linearly independent.*

Theorem 3.2 ([12], p. 612). *There exists a continuous linear extension operator*

$$E_K:\ L^2(\Omega) \to X_K(G).$$

In the proof of Theorem 3.2 presented in [12], E_K is defined by

$$(E_K v)(x) = \begin{cases} v(x), & x \in \Omega, \\ \displaystyle\sum_{j=1}^{K} c_j\,\varepsilon_j(x), & x \in \omega \equiv G \setminus \Omega, \end{cases} \tag{3.6}$$

where the coefficients c_j are determined from the equations

$$\sum_{j=1}^{K} c_j \int_\omega \varepsilon_j(x)\,\varepsilon_k(x)\,dx = -\int_\Omega v(x)\,\varepsilon_k(x)\,dx, \quad k = 1, 2, \ldots, K. \tag{3.7}$$

Obviously, the system of linear equations (3.7) ensures that the function (3.6) is in the space $X_K(G)$, and it is clearly an extension of v, i.e.,

$$(E_K v)\big|_\Omega = v.$$

The existence result of Theorem 3.2 relies on the existence of a solution of (3.7). Let

$$a_{ij} = \int_\omega \varepsilon_i(x)\,\varepsilon_j(x)\,dx \quad \text{and} \quad b_i = -\int_\Omega v(x)\,\varepsilon_i(x)\,dx,$$

then $c = (c_1, c_2, \ldots, c_K)^T$ is a solution of the system of linear equations $Ax = b$, where $A = (a_{ij})_{i,j=1}^K$ and $b = (b_1, b_2, \ldots, b_K)^T$. One can easily prove that the matrix A is positive definite. Therefore, the system of linear equations $Ax = b$ has a unique solution $A^{-1}b$. A numerical method can be directly derived from this approach toward the computation of the extended initial value (see Section 3.2).

Theorem 3.3 ([12], p. 605). *For any $v_0 \in X_K(G)$ there holds for the solution v of (3.2)–(3.4)*

$$\|v(\cdot, t)\|_{L^2(G)} \le M\, e^{-\sigma_0 t}\, \|v_0\|_{L^2(G)} \qquad for \quad t \ge 0. \tag{3.8}$$

Theorem 3.4 ([12], p. 614). *For any initial value $u_0 \in L^2(\Omega)$ and any rate $\sigma_0 > 0$, there is a control q defined on $\Gamma \times \mathbb{R}_+$ such that the solution $u(x, t)$ of problem (2.1)–(2.4) satisfies the condition*

$$\|u(\cdot, t)\|_{L^2(\Omega)} \le c\, e^{-\sigma_0 t} \qquad as \quad t \to \infty. \tag{3.9}$$

In the proof of Theorem 3.4 presented in [12], the following mapping is constructed:

$$E_K : L^2(\Omega) \to X_K(G),$$

which is an extension operator according to Theorem 3.2, where G is an extension of Ω. Let $v(x, t)$ be the solution of problem (3.2)–(3.3) satisfying

$$v(x, t)\big|_{t=0} = v_0(x) := (E_K u_0)(x) \tag{3.10}$$

and define

$$q(\cdot, t) = v(\cdot, t)|_\Gamma.$$

It follows from Theorem 3.2 and Theorem 3.3 that the solution u of problem (2.1)–(2.4) corresponding to q fulfills condition (3.9). For nonlinear problems, the same approach is applied to the linearization at u_s of the partial differential equations.

3.2. Calculating an extended initial value

The main difficulty in the extension method is the construction of an adequate extended initial value v_0. In the following we introduce two numerical approaches for this step. The overall goal is to construct v_0 satisfying $v_0|_\Omega = u_0$ and

$$\int_G v_0(x)\, \varepsilon_j(x)\, dx = 0, \quad j = 1, 2, \ldots, K. \tag{3.11}$$

a) Eigenfunction method: In the eigenfunction method (EFM) we use the extension operator E_K defined in (3.6) to construct v_0. Numerically this leads to the solution of the system of linear equations

$$\sum_{j=1}^K \xi_j \int_\omega \varepsilon_j(x)\, \varepsilon_i(x)\, dx = -\int_\Omega u_0(x)\, \varepsilon_i(x)\, dx, \quad i = 1, 2, \ldots, K, \tag{3.12}$$

with $\omega := G \setminus \Omega$. The extension v_0 is then defined in the following way

$$
v_0(x) := \begin{cases} u_0(x), & x \in \Omega, \\ \sum_{j=1}^{K} \xi_j \, \varepsilon_j(x), & x \in \omega. \end{cases}
$$

Despite its simplicity, the main drawback of this approach is related to the fact that the condition number of the resulting linear system increases extremely rapidly for small extension domain and for an increasing number of unstable modes (see Table 1).

b) Least squares method: The main difference between the least squares method (LSM) and the eigenfunction method is that in the least squares method we do not use $\varepsilon_1, \varepsilon_2, \ldots, \varepsilon_K$ as a basis to determine v_0, but a basis of a finite element space V_h defined by

$$
V_h = \mathrm{span}\,\{\varphi_1, \cdots, \varphi_l\} \oplus \mathrm{span}\,\{\tilde{\varphi}_1, \cdots, \tilde{\varphi}_l\},
$$

which contains functions whose traces lie on $G \setminus \bar{\Omega}$. In order to attain this goal we assume the following expression for v_0:

$$
v_0(x) = u_0(x), \qquad x \in \Omega, \tag{3.13}
$$

$$
v_0(x) = \sum_{j=1}^{l} \xi_j \varphi_j + \sum_{j=l+1}^{r} \tilde{\xi}_j \tilde{\varphi}_j, \qquad x \in G \setminus \Omega, \tag{3.14}
$$

where we assume for the ansatz functions

$$
\varphi_j(x) = 0, \qquad x \in \partial(G \setminus \Omega), \quad j \in [1, l], \tag{3.15}
$$

$$
\tilde{\varphi}_j(x) \neq 0, \qquad x \in \partial(G \setminus \Omega), \quad j \in [l+1, r]. \tag{3.16}
$$

At the interface between Ω and $G \setminus \Omega$ the values $\tilde{\xi}_j$ are set to ensure a continuity condition with v_0. On the non interfacing part of the boundary of $G \setminus \Omega$ the extension is assumed to be equal to 0. Therefore the components $\tilde{\xi}_j$ for $j \in [l+1, r]$ are given by the problem setting. The orthogonality condition (3.11) leads to the following constraints

$$
\int_{G \setminus \Omega} \left(\sum_{j=1}^{l} \xi_j \varphi_j(x) + \sum_{j=l+1}^{r} \tilde{\xi}_j \tilde{\varphi}_j \right) \varepsilon_i(x) dx + \int_{\Omega} v_0(x) \varepsilon_i(x) dx = 0, \quad i \in [1, n]. \tag{3.17}
$$

The resulting system of n linear equations is used to determine the $l \geq n$ unknowns $\{\xi_j\}_{1 \leq j \leq l}$ by means of a least square approach, i.e., we solve the following problem:

$$
\min_{\xi} \frac{1}{2} \xi^T Q \xi + \tilde{c}^T \xi \qquad \text{s.t.} \quad T\xi = \tilde{b}, \tag{3.18}
$$

where $Q = (q_{ij}) \in \mathbb{R}^{l \times l}$, $\tilde{c} = (\tilde{c}_i) \in \mathbb{R}^l$, $T = (t_{ij}) \in \mathbb{R}^{K \times l}$, $\tilde{b} = (\tilde{b}_i) \in \mathbb{R}^l$ with

$$q_{ij} = \int_{G \setminus \Omega} \nabla \varphi_i(x) \cdot \nabla \varphi_j(x) dx, \tag{3.19}$$

$$\tilde{c}_i = \int_{G \setminus \Omega} \nabla \varphi_i(x) \big(\sum_{j=l+1}^{r} \tilde{\xi}_j \nabla \tilde{\varphi}_j \big) dx, \tag{3.20}$$

$$t_{ij} = \int_{G \setminus \Omega} \varphi_j(x) \varepsilon_i(x) dx, \tag{3.21}$$

$$\tilde{b}_i = - \int_{\Omega} v_0(x) \varepsilon_i(x) dx - \int_{G \setminus \Omega} \big(\sum_{j=l+1}^{r} \tilde{\xi}_j \tilde{\varphi}_j \big) \varepsilon_i(x) dx. \tag{3.22}$$

Solving the problem (3.18) corresponds to minimize the semi-norm $|\cdot|_1$ on G of the extension v_0 under the orthogonality constraints (3.11). The optimization problem (3.18) can be solved by means of the standard Euler-Lagrange formulation which leads to the following two steps:

$$\text{step 1:} \qquad \eta = (TQ^{-1}T^T)^{-1}[\tilde{b} + TQ^{-1}\tilde{c}], \tag{3.23}$$

$$\text{step 2:} \qquad \xi = Q^{-1}(-\tilde{c} + T^T \eta). \tag{3.24}$$

Here η denotes the Lagrange multiplier.

Remark 3.5. Instead of solving the problem (3.18) which aims at minimizing the H_1-seminorm one could consider the following minimization problem

$$\min_{\xi} \frac{1}{2} \xi^T M \xi + \hat{c}^T \xi \qquad \text{s.t.} \quad T\xi = \tilde{b}, \tag{3.25}$$

where $M = (m_{ij}) \in \mathbb{R}^{l \times l}$, $\hat{c} = (\hat{c}_i) \in \mathbb{R}^l$ with

$$m_{ij} = \int_{G \setminus \Omega} \varphi_i(x) \varphi_j(x) dx \tag{3.26}$$

$$\hat{c}_i = \int_{G \setminus \Omega} \varphi_i(x) \big(\sum_{j=l+1}^{r} \tilde{\xi}_j \tilde{\varphi}_j \big) dx, \tag{3.27}$$

In this second approach the goal is to minimize the L_2-norm on the extension. In practice however this approach leads to highly oscillating solutions at the interface and as a consequence to numerical instabilities (see Figure 1).

3.3. Extension method

By solving the problem on computer, rounding errors cannot be avoided. Hence, the computed extension initial value may have some components in the unstable manifold. At some point this may lead to a blow up of the computed solution in the extended domain. In order to avoid this phenomenon, we check at which stage the state solution stops to decrease in time. Let t_1 be a such a time step. In order

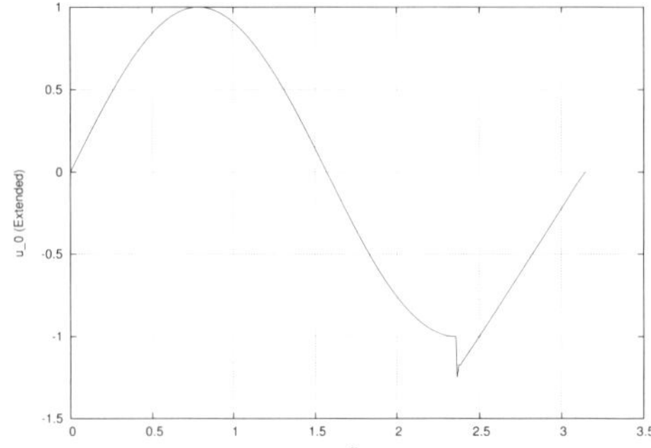

FIGURE 1. Extension of the function $v_0(x) = \sin(2x)$ from $\Omega = [0, 3\pi/4]$ to $G = [0, \pi]$ assuming $\gamma = 3$ in (5.5). This configuration leads to one unstable mode $\varepsilon_1(x) = \sin(x)$. Here we consider the extension approach based on (3.23), i.e., on the minimization of L_2 norm of the extension. This approach is numerically not adequate since it leads to a solution which is highly oscillating at the interface between Ω and $G\backslash\Omega$.

to damped the unwanted component lying in the unstable manifold we restart the whole extension process at $t = t_1$, i.e., the operator E_K is applied to $u(x, t_1)$:

$$v_1(x) := E_K u(x, t_1).$$

Let $v(x, t)$ be the solution of the system (3.2)–(3.3) with $t \in (t_1, T)$ and the initial state $v(x, t_1) = v_1(x)$, and

$$(u(\cdot, t),\, q(\cdot, t)) = (v|_\Omega(\cdot, t),\, v|_\Gamma(\cdot, t)), \quad t \in (t_1, T).$$

Let $t_2 > t_1$ be the next exploding point, we then restart the stabilization process again at $t = t_2$ and so on. By combining the controls in subintervals we obtain the control in (0,T). The overall process leads to the following algorithm:

Algorithm (Stabilization algorithm for linear problems). Given an initial value u_0, an extended domain G, a decreasing rate σ_0, a convergence constant c, and constants k_1 and k_2, e.g., $k_1 = 1$ and $k_2 = 3$. Let V_h be a given finite element space and $\{t_0, \ldots, t_m\}$ be a grid in time.

- Compute an extended initial value v_0 by using one of the algorithms (EFM) or (LSM).
- Set $i = 1$ and $k = 0$.
- **do while** $i \leq m$
    1. Solve the discrete variational problem of the extended state equation at $t = t_i$ and obtain $v_h^i(x)$.
    2. If $k \geq k_1$ and $\|v_h^i\|_\Omega \geq \|v_h^{i-1}\|_\Omega$ then go to 5.
    3. If $k \geq k_2$ and $\|v_h^i\|_\Omega > c\, e^{-\sigma_0 t_i}$ then go to 5.

4. Go to 7.
5. Set $i = i - 1$ and $k = -1$.
6. Let v_h^i be an extension of $v_h^i|_\Omega$ calculated by using one of the algorithms (EFM) or (LSM).
7. Set $i = i + 1$ and $k = k + 1$.

end do

- Set $q = v_h|_\Gamma$.

4. Optimal control based solution process

4.1. Formulation

We consider the system (2.1)–(2.4) with

$$F(u) = \nu\Delta u + H(u),$$

where H is a functional of u and first order derivatives of u. Let $(Q, \|\cdot\|_Q)$ be the space $(H_0^{1/2}(\Gamma), \|\cdot\|_{L^2(\Gamma)})$ or $(H_0^1(\Gamma), \|\cdot\|_{H^1(\Gamma)})$ depending on the considered formulation of the optimization problem. We denote further $Q_{ad} := L^\infty(0, T; Q)$. Assume that there exists for each $q \in Q_{ad}$ a unique weak solution u of the system (2.1)–(2.4), which is continuously depending on the control $q \in Q_{ad}$.

Our problem reads as follows: Find $q^* \in Q_{ad}$ such that

$$J(q^*) \leq J(q), \quad \forall q \in Q_{ad}, \tag{4.1}$$

where

$$J(q) = \frac{K_1}{2} \int_0^T \|q(\cdot, t)\|_Q^2 \, dt + \frac{1}{2} \int_0^T K_2(t) \|u(\cdot, t) - u_s\|_\Omega^2 \, dt, \tag{4.2}$$

$K_1 \geq 0$, $0 < K_2^0 \leq K_2(t) \leq K_2^1$ for all $t \in [0, T]$, and u is the weak solution of the system (2.1)–(2.4). We should set K_2 as a function, which increases in time, and K_1 should be sufficiently small compared with K_2.

We describe the optimization problem in case $\|\cdot\|_Q = \|\cdot\|_{L^2(\Gamma)}$ as an L^2-*optimization problem*, and in case of $\|\cdot\|_Q = \|\cdot\|_{H^1(\Gamma)}$ as an H^1-*optimization problem*.

Since $(H_0^{1/2}(\Gamma), \|\cdot\|_{L^2(\Gamma)})$ is not complete the standard existence theory can not be directly applied while the existence of an optimal solution for the H^1-optimization problem at least for linear parabolic problems is guaranteed. In that last case, the regularity requirements for q are however usually to high in practice.

4.2. First order optimality condition

We first consider the L^2-optimization formulation, that means

$$(Q, \|\cdot\|_Q) = (H_0^{1/2}(\Gamma), \|\cdot\|_\Gamma).$$

The case of the H^1-optimization formulation can be treated similarly.

Theorem 4.1. *For the L^2-optimization problem there holds for each $q \in Q_{ad}$*

$$\nabla J(q) = K_1 q - \nu \partial_n z, \tag{4.3}$$

where $z \in V_{ad}^0$ solves the following adjoint system

$$
\begin{aligned}
-\partial_t z - \nu \Delta z - H'(u)^*(z) &= K_2(u - u_s) \quad &&in \quad \Omega \times (0, T), &&(4.4)\\
z(x, T) &= 0 &&for \quad x \in \Omega, &&(4.5)\\
z(x, t) &= 0 &&on \quad \partial\Omega \times (0, T), &&(4.6)
\end{aligned}
$$

with $H'(u)^*$ is the adjoint operator of $H'(u)$ and

$$V_{ad}^0 := H^1(0, T; L_0^2(\Omega)) \cap L^2(0, T; H_0^1(\Omega)).$$

Proof. From the definition of the functional F we have

$$F'(u)(v) = \nu \Delta v + H'(u)(v).$$

Let $\Sigma = \Omega \times (0, T)$. With every function z belonging to the space V_{ad}^0 we have

$$\int_\Sigma H'(u)(v) \cdot z \, dx \, dt = \int_\Sigma v \cdot H'(u)^*(z) \, dx \, dt. \tag{4.7}$$

In order to compute $\nabla J(q)$, we shall use a formal perturbation analysis. Let us consider $q \in Q_{ad}$ and a small perturbation δq of q. We then have

$$
\begin{aligned}
\delta J(q) &= \int_0^T \!\! \int_\Gamma \nabla J(q) \cdot \delta q \, dO \, dt \\
&= K_1 \int_0^T \!\! \int_\Gamma q \cdot \delta q \, dO \, dt + \int_0^T \!\! \int_\Omega K_2(t)(u - u_s) \cdot \delta u \, dx \, dt, \tag{4.8}
\end{aligned}
$$

where δu satisfies

$$
\begin{aligned}
\partial_t \delta u - \nu \Delta \delta u - H'(u)(\delta u) &= 0 \quad &&in \quad \Omega \times (0, T), &&(4.9)\\
\delta u(x, 0) &= 0 &&for \quad x \in \Omega, &&(4.10)\\
\delta u(x, t) &= 0 &&on \quad (\partial\Omega \backslash \Gamma) \times (0, T), &&(4.11)\\
\delta u(x, t) &= \delta q(x, t) &&on \quad \Gamma \times (0, T). &&(4.12)
\end{aligned}
$$

Let us consider $z \in V_{ad}^0$. Multiplying both sides of (4.9) by z and integrating over Σ, we obtain

$$\int_\Sigma \partial_t \delta u \cdot z \, dx \, dt - \int_\Sigma \nu \Delta \delta u \cdot z \, dx \, dt - \int_\Sigma H'(u)(\delta u) \cdot z \, dx \, dt = 0.$$

Integrating by parts over Σ and taking into account relations (4.10)–(4.12) and (4.7) gives

$$
\begin{aligned}
0 &= \int_\Omega \delta u(x,T) \cdot z(x,T)\,dx - \int_\Omega \delta u(x,0) \cdot z(x,0)\,dx - \int_\Sigma \delta u \cdot \partial_t z\,dx\,dt \\
&\quad - \int_{\partial\Omega\times(0,T)} \nu \partial_n \delta u \cdot z\,dO\,dt + \int_{\partial\Omega\times(0,T)} \nu \delta u \cdot \partial_n z\,dO\,dt \\
&\quad - \int_\Sigma \nu \delta u \cdot \Delta z\,dx\,dt - \int_\Sigma \delta u \cdot H'(u)^*(z)\,dx\,dt \\
&= \int_\Omega \delta u(x,T) \cdot z(x,T)\,dx + \int_{\Gamma\times(0,T)} \nu \delta q \cdot \partial_n z\,dO\,dt \\
&\quad + \int_\Sigma \big(-\partial_t z - \nu \Delta z - H'(u)^*(z) \big) \cdot \delta u\,dx\,dt.
\end{aligned}
\tag{4.13}
$$

Suppose that, in addition to the condition $z \in V_{ad}^0$, the function z satisfies the adjoint system (4.4)–(4.6). Then from (4.13) we have

$$
\int_\Sigma K_2(u - u_s) \cdot \delta u\,dx\,dt = -\int_{\Gamma\times(0,T)} \nu \delta q \cdot \partial_n z\,dO\,dt.
\tag{4.14}
$$

It follows from (4.8) and (4.14) that

$$
\begin{aligned}
\int_0^T \int_\Gamma \nabla J(q) \cdot \delta q\,dO\,dt &= K_1 \int_0^T \int_\Gamma q \cdot \delta q\,dO\,dt - \int_0^T \int_\Gamma \nu \delta q \cdot \partial_n z\,dO\,dt \\
&= \int_0^T \int_\Gamma (K_1 q - \nu \partial_n z) \cdot \delta q\,dO\,dt.
\end{aligned}
\tag{4.15}
$$

Since (4.15) is valid for all (small) perturbation δq, we finally obtain

$$
\nabla J(q) = K_1 q - \nu \partial_n z. \qquad \square
$$

The first order necessary optimality condition of the L^2-optimization problem reads as follows.

Theorem 4.2. *If $q^* \in Q_{ad}$ is a local minimizer of the L^2-optimization problem, then*

$$
K_1 q^* = \nu \partial_n z^*,
$$

where z^* is the adjoint solution corresponding to the control q^*.

With an analogous argument we obtain the optimality condition of the H^1-optimization problem.

Theorem 4.3. *For the H^1-optimization problem the gradient of the objective functional solves*

$$
\int_0^T \int_\Gamma \nabla J(q) \cdot \varphi\,dO\,dt = K_1 \int_0^T (q,\varphi)_{H^1(\Gamma)}\,dt - \int_0^T \int_\Gamma \nu \partial_n z \cdot \varphi\,dO\,dt,
$$

for all $\varphi \in L^\infty(0,T; H^1(\Gamma))$, where z is the solution of the adjoint system (4.4)–(4.6).

Theorem 4.4. *The first order necessary optimality condition of the H^1-optimization problem reads as follows:*

$$K_1 \int_0^T (q^*, \varphi)_{H^1(\Gamma)} \, dt - \int_0^T \int_\Gamma \nu \partial_n z^* \cdot \varphi \, dO \, dt = 0, \quad \forall \varphi \in L^\infty(0, T; H^1(\Gamma)),$$

where z^* is the adjoint solution corresponding to the control q^*.

4.3. Solution process

In order to solve the optimization problem we use the L-BFGS method, which needs only the gradient of the objective functional. The advantage of the L-BFGS method in the comparison with the BFGS method is that the approximation of the Hessian matrix does not need to be stored (see, e.g., [15])

At each step of the L-BFGS method we have to calculate the gradient of the objective functional, this means the primal and adjoint state solution has be available. To solve the primal and adjoint equations numerically we apply the implicit Euler method for discretization in time and finite element method for discretization in space.

Let $0 = t_0 < t_1 < \cdots < t_m = T$ be a grid in time, $k_i := t_i - t_{i-1}$ ($i = 1, 2, \ldots, m$), and V_h and V_h^0 be given finite element spaces

$$V_h = \left\{ v_h \in C(\Omega) \,\middle|\, v_h|_K \in P(K), \forall K \in \mathcal{T}_h \right\},$$
$$V_h^0 = \left\{ v_h \in V_h \,\middle|\, v_h|_{\partial\Omega} = 0 \right\}.$$

We denote

$$Q_h := \left\{ v_h|_\Gamma \,\middle|\, v_h \in V_h \right\} \cap H_0^s(\Gamma),$$

for some suitable $s > 0$. Let q_h^i ($i = 1, 2, \ldots, m$) be the interpolation of $q(\cdot, t_i)$ in Q_h. Then the discrete variational problem of the primal equation (2.1)–(2.4) at the ith time step ($i = 1, 2, \ldots, m$) is the following: Find $u_h^i \in q_h^i + V_h^0$ such that

$$(u_h^i, \varphi)_\Omega + k_i \, a(u_h^i, \varphi) = (u_h^{i-1}, \varphi)_\Omega, \quad \forall \varphi \in V_h^0, \tag{4.16}$$
$$u_h^0(x) = u_0(x) \quad \text{in } \Omega, \tag{4.17}$$

where the bilinear form $a : H^1(\Omega) \times H_0^1(\Omega) \to \mathbb{R}$ is defined by

$$a(v, w) := \nu(\nabla v, \nabla w)_\Omega - (H(v), w)_\Omega.$$

The discrete variational formulation of the adjoint problem (4.4)–(4.6) at the ith time step ($i = m - 1, m - 2, \ldots, 0$) reads: Find $z_h^i \in V_h^0$ such that there holds for all $\varphi \in V_h^0$

$$(z_h^i, \varphi)_\Omega + k_{i+1} a^*(z_h^i, \varphi; u_h^i) = (z_h^{i+1}, \varphi)_\Omega + k_{i+1}\big(K_2(t_i)(u_h^i - u_s), \varphi\big)_\Omega, \tag{4.18}$$
$$z_h^m \equiv 0, \tag{4.19}$$

where

$$a^*(v, \phi; w) := \nu(\nabla v, \nabla \phi)_\Omega - \big(H'(w)^*(v), \phi\big)_\Omega.$$

Discretizing the gradient of the objective functional at the continuous level, we obtain that the ith coordinate of the discrete gradient $\nabla J_{h,\Delta t}(q_{h,\Delta t})$ solves the following problem: Find $\nabla J_{h,\Delta t}(q_{h,\Delta t})_i \in Q_h$ such that

$$\big(\nabla J_{h,\Delta t}(q_{h,\Delta t})_i, \chi\big)_\Gamma = \big(K_1 q_h^i - \nu \partial_n z_h^i, \chi\big)_\Gamma, \quad \forall \chi \in Q_h. \tag{4.20}$$

5. Numerical experiments

5.1. Unstable linear parabolic problem in 1D

First, we consider the following one-dimensional linear parabolic problem:

$$\partial_t u - \Delta u - \gamma u \;=\; 0 \qquad \text{in} \quad [0, \tfrac{3\pi}{4}] \times (0, T), \tag{5.1}$$

$$u(x, 0) \;=\; u_0(x) \quad \text{for} \quad x \in [0, \tfrac{3\pi}{4}], \tag{5.2}$$

$$u(0, t) \;=\; 0 \qquad \text{for} \quad t \in (0, T), \tag{5.3}$$

$$u(\tfrac{3\pi}{4}, t) \;=\; q(t) \qquad \text{for} \quad t \in (0, T). \tag{5.4}$$

Our goal is to stabilize the problem toward the stationary solution $u_s = 0$. For the extension method we assume for the extended domain $G = (0, g_{\max})$ where obviously $g_{\max} > \frac{3\pi}{4}$. Note that the eigenvalues of the operator $\Delta v + \alpha v$ with $v : G \to \mathbb{R}$ and $v|_{\partial G} = 0$ depend on γ and $g_{\max}$. For the discretization we use uniform grids in time and space with $\Delta t = 10^{-3}$ and $h = 2^{-8}$.

Table 1 shows the condition number of the linear system (3.12) for various values of $g_{\max}$ and γ leading to different number of unstable modes. The solution of the linear system (3.12) is needed to compute the extended initial value by means of the numerical approach EFM (see Section 3.2). Table 1 clearly shows that this condition number drastically increase for a decreasing size of the extended domain and for an increasing number number of unstable modes. For a small extended domain, the linear system is badly conditioned even if a few number of unstable modes has to be filtered. The method LSM should be therefore preferred in that context and is used for all presented numerical tests.

The extension method performs efficiently and robustly for the problem (5.1–5.4). Corresponding results are depicted on Figures 2–7. Especially for an increasing number of unstable modes no special tuning of the method is needed to obtain the stabilizing control variable. The optimal control based approach describes in Section 4 behaves well for a small number of unstable modes (see Figure 8). For an increasing number of unstable modes a fine tuning of the parameters K_1 and K_2 as well as a feedback control approach are needed to attain convergence if possible at all.

g_{max} / NEV	1.1	1.2	1.5	2
2	5.13e+04	3.90e+03	1.33e+02	1.22e+01
3	2.66e+09	1.53e+07	1.88e+04	1.97e+02
4	5.04e+12	5.39e+10	2.55e+06	4.08e+03
5	1.59e+13	3.41e+13	3.55e+08	9.63e+04
6	7.67e+12	5.89e+13	5.18e+10	2.44e+06
7	9.71e+12	5.06e+13	6.11e+12	6.51e+07
8	1.34e+13	2.83e+14	2.53e+13	1.79e+09
9	2.89e+13	1.04e+14	2.54e+13	5.05e+10
10	1.70e+13	1.71e+14	2.34e+13	1.31e+12

TABLE 1. Condition number of the linear system (3.12) for various values of g_{max} and γ leading to different number of unstable modes NEV. The solution of the linear system (3.12) is needed to compute the extended initial value by means of the numerical approach EFM (see Section 3.2).

5.2. Unstable nonlinear parabolic problem in 1D

The extension method presented in Section 3 relies on a linearization at u_s of the considered partial differential equations. Our goal in this section is to exhibit some of the properties of this approach for a nonlinear problem. We consider the following one-dimensional nonlinear parabolic problem:

$$\partial_t u - \Delta u - \gamma u - u^2 = 0 \qquad \text{in} \quad [0, \frac{3\pi}{4}] \times (0, T), \qquad (5.5)$$

$$u(x, 0) = u_0(x) \quad \text{for} \quad x \in [0, \frac{3\pi}{4}], \qquad (5.6)$$

$$u(0, t) = 0 \quad \text{for} \quad t \in (0, T), \qquad (5.7)$$

$$u(\frac{3\pi}{4}, t) = q(t) \quad \text{for} \quad t \in (0, T). \qquad (5.8)$$

Our goal is to stabilize the problem toward the stationary solution $u_s = 0$. We prescribe $G = (0, \pi)$. In the range $\gamma < 8$, the extension method performs robustly and efficiently toward a stabilization of this unstable problem (see Figure 8). For more than two unstable modes the stabilization of the equations could not be obtained even for small initial data. The optimal control based approach shows for this problem similar stabilizing properties in comparison to the linear case (5.1)–(5.4).

5.3. Unstable linear parabolic problem in 2D

Let $\Omega = (0,1) \times (0,1)$, $\Gamma = (0,1) \times \{1\}$, and $T = 1$. Consider the following two-dimensional linear parabolic problem:

$$
\begin{aligned}
\partial_t u - \Delta u - 25\,u &= 0 && \text{in } \Omega \times (0,T), & (5.9)\\
u(x,0) &= u_0(x) && \text{for } x \in \Omega, & (5.10)\\
u(x,t) &= 0 && \text{for } (x,t) \in (\partial\Omega \backslash \Gamma) \times (0,T), & (5.11)\\
u(x,t) &= q(t) && \text{for } (x,t) \in \Gamma \times (0,T), & (5.12)
\end{aligned}
$$

where

$$
u_0(x) = \sin(\pi x_1)\,\sin(\pi x_2) \qquad \text{for } x = (x_1, x_2) \in \Omega. \tag{5.13}
$$

Without control, i.e., $q = 0$, the state solution blows up (see Figure 9). Our goal is to stabilize the problem toward the stationary solution $u_s = 0$. For the discretization we consider a uniform grid in time and space with $\Delta t = 0.01$ and $h = 2^{-6}$ and discretize the problem in space by means of Q_1 finite elements. We consider the optimization based approach described in Section 4 for different coefficients K_1 and $K_2(t)$ (see Table 2).

	Case 1	Case 2	Case 3	Case 4
K_1	10^{-3}	10^{-3}	10^{-3}	10^{-3}
$K_2(t)$	t	e^{4t}	e^{6t}	$0.1e^{8t}$

TABLE 2. Setup of the coefficients for the optimization based approach (see Section 4) used to solve the problem (5.9)–(5.12).

The obtained optimal controls and the corresponding state solutions and there norms are presented in Figures 10–15. These results clearly show that for a wide range of parameters K_2 this numerical approach allows to obtain the needed stabilizing properties.

6. Conclusion

In the extension method, the computational costs are mainly related to the computation of the unstable eigenfunctions as well as one forward solution of the considered partial differential equations. The costs related to the computation of the extension operator are almost negligible. The delicate and still open issue for this approach is the adequate choice of the extended domain especially for higher-dimensional problems. The optimal control based approach is more straightforward to apply since it does not rely on the definition of an extended domain. The main computational costs are related to the calculation of the gradient, i.e., the computation of the primal and adjoint state solution. For highly unstable problems, a feedback control approach has to be considered for both methods.

 V. Heuveline and H. Nam-Dung

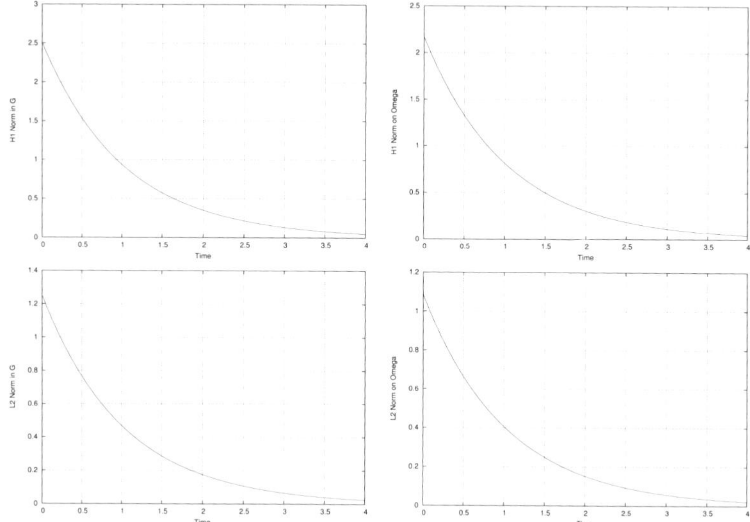

FIGURE 2. Time variation of (upper left) $\|v(\cdot,t)\|_{H_1(G)}$, (upper right) $\|u(\cdot,t)\|_{H_1(\Omega)}$, (lower left) $\|v(\cdot,t)\|_{L_2(G)}$, (lower right) $\|u(\cdot,t)\|_{L_2(\Omega)}$. for problem (5.1)-(5.4) with $\gamma = 3$ (one unstable mode).

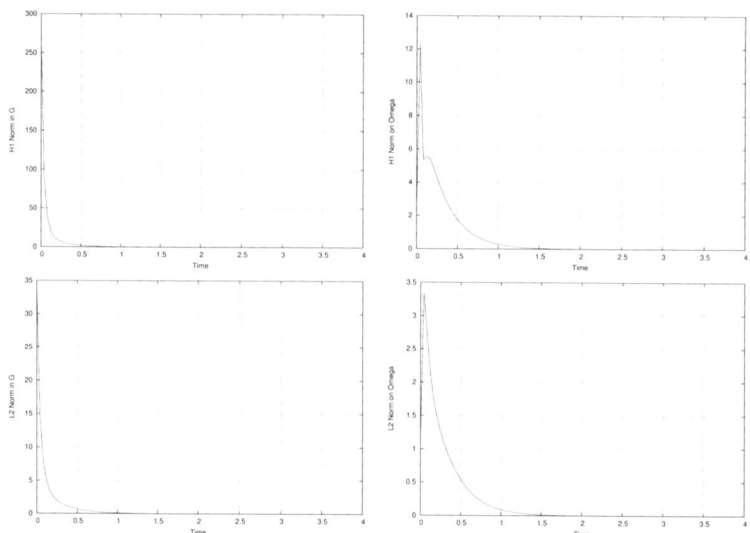

FIGURE 3. The description of the plots is identical to Figure 2. Problem (5.1)–(5.4) for $\gamma = 5$ (two unstable modes).

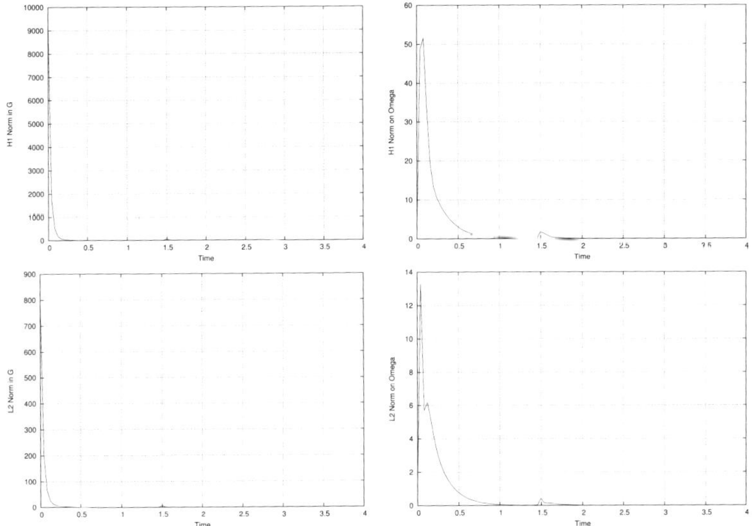

FIGURE 4. Time variation of (upper left) $\|v(\cdot,t)\|_{H_1(G)}$, (upper right) $\|u(\cdot,t)\|_{H_1(\Omega)}$, (lower left) $\|v(\cdot,t)\|_{L_2(G)}$, (lower right) $\|u(\cdot,t)\|_{L_2(\Omega)}$. for problem (5.1)–(5.4) with $\gamma = 10$ (three unstable modes).

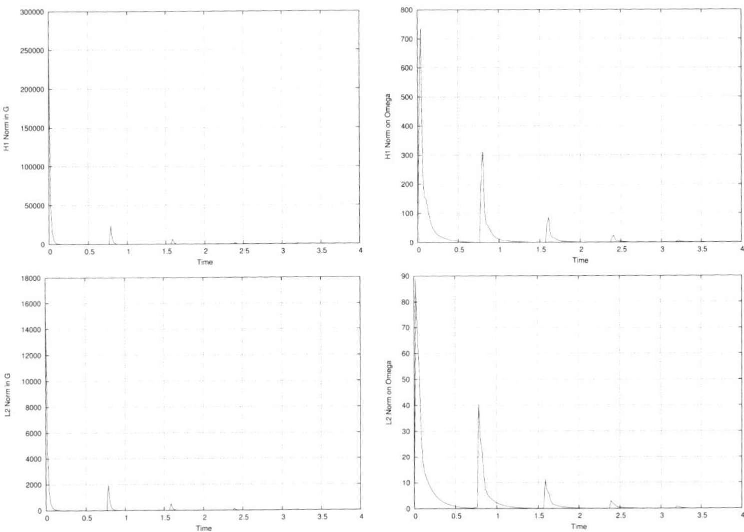

FIGURE 5. The description of the plots is identical to Figure 4. Problem (5.1)–(5.4) for $\gamma = 17$ (four unstable modes).

                V. Heuveline and H. Nam-Dung

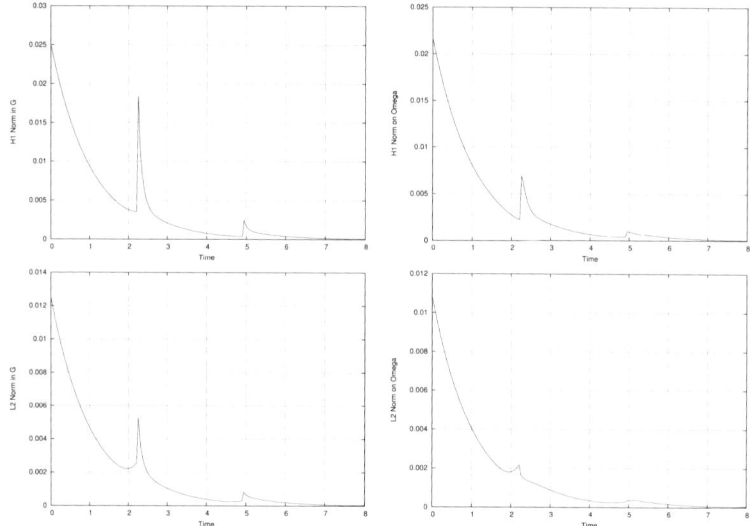

FIGURE 6. Time variation of (upper left) $\|v(\cdot,t)\|_{H_1(G)}$, (upper right) $\|u(\cdot,t)\|_{H_1(\Omega)}$, (lower left) $\|v(\cdot,t)\|_{L_2(G)}$, (lower right) $\|u(\cdot,t)\|_{L_2(\Omega)}$ for problem (5.5)–(5.8) with $\gamma = 3$, $u_0(x) = 0.01\sin(2x)$.

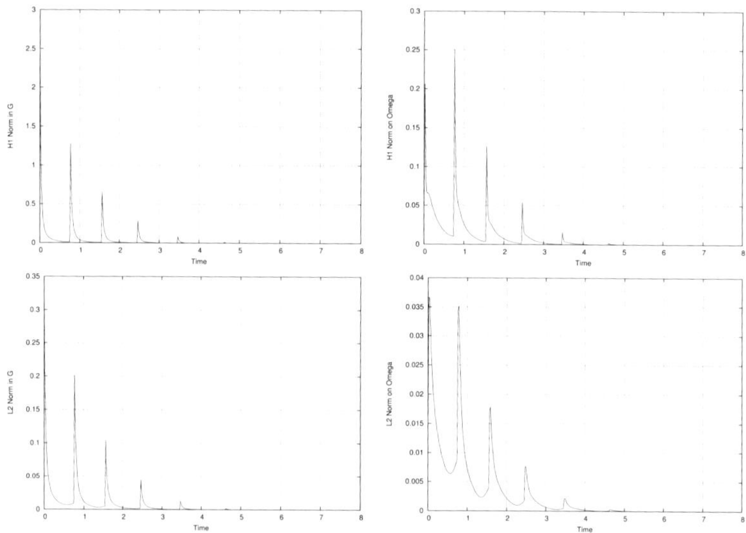

FIGURE 7. The description of the plots is identical to Figure 6. Problem (5.5)–(5.8) for $\gamma = 5$, $u_0(x) = 0.01\sin(2x)$.

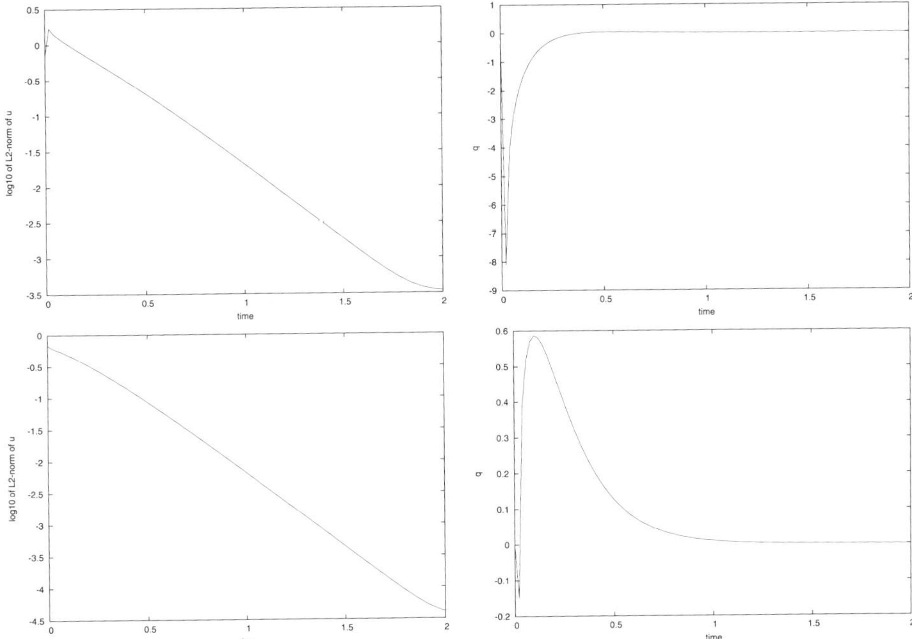

FIGURE 8. Results obtained by means of the optimization based approach depicted in Section 4 for the stabilization problem (5.1)–(5.4) assuming $K_1 = 10^{-3}$ and $K_2(t) = e^{4t}$. For the upper (resp. lower) plots we imposed $\gamma = 3$ (resp. $\gamma = 5$). The plots on the left column describe the time variation of $\log_{10}(\|u(\cdot, t)\|_{L^2(\Omega)})$. On the right column the time behavior of the control q is plotted.

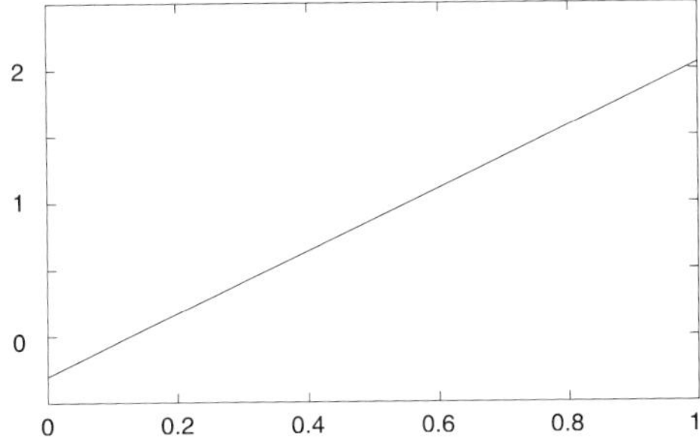

FIGURE 9. The function $\log_{10}\|u(\cdot, t)\|_\Omega$ assuming no control for the stabilization problem (5.9)–(5.12).

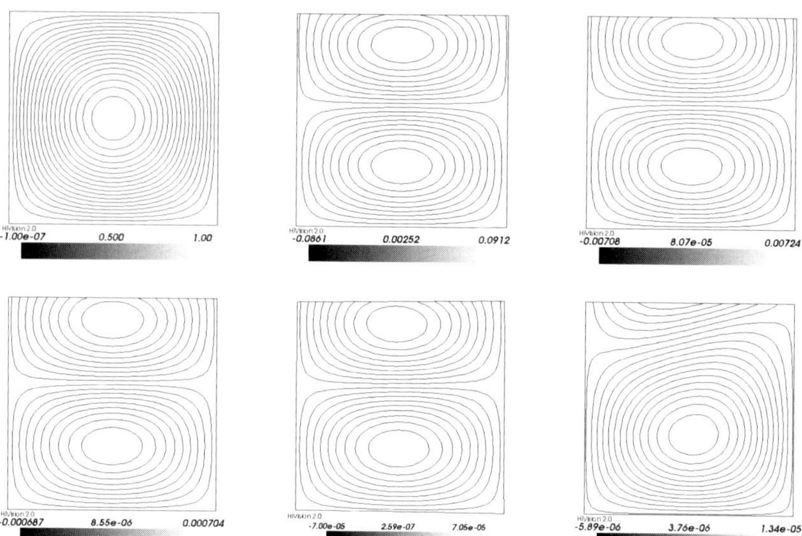

FIGURE 10.   Isolines of the computed stabilized solution of the problem (5.9)–(5.12) for $t = 0, 0.2, 0.4, 0.6, 0.8, 1$. (case 1)

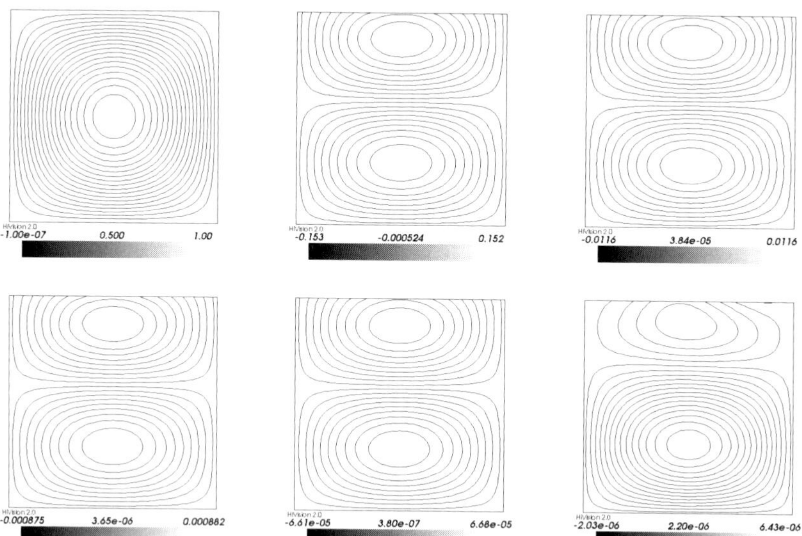

FIGURE 11.   Isolines of the computed stabilized solution of the problem (5.9)–(5.12) for $t = 0, 0.2, 0.4, 0.6, 0.8, 1$. (case 2)

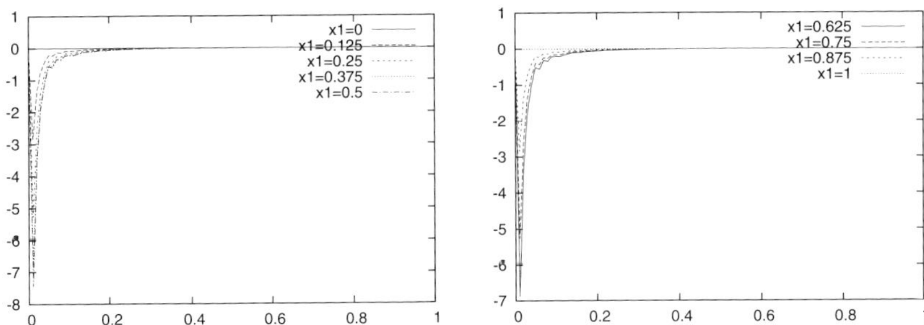

FIGURE 12. Optimal control for problem (5.9)–(5.12) and case 1.

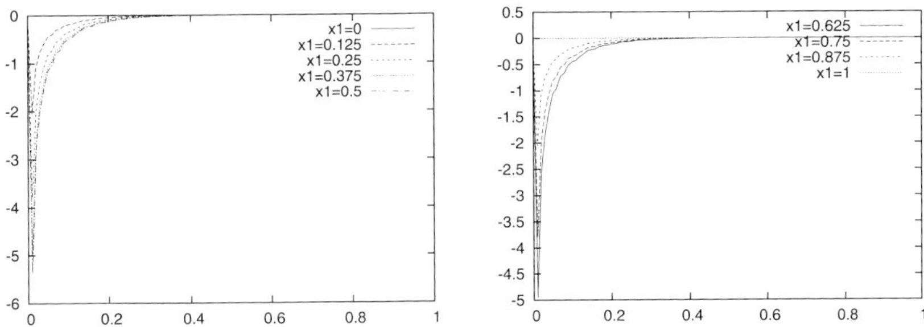

FIGURE 13. Optimal control for problem (5.9)–(5.12) and case 2.

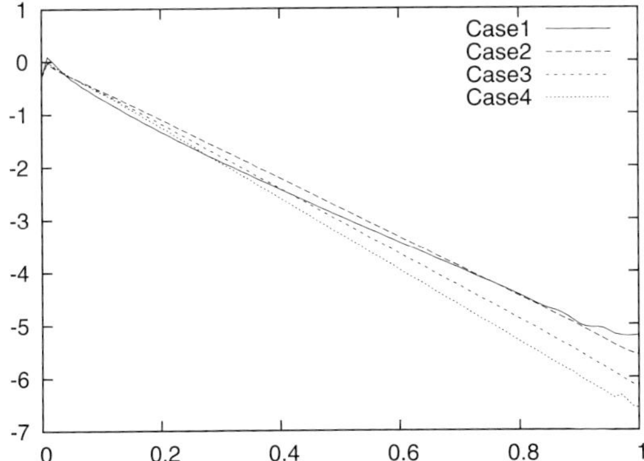

FIGURE 14. The function $\log_{10}\|u(\cdot,t)\|_\Omega$ for problem (5.9)–(5.12) assuming corresponding optimal Dirichlet control.

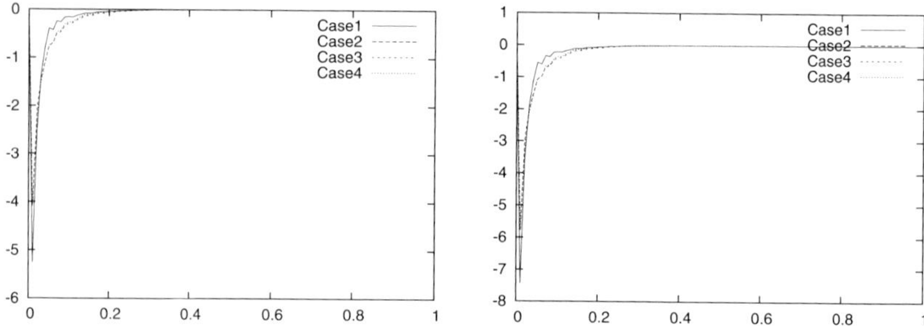

FIGURE 15. Comparison between optimal controls in the 4 cases (see Table 2) at $x_1 = 0.25$ and $x_1 = 0.5$.

Acknowledgments

The first author greatly thanks Prof. Andrei V. Fursikov for fruitful discussions on the stabilization problem as well as for important suggestions with respect to the extension method.

References

[1] B.S. Mordukhovich and J.-P. Raymond, *Neumann boundary control of hyperbolic equations with pointwise state constraints.* SIAM J. Control Optimization **43(4)** (2005), 1354–1372.

[2] V. Barbu, D. Coca, and Y. Yan, *Stabilizing semilinear parabolic equations.* Numer. Funct. Anal. Optimization **26(4-5)** (2005), 449–480.

[3] A.V. Fursikov, M.D. Gunzburger, and L.S. Hou, *Optimal boundary control for the evolutionary Navier-Stokes system: The three-dimensional case.* SIAM J. Control Optimization **43(6)** (2005), 2191–2232.

[4] A.V. Fursikov, *Stabilization for the 3D Navier-Stokes system by feedback boundary control.* Discrete Contin. Dyn. Syst. **10(1-2)** (2004), 289–314.

[5] S. Anita and J.-P. Raymond, *Positive stabilization of a parabolic equation by controls localized on a curve.* J. Math. Anal. Appl. **286(1)** (2003), 107–115.

[6] V. Barbu, *Feedback stabilization of Navier-Stokes equations.* ESAIM, Control Optim. Calc. Var. **9** (2003), 197–206.

[7] V. Barbu and G. Wang, *Feedback stabilization of semilinear heat equations.* Abstr. Appl. Anal. **12** (2003), 697–714.

[8] N. Arada and J.-P. Raymond, *Dirichlet boundary control of semilinear parabolic equations. II: Problems with pointwise state constraints.* Appl. Math. Optimization **45(2)** (2002), 145–167.

[9] N. Arada and J.-P. Raymond, *Dirichlet boundary control of semilinear parabolic equations. I: Problems with no state constraints.* Appl. Math. Optimization **45(2)** (2002), 125–143.

[10] A.V. Fursikov, M.D. Gunzburger, and L.S. Hou, *Inhomogeneous boundary value problems for the three-dimensional evolutionary Navier-Stokes equations.* J. Math. Fluid Mech. **4(1)** (2002), 45–75.

[11] P.A. Nguyen and J.-P. Raymond, *Control problems for convection-diffusion equations with control localized on manifolds.* ESAIM, Control Optim. Calc. Var. **6** (2001), 467–488.

[12] A.V. Fursikov, *Stabilizability of a quasi-linear parabolic equation by means of a boundary control with feedback.* Sbornik Mathematics **192(4)** (2001), 593–639.

[13] A.V. Fursikov, *Stabilizability of two-dimensional Navier-Stokes equations with help of a boundary feedback control.* J. Math. Fluid Mech. **3(3)** (2001), 259–301.

[14] A.V. Fursikov, *Optimal control of distributed systems. Theory and applications.*, Transl. from the Russian by Tamara Roszkovskaya, Translations of Mathematical Monographs. 187. Providence, RI: AMS, American Mathematical Society. xiv, 2000.

[15] J. Nocedal and S. J. Wright, *Numerical Optimization.*, Springer, New York, 1999.

[16] A.V. Fursikov and O.Yu. Imanuvilov, *Exact controllability of the Navier-Stokes and Boussinesq equations.* Russian Math. Surveys **54(3)** (1999), 565–618.

[17] A.V. Fursikov, M.D. Gunzburger, and L.S. Hou, *Boundary value problems and optimal boundary control for the Navier-Stokes system: The two-dimensional case.* SIAM J. Control Optimization **36(3)** (1998), 852–894.

[18] J.W. He, R. Glowinski, R. Metacalfe, and J. Periaux, *A numerical approach to the control and stabilization of advection-diffusion systems: Application to viscous drag reduction.* Int. J. Comput. Fluid Dyn. **11(1-2)** (1998), 131–156.

[19] J.W. He and R. Glowinski, *Neumann control of unstable parabolic systems: Numerical approach.* J. Optimization Theory Appl. **96(1)** (1998), 1–55.

[20] J.L. Lions, *Optimal control of Systems Governed by Partial Differential Equations.*, Springer, Berlin, 1971.

Vincent Heuveline
Institute for Applied Mathematics and Computing Center
University Karlsruhe (TH)
Zirkel 2
D-76128 Karlsruhe, Germany
e-mail: `vincent.heuveline@math.uni-karlsruhe.de`

Hoang Nam-Dung
Zuse Institute Berlin (ZIB)
Takustrasse 7
D-14195 Berlin, Germany
e-mail: `hoang@zib.de`

International Series of Numerical Mathematics, Vol. 155, 119–147
© 2007 Birkhäuser Verlag Basel/Switzerland

Fast Solution Techniques in Constrained Optimal Boundary Control of the Semilinear Heat Equation

M. Hintermüller, S. Volkwein and F. Diwoky

Abstract. Efficient numerical techniques for the solution of constrained optimal control problems for the nonlinear heat equation are considered. The nonlinearity in the governing equation is due to the boundary conditions which cover the Boltzmann radiation boundary condition. With respect to numerical algorithms, variants of semismooth Newton methods are proposed which allow a convergence analysis in function space. For the latter aspect the concept of generalized (Newton, or slant) differentiability is invoked. The paper ends with a comparison of the proposed algorithms among each other and with a sequential quadratic programming method.

Mathematics Subject Classification (2000). Primary 49K20, 65K05; Secondary 35Kxx.

Keywords. Constrained optimal control, integral constraints, nonlinear heat equation, pointwise constraints, semismooth Newton.

1. Introduction

In this paper we study methods for the numerical solution of constrained optimal control problems for thermal processes. The governing time-dependent partial differential equation is given by the heat equation

$$y_t - \alpha \Delta y = f \quad \text{in } Q, \quad y(0) = y_\circ \quad \text{on } \Omega, \tag{1.1}$$

where Ω represents the domain of interest and Q is the space-time cylinder. Further, f is a given source, $y_\circ$ denotes the initial temperature, and $\alpha > 0$ is a given

The authors gratefully acknowledge support by the Austrian Science Fund FWF under grant no. P16760-N12. The first author (M.H.) also acknowledges support via the START-program Y305 "Interfaces and Free Boundaries".

constant reflecting heat conduction properties. We consider (1.1) together with nonlinear boundary conditions of the type

$$\frac{\partial y}{\partial n} + b(y) = \beta u, \tag{1.2}$$

with a given possibly nonlinear function b. Above, u denotes the control, and $\beta \neq 0$ is a given constant. Note that (1.2) covers locally radiative heat transfer across the boundary which is modelled by the Stefan-Boltzmann-radiation boundary condition

$$\frac{\partial y}{\partial n} + y^3|y| = \beta u.$$

The latter law is of interest, e.g., in crystal growth phenomena [21].

In many practically relevant situations the control u has to obey certain restrictions. For instance, due to technical limitations and/or to prevent phase changes it is required that the control only acts within certain bounds. This can be modelled by bilateral pointwise inequality constraints of the type

$$u_a \leq u \leq u_b \quad \text{almost everywhere in } Q.$$

Also, it might be of interest to constrain the available "volume" of the control action. This results in inequality constraints of integral type. The control problem stated in the next section contains all these constraints simultaneously. In addition, it also covers integral state constraints.

For theoretical investigations of constrained optimal control problems involving parabolic partial differential equations we refer to, e.g., [7, 21, 23] and the many references therein. These sources essentially focus on the derivation of practically relevant first or second order optimality conditions. This is a complex task due to the presence of the inequality constraints, and requires a careful investigation of the solvability of the state equation, the adjoint system, and of the choice of the involved function spaces. We point out that the first order optimality system is usually the starting point for the development of numerical solution algorithms.

Besides the modelling issues and the theoretical aspects mentioned above, in practical applications the efficient numerical solution of the underlying control problem is of utmost importance. In the present work we emphasize variants of semismooth Newton techniques. The latter algorithm class is based on a generalized differentiability concept in function space [18] and turned out to be extremely efficient for classes of elliptic control problems; see some of the references in [18]. For its globalization we propose a combination of problem dependent modifications of the Hessian of the Lagrangian (see [15, 17]) and a line search procedure. Let us point out that an alternative trust-region globalization for a related problem class was considered in [11]. The resulting algorithm is implemented in the `trice`-project (see `www.caam.rice.edu/~trice`).

The subsequent sections are organized as follows: In Section 2 we give a brief introduction highlighting the model problem together with the involved function spaces. Further we state the first order optimality system for characterizing (local) solutions. In order to avoid conflicting situations due to constraint requirements, we introduce an augmented Lagrangian penalization approach for the integral constraint. For later purposes we also study the reduced problem and differentiability properties of the control-to-state mapping. Section 3 is concerned with the development of algorithms for the efficient solution of the augmented Lagrangian subproblems. We introduce a *partial* semismooth Newton method, which only linearizes some parts of the first order system and leaves the state equation nonlinear, and we study classical semismooth Newton methods. In either case we consider two subtypes of methods: The first variant considers the state equation as an explicit constraint, while the second one performs a reduction process via the state equation by expressing the state as a function of the control variable. Finally, in Section 4 we report on numerical results, and we provide a comparison with a sequential quadratic programming method. Due to the augmentation of the integral constraint, the latter method is also a variant of semismooth Newton algorithms.

2. Optimal control problem and augmented Lagrangian formulation

In this section we formulate the optimal control problem with integral constraints and review first order necessary optimality conditions. Moreover, to handle the integral constraints, we propose an augmented Lagrangian approach.

2.1. The constrained optimal control problem

Suppose that Ω is an open and bounded subset of $\mathbb{R}^d$, with $d \in \{2,3\}$, with Lipschitz boundary $\Gamma = \partial\Omega$. For $T > 0$ we set $Q = (0,T) \times \Omega$ and $\Sigma = (0,T) \times \Gamma$. Moreover, by $L^2(0,T;H^1(\Omega))$ we denote the space of (equivalence classes) of measurable abstract functions $\varphi : [0,T] \to H^1(\Omega)$, which are square integrable, i.e.,

$$\int_0^T \|\varphi(t)\|^2_{H^1(\Omega)}\, dt < \infty.$$

For the definition of Sobolev spaces we refer the reader, e.g., to [1, 14]. When t is fixed, the expression $\varphi(t)$ stands for the function $\varphi(t,\cdot)$ considered as a function in Ω only. Recall that

$$W(0,T) = \left\{\varphi \in L^2(0,T;H^1(\Omega)) : \varphi_t \in L^2(0,T;H^1(\Omega)')\right\}$$

is a Hilbert space supplied with its common inner product; see [10, p. 473], for instance. Notice that $L^2(0,T;L^2(\Omega))$ can be identified with $L^2(Q)$.

We consider an optimal boundary control problem for the heat equation with pointwise control constraints and integral constraints. The goal is to minimize the cost function $J : W(0,T) \times L^2(\Sigma) \to [0,\infty)$ given by

$$
\begin{aligned}
J(y,u) \;=\; &\frac{1}{2}\int_0^T\!\!\int_\Omega \alpha_Q\, |y - z_Q|^2\, d\mathbf{x}dt + \frac{1}{2}\int_\Omega \alpha_\Omega\, |y(T) - z_\Omega|^2\, d\mathbf{x} \\
&+ \frac{\kappa}{2}\int_0^T\!\!\int_\Gamma |u|^2\, dsdt,
\end{aligned}
\tag{2.1}
$$

where the state y and the control u are coupled by the following semilinear boundary value problem

$$y_t(t, \mathbf{x}) - \alpha \Delta y(t, \mathbf{x}) = f(t, \mathbf{x}) \qquad \text{for all } (t, \mathbf{x}) \in Q, \tag{2.2a}$$

$$\frac{\partial y}{\partial n}(t, s) + b(y(t, s)) = \beta u(t, s) \qquad \text{for all } (t, s) \in \Sigma, \tag{2.2b}$$

$$y(0, \mathbf{x}) = y_\circ(\mathbf{x}) \qquad \text{for all } \mathbf{x} \in \Omega. \tag{2.2c}$$

In (2.1) we assume that α_Q and α_Ω are non-negative weights satisfying $\alpha_Q \in L^\infty(Q)$ and $\alpha_\Omega \in L^\infty(\Omega)$, respectively. The desired states $z_Q \in L^\infty(Q)$ and $z_\Omega \in L^\infty(\Omega)$ are given, and $\kappa > 0$ denotes a regularization parameter. For the data in (2.2) we suppose that the inhomogeneity f belongs to $L^\infty(Q)$, $\alpha > 0$ and $\beta \neq 0$ hold true, and the initial state satisfies $y_\circ \in C(\overline{\Omega})$. Moreover, the nonlinear function $b : \mathbb{R} \to \mathbb{R}$ is twice continuously differentiable, and there exist real constants C_1, C_2, C_3 with $C_1, C_3 \geq 0$ such that

$$|b(0)| \leq C_1 \quad \text{and} \quad C_2 \leq b'(r) \leq C_3\, \eta(|r|) \text{ for } r \in \mathbb{R}, \tag{2.3}$$

where $\eta : [0, \infty) \to [0, \infty)$ is a non-decreasing function. It was proven in [7] that for any $u \in L^{\sigma_1}(\Sigma)$, with $\sigma_1 > d+1$, there exists a unique solution $y \in W(0, T) \cap C(\overline{Q})$ to the state equation (2.2). Moreover, the mapping $u \mapsto y(u)$ is continuous from $L^{\sigma_1}(\Sigma)$ to $C(\overline{Q})$.

We also impose bilateral control constraints on the control variable u. For that purpose let $u_a, u_b \in L^{\sigma_2}(\Sigma)$, with $\sigma_2 \geq \sigma_1$, be given lower and upper bounds, respectively. Then the admissible controls u are required to belong to the closed convex set

$$U_{\mathsf{ad}} = \big\{ u \in L^2(\Sigma) \,\big|\, u_a \leq u \leq u_b \text{ on } \Sigma \text{ almost everywhere (a.e.)} \big\} \subset L^{\sigma_2}(\Sigma). \tag{2.4}$$

In addition, an integral constraint has to be satisfied, i.e., we assume that

$$\int_0^T \int_\Gamma (1 - \tau)u + \tau y \,\mathrm{d}s\mathrm{d}t + \zeta \int_Q y \,\mathrm{d}\mathbf{x}\mathrm{d}t \leq C_I \tag{2.5}$$

where $C_I \in \mathbb{R}$, and $\tau, \zeta \in [0, 1]$ are fixed. Note that for $\tau = \zeta = 0$ we have an integral control constraint, whereas for $\tau = 1$ condition (2.5) becomes an integral state constraint.

For a compact formulation of the optimal control problem we introduce the two Banach spaces $Y = W(0, T) \cap C(\overline{Q})$, $X = Y \times L^{\sigma_1}(\Sigma)$, and the nonlinear mapping $e : X \to L^2(0, T; H^1(\Omega)')$ by

$$\langle e(y, u), \varphi \rangle_{L^2(0,T;H^1(\Omega)'),\, L^2(0,T;H^1(\Omega))} = \int_0^T \langle y_t(t), \varphi(t) \rangle_{H^1(\Omega)',\, H^1(\Omega)} \,\mathrm{d}t$$

$$+ \int_0^T \int_\Omega \alpha \nabla y \cdot \nabla \varphi - f\varphi \,\mathrm{d}\mathbf{x}\mathrm{d}t + \int_0^T \int_\Gamma \alpha\big(b(y(\cdot)) - \beta u\big)\varphi \,\mathrm{d}s\mathrm{d}t$$

for $\varphi \in L^2(0, T; H^1(\Omega))$, where $\langle \cdot, \cdot \rangle_{H^1(\Omega)', H^1(\Omega)}$ denotes the duality pairing between $H^1(\Omega)$ and the dual $H^1(\Omega)'$. The feasible set is given by

$$\mathcal{F}(\mathbf{P}) = \left\{ x = (y, u) \in X \,\middle|\, e(x) = 0, \ y(0) = y_\circ, \ u \in U_{\mathsf{ad}}, \ (2.5) \text{ is satisfied} \right\}.$$

Throughout the paper we assume that $\mathcal{F}(\mathbf{P}) \neq \emptyset$. Note that in the special case $\tau = \zeta = 0$ this can be guaranteed by the requirement

$$\int_0^T \int_\Gamma u_a \, \mathrm{d}s\mathrm{d}t \leq C_I. \tag{2.6}$$

Our infinite-dimensional optimal control problem now reads

$$\min J(x) \quad \text{subject to (s.t.)} \quad x \in \mathcal{F}(\mathbf{P}). \tag{P}$$

Since $\mathcal{F}(\mathbf{P}) \neq \emptyset$ by assumption, there exists at least one (global) solution $x^* = (y^*, u^*)$ to $(\mathbf{P})$. For a proof we refer to [4, 7].

Next we introduce the Lagrange functional $\mathcal{L} : X \times L^2(0, T; H^1(\Omega)) \to \mathbb{R}$ by

$$\mathcal{L}(x, p) = J(x) + \langle e(x), p \rangle_{L^2(0,T;H^1(\Omega)'), L^2(0,T;H^1(\Omega))}$$

for the primal variable $x = (y, u) \in X$ and the dual variable $p \in L^2(0, T; H^1(\Omega))$. The first order necessary optimality conditions in the next theorem follow, e.g., from the results in [7] and [23].

Theorem 2.1. *Suppose that* $(y^*, u^*) = x^* \in \mathcal{F}(\mathbf{P})$ *is a local solution to* $(\mathbf{P})$. *Then there exist unique Lagrange multipliers* (p^*, λ^*, ξ^*) *in* $W(0, T) \times L^2(\Sigma) \times \mathbb{R}$ *satisfying, together with* (y^*, u^*), *the dual system (here written in its strong form)*

$$-p_t^* - \alpha \Delta p^* = -\alpha_Q (y^* - z_Q) - \zeta \xi^* \qquad\qquad in\ Q, \tag{2.7a}$$

$$\frac{\partial p^*}{\partial n} + b'(y^*(\cdot))p^* + \tau \xi^* = 0 \qquad\qquad on\ \Sigma, \tag{2.7b}$$

$$p^*(T) = -\alpha_\Omega (y^*(T) - z_\Omega) \qquad\qquad in\ \Omega, \tag{2.7c}$$

$$\kappa u^* - \alpha \beta p^* + (1 - \tau)\xi^* + \lambda^* = 0 \qquad\qquad on\ \Sigma, \tag{2.7d}$$

$$\lambda^* = \max\left(0, \lambda^* + \sigma(u^* - u_b)\right) + \min\left(0, \lambda^* + \sigma(u^* - u_a)\right) \qquad on\ \Sigma, \tag{2.7e}$$

$$\xi^* = \max\left(0, \xi^* + \varrho\left(\int_0^T \int_\Gamma (1 - \tau)u^* + \tau y^* \, \mathrm{d}s\mathrm{d}t \right.\right. \tag{2.7f}$$

$$\left.\left. + \zeta \int_Q y^* \mathrm{d}x\mathrm{d}t - C_I \right)\right) \qquad\qquad in\ \mathbb{R}$$

for any $\sigma, \varrho > 0$, *where in (2.7e) the functions* min *and* max *are interpreted in the pointwise almost everywhere sense. In particular, if* $z_\Omega \in C(\overline{\Omega})$, $p^* \in C(\overline{Q})$ *holds true.*

In (2.7b) and (2.7d) we identify $\xi^*(s) = \xi^*$ for all $s \in \Sigma$; analogously for ξ^* in (2.7a). Further, (2.7e) and (2.7f) are nonlinear complementarity problem (NCP) function based reformulations of the complementarity systems

$$\lambda_a^* \geq 0, \quad u_a - u^* \leq 0, \quad \lambda_a^*(u_a - u^*) = 0, \quad \text{a.e. on } \Sigma, \tag{2.8}$$

$$\lambda_b^* \geq 0, \quad u^* - u_b \leq 0, \quad \lambda_b^*(u^* - u_b) = 0, \quad \text{a.e. on } \Sigma, \tag{2.9}$$

with $\lambda^* = \lambda_b^* - \lambda_a^*$, and

$$\begin{cases} \xi^* \geq 0, \quad \int_0^T \int_\Gamma (1-\tau)u^* + \tau y^* \, dsdt + \zeta \int_Q y^* \, d\mathbf{x}dt - C_I \leq 0, \\ \xi^*\left(\int_0^T \int_\Gamma (1-\tau)u^* + \tau y^* \, dsdt + \zeta \int_Q y^* \, d\mathbf{x}dt - C_I \right) = 0. \end{cases} \tag{2.10}$$

2.2. Augmentation of the integral constraint

In our approach we handle the integral constraint (2.5) by an augmented Lagrangian penalization. See, e.g., [5, 6] for a detailed account of this technique. In fact, for $\varrho > 0$ and $\hat{\xi} \in \mathbb{R}_0^+$ we introduce the modified cost functional

$$J_{\hat{\xi}}^\varrho(y, u) = J(y, u) + \frac{1}{2\varrho} \max\left\{0, \hat{\xi} + \varrho\left(\int_0^T \int_\Gamma (1-\tau)u + \tau y \, dsdt \right.\right.$$
$$\left.\left. + \zeta \int_Q y \, d\mathbf{x}dt - C_I \right)\right\}^2.$$

Then, we consider the optimal control problem

$$\min J_{\hat{\xi}}^\varrho(x) \quad \text{s.t.} \quad e(x) = 0, \ y(0) = y_\circ \text{ and } x = (y, u) \in Y \times U_{\text{ad}} \tag{$\mathbf{P}^\varrho$}$$

for fixed $\varrho > 0$ and $\hat{\xi} \in \mathbb{R}_+$ instead of $(\mathbf{P})$. Using analogous arguments as for $(\mathbf{P})$ one can prove that $(\mathbf{P}^\varrho)$ has at least one global solution for arbitrarily $\varrho > 0$ and $\hat{\xi} \in \mathbb{R}_0^+$. Notice that $(\mathbf{P}^\varrho)$ does not involve the integral constraint explicitly. Rather this constraint is realized by adding an augmented Lagrangian-type penalty term to the original objective function. It is well known that the augmented Lagrangian penalization is exact for sufficiently large $\rho > 0$, i.e., a local solution $x^* \in Y \times U_{\text{ad}}$ to $(\mathbf{P})$ is also a local solution to $(\mathbf{P}^\varrho)$.

The corresponding augmented Lagrange function

$$\mathcal{L}_{\hat{\xi}}^\varrho : X \times L^2(0, T; H^1(\Omega)) \to \mathbb{R}$$

associated with $(\mathbf{P}^\varrho)$ is

$$\mathcal{L}_{\hat{\xi}}^\varrho(x, p) = J_{\hat{\xi}}^\varrho(x) + \langle e(x), p \rangle_{L^2(0,T;H^1(\Omega)'), L^2(0,T;H^1(\Omega))}$$

for $x = (y, u) \in X$, $p \in L^2(0, T; H^1(\Omega))$, $\varrho > 0$, and $\hat{\xi} \in \mathbb{R}_0^+$.

Suppose that $x^\varrho = (y^\varrho, u^\varrho) \in X$ denotes a solution to $(\mathbf{P}^\varrho)$ for fixed $\varrho > 0$ and $\hat{\xi} \in \mathbb{R}_0^+$. The first order necessary optimality conditions of $(\mathbf{P}^\varrho)$ involve the dual system

$$-p_t^\varrho - \alpha \Delta p^\varrho = -\alpha_Q(y^\varrho - z_Q) - \zeta \xi^\varrho \qquad \text{in } Q, \qquad (2.11\text{a})$$

$$\frac{\partial p^\varrho}{\partial n} + b'(y^\varrho(\cdot))p^\varrho + \tau \xi^\varrho = 0 \qquad \text{on } \Sigma, \qquad (2.11\text{b})$$

$$p^\varrho(T) = -\alpha_\Omega(y^\varrho(T) - z_\Omega) \qquad \text{in } \Omega, \qquad (2.11\text{c})$$

$$\kappa u^o - \alpha\beta p^\varrho + (1 - \tau)\xi^\varrho + \lambda^\varrho = 0 \qquad \text{on } \Sigma, \qquad (2.11\text{d})$$

$$\lambda^\varrho = \max\left(0, \lambda^\varrho + \sigma(u^\varrho - u_b)\right) + \min\left(0, \lambda^\varrho + \sigma(u^\varrho - u_a)\right) \qquad \text{on } \Sigma, \qquad (2.11\text{e})$$

$$\xi^\varrho = \max\left(0, \hat{\xi} + \varrho\left(\int_0^T \int_\Gamma (1 - \tau)u^\varrho + \tau y^\varrho \, dsdt \right.\right. \qquad (2.11\text{f})$$

$$\left.\left. + \zeta \int_Q y^\varrho \, \mathbf{dx}dt - C_I\right)\right) \qquad \text{in } \mathbb{R}.$$

In contrast to (2.7f) the scalar ξ^ϱ is now given explicitly by (2.11f) since $\hat{\xi}$ is fixed.

Let us review the augmented Lagrangian algorithm, which can be interpreted as a combination of penalty functions and local duality methods.

Algorithm 1 (Augmented Lagrangian algorithm).

1. Choose a starting value $\xi^0 \in \mathbb{R}_+$ for the Lagrange multiplier associated to the integral constraint (2.5), the initial parameter $\varrho_0 > 0$ for the augmentation, a factor $\beta^\varrho > 1$ and a stopping criterion; set $n = 0$.
2. Determine a (local) solution $x^{n+1} = (y^{n+1}, u^{n+1}) \in Y \times U_{\text{ad}}$ of $(\mathbf{P}^\varrho)$ with $\rho = \rho_n$ and $\hat{\xi} = \xi^n$.
3. Set

$$\xi^{n+1} = \max\left(0, \xi^n + \varrho_n\left(\int_0^T \int_\Gamma (1 - \tau)u^{n+1} + \tau y^{n+1} \, dsdt + \right.\right.$$

$$\left.\left. \zeta \int_Q y^{n+1}\mathbf{dx}dt - C_I\right)\right).$$

4. Unless the stopping rule is satisfied, set $\varrho_{n+1} = \beta^\varrho \varrho_n$, $n = n+1$, and continue with step (2).

Remark 2.2.

1. Other augmentation rules for the parameter ϱ than the one realized in step (4) can be found, e.g., in [6, p. 405].
2. In the process of solving $(\mathbf{P})$ the augmented Lagrangian algorithm acts as the outer iteration of our whole optimization method, whereas at each level of Algorithm 1 the solution of $(\mathbf{P}^\varrho)$ is computed by an inner iteration method. In section 3 we propose variants of semismooth Newton techniques for solving $(\mathbf{P}^\varrho)$ with $\rho = \rho_n$. $\diamond$

2.3. The reduced optimal control problem

Since for any $u \in L^{\sigma_1}(\Sigma)$ there exists a unique solution $y \in Y$ to the state equation (2.2), we can define the nonlinear solution operator

$$\mathcal{S} : L^{\sigma_1}(\Sigma) \to Y$$

by $y = \mathcal{S}(u)$ for $u \in L^{\sigma_1}(\Sigma)$. Introducing the so-called reduced cost functional

$$\hat{J}^{\varrho}_{\hat{\xi}}(u) = J^{\varrho}_{\hat{\xi}}(\mathcal{S}(u), u),$$

the problem ($\mathbf{P}^{\varrho}$) can be equivalently expressed as

$$\min \hat{J}^{\varrho}_{\hat{\xi}}(u) \quad \text{s.t.} \quad u \in U_{\mathrm{ad}} \tag{$\hat{\mathbf{P}}^{\varrho}$}$$

with $\varrho > 0$ and $\hat{\xi} \in \mathbb{R}^+_0$ fixed. Notice that ($\hat{\mathbf{P}}^{\varrho}$) is a minimization problem with bilateral control constraints, but with no equality constraints. For accessing the gradient of $\hat{J}^{\varrho}_{\hat{\xi}}$ we have to guarantee differentiability of $\mathcal{S}(u)$ with respect to u. This is the content of Theorem 2.3. For its proof we refer to [24, Theorem 5.10].

Theorem 2.3. *The solution operator $\mathcal{S}$ of the state equation is continuously differentiable as a mapping from $L^{\sigma_1}(\Sigma)$ to Y. The action of the derivative $\mathcal{S}'(u)$ (we also write $y'(u)$) on some $v \in L^{\sigma_1}(\Sigma)$, i.e., $\mathcal{S}'(u)v = w$, is characterized by the solution w to the initial-boundary value problem*

$$w_t - \alpha \Delta w = 0 \quad in \ Q, \tag{2.12}$$

$$\frac{\partial w}{\partial n} + b'(\mathcal{S}(u))w = \beta v \quad in \ \Sigma, \tag{2.13}$$

$$w(0) = 0. \tag{2.14}$$

We obtain an analogous differentiability result for the adjoint state p^{ϱ} satisfying (2.11a)–(2.11c) and considered as a function of u^{ϱ} and ξ^{ϱ}.

The derivative of $\hat{J}^{\varrho}_{\hat{\xi}}$ at a point $u \in L^{\sigma_1}(\Sigma)$ is represented by

$$(\hat{J}^{\varrho}_{\hat{\xi}})'(u) = \frac{\partial J^{\varrho}_{\hat{\xi}}(\mathcal{S}(u), u)}{\partial y}\mathcal{S}'(u) + \frac{\partial J^{\varrho}_{\hat{\xi}}(\mathcal{S}(u), u)}{\partial u} = -\alpha\beta p + \kappa u + (1 - \tau)\xi \quad \text{on } \Sigma$$

(compare (2.11d)), where $p \in W(0, T)$ solves the adjoint system (2.11a)–(2.11c) for the state $y = \mathcal{S}(u)$. Further, ξ satisfies

$$\xi = \max\left(0, \hat{\xi} + \varrho\left(\int_0^T \int_\Gamma (1 - \tau)u + \tau y \, dsdt + \zeta \int_Q y \, dxdt - C_I\right)\right). \tag{2.15}$$

Recall that the quantity $\hat{\xi}$ is fixed by the outer iteration, i.e., the augmented Lagrangian algorithm. From (2.11) we derive that the first order necessary optimality condition

$$\langle (\hat{J}^{\varrho}_{\hat{\xi}})'(u^{\varrho}), u - u^{\varrho} \rangle_{L^2(\Sigma)} \geq 0 \quad \text{for all } u \in U_{\mathrm{ad}}$$

for ($\hat{\mathbf{P}}^{\varrho}$) is equivalent to

$$\kappa u^{\varrho} - \alpha\beta p^{\varrho} + (1 - \tau)\xi^{\varrho} + \lambda^{\varrho} = 0 \quad \text{on } \Sigma, \tag{2.16a}$$

where the Lagrange multiplier λ^ϱ associated with the bilateral control constraints satisfies

$$\lambda^\varrho = \max\left(0, \lambda^\varrho + \sigma(u^\varrho - u_b)\right) + \min\left(0, \lambda^\varrho + \sigma(u^\varrho - u_a)\right) \quad \text{on } \Sigma \qquad (2.16b)$$

for some arbitrarily fixed $\sigma > 0$. Thus, the first order necessary optimality conditions for $(\hat{\mathbf{P}}^\varrho)$ are given by (2.16) together with (2.11a)–(2.11c) and (2.15) with $\xi = \xi^\varrho$ and $u = u^\varrho$, $y = \mathcal{S}(u^\varrho)$.

3. Solution methods for $(\mathbf{P}^\varrho)$

In this section we describe different strategies for solving $(\mathbf{P}^\varrho)$, which is a nonlinear optimal control problem with bilateral pointwise control constraints. The efficient solution of $(\mathbf{P}^\varrho)$ is the key part in our augmented Lagrangian method. Here, we explore several variants of semismooth Newton techniques and compare them with a (reduced) sequential quadratic programming (SQP) method.

3.1. Partial semismooth Newton methods

Our first method aims at solving $(\mathbf{P}^\varrho)$ by a Newton-type algorithm applied to the first order necessary optimality conditions of either the full problem or its reduced variant $(\hat{\mathbf{P}}^\varrho)$. In this context, the non-differentiability of the min- and max-functions in (2.16b) complicates the treatment. In order to cope with this difficulty, we recall the notion of Newton differentiability, which holds true in finite as well as in infinite-dimensional spaces; see, e.g., [8, 18]. It generalizes the classical Fréchet differentiability concept, and it allows to formulate a generalized variant of Newton's method for solving $(\mathbf{P}^\varrho)$, respectively $(\hat{\mathbf{P}}^\varrho)$.

Definition 3.1. Let V, W be two Banach spaces, $S \subset V$ a nonempty open set, $F : S \to W$ a given mapping, and $v^* \in S$. If there exists a neighborhood $N(v^*) \subset S$ and a family of mappings $G : N(v^*) \to L(V, W)$ such that

$$\lim_{\|h\|_V \downarrow 0} \frac{1}{\|h\|_V} \|F(v^* + h) - F(v^*) - G(v^* + h)(h)\|_W = 0, \qquad (3.1)$$

then F is called *Newton-differentiable at* v^*, and $G(v^*)$ is said to be a *generalized derivative (or Newton map) for* F *at* v^*. Here, $L(V, W)$ denotes the Banach space of all bounded and linear operators from V to W endowed with the common norm.

Remark 3.2. The function $\max : L^p(\Sigma) \to L^q(\Sigma)$ is Newton differentiable for $1 \leq q < p \leq \infty$ (see [18]). If $F : L^r(\Sigma) \to L^p(\Sigma)$ is Fréchet differentiable for some $1 \leq r \leq \infty$, then the function

$$(t, s) \mapsto \chi_{\mathcal{A}}(t, s) \cdot \nabla F(u(t, s)), \quad (t, s) \in \Sigma, \qquad (3.2)$$

is a generalized derivative of $\max(0, F(\cdot)) : L^r(\Sigma) \to L^q(\Sigma)$. Here, $\chi_{\mathcal{A}}$ denotes the characteristic function of the set $\mathcal{A} \subset \Sigma$, where $F(u(\cdot))$ is positive, i.e., $\chi_{\mathcal{A}}(t, s) = 1$ if $F(u(t, s)) > 0$ and $\chi_{\mathcal{A}}(t, s) = 0$ otherwise. From $\min(0, F(\cdot)) = -\max(0, -F(\cdot))$, we see that an analogous differentiation formula holds true for the min-function. $\diamond$

Using this concept of generalized differentiability, we can formulate semis-mooth Newton variants for the solution of $(\mathbf{P}^\varrho)$. In this section we realize a *partial* semismooth Newton method; compare [20] for a related approach. This method operates with a linearization of the nonlinearity due to the min- and max-operations, whereas $u \mapsto \mathcal{S}(u)$ is kept nonlinear. Choosing $\sigma = \kappa$ in (2.16b), a prerequisite for proving locally superlinear rate of convergence of semismooth Newton methods [18], and taking into account (2.16a), we find

$$- \kappa u^\varrho + \alpha\beta p^\varrho + (\tau - 1)\xi^\varrho - \max(0, \alpha\beta p^\varrho + (\tau - 1)\xi^\varrho - \kappa u_b)$$
$$- \min(0, \alpha\beta p^\varrho + (\tau - 1)\xi^\varrho - \kappa u_a) = 0. \tag{3.3}$$

Now suppose $(u, \xi) \in L^{\sigma_1}(\Sigma) \times \mathbb{R}$ is some given approximation of (u^ϱ, ξ^ϱ). Let $\mathcal{S}(u)$ be the corresponding state, and let $p(u, \xi)$ be the pertinent adjoint state satisfying (2.11a)–(2.11c) with (y^ϱ, ξ^ϱ) replaced by $(\mathcal{S}(u), \xi)$. Further assume that $\lambda = \lambda(u)$ satisfies (2.11d) with $(u^\varrho, p^\varrho, \xi^\varrho)$ replaced by $(u, p(u, \xi), \xi)$. Then, since $p(u, \xi)|_\Sigma$ is continuously differentiable from $L^{\sigma_1}(\Sigma) \times \mathbb{R}$ to $L^{\sigma_3}(\Sigma)$ with $\sigma_3 > \sigma_1$, Remark 3.2 provides Newton differentiability of the min- and max-terms in (3.3), respectively. Using (3.2) and defining

$$\begin{aligned}
\mathcal{A}_b &= \big\{(t, s) \in \Sigma \,\big|\, \lambda + \kappa(u - u_b) > 0 \text{ a.e.}\big\}, \\
\mathcal{A}_a &= \big\{(t, s) \in \Sigma \,\big|\, \lambda + \kappa(u - u_a) < 0 \text{ a.e.}\big\}, \\
\mathcal{I} &= \Sigma \setminus \mathcal{A}, \quad \text{with } \mathcal{A} = \mathcal{A}_b \cup \mathcal{A}_a,
\end{aligned} \tag{3.4}$$

we obtain the following linearization of (3.3) at $(u, p(u, \xi), \xi)$ with respect to the independent variables u and ξ:

$$\begin{aligned}
&- \kappa(u + \delta u) + \alpha\beta\,(p(u, \xi) + p_u(u, \xi)\delta u + p_\xi(u, \xi)\delta\xi) - (1 - \tau)(\xi + \delta\xi) \\
&- \chi_{\mathcal{A}_b}\,(\alpha\beta(p(u, \xi) + p_u(u, \xi)\delta u + p_\xi(u, \xi)\delta\xi) - (1-\tau)(\xi + \delta\xi) - \kappa u_b) \\
&- \chi_{\mathcal{A}_a}\,(\alpha\beta(p(u, \xi) + p_u(u, \xi)\delta u + p_\xi(u, \xi)\delta\xi) - (1-\tau)(\xi + \delta\xi) - \kappa u_a) = 0.
\end{aligned} \tag{3.5}$$

Here $\delta u \in L^{\sigma_1}(\Sigma)$ and $\delta\xi \in \mathbb{R}$ represent the increments. A closer look reveals:

$$u^+ = a \text{ on } \mathcal{A}_a, \tag{3.6a}$$

$$u^+ = b \text{ on } \mathcal{A}_b, \tag{3.6b}$$

$$-\kappa u^+ + \alpha\beta\,(p(u, \xi) + p_u(u, \xi)\delta u + p_\xi(u, \xi)\delta\xi) - (1 - \tau)\xi^+ = 0 \text{ on } \mathcal{I}, \tag{3.6c}$$

where $u^+ = u + \delta u$ and $\xi^+ = \xi + \delta\xi$. Note that (3.6c) can be viewed as

$$\lambda + \delta\lambda = 0 \text{ on } \mathcal{I}, \tag{3.6d}$$

with

$$\delta\lambda = -\kappa\delta u + \alpha\beta\,(p_u(u, \xi)\delta u + p_\xi(u, \xi)\delta\xi) + (\tau-1)\delta\xi.$$

Define

$$\gamma_{\hat{\xi}}^\varrho(y, u) = \hat{\xi} + \rho\left(\int_0^T \int_\Gamma (1 - \tau)u + \tau y(u)\,\mathrm{d}s\mathrm{d}t + \zeta \int_Q y\,\mathrm{d}\mathbf{x}\mathrm{d}t - C_I\right).$$

The linearization of (2.15) in the generalized sense yields

$$\xi^+ = 0 \quad \text{if} \quad \gamma_{\hat{\xi}}^{\varrho}(y, u) \leq 0, \tag{3.7a}$$

and

$$\xi^+ = \hat{\xi} + \rho\left(\int_0^T \int_\Gamma (1 - \tau)u^+ + \tau(y(u) + y'(u)\delta u)\, ds dt \right. \tag{3.7b}$$

$$\left. + \zeta \int_Q y(u) + y'(u)\delta u\, d\mathbf{x}dt - C_I \right) \quad \text{else.}$$

The active respectively inactive set behavior of the variables in (3.6) motivates the following algorithm.

Algorithm 2 (Partial semismooth Newton (pSSN) method).

(0) Inputs are $\varrho > 0$ and $\hat{\xi} \in \mathbb{R}$ (from Algorithm 1).

1. Choose starting values λ_0^ϱ and u_0^ϱ, compute $y_0^\varrho = \mathcal{S}(u_0^\varrho)$ and $(p_0^\varrho, \xi_0^\varrho)$ satisfying (2.11a)–(2.11d). Set $k := 0$.

2. Unless some stopping rule is satisfied, determine the active and inactive sets

$$\mathcal{A}_a^k = \left\{ (t, s) \in \Sigma \,\middle|\, \lambda_k^\varrho + \kappa(u_k^\varrho - u_a) < 0 \text{ a.e.} \right\},$$

$$\mathcal{A}_b^k = \left\{ (t, s) \in \Sigma \,\middle|\, \lambda_k^\varrho + \kappa(u_k^\varrho - u_b) > 0 \text{ a.e.} \right\},$$

$$\mathcal{I}^k = \Sigma \setminus \mathcal{A}^k, \quad \text{with } \mathcal{A}^k = \mathcal{A}_b^k \cup \mathcal{A}_a^k.$$

3. Compute a local solution $x_k^\varrho = (y_k^\varrho, u_k^\varrho)$ with pertinent multiplier λ_k^ϱ and adjoint state p_k^ϱ of

$$\begin{cases} \min\limits_{\hat{\xi}} \quad J_{\hat{\xi}}^\varrho(x) \quad \text{over} \quad x = (y, u) \in Y \times L^{\sigma_1}(\Sigma) \\ \text{s.t.} \quad e(x) = 0,\ y(0) = y_\circ, \\ \qquad u = u_a \text{ on } \mathcal{A}_a^k, \\ \qquad u = u_b \text{ on } \mathcal{A}_b^k. \end{cases} \tag{3.8}$$

Set $k = k + 1$, and return to (2).

In our numerics Section 4 we report on two variants of Algorithm 2 which differ in the way how the subproblems (3.8) are solved:

pSSN-SQP. This variant achieves the solution of the nonlinearly constrained problem (3.8) by a SQP method. In every iteration of the SQP iteration a quadratic program of the type

$$\begin{cases} \min \quad \nabla J_{\hat{\xi}}^\varrho(x_{k-1}^\varrho)\delta x + \frac{1}{2}\langle H_{\hat{\xi}}^{\varrho,k-1}\delta x, \delta x\rangle \quad \text{over } \delta x = (\delta y, \delta u) \\ \text{s.t.} \quad e(x_{k-1}^\varrho) + \nabla e(x_{k-1}^\varrho)\delta x = 0,\ \delta y(0) = 0, \\ \qquad u_{k-1}^\varrho + \delta u = u_a \text{ on } \mathcal{A}_a^k, \\ \qquad u_{k-1}^\varrho + \delta u = u_b \text{ on } \mathcal{A}_b^k. \end{cases} \tag{3.9}$$

has to be solved. Above, $H_{\hat{\xi}}^\varrho$ denotes the generalized Hessian of $\mathcal{L}_{\hat{\xi}}^\varrho$, the Lagrange function of

$$\min \quad J_{\hat{\xi}}^\varrho(x) \quad \text{s.t.} \quad e(x) = 0$$

defined by

$$\mathcal{L}_{\xi}^{\varrho}(x,p) = J_{\xi}^{\varrho}(x) + \langle e(x),p\rangle_{L^2(0,T;H^1(\Omega)'),L^2(0,T;H^1(\Omega))}.$$

The use of the generalized derivative in the computation of the Hessian is necessitated by the augmented Lagrangian penalty term, which is only once continuously Fréchet differentiable. Due to the integration process inside the max-operation of the penalty term, the non-differentiability is on the level of the reals only. Hence, standard non-smooth theory (see [9]) applies when differentiating the objective function in $(\mathbf{P}^{\varrho})$ and when analyzing the algorithm. In fact, the SQP method for computing a solution of (3.8) can be interpreted as a semismooth Newton method for solving the firstorder optimality system of (3.8). In our implementation we use the following generalized derivative of $\max(0,\cdot) : \mathbb{R} \to \mathbb{R}_0^+$:

$$g_{\max}(z) = 0 \text{ if } z \leq 0, \text{ and } g_{\max}(z) = 1 \text{ else.}$$

With this definition, it is easy to check that $F : L^{\sigma_1}(\Sigma) \to \mathbb{R}_0^+$,

$$F(u) = \max\left(0, \int_0^T \int_\Gamma (1-\tau)u + \tau y\,ds dt + \zeta \int_Q y\,d\mathbf{x}dt - C_I\right)$$

satisfies (3.1). Hence, from [18, Theorem 1.1] it follows that the SQP method with subproblems (3.9) converges at a superlinear rate, provided that $(H_\xi^\varrho)^{-1}$ exists and is bounded in a neighborhood of a (local) solution x_k^ϱ and the SQP starting point $(x_{k,0}^\varrho, p_{k,0}^\varrho, \lambda_{k,0}^\varrho)$ is sufficiently close to $(x_k^\varrho, p_k^\varrho, \lambda_k^\varrho)$.

pSSN-Newton. In our second approach, (3.8) is solved by a reduction process and a subsequent application of a the semismooth Newton method: First, the state equation is used to obtain $y = y(u) = \mathcal{S}(u)$. This allows us to consider the reduced problem

$$\min \quad \hat{J}_\xi^\varrho(u) \quad \text{s.t.} \quad u = u_a \text{ on } \mathcal{A}_a^k \quad \text{and} \quad u = u_b \text{ on } \mathcal{A}_b^k \tag{3.10}$$

instead of (3.8). Its first order necessary optimality conditions are given by

$$\lambda = 0 \quad \text{on } \mathcal{I}^k, \tag{3.11a}$$

$$u = u_b \quad \text{on } \mathcal{A}_b^k, \tag{3.11b}$$

$$u = u_a \quad \text{on } \mathcal{A}_a^k, \tag{3.11c}$$

$$y_t - \alpha\Delta y = f \quad \text{in } Q, \tag{3.11d}$$

$$\frac{\partial y}{\partial n} + b(y(\cdot)) = \beta u \quad \text{on } \Sigma, \tag{3.11e}$$

$$y(0) = y_\circ \quad \text{in } \Omega, \tag{3.11f}$$

$$-p_t - \alpha\Delta p = -\alpha_Q(y - z_Q) - \zeta\xi \quad \text{in } Q, \tag{3.11g}$$

$$\frac{\partial p}{\partial n} + b'(y)p + \tau\xi = 0 \quad \text{on } \Sigma, \tag{3.11h}$$

$$p(T) = -\alpha_\Omega(y(T) - z_\Omega) \quad \text{in } \Omega, \tag{3.11i}$$

$$\kappa u - \alpha\beta p + (1-\tau)\xi + \lambda = 0 \quad \text{on } \Sigma, \tag{3.11j}$$

$$\xi = \max\left(0, \hat{\xi} + \rho\left(\int_0^T\int_\Gamma (1-\tau)u + \tau y \, dsdt \right.\right. \tag{3.11k}$$
$$\left.\left. + \zeta\int_Q y \, \mathbf{dx}dt - C_I\right)\right).$$

As for pSSN-SQP, we obtain generalized differentiability of the max-operation in the last equation of (3.11). Hence, we can again apply a semismooth Newton method for computing a solution to (3.11). Also, we obtain a locally superlinear rate of convergence, provided that the generalized Jacobian of the system has a bounded inverse in a neighborhood of a solution to (3.10).

3.2. The semi-smooth Newton method

Next we turn to our second class of methods, which also linearizes the nonlinear mapping $u \mapsto \mathcal{S}(u)$ when computing the new iterate u^ϱ_{k+1}. Again, we study two variants, one operating on the full first order system (2.11) and the other one utilizing a reduction process similar to the one in Section 3.1.

SSN. We first describe the full semismooth Newton method. For this purpose recall that the choice $\sigma = \kappa$ and inserting (2.16a) in (2.16b) yield (3.3). Hence, the first order system (2.11) is equivalent to

$$y^\varrho_t - \alpha\Delta y^\varrho = f \qquad\qquad\qquad\qquad \text{in } Q, \tag{3.12a}$$

$$\frac{\partial y^\varrho}{\partial n} + b(y^\varrho) = \beta u^\varrho \qquad\qquad\qquad \text{on } \Sigma, \tag{3.12b}$$

$$y^\varrho(0) = y_\circ \qquad\qquad\qquad\qquad\qquad \text{in } \Omega, \tag{3.12c}$$

$$-p^\varrho_t - \alpha\Delta p^\varrho = -\alpha_Q(y^\varrho - z_Q) - \zeta\xi^\varrho \qquad \text{in } Q, \tag{3.12d}$$

$$\frac{\partial p^\varrho}{\partial n} + b'(y^\varrho)p^\varrho + \tau\xi^\varrho = 0 \qquad\qquad \text{on } \Sigma, \tag{3.12e}$$

$$p^\varrho(T) = -\alpha_\Omega(y^\varrho(T) - z_\Omega) \qquad\qquad \text{in } \Omega, \tag{3.12f}$$

$$-\kappa u^\varrho + \alpha\beta p^\varrho + (\tau - 1)\xi^\varrho - \max(0, \alpha\beta p^\varrho + (\tau-1)\xi^\varrho - \kappa u_b) \tag{3.12g}$$
$$- \min(0, \alpha\beta p^\varrho + (\tau-1)\xi^\varrho - \kappa u_a) = 0, \qquad\qquad \text{on } \Sigma,$$

$$\xi^\varrho = \max\left(0, \hat{\xi} + \varrho\left(\int_0^T\int_\Gamma (1-\tau)u^\varrho + \tau y^\varrho \, dsdt \right.\right. \tag{3.12h}$$
$$\left.\left. + \zeta\int_Q y^\varrho \mathbf{dx}dt - C_I\right)\right) \qquad\qquad \text{in } \mathbb{R},$$

Let $(x^\varrho_k, \xi^\varrho_k) \in X \times \mathbb{R}$ be some guess of $(x^\varrho_*, \xi^\varrho_*)$ with y^ϱ_k satisfying either (3.12a)–(3.12c) at u^ϱ_k or a linearization thereof at some reference point. Note that in either case $y_k^\varrho = y_k^\varrho(u_k^\varrho)$ is differentiable as a mapping from $L^{\sigma_1}(\Sigma)$ to $W(0,T) \cap C(\overline{Q})$. Further let p_k^ϱ satisfy (3.12d)–(3.12f) with (y^ϱ, ξ^ϱ) replaced by $(y_k^\varrho, \xi_k^\varrho)$. Then the max- and min-expressions in the corresponding equation (3.12g) are Newton-differentiable (consider p_k^ϱ as a function of $(u_k^\varrho, \xi_k^\varrho)$). Further, (3.12h) is generalized differentiable as it was the case in Section 3.1. Hence, the semismooth Newton system, which aims at finding a zero of the generalized linearization of the nonlinear system (3.12) at $(x_k^\varrho, p_k^\varrho, \xi_k^\varrho)$, is well defined. From the linearization of (3.12g) we obtain

$$u_k^\varrho + \delta u_k^\varrho = u_a \ \text{ on } \mathcal{A}_a^k \quad \text{and} \quad u_k^\varrho + \delta u_k^\varrho = u_b \ \text{ on } \mathcal{A}_b^k, \tag{3.13}$$

where δu_k^ϱ denotes the semismooth Newton update direction for u_k^ϱ and $\mathcal{A}^k = \mathcal{A}_a^k \cup \mathcal{A}_b^k$ is determined as in step (2) of Algorithm 2 with λ_k^ϱ given by (2.11d) at $(u_k^\varrho, p_k^\varrho, \xi_k^\varrho)$. Summarizing, the linear system which has to be solved in every step of the (full) semismooth Newton method is given by

$$\lambda_k^\varrho + \delta\lambda = 0 \quad \text{on } \mathcal{I}^k, \tag{3.14a}$$

$$\delta u = u_a - u_k^\varrho \quad \text{on } \mathcal{A}_a^k, \tag{3.14b}$$

$$\delta u = u_b - u_k^\varrho \quad \text{on } \mathcal{A}_b^k, \tag{3.14c}$$

$$\delta y_t - \alpha\Delta\delta y = 0 \quad \text{in } Q, \tag{3.14d}$$

$$\frac{\partial\delta y}{\partial n} + b'(y_k^\varrho)\delta y - \beta\delta u = -\left(\frac{\partial y_k^\varrho}{\partial n} + b(y_k^\varrho) - \beta u_k^\varrho\right) \quad \text{on } \Sigma, \tag{3.14e}$$

$$\delta y(0) = y_\circ - y_k^\varrho(0) \quad \text{in } \Omega, \tag{3.14f}$$

$$-\delta p_t - \alpha\Delta\delta p + \alpha_Q\delta y + \zeta\delta\xi = 0 \quad \text{in } Q, \tag{3.14g}$$

$$\frac{\partial\delta p}{\partial n} + b'(y_k^\varrho)\delta p + \langle b''(y_k^\varrho)p_k^\varrho, \delta y\rangle + \tau\delta\xi = -\left(\frac{\partial p_k^\varrho}{\partial n} + b'(y_k^\varrho)p_k^\varrho + \tau\xi_k^\varrho\right) \quad \text{on } \Sigma, \tag{3.14h}$$

$$\delta p(T) + \alpha_\Omega\delta y(T) = 0 \quad \text{in } \Omega, \tag{3.14i}$$

$$\kappa\delta u - \alpha\beta\delta p + (1-\tau)\delta\xi + \delta\lambda = 0 \quad \text{on } \Sigma, \tag{3.14j}$$

$$\delta\xi = -\xi_k^\varrho \text{ if } \gamma_k^\varrho \le 0, \tag{3.14k}$$

$$\delta\xi = -\xi_k + \gamma_k^\varrho \tag{3.14l}$$

$$+ \varrho\left(\int_0^T \int_\Gamma (1-\tau)\delta u + \tau\delta y \, ds dt + \zeta\int_Q \delta y \, d\mathbf{x}dt\right) \text{ if } \gamma_k^\varrho > 0.$$

Above, γ_k^ϱ is given by

$$\gamma_k^\varrho = \hat{\xi} + \varrho\int_0^T \int_\Gamma (1-\tau)u_k^\varrho + \tau y_k^\varrho \, ds dt + \zeta\int_Q y_k^\varrho \, d\mathbf{x}dt - C_I.$$

The next iterates are then defined by $x_{k+1}^\varrho = x_k^\varrho + \delta x_k^\varrho$, $p_{k+1}^\varrho = p_k^\varrho + \delta p_k^\varrho$, $\xi_{k+1}^\varrho = \xi_k^\varrho + \delta \xi_k^\varrho$, and $\lambda_{k+1}^\varrho = \lambda_k^\varrho + \delta \lambda_k^\varrho$. The relations (3.13) and $\kappa \delta u = \alpha \beta \delta p$ on $\mathcal{I}^k$, as well as regularity theory for parabolic equations yield that $u_{k+1}^\varrho \in L^{\sigma_1}(\Sigma)$ provided that $u_k^\varrho \in L^{\sigma_1}(\Sigma)$. Further, from parabolic regularity theory we obtain $p_{k+1}^\varrho = p_{k+1}^\varrho(u_{k+1}^\varrho, \xi_{k+1}^\varrho) \in W(0,T) \cap C(\overline{Q})$ for $z_Q \in C(\overline{Q})$. Hence, the left-hand side of (3.12g) is Newton differentiable at $(u_{k+1}^\varrho, \xi_{k+1}^\varrho)$, and the semismooth Newton system is again well defined. As a consequence, under a bounded invertibility assumption in a neighborhood of a solution to (3.12) the resulting semismooth Newton method converges at a superlinear rate provided that the initial point x_0^ϱ is sufficiently regular, and p_0^ϱ satisfies the adjoint system for given $\xi_0^\varrho \in \mathbb{R}$.

We summarize the full semismooth Newton method in the following algorithm.

Algorithm 3 (Semismooth Newton method).

(0) Inputs are $\varrho > 0$ and $\hat{\xi} \in \mathbb{R}$ (from Algorithm 1).
1. Choose starting values u_0^ϱ, λ_0^ϱ, and ξ_0^ϱ and compute $y_0^\varrho = \mathcal{S}(u_0^\varrho)$, and p_0^ϱ by solving the adjoint system (2.11a)–(2.11c). Set $k := 0$.
2. Unless some stopping rule is satisfied, determine the active and inactive sets

$$\mathcal{A}_a^k = \left\{ (t,s) \in \Sigma \,\middle|\, \lambda_k^\varrho + \kappa(u_k^\varrho - u_b) > 0 \text{ a.e.} \right\},$$
$$\mathcal{A}_b^k = \left\{ (t,s) \in \Sigma \,\middle|\, \lambda_k^\varrho + \kappa(u_k^\varrho - u_a) < 0 \text{ a.e.} \right\},$$
$$\mathcal{I}^k = \Sigma \setminus \mathcal{A}^k, \quad \mathcal{A}^k = \mathcal{A}_a^k \cup \mathcal{A}_b^k.$$

3. Solve the Newton system (3.14), and compute the new iterates

$$(x_{k+1}^\varrho, p_{k+1}^\varrho, \lambda_{k+1}^\varrho, \xi_{k+1}^\varrho) = (x_k^\varrho, p_k^\varrho, \lambda_k^\varrho, \xi_k^\varrho) + (\delta x_k^\varrho, \delta p_k^\varrho, \delta \lambda_k^\varrho, \delta \xi_k^\varrho).$$

Set $k := k + 1$, and return to (2).

SSN-Newton. The second variant performs first a reduction by expressing y as $y(u) = \mathcal{S}(u)$ through the solution of the state equation, and then it utilizes a semismooth Newton method for the solution of the resulting minimization problem. Note that in contrast to SSN for SSN-Newton the nonlinear state equation is always satisfied. Thus, in every step of SSN-Newton the boundary condition of the state equation reads

$$\frac{\partial \delta y}{\partial n} + b'(y_k^\varrho)\delta y - \beta \delta u = 0. \tag{3.15}$$

The rest of (3.14) remains valid. When computing the next iterates, one first updates u_k^ϱ, λ_k^ϱ, and ξ_k^ϱ as above, and then computes $y_{k+1}^\varrho = \mathcal{S}(u_{k+1}^\varrho)$ and p_{k+1}^ϱ by solving the adjoint system at $(y_{k+1}^\varrho, \xi_{k+1}^\varrho)$. Arguments similar to the ones given for SSN yield a locally superlinear rate of convergence of SSN-Newton.

3.3. Reduced SQP methods

In order to compare the methods in Sections 3.1 and 3.2 with standard optimization methods for nonlinear problems, in Section 4 we also report on results obtained by a reduced sequential quadratic programming (rSQP) algorithm, where the reduction onto the control space takes place when solving the QP-subproblems. For details on rSQP methods we refer the reader to, e.g., [22]. In our case, in every iteration of the rSQP method we have to solve a linear-quadratic optimal control problem with bilateral control constraints. Its solution is computed by a variant of the primal-dual active set strategy, which is based on a generalized Moreau-Yosida approximation of the indicator function of the set of admissible controls. The method was developed in [3] and extended in [16]. Its local convergence is q-superlinear (the analysis is similar to the one in [17]), and it exhibits a mesh independent behavior. The latter aspect can be made rigorous following the analysis in [19] and is the subject of a subsequent paper.

3.4. Globalization strategy

The globalization of our SQP method is achieved by applying a line search strategy. It is realized with the backtracking Armijo-rule for the L^1-merit functional

$$\Phi(x; \mu_1, \mu_2) = J^\varrho_\xi(x) + \mu_1 \|e(x)\|_{L^2(0,T;H^1(\Omega)')}$$
$$+ \mu_2 \left(\|\min(0, u - u_a)\|_{L^2(\Sigma)} + \|\max(0, u - u_b)\|_{L^2(\Sigma)}\right)$$

with penalty parameters $\mu_1, \mu_2 > 0$, see, e.g., [22]. In Section 4 we also report on line search results for the augmented Lagrangian functional

$$\Psi(x, p, \lambda_a, \lambda_b; c) = J^\varrho_\xi(x) + \langle e(x), p \rangle_{L^2(0,T;H^1(\Omega)'), L^2(0,T;H^1(\Omega))}$$
$$+ \frac{c}{2} \|e(x)\|^2_{L^2(0,T;H^1(\Omega)')} + \frac{1}{2c} \left(\|\max(0, \lambda_b + c(u - u_b))\|^2_{L^2(\Sigma)} - \|\lambda_b\|^2_{L^2(\Sigma)}\right.$$
$$\left. + \|\min(0, \lambda_a + c(u - u_a))\|^2_{L^2(\Sigma)} - \|\lambda_a\|^2_{L^2(\Sigma)}\right)$$

with penalty parameter $c > 0$.

To ensure that the search direction has a descent property we employ a technique that modifies the Hessian of the Lagrange function in an appropriate manner. In fact, it utilizes the particular problem structure, i.e., the convexity of the objective function and the way how the nonlinearity enters into the state equation. If a lack of positive definiteness occurs while computing the new update direction, the nonlinear term in the state equation and correspondingly its derivatives occurring in the Hessian are dampened by a parameter $\gamma \in [0, 1]$. Since this modifies the (QP)-subproblem, the computation of the search direction is restarted. In Section 4 we report on two different implementations of this strategy: The γ-*strategy* adjusts $\gamma \in [0, 1]$ by a backtracking technique starting with $\gamma = 1$. Hence, $0 < \gamma < 1$ is possible. The second technique sets $\gamma = 0$ whenever lack of positive definiteness of the Hessian of the Lagrangian in the actual direction is detected. In this second case, positive definiteness can be guaranteed after the γ-reduction step. We refer to this strategy as 0/1-*strategy*. For more details on these strategies we refer to [15, 17].

4. Numerical tests

This section is devoted to numerical test examples. We consider three different problems including integral control as well as integral state constraints, terminal observations in Ω and partial observation in Q. All coding is done in MATLAB using routines from the FEMLAB package for the finite element discretization. All computations are performed on a standard 1.7 GHz desktop PC. We stop the respective algorithm as soon as the relative violation of the first order necessary conditions drops below 10^{-12}.

Example 4.1. In our first example we consider the problem

$$\min J(y, u) = \frac{1}{2} \int_\Omega |y(T) - y_d|^2 \, \mathrm{d}x + \frac{\kappa}{2} \int_0^T \int_\Gamma |u|^2 \, \mathrm{d}s$$

subject to the non-linear heat equation

$$\begin{aligned}
y_t - \alpha \Delta y &= 0 &&\text{in } Q = (0, T) \times \Omega, \\
\frac{\partial y}{\partial n} + y^3 |y| &= u &&\text{on } \Sigma = (0, T) \times \Gamma, \\
y(0) &= 8 &&\text{in } \Omega
\end{aligned}$$

and the inequality constraints

$$-15 \leq u(t, s) \leq -5 + 15t \quad \text{for } (t, s) \in \Sigma,$$

$$\int_0^T \int_\Gamma u \, \mathrm{d}s \mathrm{d}t \leq -20.$$

We choose the heat conduction parameter $\alpha = 0.1$ and the terminal time $T = 1$. The spatial domain $\Omega \subset \mathbb{R}^2$ is the unit circle with center at the origin. Moreover, $y_d(x_1, x_2) = 2 - 2\mathrm{sign}(x_1)x_1$ is the desired state (see Figure 1) and $\kappa = 10^{-4}$ the regularization parameter. Setting $z_\Omega = y_d$, $\alpha_Q = 0$, $\alpha_\Omega = 1$ the cost functional coincides with the one in (2.1). The spatial domain is discretized with triangular finite elements with a maximal edge length of $h = 0.064$. The time interval is discretized uniformly with stepsize $\Delta t = 0.01$ in the time interval $[0, T]$. This yields $3.58 \cdot 10^4$ degrees of freedom (d.o.f.) for the state y as well as for the adjoint state p, $5.2 \cdot 10^3$ d.o.f. for the control u and the dual λ, respectively. The geometry together with the given data yield a symmetric solution. Up to the discretization error, this solution symmetry should be preserved in the discretized setting. Notice that the influence of the non-linear radiation term $b(y) = y^3 |y|$ is 10 times larger than that of the diffusion term $\alpha \Delta y$. In Figure 1 the desired terminal state is compared with the discrete optimal temperature y at $t = T$. The discrete optimal control u together with the associated adjoint p at time $t = T$ are shown in Figure 2. One can clearly identify the nonempty upper and lower active sets from the shaded region in the left plot. Also, the integral constraint is active at the discrete solution. The multiplier ξ^* associated to the integral constraint has the value $\xi^* = 4.27 \cdot 10^{-4} > 0$.

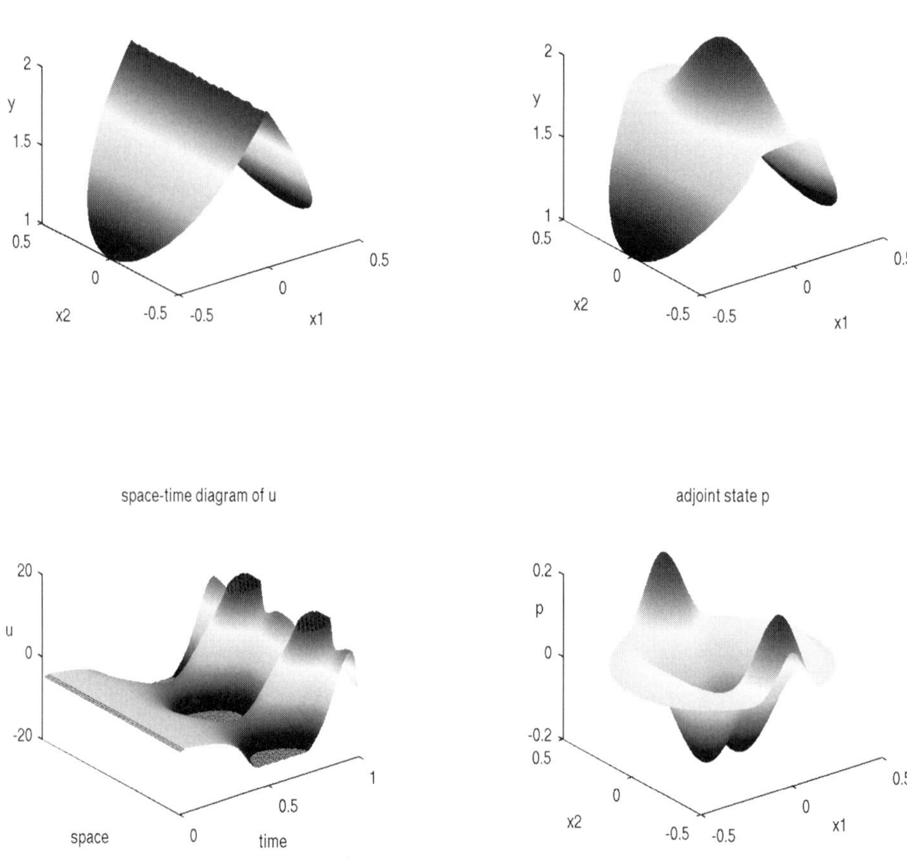

FIGURE 2. Example 4.1: perimeter-time diagram of the discrete optimal control u (left) and discrete optimal adjoint p at $t = T$ (right).

Next we compare the performance of our algorithms. First we choose rSQP, where each QP subproblem is solved by the primal-dual active set strategy (PDA), see [16, 18]. In Table 1 we present the results if we apply the γ-strategy for the globalization of the Hessian and the augmented Lagrange merit function with n_{aug} denoting the iteration of the augmented Lagrange method, n_{rSQP} the iteration of the reduced SQP algorithm and $\#n_{\mathrm{PDA}}$ total number of PDA steps for each rSQP iteration. Only in the first rSQP step a modification of the Hessian is required. The corresponding γ-value is $\gamma = 0.09987$. It turns out that this method needs two outer augmented Lagrange iterations, and fast local convergence is observed for the inner rSQP algorithm. Furthermore, the number of inner PDA steps decreases as the outer rSQP steps increases. This behavior can also be seen in the following

n_{aug}	n_{rSQP}	$\#n_{\mathrm{PDA}}$	$\|\nabla_y \mathcal{L}_\xi^\varrho(y,u,p)\|$	$\|e(y,u)\|$
1	1	5	1.901e-03	1.111e-00
	2	5	1.192e-03	1.730e-01
	3	5	1.780e-04	5.403e-02
	4	4	3.669e-06	2.074e-03
	5	2	2.952e-09	3.081e-06
	6	1	3.854e-11	7.004e-12
2	1	1	2.135e-11	7.362e-10
	2	1	2.243e-11	6.075e-15

TABLE 1. Example 4.1: performance of rSQP with augmented Lagrangian line search combined with the γ-strategy for $h_{\max} = 0.064$ and $\Delta t = 0.01$.

n_{aug}	n_{rSQP}	$\#n_{\mathrm{PDA}}$	$\|\nabla_y \mathcal{L}_\xi^\varrho(y,u,p)\|$	$\|e(y,u)\|$
1	1	4	3.234e-04	8.363e-01
	2	4	1.859e-04	1.173e-01
	3	4	8.884e-05	5.885e-03
	4	3	5.394e-06	4.997e-03
	5	3	8.946e-08	1.634e-04
	6	1	6.532e-11	1.208e-07
	7	1	3.539e-11	4.506e-14
2	1	1	2.097e-11	7.363e-10
	2	1	2.197e-11	5.567e-15

TABLE 2. Example 4.1: performance of rSQP with augmented Lagrangian line search combined with the 0/1-strategy for $h_{\max} = 0.064$ and $\Delta t = 0.01$.

tests. If we utilize the 0/1-strategy to modify the Hessian then we obtain the results presented in Table 2. Compared to the previous variant with γ-strategy two modification of the Hessian are necessary ($n_{\mathrm{rSQP}} = 1$ and $n_{\mathrm{rSQP}} = 2$). The behavior of the algorithm is similar in both cases. Next we combine rSQP with the line search based on the L^1 merit function Φ and the γ-strategy; see Table 3. The performance of rSQP with the L^1-merit function and 0/1-strategy is shown in Table 4. While the total number of PDA iterations is comparable for the γ-strategies in case of both line search functions and for the 0/1-strategy with the augmented Lagrangian line search, the combination of the 0/1-strategy with the L^1 line search requires significantly more PDA steps. The difference in computational effort with respect to cg iterations when solving the linear systems in each method is exhibited in Table 5. It turns out that rSQP combined with the 0/1 strategy and the line search

n_{aug}	n_{rSQP}	$\#n_{\text{PDA}}$	$\|\nabla_y \mathcal{L}^{\varrho}_{\xi}(y,u,p)\|$	$\|e(y,u)\|$
1	1	5	1.256e-03	9.679e-01
	2	6	1.377e-04	1.973e-01
	3	5	7.973e-06	1.255e-02
	4	3	1.006e-07	9.385e-05
	5	1	1.292e-11	7.249e-09
	6	1	6.794e-13	5.499e-15
2	1	1	6.958e-13	7.363e-10
	2	1	8.124e-13	6.090e-15

TABLE 3. Example 4.1: performance of rSQP with L^1 line search combined with the γ-strategy for $h_{\max} = 0.064$ and $\Delta t = 0.01$.

n_{aug}	n_{rSQP}	$\#n_{\text{PDA}}$	$\|\nabla_y \mathcal{L}^{\varrho}_{\xi}(y,u,p)\|$	$\|e(y,u)\|$
1	1	5	4.408e-02	3.624e-02
	2	4	8.472e-03	1.402e-01
	3	6	3.311e-04	3.738e-01
	4	4	2.246e-05	4.191e-02
	5	3	8.172e-07	6.982e-04
	6	2	3.830e-10	3.325e-07
	7	1	3.662e-11	9.693e-14
	1	1	2.072e-11	7.363e-10
	2	1	2.213e-11	6.168e-15

TABLE 4. Example 4.1: performance of rSQP with L^1 line search globalization and 0/1-strategy for $h_{\max} = 0.064$ and $\Delta t = 0.01$.

Modification of Hessian	Merit function	relative amount
γ-strategy	augmented Lagrangian	0.73
0/1-strategy	augmented Lagrangian	0.68
γ-strategy	L^1 function	0.77
0/1-strategy	L^1 function	1.00

TABLE 5. Example 4.1: relative amount of cg iterations for rSQP with different modification strategies for the Hessian and different line search methods.

based on the L^1 function requires more cg iterations than all other three variants, which, among themselves, do not differ significantly. Therefore, we continue our numerical comparison by choosing the γ-strategy combined with a line search based

n_{aug}	n_{pSSN}	$\#n_{\text{rSQP}}$	$\|\nabla_y \mathcal{L}_\xi^\varrho(y,u,p)\|$	$\|e(y,u)\|$
1	1	9	8.565e-13	2.464e-11
	2	5	9.977e-13	5.430e-15
	3	4	9.580e-13	6.111e-15
	4	3	8.811e-13	5.477e-15
	5	2	6.856e-13	1.990e-13
	6	1	7.284e-13	4.965e-13
2	1	2	7.566e-13	5.489e-15

TABLE 6. Example 4.1: performance of pSSN-SQP with line search globalization for $h_{\max} = 0.064$ and $\Delta t = 0.01$.

n_{aug}	n_{pSSN}	$\#n_{\text{Newton}}$	$\|(\hat{J}_\xi^\varrho)'(u)|_\mathcal{I}\|$
1	1	8	2.933e-12
	2	5	2.041e-11
	3	4	9.332e-13
	4	3	8.745e-13
	5	2	7.842e-13
	6	1	7.271e-13
2	1	1	6.089e-12

TABLE 7. Example 4.1: performance of pSSN-Newton with line search globalization for $h_{\max} = 0.064$ and $\Delta t = 0.01$.

on the L^1-merit function. Next we turn to the pSSN-SQP variant that also needs 2 outer augmented Lagrange iterations; see Table 6. Further, it turns out that the number of inner SQP iterations per pSSN iteration decreases as the number of pSSN iterations increases. In particular, in the sixth pSNN iteration only one SQP iteration is necessary to achieve convergence of the first augmented Lagrange iteration. To avoid negative curvature, the γ-strategy is applied twice at the first two pSSN iterations to modify the Hessian. Now we test pSSN-Newton, where in contrast to pSSN-SQP problem (3.11) is not solved by applying the reduced SQP method, but the Newton algorithm. Analogously to pSSN-SQP the number of inner Newton iterations per pSSN iteration decreases as the number of pSSN iterations increases, see Table 7. Since the equality constraint $e(y,u) = 0$ is always fulfilled within the reduced problem $(\hat{\mathbf{P}}^\varrho)$, we do not present the norms $\|e(y,u)\|$ in Table 7. The γ-strategy is applied twice at the first two pSSN iterations to modify the Hessian. Note that both variants, pSSN and pSSN-Newton, behave similarly. Next we apply SSN-Newton to our example, i.e., the semismooth Newton method for $(\hat{\mathbf{P}}^\varrho)$. As for the previous variants we need two outer augmented Lagrange

n_{aug}	n_{SSN}	$\|(\hat{J}_\xi^\varrho)'(u)|_{\mathcal{I}}\|$
1	1	1.103e-02
	2	2.822e-03
	3	2.706e-03
	4	2.708e-04
	5	1.016e-04
	6	5.490e-05
	7	1.395e-05
	8	6.387e-07
	9	1.660e-09
	10	7.232e-12
2	1	7.986e-12

TABLE 8. Example 4.1: performance of SSN-Newton with line search globalization for $h_{\mathrm{max}} = 0.064$ and $\Delta t = 0.01$.

n_{aug}	n_{SSN}	$\|\nabla_y \mathcal{L}_\xi^\varrho(y,u,p)\|$	$\|e(y,u)\|$
1	1	1.286e-03	1.172e-00
	2	1.230e-03	2.608e-00
	3	6.002e-04	9.327e-01
	4	1.552e-04	2.372e-01
	5	6.744e-05	1.038e-01
	6	1.191e-05	1.448e-02
	7	6.238e-07	8.967e-04
	8	2.194e-09	3.791e-06
	9	8.391e-13	4.902e-11
2	1	8.537e-13	7.363e-10
	2	8.025e-13	5.548e-15

TABLE 9. Example 4.1: performance of SNN with line search globalization for $h_{\mathrm{max}} = 0.064$ and $\Delta t = 0.01$.

iterations; see Table 8. In the first augmented Lagrange iteration we observe fast convergence for $n_{\mathrm{SSN}} \geq 7$. Let us mention that in the first and sixth iteration, the Hessian has to be modified due to negative curvature. As for pSSN-Newton the equality constraint $e(y,u) = 0$ is satisfied for each iteration so that we do not present the norms $\|e(y,u)\|$. Finally, we test the semismooth Newton method SSN. The results are presented in Table 9. In contrast to SSN-Newton the equality constraint $e(y,u) = 0$ is only approximately satisfied in the course of the iteration.

Algorithm	Relative amount
rSQP	1.00
pSSN-SQP	0.99
pSSN-Newton	0.95
SSN-Newton	0.56
SSN	0.50

TABLE 10. Example 4.1: relative amount of cg iterations for all variants with γ-strategy and L^1 line search globalization for $h_{\max} = 0.064$ and $\Delta t = 0.01$.

$h_{\max}$	Δt	$\#Q$	$\#\Sigma$	$\#\mathcal{A}_a(\%)$	$\#\mathcal{A}_b(\%)$	$n_{\mathrm{PDA}}(n_{\mathrm{aug}}=1)$						$n_{\mathrm{aug}}=2$
0.064	0.010	3.58e04	5.20e03	10.69	5.62	5	5	5	4	2	1	1	1
0.032	0.010	1.39e05	1.00e04	11.18	6.60	5	6	5	3	2	1	1	1
0.016	0.010	5.46e05	1.96e04	11.40	6.57	6	6	5	4	2	1	1	1
0.064	0.005	7.16e04	1.04e04	9.06	4.57	5	5	4	3	2	1	1	1
0.032	0.005	2.77e05	2.00e04	9.86	5.31	5	6	5	3	3	1	2	1

TABLE 11. Example 4.1: influence of mesh size on the number of PDA iterations per rSQP step for each augmented Lagrange iteration.

Comparing the pSSN-variants to the SSN-variants, we observe that the latter ones require a smaller number of iterations for successful termination.

We compare the efficiency of the algorithms in Table 10, where the relative amount of cg iterations when solving the respective linear systems to compute the next iterates is presented. It turns out that SSN is the most efficient method for our test example. The variant SSN-Newton requires only slightly more cg iterations. We observe that pSSN-SQP as well as pSSN-Newton are as efficient as rSQP. In Table 11 the degrees of freedom in Q and on Σ are shown. Moreover, $\#\mathcal{A}_a(\%)$ and $\#\mathcal{A}_b(\%)$ denote the relative amount of active points, e.g., $\#\mathcal{A}_a(\%) = 100\% \cdot (\#\mathcal{A}_a)/(\#\Sigma)$. Finally, we observe strong mesh-independence for rSQP; compare Table 11. Let us mention that it is well known that the SQP method is mesh-independent for twice continuously Fréchet differentiable cost functional and constraints; see, e.g., [2, 12, 25]. Moreover, the primal-dual active set strategy satisfies a mesh-independent principle; see [19]. From Table 11 we can see that the combination of the augmented Lagrangian method with the rSQP (with augmented Lagrangian line search) and the primal-dual active set strategy has mesh-independent behavior. Notice that in our case, the cost functional possesses only a generalized second derivative; compare Section 3.1. We point out that L^1 line search globalization yields also a mesh-independent algorithm. $\Diamond$

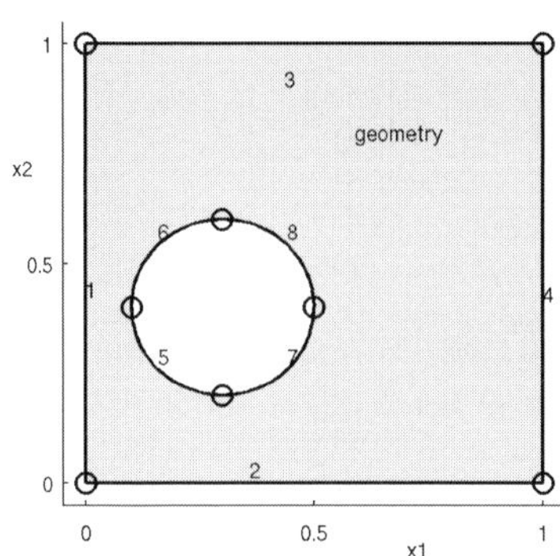
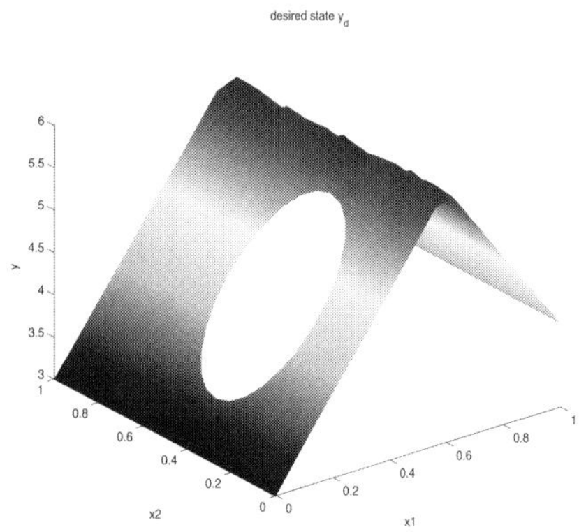

FIGURE 3. Example 4.2: spatial domain Ω with edge ordering (left) and desired state (right).

Example 4.2 (Integral state constraints). In our second example we consider the problem

$$\min J(y,u) = \frac{1}{2} \int_{0.2}^{T} \int_{\Omega} |y - y_d|^2 \,\mathrm{d}x\mathrm{d}t + \frac{\kappa}{2} \int_{0}^{T} \int_{\Gamma} |u|^2 \,\mathrm{d}s\mathrm{d}t$$

$$y_t - \alpha\Delta y = 0 \qquad\qquad \text{in } Q = (0,T) \times \Omega,$$

$$\frac{\partial y}{\partial n} + y^3|y| = \beta u \qquad\qquad \text{on } \Sigma = (0,T) \times \Gamma,$$

$$y(0) = 4 \qquad\qquad \text{in } \Omega$$

and the inequality constraints

$$0 \leq u(t,s) \quad \text{for } (t,s) \in \Sigma,$$

$$\int_{0}^{T} \int_{\Omega} y \,\mathrm{d}s\mathrm{d}t \leq \frac{18}{5}.$$

We choose the heat conduction parameter $\alpha = 0.1$, $\beta = 100$ and terminal time $T = 1$. The spatial domain $\Omega \subset \mathbb{R}^2$ is the unit square with a circular hole of radius 0.2 centered at the point $(0.3, 0.4)$; see left plot in Figure 3. Moreover, $y_d(x_1, x_2, t) = 6 - 5(x_1 - 0.6)\mathrm{sign}\,(x_1 - 0.6)$ is the desired state (compare right plot in Figure 3) and $\kappa = 10^{-3}$ the regularization parameter. Setting $z_Q = y_d$, $\alpha_\Omega = 0$, $\alpha_Q = 1$ in $(0.2, T) \times \Omega$ and $\alpha_Q = 0$ in $(0, 0.2] \times \Omega$ the cost functional coincides with the one in (2.1). The spatial domain is discretized with triangular finite elements with an maximal edge length of $h = 5.3 \cdot 10^{-2}$. The time interval was discretized non-uniformly with stepsize $\Delta t = 0.01$ in the time interval $[0, 0.3]$ and $\Delta t = 0.02$ in the time interval $(0.3, 1]$. This reflects the dynamics of the control as well as the temperature, i.e., the state y. In total, we have $2.68 \cdot 10^4$ d.o.f. for the state y and the dual p and about $5.7 \cdot 10^3$ d.o.f. for the control u and the multiplier λ. In

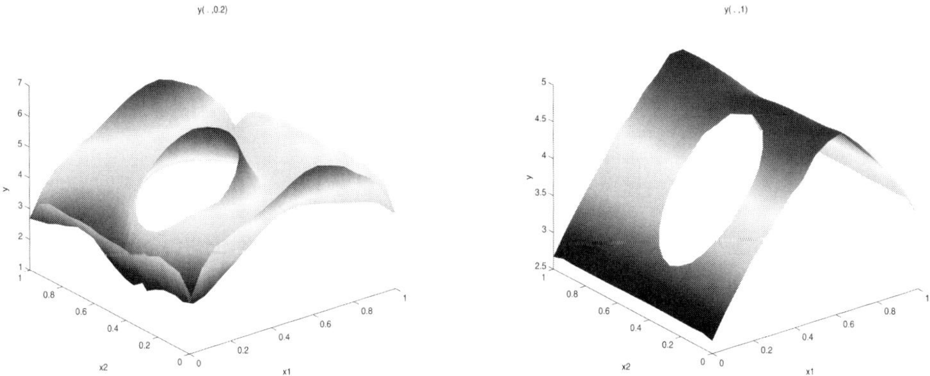

FIGURE 4. Example 4.2: optimal temperature at $t = 0.2$ (left) and at $t = T$ (right)

	$n_{\mathrm{aug}} = 1$	$n_{\mathrm{aug}} = 2$	$n_{\mathrm{aug}} = 3$	$n_{\mathrm{aug}} = 4$	total
rSQP	6	11	15	11	43
pSSN-SQP	6	32	94	56	188
SSN-SQP	6	6	16	11	39

TABLE 12. Example 4.2: Number of KKT solves for SQP, pSSN-SQP and SSN-SQP.

fact, the boundary of the domain, where the control action occurs, consists of two disjoint components. In Figure 4 (left plot) the optimal temperature at $t = 0.2$ (the beginning of the observation interval) and $t = T$ (right plot) are shown. In this example, the pointwise control constraint from below is only active at the circular hole. In Figure 5 we present the active sets for the unilateral control constraint $u \geq 0$ on Σ, where the outer boundary is revolved onto the larger block with edge ordering: 2, 4, 3, 1 (see Figure 3), the circular boundary is represented above with edge ordering: 7, 8, 6, 5. Moreover, the integral constraint is active. In Table 12 we compare the variants rSQP, pSSN-SQP and SSN-SQP. We observe that SSN is the fastest variant and that rSQP is nearly as efficient as SSN. On the other hand, pSSN-SQP needs five times more systems solves than SSN. $\diamond$

Example 4.3 (Rail profile example). We set $T = 600$, $\alpha = 2.27 \cdot 10^{-5}$, $y_d = 800$ and $\kappa = 10^{-6}$. Consider the problem

$$\min J(y, u) = \frac{1}{2} \int_{\Omega} |y(T) - y_d|^2 \, \mathrm{d}x + \frac{\kappa}{2} \int_0^T \int_{\Gamma} |u|^2 \, \mathrm{d}s \mathrm{d}t$$

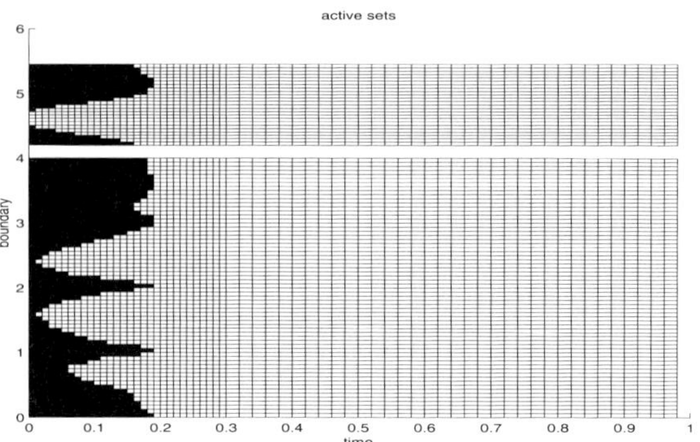

FIGURE 5. Example 4.2: distribution of the active set (black color) for the unilateral bound $u \geq 0$ in a boundary-time diagram of the control, where the boundary is projected onto the x_2-axis (bottom). The outer boundary is revolved onto the larger block with edge ordering: 2, 4, 3, 1, the circular boundary is represented above with edge ordering: 7, 8, 6, 5, compare Figure 3.

subject to the non-linear equation

$$y_t - \alpha \Delta y = 0 \qquad \text{in } Q = (0, T) \times \Omega,$$

$$\frac{\partial y}{\partial n} + 7.09 \cdot 10^{-11} y^3 |y| = 0.522 + u \qquad \text{on } \Sigma = (0, T) \times \Gamma,$$

$$y(0) = 300 \qquad \text{in } \Omega$$

and to the inequality constraint

$$8 \cdot 10^3 \leq u(t, s) \leq 3.2 \cdot 10^4 \quad \text{for } (t, s) \in \Sigma,$$

$$\int_0^T \int_\Gamma u \, ds dt \leq 6 \cdot 10^6$$

Here, the spatial domain corresponds to a rail profile; see Figure 6. The material parameters reflect the thermal properties of steel; see [13]. All units are given in SI, e.g., space is given in $[x_i] = m$, time in $[t] = s$, the state in $[y] = K$, the control in $[u] = W/m^2$, the cost parameter in $[\kappa] = K^2 m^4/W^2$, and the integral constraint in $[\int_0^T \int_\Gamma u \, ds dt] = J/m$. The parameters are assumed to be independent of temperature, which is a simplification in the temperature range of interest. The ambient temperature is assumed to be $300K$. The boundary nonlinearity is weak due to the small value of the radiation coefficient $7.09 \cdot 10^{-11}$. We apply the variant rSQP and point out that – due to the smallness of the nonlinear term – all variants have similar performance. Already the initial guess lies within the attraction basin of fast local convergence. Hence, in this example only 4 rSQP steps are needed for the first augmentation step to converge, and a q-quadratic

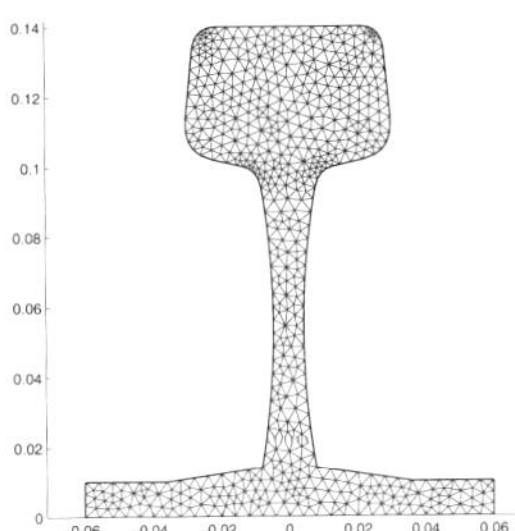 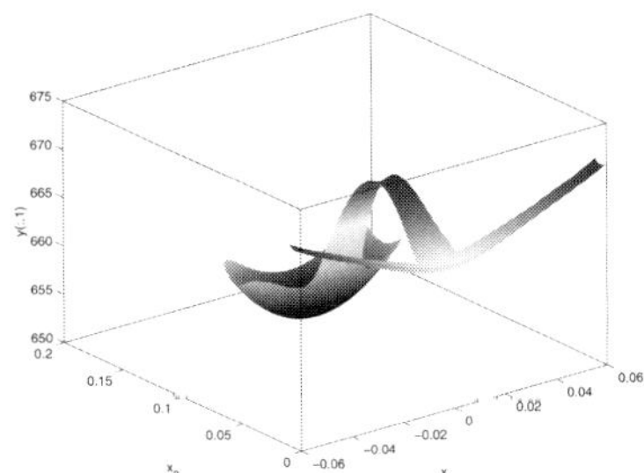

FIGURE 6. Example 4.3 finite element mesh of the rail's profile (left) and discrete optimal temperature a time $t = T$ (right).

n_{aug}	n_{rSQP}	$\#n_{\mathrm{PDA}}$	$\|\nabla_y \mathcal{L}_\xi^\varrho(y,u,p)\|$	$\|e(y,u)\|$	$J(y,u)$
1	1	5	5.386e-04	4.365e-03	53.58
	2	4	8.893e-07	5.499e-06	54.15
	3	2	7.338e-12	1.268e-10	54.15
	4	1	1.499e-12	3.590e-14	54.15
2	1	2	1.102e-09	7.246e-09	54.19
	2	1	4.947e-12	3.377e-14	54.19
3	1	1	6.383e-11	2.843e-14	54.19

TABLE 13. Example 4.3: Iteration results with $h_{\mathrm{max}} = 0.004$ and $\Delta t = 0.02$.

h_{max}	Δt	$\#Q$	$\#\Sigma$	$\#\mathcal{A}_a(\%)$	$\#\mathcal{A}_b(\%)$	$n_{\mathrm{PDA}}(n_{\mathrm{aug}} = 1)$				2		3
0.004	0.020	3.16e04	8.10e03	9.037	6.037	5	4	2	1	2	1	1
0.004	0.010	6.32e04	1.62e04	8.907	5.932	6	4	2	1	2	1	1
0.004	0.005	1.26e05	3.24e04	8.821	5.975	6	4	2	1	2	1	1
0.002	0.020	1.07e05	1.48e04	9.662	5.101	5	4	2	1	3	1	1
0.002	0.010	2.13e05	2.96e04	9.500	5.068	5	4	2	1	3	1	1
0.002	0.005	4.26e05	5.92e04	9.449	5.081	6	4	2	1	3	1	1
0.001	0.020	4.13e05	2.91e04	9.973	5.034	6	4	3	1	3	1	1
0.001	0.010	8.27e05	5.82e04	9.773	4.973	6	4	3	1	3	1	1
0.001	0.005	1.65e06	1.16e05	9.680	4.974	6	4	2	1	3	1	1

TABLE 14. Example 4.3: Influence of the mesh fineness on the power of the active sets, the cost functional and the overall number of cg-iterations.

rate of convergence is achieved from iteration one on. Also, we observe that the amount of cg-iterations is significantly lower than the one in the previous examples. The corresponding convergence behavior is shown in Table 13. In Table 14 the degrees of freedom as well as the relative amount of active points are presented for different discretizations. From the results in Table 14 we observe (strong) mesh independence of rSQP already on the coarsest mesh considered.   ◊

References

[1] R.A. Adams. *Sobolev Spaces*. Academic Press, New York-London, 1975. Pure and Applied Mathematics, Vol. 65.

[2] W. Alt. Discretization and mesh-independence of Newton's method for generalized equations. In *Mathematical Programmierung with Data Perturbation*. Lecture Notes in Pure and Appl. Math., 195:1–30, 1996.

[3] M. Bergounioux, K. Ito, and K. Kunisch. Primal-dual strategy for constrained optimal control problems. *SIAM J. Contr. Optim.*, 35:1524–1543, 1997.

[4] M. Bergounioux and F. Tröltzsch. Optimal control of semilinear parabolic equations with state-constraints of bottleneck type. *ESAIM: Control, Optimisation and Calculus of Variations*, 4:595–608, 1999.

[5] D.P. Bertsekas. *Constrained Optimization and Lagrange Multiplier Methods*. Academic Press, New York, 1982

[6] D.P. Bertsekas. *Nonlinear Programming*. Athena Scientific, 1995.

[7] E. Casas, J.-P. Raymond, and H. Zidani. Pontryagin's principle for local solutions of control problems with mixed control-state constraints. *SIAM J. Control and Optimization*, 39:1182–1203, 2000.

[8] X. Chen, Z. Nashed, and L. Qi. Smoothing methods and semismooth methods for nondifferentiable operator equations. *SIAM J. Numer. Anal.*, 38: 1200–1216, 2000.

[9] F.H. Clarke. *Optimization and Nonsmooth Analysis*. Wiley, New York, 1983.

[10] R. Dautray and J.-L. Lions. *Evolution Problems I*. Mathematical Analysis and Numerical Methods for Science and Technology. Volume 5, Springer-Verlag, Berlin, 1992.

[11] J.E. Dennis, M. Heinkenschloss, and L.N. Vicente. Trust-region interior-point SQP algorithms for a class of nonlinear programming problems. *SIAM J. Control Optimization*, 36(5):1750–1794, 1998.

[12] A.L. Dontchev, W.W. Hager, and V.M. Veliov. Uniform convergence and mesh-independence of Newton's method for discretized variational problems. *SIAM J. Control Optim.*, 39(3):961–980, 2000.

[13] H. Dubbel and K.H. Grote. *Taschenbuch für den Maschinenbau*. 21., neubearb. und erw. Aufl., Springer, Berlin, 2005.

[14] L.C. Evans. *Partial Differential Equations*. Graduate Studies in Mathematics, vol. 19, American Mathematical Society, Providence, Rhode Island, 1998.

[15] M. Hintermüller. On a globalized augmented Lagrangian-SQP algorithm for nonlinear optimal control problems with box constraints. Hoffmann, Karl-Heinz (ed.) et

al., Fast solution of discretized optimization problems. Workshop held at the Weierstrass Institute for Applied Analysis and Stochastics, Berlin, Germany, May 8-12, 2000. Basel: Birkhäuser. ISNM, Int. Ser. Numer. Math. 138, 139–153, 2001.

[16] M. Hintermüller. A primal-dual active set algorithm for bilaterally control constrained optimal control problems. *Q. Appl. Math.*, 61(1):131–160, 2003.

[17] M. Hintermüller and M. Hinze. Globalization of SQP-methods in control of the instationary Navier-Stokes equations. *M2AN, Math. Model. Numer. Anal.* 36(4):725–746, 2002.

[18] M. Hintermüller, K. Ito, and K. Kunisch. The primal-dual active set strategy as a semi-smooth Newton method. *SIAM J. Optimization*, 13:865–888, 2003.

[19] M. Hintermüller, and M. Ulbrich. A mesh-independence result for semismooth Newton methods. *Math. Program.*, Ser. B, 101: 151–184, 2004.

[20] K. Ito and K. Kunisch. The primal-dual active set method for nonlinear optimal control problems with bilateral constraints. *SIAM J. Control Optimization*, 43(1):357–376, 2004.

[21] C. Meyer, P. Philip, and F. Tröltzsch. Optimal control of a semilinear PDE with nonlocal radiation interface conditions. Preprint, TU-Berlin, Department of Mathematics, 2005.

[22] J. Nocedal and S.J. Wright. *Numerical Optimization*. Springer Series in Operation Research, Springer-Verlag, New York, 1999.

[23] J.-P. Raymond and H. Zidani. Hamiltonian Pontryagin's principles for control problems governed by semilinear parabolic equations. *Appl. Math. Optim.*, 39:143–177, 1999.

[24] F. Tröltzsch. *Optimale Steuerung partieller Differentialgleichungen*. Vieweg Verlag, Wiesbaden, 2005.

[25] S. Volkwein. *Mesh-independence for an augmented Lagrangian-SQP method in Hilbert spaces. SIAM J. Control Optim.*, 38:1938–1984, 2000.

M. Hintermüller
Karl-Franzens-University of Graz
Department of Mathematics and Scientific Computing
Heinrichstrasse 36
A-8010 Graz, Austria
e-mail: `michael.hintermueller@uni-graz.at`

S. Volkwein
Karl-Franzens-University of Graz
Department of Mathematics and Scientific Computing
Heinrichstrasse 36
A-8010 Graz, Austria
e-mail: `stefan.volkwein@uni-graz.at`

F. Diwoky
Karl-Franzens-University of Graz
Department of Mathematics and Scientific Computing
Heinrichstrasse 36
A-8010 Graz, Austria

International Series of Numerical Mathematics, Vol. 155, 149–174
© 2007 Birkhäuser Verlag Basel/Switzerland

Optimal and Model Predictive Control of the Boussinesq Approximation

Michael Hinze and Ulrich Matthes

Abstract. We discuss optimal and model predictive control techniques applied to the Boussinesq approximation of the Navier-Stokes system. We focus on mathematical modeling, discuss possible control scenarios, and provide a concise description of the numerical implementation. Furthermore, several numerical examples are provided.

Mathematics Subject Classification (2000). 35Q30,49J20,49K20.

Keywords. Boussinesq approximation, model-predictive control, optimal control,numerics.

1. Introduction

The Boussinesq approximation of the Navier-Stokes system is frequently used as mathematical model for fluid flow in semiconductor melts. In many crystal growth technics, such as Czochralski growth, and zone-melting technics the behavior of the flow has considerable impact on the crystal quality. It is therefore quite natural to establish flow conditions which guarantee desired crystal properties.

As a first step towards controlling the crystal-melt complex in Czochralski growth we study in the present paper optimal and model predictive control technics for the Boussinesq approximation. As control actions we consider distributed forcing, distributed heating, and boundary heating, as well as its combinations.

To the best of the authors knowledge up to now there are no contribution to model predictive control for the Boussinesq approximation. However, in the past decade considerable progress has been made in the field of flow control, see [5] for a comprehensive overview and further literature in the field. In the literature also

The authors acknowledge financial support of the Collaborative Research Grant SFB 609 *Elektromagnetische Strömungsbeeinflussung in Metallurgie, Kristallzüchtung und Elektrochemie*, sponsored by the Deutsche Forschungsgemeinschaft.

contributions to optimal control of the Boussinesq approximation can be found. Here we mention the works [1] and [12].

The paper is organized as follows. In Section 2 the variational form of the Boussinesq approximation is introduced and the time discretization scheme is presented. In Section 3 model predictive control is introduced, and in Section 4 numerical results are given. In Section 5 we summarize the numerical results and give some conclusions.

2. Mathematical model

2.1. Boussinesq approximation

The Boussinesq approximation of the Navier-Stokes system in the primitive setting is given by

$$
\begin{aligned}
\tfrac{\partial y}{\partial t} - \nu \Delta y + \nabla p &= -(y\nabla)y - \gamma\, g\, \tau + u_F & \text{in } Q, \\
-\operatorname{div} y &= 0 & \text{in } Q, \\
y &= 0 & \text{on } \Sigma, \\
y(0) &= y_0 & \text{in } \Omega, \\
\tfrac{\partial \tau}{\partial t} - a\Delta\tau &= -(y\nabla)\tau + u_Q & \text{in } Q, \\
a\partial_\eta \tau &= \alpha(u - \tau) & \text{on } \Sigma, \\
\tau(0) &= \tau_0 & \text{in } \Omega,
\end{aligned}
\tag{1}
$$

were y, p, τ denote the velocity, pressure and temperature field, respectively. Further a denotes the thermal diffusivity, ν the kinematic viscosity, $g \in \mathbf{R}^2$ the acceleration of gravity, γ the coefficient of volume expansion, and α a positive number. Here $\Omega \subset \mathbf{R}^2$ denotes an open, bounded domain, with boundary $\Gamma = \partial\Omega$ which is assumed to be sufficiently smooth. We set $Q := (0, T) \times \Omega$ and $\Sigma := (0, T) \times \Gamma$ with T denoting the time horizon.

The variables u, u_F, u_Q denote the control actions; u the boundary temperature, u_F distributed force, and u_Q distributed heating.

To prepare for the variational formulation of (1) we further introduce the solenoidal spaces

$$
H = \left\{ v \in C_0^\infty(\Omega)^2 : \ \operatorname{div} v = 0 \right\}^{-\|\cdot\|_{L^2(\Omega)^2}}
$$

and

$$
V = \left\{ v \in C_0^\infty(\Omega)^2 : \ \operatorname{div} v = 0 \right\}^{-\|\cdot\|_{H^1(\Omega)^2}}.
$$

Also if X is a Banach space, $L^p(0, T; X)$ denotes the space of L^p-integrable functions from $(0, T)$ into X, which itself is a Banach space.

2.1.1. Variational formulation. Following [1] and [15], the variational formulation of (1) reads: Given $f = (u_F, u_Q)^T \in L^2(0, T; V^* \times H^1(\Omega)^*), u \in L^2(0, T; L^2(\Gamma))$

and $Y_0 \in H \times L^2(\Omega)$, find $Y \in L^2(0,T; V \times H^1(\Omega))$ satisfying

$$\frac{d}{dt}(Y,U) + a(Y,U) + b(y,Y,U) + (\gamma g\tau, v)_{L^2(\Omega)^2} + (\alpha\tau, \eta)_{L^2(\Gamma)}$$
$$= (f,U)_{(V \times H^1(\Omega))^* (V \times H^1(\Omega))} + (\alpha u, \eta)_{L^2(\Gamma)}$$
$$\forall U \in V \times H^1(\Omega), \text{ and almost all } t \in (0,T), \tag{2}$$

and

$$Y(0) = Y_0 := \begin{pmatrix} y(0) \\ \tau(0) \end{pmatrix}. \tag{3}$$

Here we use the notation

$$Y := \begin{pmatrix} y \\ \tau \end{pmatrix}, \quad U := \begin{pmatrix} v \\ \eta \end{pmatrix}, \quad W := \begin{pmatrix} w \\ \kappa \end{pmatrix},$$

and forms $a(\cdot,\cdot)$ and $b(\cdot,\cdot,\cdot)$ are defined by

$$a(Y,U) := \nu \int_\Omega \nabla y \nabla u \, dx + a \int_\Omega \nabla\tau\nabla\eta \, dx \quad \forall Y, U \in V,$$
$$b(U,Y,W) := \int_\Omega (v\nabla)yw \, dx + \int_\Omega (v\nabla)\tau\kappa \, dx \quad \forall U, Y, W \in V \times H^1(\Omega),$$

and

$$(\alpha\tau, \cdot) := (S(\alpha\tau), S\cdot)_{L^2(\Gamma)} \in V^* \text{ with } S \text{ denoting the trace operator.}$$

2.1.2. Existence and uniqueness. Analogously to [15, Chap. III] we can prove existence and uniqueness of solutions to (2)–(3).

Theorem 2.1. *Let* $u_F \in L^2(0,T; V^*), u_Q \in L^2(0,T, H^1(\Omega)^*), u \in L^2(0,T; L^2(\Gamma))$ *and* $y_0 \in H, \tau_0 \in L^2(\Omega)$. *Then there exists an unique solution* Y *of* (2)–(3) *which satisfies* $Y \in L^2(0,T; V \times H^1(\Omega)), Y' \in L^2(0,T; V^* \times H^1(\Omega)^*)$. *Moreover,* $Y \in C([0,T]; H \times L^2(\Omega))$ *and*

$$Y(t) \to Y_0, \text{ in } H \times L^2(\Omega), \text{ as } t \to 0. \tag{4}$$

For the convenience of the reader a proof of this theorem is provided in the Appendix 6.

2.2. Time discretization

As time discretization scheme for (1) we use a semi-implicit Euler with time step size dt. Semi-implicit here means that the convective parts are discretized explicitly.

Giving y^i, τ^i at time instance t_i the resulting system for y^{i+1} and τ^{i+1} at time instance t_{i+1} in the primitive setting reads:

$$\frac{y^{i+1} - y^i}{dt} - \nu \Delta y^{i+1} + \nabla p^{i+1} = -(y^i \nabla) y^i - \tau^{i+1} \gamma g + u_F^{i+1} \qquad \text{in } \Omega, \qquad (5)$$

$$-\operatorname{div} y^{i+1} = 0 \qquad \text{in } \Omega, \qquad (6)$$

$$y^{i+1} = 0 \qquad \text{on } \Gamma, \qquad (7)$$

$$\frac{\tau^{i+1} - \tau^i}{dt} - a \Delta \tau^{i+1} = -(y^i \nabla) \tau^i + u_Q^{i+1} \qquad \text{in } \Omega, \qquad (8)$$

$$a \partial_\eta \tau^{i+1} = \alpha(u^{i+1} - \tau^{i+1}) \qquad \text{on } \Gamma, \qquad (9)$$

where $y^0 := y_0$ and $\tau^0 = \tau_0$ with y_0, τ_0 from (1).

The treatment of the convection term in (8) allows to compute the temperature τ^{i+1} by solving (8),(9), and subsequently the velocity y^{i+1} and p^{i+1} by (5)–(7). To anticipate the discussion, this coupling is also advantageous for the evaluation of descent directions in the formulation of the instantaneous control method.

It is worth noting that for given $y^i, \tau^i, u_F^{i+1}, u_Q^{i+1}, u^{i+1}$ in $V \times H^1(\Omega) \times V^* \times (H^1)^* \times L^2(\Gamma)$ the system (5)–(9) admits a unique weak solution $y^{i+1} \in V, \tau^{i+1} \in H^1(\Omega)$, compare [3].

3. Model predictive control

For an integer $M \geq 1$ given, model predictive control, frequently also called receding horizon control, applies repeatedly optimal control on a finite discrete time horizon containing M time steps, and uses the optimal control action associated to the first time step to steer the system towards a prescribed desired state $(z, S) = (z(t, x), S(t, z))$. In the present work the optimization problem associated to time step i is given by:

$$\min J(y, \tau, u, u_F, u_Q) = \sum_{j=i+1}^{i+M} \left(\frac{c_0}{2} \int_\Omega (y^j - z^j)^2 dx + \frac{c_1}{2} \int_\Omega (\tau^j - S^j)^2 dx \right.$$

$$\left. + \frac{c_2}{2} \int_\Gamma {u^j}^2 dx + \frac{c_3}{2} \int_\Omega {u_F^j}^2 dx + \frac{c_4}{2} \int_\Omega {u_Q^j}^2 dx \right) \qquad (10)$$

for $(y, \tau, u, u_F, u_Q) \in V^M \times H^1(\Omega)^M \times L^2(\Gamma)^M \times H^M \times (L^2)^M$, subject to:

$$\tau^{j+1} - dt\, a\, \Delta \tau^{j+1} = dt c_Q\, u_Q^{j+1} + \tau^j - dt\, (y^j \nabla) \tau^j \quad \text{in } \Omega$$

$$a \partial_\eta \tau^{j+1} = \alpha(u^{j+1} - \tau^{j+1}) \qquad \text{on } \Gamma$$

$$y^{j+1} - dt\nu \Delta y^{j+1} + \nabla(dt p^{j+1}) = -dt\gamma g \tau^{j+1} + dt c_F u_F^{j+1} + y^j$$

$$- dt(y^j \nabla) y^j \qquad \text{in } \Omega \qquad (11)$$

$$-\operatorname{div} y^{j+1} = 0 \qquad \text{in } \Omega$$

$$y^{j+1} = 0 \qquad \text{on } \Gamma$$

with $j = i, \ldots, i + M - 1$. In particular in this setting we assume that controls are at least square integrable functions.

Since the transition constraints (11) for given u_Q, u_F, u admit an unique solution, we may introduce the reduced functional

$$\hat{J}(u, u_F, u_Q) := J(y(u, u_F, u_Q), \tau(u, u_f, u_Q), u, u_F, u_Q).$$

Problem (10),(11) then is equivalent to

$$\min \hat{J}(u, u_F, u_Q), \quad \text{for } (u, u_F, u_Q) \in L^2(\Gamma)^M \times H^M \times (L^2)^M. \qquad (12)$$

Since $\hat{J}$ is a quadratic functional and the constraints (11) are linear, problem (12) admits a unique solution.

It is well known, that the gradient $\hat{J}'(u, u_F, u_Q)$ can be expressed with the help of the adjoint variables associated to (10),(11). Let us discuss the details for the case $M = 1$ which also forms the starting point of our investigations of the instantaneous control strategy. In the following, superscripts are dropped. The adjoint equations associated to problem (11) for $M = 1$ are given by

$$
\begin{aligned}
p^y - dt\,\nu\Delta p^y + \nabla p^p &= c_0(y - z) && \text{in } \Omega, \\
p^y &= 0 && \text{on } \Gamma, \\
-\text{div }p^y &= 0 && \text{in } \Omega, \\
p^\tau - dt\,a\Delta p^\tau &= c_1(\tau - S) - dt\,\gamma g p^y && \text{in } \Omega, \\
a\partial_\eta p^\tau &= -\alpha p^\tau && \text{on } \Gamma,
\end{aligned}
$$

where p^y, p^p denote the adjoint velocity field and pressure, respectively, and p^τ the adjoint temperature field. With the adjoint variables available, there holds

$$
\begin{aligned}
\left(\hat{J}'(u, u_F, u_Q), (v, v_f, v_Q) \right) &= (c_2 u - dt\,a\partial_\eta p^\tau, v)_{L^2(\Gamma)^2} \\
&\quad + (c_3 u_F + dt p^y, v_F)_H + (c_4 u_Q + dt p^\tau, v_Q)_{L^2} \qquad (13)
\end{aligned}
$$

In the instantaneous control approach the reduced optimization problem (12) is solved approximately, by applying only one steepest descent step to obtain an approximate solution [2, 7, 8, 9, 10, 11, 13, 16]. Instantaneous control therefore may be regarded as an inexact variant of MPC for $M = 1$.

To compute the gradient $\hat{J}'(u, u_F, u_Q)$ for given u, u_F, u_Q the coupled system of equations (11) and (13) has to be solved for p^y, p^p, p^τ. This is accomplished by using a preconditioned conjugate gradient method for the associated Schur-complement as proposed in [4] and [6], say.

In system (1) different control actions are possible. To optimize all of them simultaneously a suitable scaling of the gradient $\hat{J}'(u, u_F, u_Q)$ in the steepest descent method has to be introduced. This may be regarded as preconditioning and is achieved by replacing $\hat{J}'$ by $D\hat{J}'$ with D denoting as suitable 3×3 diagonal matrix. For more details see Section 4.3.2. The step size in the steepest descent method for $\hat{J}$ is computed exactly for the instantaneous control (IC) method. This is possible since $\hat{J}$ is quadratic in its arguments u, u_F and u_Q, compare [6]. The optimal step size in direction d is computed via a steepest descent step with trial step size ρ_p and calculating the minimum of the parabola defined by $\hat{J}(u_*), \hat{J}'(u_*)$ and $\hat{J}(u_* + \rho_p d)$, where $u_* = (u, u_F, u_Q)^T$. For more details we also refer to Section 4.3.4.

4. Numerical results

4.1. Introduction

We test IC and MPC for two numerical examples. The control goal in both examples consists in tracking of a desired velocity and a desired temperature field. In Example 1 the desired (normalized) temperature is zero and the desired velocity field is obtained from a forward simulations with pre-specified boundary temperature, see Section 4.3 for details. We investigate the performance of IC for all three control actions. It turns out that IC performs very well. These control results can be improved by applying MPC to larger time horizons, i.e., for $M > 1$. This is illustrated for boundary control in Section 4.4.1.

In Example 2 the desired velocity field is zero, and a non-trivial temperature field is desired. The control action here is given by boundary control alone. The control aim is to find a good trade-off between reaching the two aims in the case of boundary control. Difficulties are introduced into this control problem through the different time scales of steering the temperature vs. steering the velocity field. The IC fails, see Section 4.4.2, but the MPC is able to compute an acceptable control.

4.2. Implementation and numerical examples

All numerical examples are computed on a 20×20 equidistant grid on $\Omega := (0,1)^2$. For the velocity-pressure discretization the related staggered grid is used. The temperature is taken on the pressure nodes. The discretization of the Laplacian is based on the 5-point star. The parameters in our computations can be found in Table 1. The resulting Reynolds-number for Example 1 is $\mathrm{Re} = \frac{L\,\|z\|}{\nu} = 33.2$ (with $L = 1$ unit square), and for Example 2 it depends on the control. The Grashof-number for Example 2 is $\mathrm{Gr} = \frac{\gamma|g|L^3(\delta S)}{\nu^3} = 2060$ (were δS denotes the maximal temperature difference in the desired temperature field S) and for Example 1 it depends on the control.

All elliptic subproblems are solved with the SSOR method, which for the numerical examples presented below converges within a few steps.

For the parameter α in the boundary condition we also choose $\alpha = \infty$, i.e., Dirichlet conditions, to test the robustness of the algorithm.

The boundary conditions are $y = 0$ on Γ, and $\tau = 0$ on Γ if no boundary temperature control is used. The initial conditions are chosen as $y_0 \equiv 0$ and $\tau_0 \equiv 0$.

The time horizon for the integrations of J is given by $[0, T]$, with $T = 360$.

4.3. IC and Example 1

In this example we present a detailed discussion of the instantaneous control strategy (IC), which proves very powerful in various applications to flow control, see [2, 6] and the literature cited there. IC is an inexact variant of MPC for $M = 1$. For the approximate solution of the optimality system in this case only one steepest descent step is applied. The parameters used in our computations for this case can be found in Table 1, left.

Parameter	Example 1	Example 2
c_0	$2 \cdot 10^6$	$2 \cdot 10^6$
c_1	2	20
c_2	$2 \cdot 10^{-4}$	$2 \cdot 10^{-3}$
c_3	$2 \cdot 10^{-2} dt$	$2 \cdot 10^{-3} dt$
c_4	$2 \cdot 10^{-2} dt$	$2 \cdot 10^{-3} dt$
a	$1.44 \cdot 10^{-4}$	$1.44 \cdot 10^{-3}$
ν	$2.5 \cdot 10^{-4}$	$1 \cdot 10^{-3}$
γ	$2.1 \cdot 10^{-4}$	$2.1 \cdot 10^{-4}$
g	$(0, -9.81)$	$(0, -9.81)$

TABLE 1. Parameters for the examples 1 and 2.

The desired state z for Example 1 is depicted in Figure 1, left. It is the stationary velocity field obtained after a time of $t = 10000$ by choosing the constant boundary temperature of 1 on the right half of the lower boundary, and 0 otherwise. As desired temperature $S \equiv 0$ is chosen but with only a small weight in the cost functional, see Table 1. For tracking of an optimal trajectory (see 4.3.8) the desired state is the solution of an optimal control problem. For further details and results see [14].

4.3.1. Control actions. Three different control actions are investigated: distributed force, distributed heat, boundary temperature, and also their combinations. In all cases the steepest descent step for IC is initialized with zero control. The optimal steepest descent step size is used and the time step is set to $dt = 0.8$. The results are shown in the Figure 1, right, and figures 2, 3, 4. This figures show the temperature and velocity field at $T = 360$. In all cases the IC performs very well and is able to reach the desired state approximately. The evolution of cost functional J is shown in Figure 5. We take up again the case of boundary temperature control in Section 4.4.1.

4.3.2. Gradient scaling. Gradient scaling is preconditioning of the steepest descent method. It needs to be applied if combinations of control actions are used. In Figure 6 the values of the time integrated cost functional are presented for the parameter range $c_F, c_Q \in [10^{-7}, 10^{-1}] \times [10^{-5}, 10]$. As expected, for this example small values of c_Q give the best reduction of the cost functional. Here we apply diagonal scaling with the diagonal matrix

$$D = \begin{pmatrix} 1 & 0 & 0 \\ 0 & c_F & 0 \\ 0 & 0 & c_Q \end{pmatrix}.$$

In all numerical computations presented (except those of Figure 6) we set $c_F = 10^{-3}, c_Q = 0.3$. For this choice of parameters IC also performs very well on the other three examples investigated in [14]. However, we note no general rule for choosing the parameters c_F, c_Q is known yet.

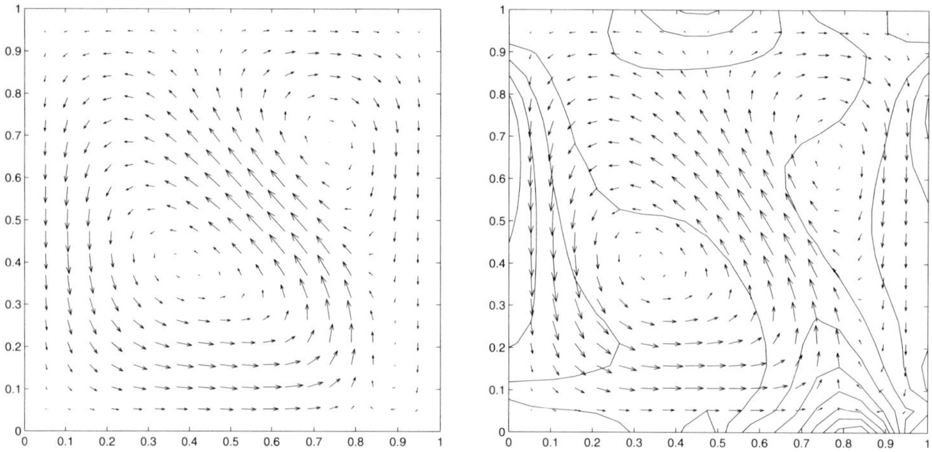

FIGURE 1. Left: Desired state. Right: Flow controlled by boundary temperature.

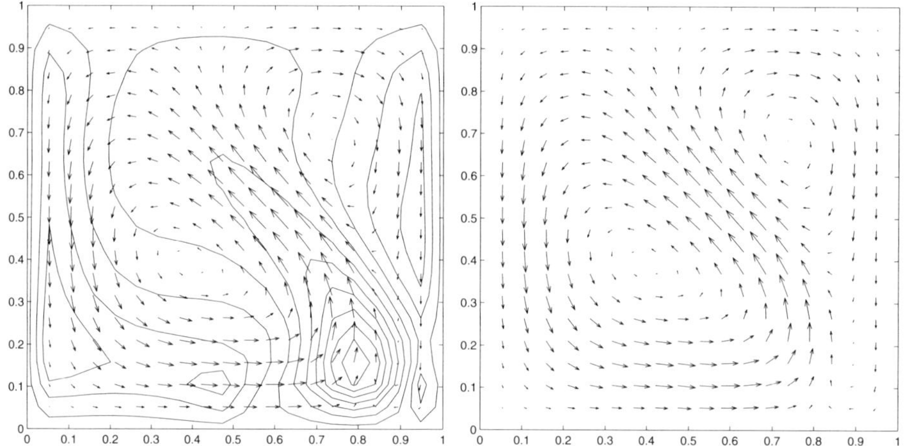

FIGURE 2. Left: Flow controlled by distributed heat. Right: Flow controlled by distributed force.

4.3.3. Initial controls in steepest descent. As initial values for the steepest descent method either the zero control or the control from the optimization at the previous time slice are chosen. It is observed that choosing zero control as initialization the performance of MPC/IC is very sensitive with regard to gradient scaling.

It is remarkable, that IC initialized with the control of the previous time slice in the long run performs similar to MPC with $M = 1$, provided the controls vary not too much between the time slices.

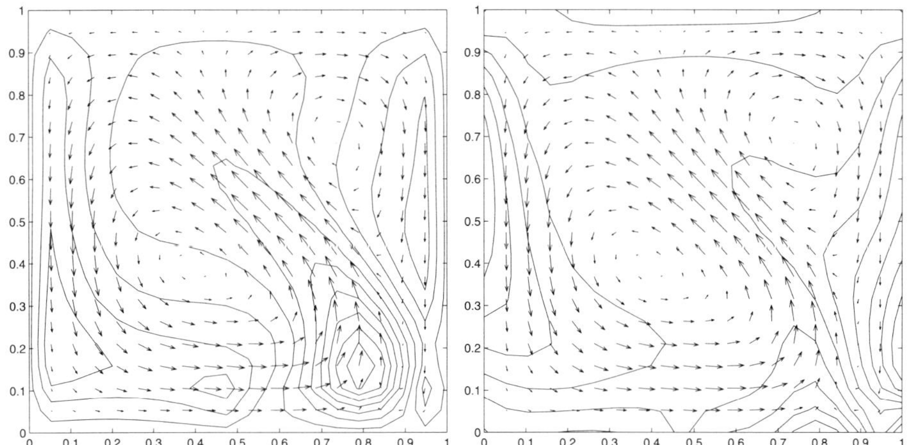

FIGURE 3. Left: Flow controlled by boundary temperature and distributed heat. Right: Flow controlled by boundary temperature and distributed force.

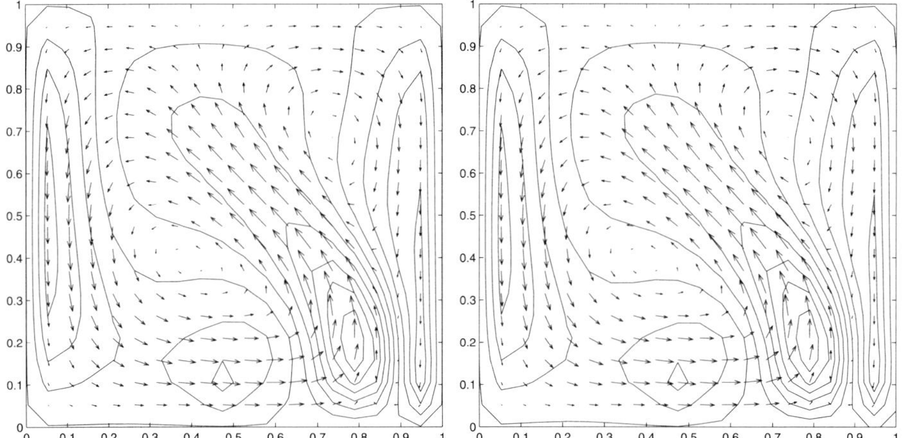

FIGURE 4. Left: Flow controlled by distributed heat and distributed force. Right: Flow controlled by all three controls.

Start with zero control is worse in cases where combinations of controls are used. On the other hand using the control from the previous time slice sometimes turns out to be less robust. This strongly depends on the dynamical behavior of the underlying physical process, see [14] for details.

4.3.4. Steepest descent step size. Since the IC control problem (10),(11) is linear-quadratic the optimal step size ρ^* in the steepest descent algorithm can be calculated exactly. In the calculations presented the value ρ^* is taken as minimum

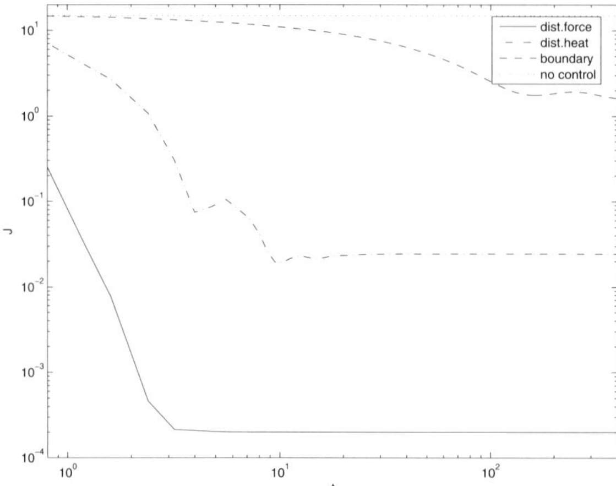

FIGURE 5. Evolution of cost functional J for some differed control actions.

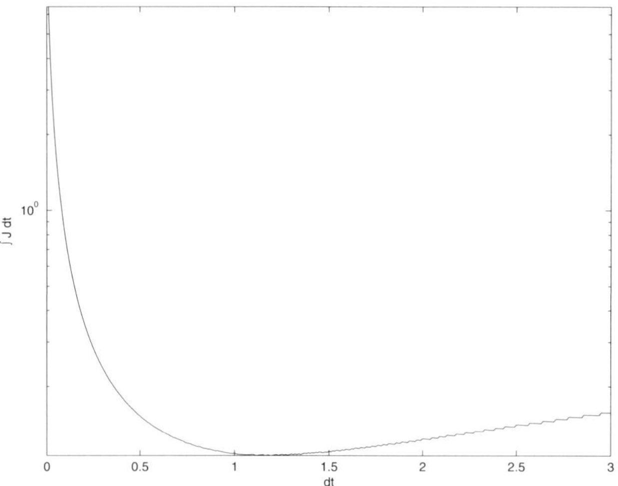

FIGURE 6. Different scaling of controls, all controls.

of the scalar parabola $h(\rho)$ defined by the function values $\hat{J}(u)$, $\hat{J}(u + \rho_p d)$ and the derivative $\hat{J}'(u)d$, where $d := -J'(u)$ and ρ_p is an estimation of the steepest descent step size taken from the optimization problem at the previous time slice. Compared to taking constant steepest descent step sizes the numerical overhead is caused by an additional function evaluation $\hat{J}(u + \rho_p d)$, which amounts to solving (11) with control $u + \rho_p d$.

Using a constant steepest descent step size ρ results in a slightly faster algorithm but requires knowledge about the magnitude of this step size. If the step size is too large the method diverges. Too short steepest descent steps lead to

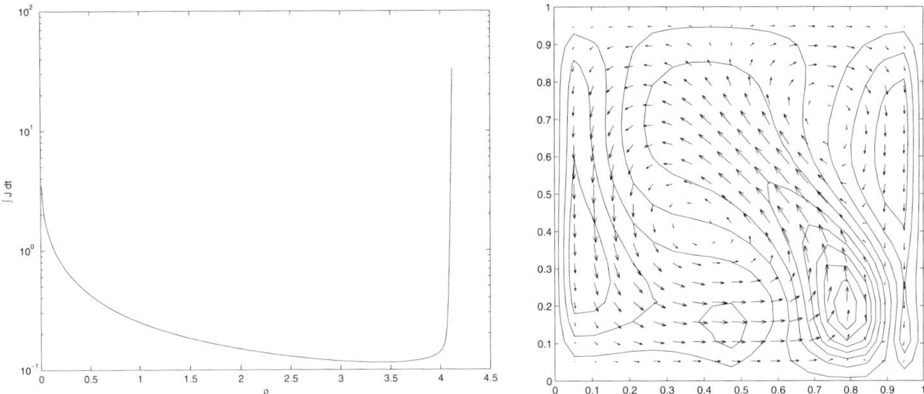

FIGURE 7. Left: Distributed heat, integrated functional for different constant steepest descent step sizes. Right: distributed heat, $dt = 0.1$.

ineffective controls, in particular when using zero initial control. In Figure 7, left, the dependence of $\int J dt$ on ρ is shown. Short steepest descent steps are worse with regard to reducing the functional J. On the other hand, if the steepest descent step is too large, the steepest descent method diverges. Because a useful steepest descent step size is not known a priori, we suggest to use the optimized steepest descent step size instead of a fixed steepest descent step.

4.3.5. Time step. The quality of controls obtained by the IC method depends on the length of the time step. Small time steps cause only weak control actions, so that time steps as large as possible, obeying the CFL conditions, should be taken. At greater time steps the control is more effective and so the IC predicts a greater win from producing stronger forces and heatings.

So the time step for reasons of control effectivity and computing time should be made as large as possible.

Now the performance of control at different time steps is investigated. For distributed heating compare the flows together with the temperature field in Figure 7, right $dt = 0.1$, Figure 2, left $dt = 0.8$, and Figure 8 $dt = 6.4$ respectively. As one can see flows and temperature distributions in all three cases look very similar. For larger time steps the controls and states are oscillating between two states, which are depicted in Figure 8. We note, that this is a purely numerical behavior caused by the large time step chosen.

The dependence of $\int J dt$ of dt is shown in Figure 9, left. As one can see cost reduction is most effective for $dt \approx 1.1$.

4.3.6. Simulation of practical control. In order to check the liability of numerically computed control procedures, the controls calculated on a 20×20 grid are applied to a discrete problem on a 39×39 grid. To extend the control, linear interpolation is used.

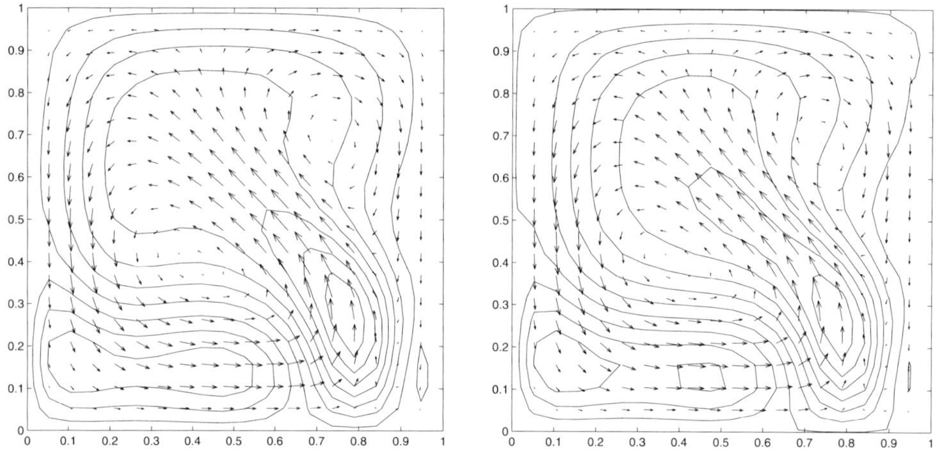

FIGURE 8. Distributed heat, $dt = 6.4$, the state and control are oscillating between these two states.

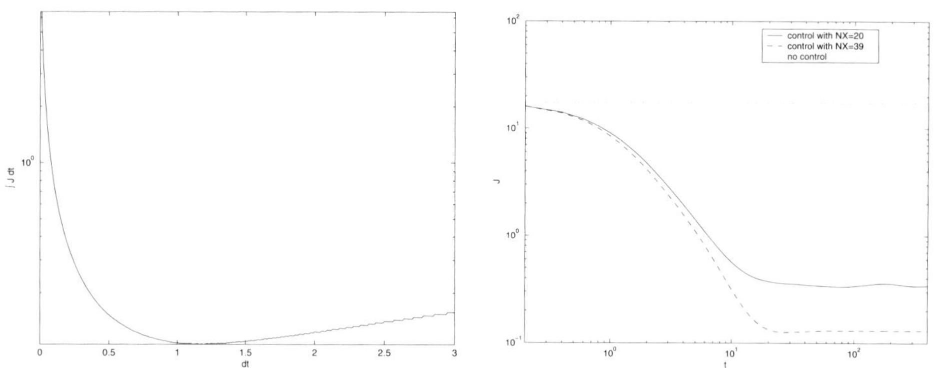

FIGURE 9. Left: Distributed heat and boundary temperature for different time steps dt. Right: distributed heat. The control from the coarser grid also works well on the finer grid.

As Figure 9, right shows, instantaneous controls obtained on the coarse grid perform pretty well also on the finer grid.

4.3.7. Comparison of IC and optimal open loop control. We now compare IC to optimal open loop control (OC). To obtain a discrete in time optimal open loop control on the time horizon $[0, T]$ the latter is divided into M time slices, and the

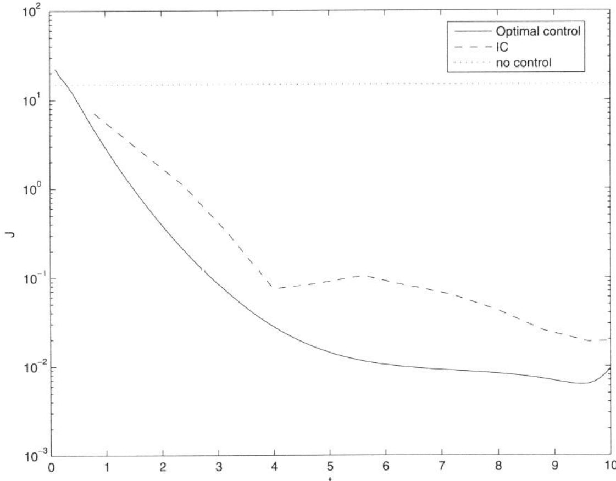

FIGURE 10. Instantaneous control vs. optimal control.

cost functional

$$J_{OC}(y,\tau,u,u_F,u_Q) = \sum_{i=1}^{M}\left(\frac{c_0}{2}\int_{\Omega}(y^i-z)^2dx + \frac{c_1}{2}\int_{\Omega}(\tau^i-S)^2dx \right. \tag{14}$$

$$\left. + \frac{c_2}{2}\int_{\Gamma}(u^i)^2dx + \frac{c_3}{2}\int_{\Omega}(u_F{}^i)^2dx + \frac{c_4}{2}\int_{\Omega}(u_Q{}^i)^2dx\right)$$

is minimized s. t. the constraints (5)–(9), i.e., we solve (10),(11) on $[0,T]$. Numerically this is performed by applying a limited memory BFGS method on the fully discrete system. We note that the discrete optimization problem contains $3 \cdot 10^5$ unknowns. The evolution of the cost functional (at each time slice) is compared to that obtained by IC in Figure 10. The control mechanism in this case is distributed heating. Parameters taken are $M = 100$ and $dt = 0.1$ for OC. For IC $dt = 0.8$ is chosen because shorter time horizons are worse, see Section 4.3.5 for the discussion of this fact. The coefficients and desired states are that of Example 1, compare Section 4.2.

4.3.8. Tracking of optimal control with IC. Once an optimal open loop trajectory is known it may serve as dynamical desired state to be tracked by the MPC strategy. In this context MPC, and in particular IC, serve as (nonlinear) closed loop control mechanisms. In Figure 11 the results for IC and varying time step sizes dt are shown, where the control mechanism is distributed heating. As can be seen, IC is able to track the optimal open loop trajectory.

We note that for IC tracking the optimal trajectory and IC applied to original desired state the cost functionals are different. In the case of IC tracking the optimal trajectory, the functional is

$$J_o(t) = \frac{c_0}{2} \int_\Omega (y(t) - y(t)^*)^2 dt + \frac{c_1}{2} \int_\Omega (\tau(t) - \tau(t)^*)^2 dt + \cdots$$

where $(y(t)^*, \tau(t)^*)$ denotes the optimal state. The original cost functional is given by

$$J(t) = \frac{c_0}{2} \int_\Omega (y(t) - z)^2 dt + \frac{c_1}{2} \int_\Omega (\tau(t) - S)^2 dt + \cdots.$$

In Figure 11, top, the evolution of J_o for tracking the optimal control for different time steps, is shown. The method works very well, especially for $dt = 0.1$ and $dt = 0.2$. In Figure 11, bottom a comparison of J for optimal control, IC for tracking of the optimal state, and IC applied to tracking of the desired state of the optimal control problem with $dt = 0.8$ is presented. Note that J is also calculated for IC tracking the optimal trajectory (whose cost functional is in fact J_o). The dashed line represents the evolution of J for IC applied to track the desired state of the optimal control problem. The dash-dotted line shows the evolution of J for tracking of the optimal trajectory y^*, τ^* obtained from the optimal control problem. The solid line shows the evolution of J for the optimal control. For IC is $dt = 0.8$ in both cases. Note that, as shown in the Figure 11, bottom, IC tracking the optimal trajectory would even perform better with shorter time steps (in contrast to IC applied to the original problem). As one can see IC in this example is well suited to track optimal trajectories (in the sense of a nonlinear closed loop controller) and also provides suboptimal controls with cost of the same magnitude as those of the optimal control procedure.

To investigate whether IC is able to stabilize a disturbed system random disturbances β are added at each time instance;

$$u_F^{dis} = u_F + \beta_F, \qquad u_Q^{dis} = u_Q + \beta_Q$$

The functions β_F, β_Q are random numbers defined in the corresponding nodes equally distributed over $[-1, 1]$. To get an impression of the size of the disturbances we mention that their size is approximately 17 times that of the control action in the undisturbed case after the initial decrease.

Figure 12 shows the same quantities as Figure 11, but for the disturbed case. In the top figure J represents the cost functional of IC with $z(t) = y^*(t)$, $S(t) = \tau^*(t)$, where y^*, τ^* denote the optimal state. In the bottom figure J represents also the cost functional of IC but z and S are the same as for the optimal control.

As one can see, IC is able to track the perturbed optimal trajectory in the sense of a closed-loop controller, whereas the unperturbed optimal control strategy seems to fail.

4.4. Model predictive control with $M > 1$

4.4.1. MPC and Example 1. Now we try to improve the boundary temperature control by using MPC. As in the case of IC, we only use one steepest descent step to solve the corresponding optimization problems approximately. The results are presented in Figure 13. To compare the performance of MPC to that of IC only

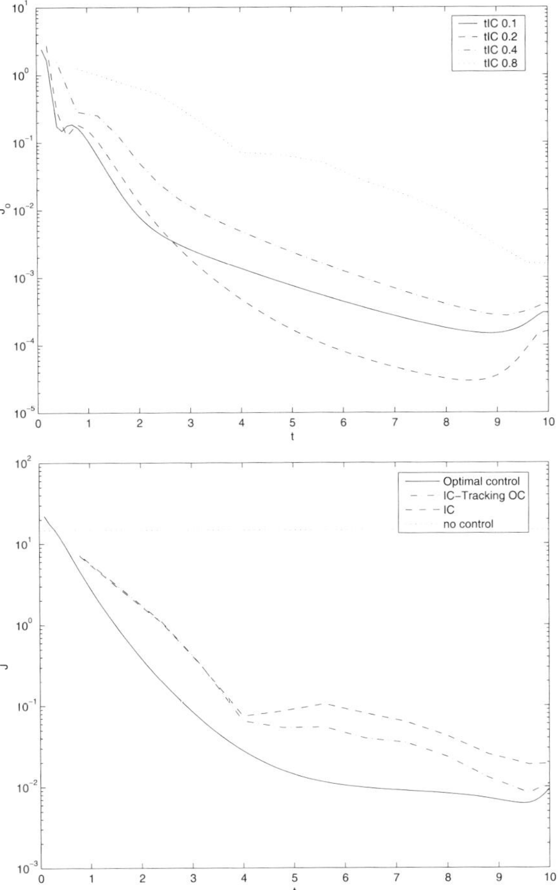

FIGURE 11. Tracking of optimal control with IC: evolution of J_o for tracking of optimal state (top), and comparison of J for optimal control, IC for tracking of the optimal state, and IC applied to tracking of the desired state of the optimal control problem with $dt = 0.8$ (bottom).

the values of the first addend in 10 are shown. Their values compare to those of the cost functional used for IC. As a result MPC with $M = 16, \ldots, 64$ and boundary heating reduces the (instantaneous) cost functional slower but in the long run as good as distributed heating with IC, see Figure 5 and Section 4.3. This is a substantial improvement compared to the control with IC.

4.4.2. MPC and Example 2. IC is not always successful in steering system states to desired states. However, as will be presented in the following, MPC on larger time horizons in general achieves this goal instead. In the present example we

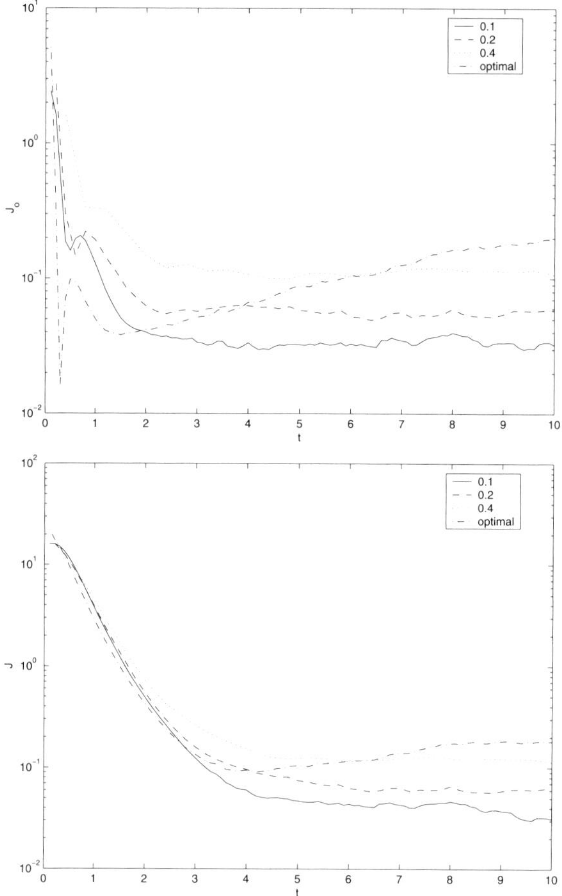

FIGURE 12. Tracking of optimal control with IC including perturbations; evolution of J for tracking of optimal state (top), and comparison of optimal control and IC tracking the optimal state (bottom).

choose $z \equiv 0$, and the desired temperature distribution is given by

$$S := \begin{cases} 1 & \text{in } [0.5, 1) \times (0, 1) \\ 0 & \text{in } (0, 0.5) \times (0, 1). \end{cases}$$

As control action boundary control is chosen. This means that the control problem consists in establishing different temperatures in the left and the right part of the domain, respectively, with velocity as small as possible. The parameters of the computation are shown in Table 1, right. We investigate (10),(11) for varying M, i.e., we vary the length of the prediction horizon in MPC.

As Figure 14 shows, MPC with $M \geq 8$ has to be applied in order to reduce the value of the cost functional. Smaller prediction horizons do not yield a reduction

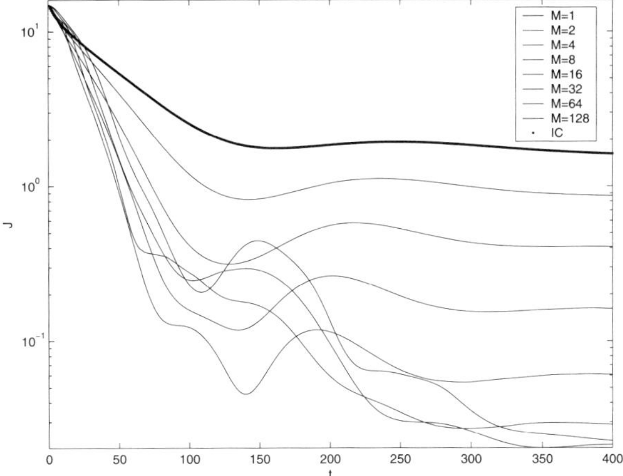

FIGURE 13. Performance of MPC in Example 1, start with control from last time slice.

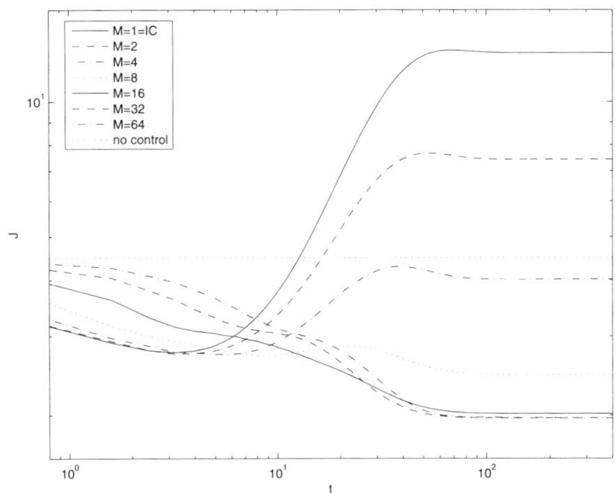

FIGURE 14. Performance of MPC in Example 2.

which is mainly caused by the fact that the velocity only has a negligible influence on the gradient of the cost functional for small time horizons. We also note that in this case the increase of $|y - z|$ is superior over the decrease of $|\tau - S|$.

5. Discussion and conclusions

Several control approaches to the Boussinesq approximation of the Navier-Stokes system are presented. Against the background of real-time control the instantaneous control method (IC) and model predictive control (MPC) mechanisms are studied in detail. As control actions, volume forces, distributed and boundary heating are considered.

IC performs very well in most of the investigated scenarios. Concerning the use of the control actions we may propose the following recipes;

- If tracking of a velocity is the control goal, either volume forces, or distributed heating, or a combination of both should be applied. Compared to their performance boundary temperature is less effective.
- If tracking of temperature distributions is the control goal, distributed heating combined with boundary heating should be applied. The influence of volume forces in this case is negligible.
- If a combination of control actions is chosen, the gradient of the cost functional has to be appropriately preconditioned in order to obtain a successful control method.

As is pointed out in Section 4.4.2, especially for tracking of temperature distributions, MPC on sufficiently large time horizons has to be applied.

IC also presents a powerful tool in the context of nonlinear closed-loop control. If an open-loop optimal control strategy for a process is given (i.e., computed a priori), IC may be used as a fast closed loop control mechanism which is capable of tracking the optimal open-loop control strategy, even in the presence of perturbations.

It is astonishing how well IC, and MPC perform in the sense of suboptimal control strategies for optimal control problems, as figures 10–12 indicate. These technics therefore also offer promising control tools for more realistic and complex configurations as they are dealt with in crystal growth, say.

6. Appendix

Proof of existence and uniqueness

Subsequently we use the notation $U = (v, \eta)$, $Y = (y, \tau)$, $W = (w, \kappa)$, and c denotes a positive generic constant. Similar to [15, Lemma 3.4]. we have

Lemma 6.1. *There holds*

$$|b(U, Y, W)| \leq c\,|v|_{L^2}\,\|Y\|_{V \times H^1(\Omega)}\,\|W\|_{V \times H^1(\Omega)} \quad \forall\, v \in V, \quad Y, W \in V \times H^1(\Omega).$$

If U belongs to $L^2(0, T; V \times H^1(\Omega)) \cap L^\infty(0, T; H \times L^2(\Omega))$ then $b(U, U, \cdot)$ belongs to $L^2(0, T; V^* \times (H^1(\Omega))^*)$ and

$$|b(U, U, \cdot)|_{L^2(0,T;V^* \times (H^1(\Omega))^*)} \leq c\,|U|_{L^\infty(0,T;H \times L^2(\Omega))}\,|U|_{L^2(0,T;V \times H^1(\Omega))}\,.$$

Proof of Lemma 6.1. By definition

$$b(U, Y, W) = \int_\Omega (v\nabla)yw \, dx + \int_\Omega (v\nabla)\tau\kappa \, dx.$$

With Hölders inequality and interpolation inequality, see [15, Lemma 3.3], we get

$$b(U, Y, W) \le c\,|v|_{L^4}\,|\nabla y|_{L^2}\,|w|_{L^4} + c\,|v|_{L^4}\,|\tau|_{H^1}\,|\kappa|_{L^4}$$

$$\le c\,|v|_{L^2}^{\frac12}\,|\nabla v|_{L^2}^{\frac12}\,|w|_{L^2}^{\frac12}\,|\nabla w|_{L^2}^{\frac12}\,|\nabla y|_{L^2} + c\,|v|_{L^2}^{\frac12}\,|\nabla v|_{L^2}^{\frac12}\,|\kappa|_{L^2}^{\frac12}\,|\kappa|_{H^1}^{\frac12}\,|\tau|_{H^1}.$$

If $U, Y, W \in V \times H^1(\Omega)$, the relation $b(U, Y, W) = -b(U, W, Y)$ gives

$$b(U, Y, W) \le c\,|v|_{L^2}^{\frac12}\,|\nabla v|_{L^2}^{\frac12}\,|y|_{L^2}^{\frac12}\,|\nabla y|_{L^2}^{\frac12}\,|\nabla w|_{L^2} + c\,|v|_{L^2}^{\frac12}\,|\nabla v|_{L^2}^{\frac12}\,|\tau|_{L^2}^{\frac12}\,|\tau|_{H^1}^{\frac12}\,|\kappa|_{H^1}.$$

This implies

$$|b(U, U, Y)| \le c\,|U|_{L^2}\,|U|_{V \times H^1}\,|Y|_{V \times H^1}.$$

If now $U \in L^2(0, T; V \times H^1(\Omega)) \cap L^\infty(0, T; H \times L^2(\Omega))$, then $b(U(t), U(t), \cdot) \in (V^* \times H^1(\Omega)^*)$ for almost every t and the estimate

$$|b(U(t), U(t), \cdot)|_{V^* \times (H^1(\Omega))^*} \le c\,|U(t)|_{L^2}\,|U(t)|_{V \times H^1}$$

implies that $b(U, U, \cdot)$ belongs to $L^2(0, T; V^* \times H^1(\Omega)^*)$. $\qquad\square$

Proof of Theorem 2.1. We begin with proving existence.

i) We apply the Galerkin procedure. Since $V \times H^1(\Omega)$ is separable and $\mathcal{V} \times \mathcal{C}^\infty(\Omega)$ is dense in $V \times H^1(\Omega)$, there exists a sequence $w_1, \ldots, w_m, \ldots$ of elements of $\mathcal{V} \times \mathcal{C}^\infty(\Omega)$, which is free and total in $V \times H^1(\Omega)$. For each $m \in \mathbb{N}$ we make the ansatz

$$Y_m = \sum_{i=1}^m g_{im}(t)w_i.$$

for an approximate solution Y_m of (2). Inserting Y_m into (2) and using w_j as test functions we obtain

$$\begin{aligned}
(Y_m'(t), w_j) &+ a(Y_m(t), w_j) + b(y_m(t), Y_m(t), w_j) \\
&+ (\gamma g \tau_m(t), w_{j\,12})_{L^2(\Omega)^2} + (\alpha \tau_m(t), w_{j\,3})_{L^2(\Gamma)} \\
&= \langle f(t), w_j \rangle + (\alpha u(t), w_{j\,3})_{L^2(\Gamma)}, \quad t \in [0, T], \ j = 1, \ldots, m,
\end{aligned} \tag{15}$$

$$Y_m(0) = Y_{0m}, \tag{16}$$

where Y_{0m} is the orthogonal projection in $H \times L^2(\Omega)$ of Y_0 onto the space spanned by $w_1, \dots, w_m$. Equations (15), (16) form a nonlinear system of differential equations for the functions $g_{1m}, \dots, g_{mm}$:

$$\sum_{i=1}^{m} (w_i, w_j) g'_{im}(t) + \sum_{i=1}^{m} a(w_i, w_j) g_{im}(t) + \sum_{i,l=1}^{m} b(w_{i12}, w_l, w_j) g_{im}(t) g_{lm}(t)$$
$$+ \sum_{i=1}^{m} (\gamma g w_{i3}, w_{j12})_{L^2(\Omega)^2} g_{im}(t) + \sum_{i=1}^{m} (\alpha w_{i3}, w_{j3})_{L^2(\Gamma)} g_{im}(t)$$
$$= \langle f(t), w_j \rangle + (\alpha u(t), w_{j3})_{L^2(\Gamma)}.$$

Since the mass matrix $(w_i, w_j)_{i,j=1}^{m}$ is nonsingular this system can be rewritten in the form

$$g'_{im}(t) + \sum_{i=1}^{m} \alpha_{ij} g_{jm}(t) + \sum_{i,k=1}^{m} \alpha_{ijk} g_{jm}(t) g_k m(t)$$
$$= \sum_{i=1}^{m} \beta_{ij} \langle f(t), w_j \rangle + \sum_{i=1}^{m} \tilde{\beta}_{ij} (au(t), w_{j3})_{L^2(\Gamma)},$$

$$\tag{17}$$

$$g_{im}(0) = (Y_{0m})_i, \tag{18}$$

with appropriate coefficients $\alpha_{ij}, \alpha_{ijk}, \beta_{ij}, \tilde{\beta}_{ij}$.

System (17), (18) admits a maximal solution defined on some interval $[0, t_m]$. If $t_m < T$, then $|Y_m(T)|$ must tend to $+\infty$ as $t \to t_m$; the a priori estimates we shall prove in ii) show that this can not happen and therefore $t_m = T$.

ii) A priori estimates.

We multiply (15) by $g_{jm}(t)$ and add the equations for $j = 1, \dots, m$. This gives

$$(Y'_m(t), Y_m(t)) + a(Y_m(t), Y_m(t)) + b(y_m(t), Y_m(t), Y_m(t))$$
$$+ (\gamma g \tau_m(t), y_m(t))_{L^2(\Omega)^2} + (\alpha \tau_m(t), \tau_m(t))_{L^2(\Gamma)}$$
$$= \langle f(t), Y_m(t) \rangle + (\alpha u(t), \tau_m(t))_{L^2(\Gamma)},$$

With [12, Lemma 2.1] and the fact that $\operatorname{div} y_m(t) = 0$ we get

$$b(y_m(t), Y_m(t), Y_m(t)) = 0,$$

and

$$(\gamma g \tau_m(t), y_m(t))_{L^2(\Omega)^2} \geq -c_1 |Y_m(t)|^2_{H \times L^2(\Omega)}.$$

We conclude now

$$\frac{d}{dt} |Y_m|^2 + 2a(Y_m(t), Y_m(t)) + 2(\alpha \tau_m(t), \tau_m(t))_{L^2(\Gamma)}$$
$$\leq 2c_1 |Y_m(t)|^2_{L^2(\Omega)} + 2 \langle f(t), Y_m(t) \rangle + 2 \langle \alpha u, \tau_m(t) \rangle_{L^2(\Gamma)},$$

which implies

$$\frac{d}{dt}\left|Y_m\right|^2 + 2\nu\left\|y_m(t)\right\|^2 + 2a\left\|\tau_m(t)\right\|^2 + 2(\alpha\tau_m(t),\tau_m(t))_{L^2(\Gamma)}$$

$$\leq 2c_1\left|Y_m(t)\right|^2_{L^2(\Omega)} + 2\left\langle u_F, y_m(t)\right\rangle + 2\left\langle u_Q, \tau_m(t)\right\rangle + 2\left\langle \alpha u, \tau_m(t)\right\rangle_{L^2(\Gamma)}.$$

Using

$$0 \leq (ca - c^{-1}b)^2 = c^2 a^2 + c^{-2}b^2 - 2ab \quad \forall a,b,c \in \mathbb{R}, c \neq 0$$

we get the estimates

$$2\left\langle u_F, y_m(t)\right\rangle \leq 2\left\|u_F(t)\right\|_{V^*}\left\|y_m(t)\right\|_V$$

$$\leq \nu\left\|y_m(t)\right\|^2_V + \frac{1}{\nu}\left\|u_F(t)\right\|^2_{V^*}, \tag{19}$$

$$2\left\langle u_Q, \tau_m(t)\right\rangle \leq 2\left\|u_Q(t)\right\|_{(H^1)^*}\left\|\tau_m(t)\right\|_{H^1}$$

$$\leq a\left\|\tau_m(t)\right\|^2_{H^1} + \frac{1}{a}\left\|u_Q(t)\right\|^2_{(H^1)^*}, \tag{20}$$

$$2\left\langle \alpha u(t), \tau_m(t)\right\rangle_{L^2(\Gamma)} \leq 2\left\|\alpha u(t)\right\|\left\|\tau_m(t)\right\|$$

$$\leq \alpha\left\|\tau_m(t)\right\|^2_{L^2(\Gamma)} + \frac{1}{\alpha}\left\|\alpha u(t)\right\|^2_{L^2(\Gamma)}, \tag{21}$$

and thus,

$$\frac{d}{dt}\left|Y_m(t)\right|^2 + \nu\left\|y_m(t)\right\|^2 + a\left\|\tau_m(t)\right\|^2 + \alpha\left|\tau_m(t)\right|^2_{L^2(\Gamma)}$$

$$\leq 2c_1\left|Y_m(t)\right|^2_{L^2(\Omega)} + \frac{1}{\nu}\left\|u_F(t)\right\|^2_{V^*} + \frac{1}{a}\left\|u_Q(t)\right\|^2_{(H^1)^*} + \left\|u(t)\right\|^2_{L^2(\Gamma)},$$

as well as

$$\frac{d}{dt}\left|Y_m(t)\right|^2 \leq 2c_1\left|Y_m(t)\right|^2_{L^2(\Omega)} + \frac{1}{\nu}\left\|u_F(t)\right\|^2_{V^*} + \frac{1}{a}\left\|u_Q(t)\right\|^2_{(H^1)^*} + \left\|u(t)\right\|^2_{L^2(\Gamma)}. \tag{22}$$

Integrating (22) from 0 to s we obtain

$$\left|Y_m(s)\right|^2 \leq \left|Y_m(0)\right|^2 + 2c_1\int_0^s\left|Y_m(t)\right|^2_{L^2(\Omega)}\,dt$$

$$+\frac{1}{\nu}\int_0^s\left\|u_F(t)\right\|^2_{V^*}\,dt + \frac{1}{a}\int_0^s\left\|u_Q(t)\right\|^2_{(H^1)^*}\,dt + \int_0^s\left\|u(t)\right\|^2\,dt.$$

Gronwall's Lemma then yields:

$$\left|Y_m(s)\right|^2 \leq \left(\left|Y_m(0)\right|^2 + \frac{1}{\nu}\int_0^s\left\|u_F(t)\right\|^2_{V^*}\,dt\right.$$

$$+ \frac{1}{a}\int_0^s\left\|u_Q(t)\right\|^2_{(H^1)^*}\,dt + \int_0^s\left\|u(t)\right\|^2\,dt\Big)e^{2c_1 s},$$

$$\left|Y_m(s)\right|^2 \leq \left(\left|Y_m(0)\right|^2 + \frac{1}{\nu}\int_0^T\left\|u_F(t)\right\|^2_{V^*}\,dt + \frac{1}{a}\int_0^T\left\|u_Q(t)\right\|^2_{(H^1)^*}\,dt\right.$$

$$+ \int_0^T\left\|u(t)\right\|^2\,dt\Big)e^{2c_1 T} \quad \forall s \in [0,T].$$

Hence,

$$\sup_{s\in[0,T]} |Y_m(s)|^2 \le (|Y_m(0)|^2 + \frac{1}{\nu}\int_0^T \|u_F(t)\|_{V^*}^2\, dt$$

$$+ \frac{1}{a}\int_0^T \|u_Q(t)\|_{(H^1)^*}^2\, dt + \int_0^T \|u(t)\|^2\, dt)e^{2c_1 T},$$

which implies that the sequence $\{Y_m\}_m$ remains in a bounded set of $L^\infty(0,T;H\times L^2(\Omega))$. Since $c_2\|\tau\|_{H^1(\Omega)} \le \|\tau\|_{H_0^1(\Omega)} + \|\tau\|_{L^2(\Gamma)}$ for some $c_2 > 0$ we get

$$c_3\|Y_m(t)\|_{H^1(\Omega)} \le \nu\|y_m(t)\|_V^2 + a\|\tau_m(t)\|_{H_0^1(\Omega)}^2 + \alpha\|\tau_m(t)\|_{L^2(\Gamma)}^2\,.$$

From (22) we now deduce

$$\frac{d}{dt}|Y_m(t)|^2 + c_3\|Y_m(t)\|_{H^1(\Omega)}$$
$$\le 2c_1|Y_m(t)|_{L^2(\Omega)}^2 + \frac{1}{\nu}\|u_F(t)\|_{V^*}^2 + \frac{1}{a}\|u_Q(t)\|_{(H^1)^*}^2 + \|u(t)\|^2 \tag{23}$$

Now we integrate (23) from 0 to T and apply Gronwall's Lemma once more to obtain the estimate

$$|Y_m(T)|^2 + c_3\int_0^T \|Y_m(t)\|_{H^1(\Omega)}\, dt \le (|Y_m(0)|^2 + \frac{1}{\nu}\int_0^T \|u_F(t)\|_{V^*}^2\, dt$$

$$+\frac{1}{a}\int_0^T \|u_Q(t)\|_{(H^1)^*}^2\, dt + \int_0^T \|u(t)\|^2\, dt)\exp(2c_1 T).$$

This implies that the sequence $\{Y_m\}_m$ remains in a bounded set of $L^2(0,T;V\times H^1(\Omega))$.

It is now straightforward to conclude that a subsequence $Y_{m'}$ exists such that $Y_{m'} \to Y$ in $L^2(0,T,V\times H^1(\Omega))$ weakly, and in $L^\infty(0,T,H\times L^2(\Omega))$ weak-star.

iii) Now we will show that $Y_{m'} \to Y$ in $L^2(0,T,H\times L^2(\Omega))$ strongly.

For this purpose we firstly show that $\{\frac{d}{dt}Y_m\} \subset L^{4/3}(V^*\times(H^1(\Omega))^*)$, see Constantin and [3]. First let us consider (15). Since $\{Y_m\}_m$ is bounded in $L^2(0,T;V\times H^1(\Omega))$ it follows that $a(Y_m,\cdot)$ is bounded in $L^2(0,T;V^*\times(H^1(\Omega))^*)$. This also holds for the forms $(\gamma g\tau_m,\cdot)_{L^2(\Omega)^2} + (a\tau_m,\cdot)_{L^2(\Gamma)}$. Note that $Y_m = (y_m,\tau_m)^T)$. By assumption f is bounded in $L^2(0,T;V^*\times(H^1(\Omega))^*)$.

It remains to investigate the term $b(y_m, Y_m,\cdot)$.
Using

$$\|v\|_{L^4(\Omega)} \le \|\nabla v\|_{L^2(\Omega)} \le \|v\|_{V\times H^1(\Omega)}\,,$$

$$\|y_m\|_{L^4(\Omega)} \le c\|y_m\|_{L^2(\Omega)}^{\frac{1}{2}}\|\nabla y_m\|_{L^2(\Omega)}^{\frac{1}{2}}\,,$$

$$\|\tau_m\|_{L^4(\Omega)} \le c\|\tau_m\|_{L^2(\Omega)}^{\frac{1}{2}}\|\tau_m\|_{H^1(\Omega)}^{\frac{1}{2}}\,,$$

and the Hölder-inequality we get

$$\int_\Omega (y_m \nabla) Y_m v \, dx = \int_\Omega (y_m \nabla) y_m v_y \, dx + \int_\Omega (y_m \nabla) \tau_m v_\tau \, dx$$

$$\leq c_1 \|y_m\|_{L^4(\Omega)} \|\nabla y_m\|_{L^2(\Omega)} \|v_y\|_{L^4(\Omega)} + c_2 \|y_m\|_{L^4(\Omega)} \|\tau_m\|_{H^1(\Omega)} \|v_\tau\|_{L^4(\Omega)}$$

$$\leq c_3 \|y_m\|_{L^2(\Omega)}^{\frac{1}{2}} \|\nabla y_m\|_{L^2(\Omega)}^{\frac{1}{2}} (\|\nabla y_m\|_{L^2(\Omega)} \|v_y\|_V + \|\tau_m\|_{H^1(\Omega)} \|v_\tau\|_{H^1(\Omega)})$$

$$\leq c_3 \|y_m\|_{L^2(\Omega)}^{\frac{1}{2}} \|\nabla y_m\|_{L^2(\Omega)}^{\frac{1}{2}} (\|\nabla y_m\|_{L^2(\Omega)} + \|\tau_m\|_{H^1(\Omega)}) \|v\|_{V \times H^1(\Omega)},$$

so that

$$\int_0^T \|b(y_m, Y_m, \cdot)\|_{V^* \times (H^1(\Omega)^*)}^{\frac{4}{3}} \, dt$$

$$\leq c_4 \int_0^T \|y_m\|_{L^2(\Omega)}^{\frac{2}{3}} (\|\nabla y_m\|_{L^2(\Omega)}^{\frac{3}{2}} + \|\nabla y_m\|_{L^2(\Omega)}^{\frac{1}{2}} \|\tau_m\|_{H^1(\Omega)})^{\frac{4}{3}} \, dt$$

$$\leq 2c_4 \int_0^T \|y_m\|_{L^2(\Omega)}^{\frac{2}{3}} \max(\|\nabla y_m\|_{L^2(\Omega)}^2, \|\nabla y_m\|_{L^2(\Omega)}^{\frac{2}{3}} \|\tau_m\|_{H^1(\Omega)}^{\frac{4}{3}}) \, dt.$$

Since $\|Y_m(t)\|_{L^2(\Omega)}$ is bounded uniformly, the right argument of the max-function can be estimated as

$$\int_0^T \|\nabla y_m\|_{L^2(\Omega)}^{\frac{2}{3}} \|\tau_m\|_{H^1(\Omega)}^{\frac{4}{3}} \, dt \leq \left(\int_0^T \|\nabla y_m\|_{L^2(\Omega)}^2 \, dt \right)^{\frac{1}{3}} \left(\int_0^T \|\tau_m\|_{H^1(\Omega)}^2 \, dt \right)^{\frac{2}{3}},$$

and since $\|\nabla y_m(t)\|_{L^2(\Omega)}^2$ and $\|T_m(t)\|_{H^1(\Omega)}^2$ are uniformly integrable with respect to m, we have $\{\frac{d}{dt} Y_m\} \subset L^{4/3}(V^* \times (H^1(\Omega))^*)$. Together with $\{Y_m\} \subset L^2(0, T; V \times H^1(\Omega))$ from ii) it follows that $\{Y_m\} \subset W_{4/3}^2(0, T; V \times H^1(\Omega))$ is bounded.

By the Aubin-Dubinskii-Lemma, see [3], $W_{4/3}^2(0, T; V \times H^1(\Omega))$ compactly embeds into $L^2(0, T; H \times L^2(\Omega))$. Therefore $Y_{m'} \to Y$ in $L^2(0, T, H \times L^2(\Omega))$ strongly for a subsequence.

iv) This convergence results enable us to pass to the limit in (15)–(16). Let ψ be a continuously differentiable function on $[0, T]$ with $\psi(T) = 0$. We multiply (15) by $\psi(t)$, and integrate by parts. This leads to

$$-\int_0^T (Y_m(t), \psi'(t) w_j) dt + \int_0^T a(Y_m(t), w_j \psi(t)) dt + \int_0^T b(y_m(t), Y_m(t), w_j \psi(t)) dt$$

$$+ \int_0^T (\gamma g \tau_m(t), w_{j_{12}} \psi(t))_{L^2(\Omega)^2} dt + \int_0^T (\alpha \tau_m(t), w_{j_3} \psi(t))_{L^2(\Gamma)} dt \tag{24}$$

$$= \int_0^T \langle f(t), w_j \psi(t) \rangle \, dt + \int_0^T (\alpha u(t), w_{j_3} \psi(t))_{L^2(\Gamma)} dt + (Y_{0m}, w_j) \psi(0).$$

Passing to the limit with the sequence m' is easy for the linear terms; for the nonlinear term we apply [15, Lemma 3.2] and obtain for every vector function w

with components in $C^1(\overline{(0,T) \times \Omega})$

$$\int_0^T b(y_\mu(t), Y_\mu(t), w(t))dt \to \int_0^T b(y(t), Y(t), w(t))dt \quad (\mu \to \infty).$$

In the limit we find that the equation

$$-\int_0^T (Y(t), \psi'(t)U)dt + \int_0^T a(Y(t), U\psi(t))dt + \int_0^T b(y(t), Y(t), U\psi(t))dt$$
$$+ \int_0^T (\gamma g\tau(t), v\psi(t))_{L^2(\Omega)^2}dt + \int_0^T (\alpha\tau(t), w\psi(t))_{L^2(\Gamma)}dt \tag{25}$$
$$= \int_0^T \langle f(t), U\psi(t)\rangle \, dt + \int_0^T (\alpha u(t), w\psi(t))_{L^2(\Gamma)}dt + (Y_0, U)\psi(0),$$

holds for $U = (v,w)^T$ in the set $\{w_1, w_2, \ldots\}$; by linearity this equation holds for U equal to any finite linear combination of the w_j, and by a continuity argument (25) is still valid for any $U \in V \times H^1(\Omega)$. Thus, Y satisfies (2) in the distributional sense.

Finally, it remains to prove that Y satisfies the initial condition (3). To show this we multiply (2) by ψ, and integrate. Integrating the first term by parts, gives

$$-\int_0^T (Y(t), \psi'(t)U)dt + \int_0^T a(Y(t), U\psi(t))dt + \int_0^T b(y(t), Y(t), U\psi(t))dt$$
$$+ \int_0^T (\gamma g\tau(t), v\psi(t))_{L^2(\Omega)^2}dt + \int_0^T (\alpha\tau(t), w\psi(t))_{L^2(\Gamma)}dt \tag{26}$$
$$= \int_0^T \langle f(t), U\psi(t)\rangle \, dt + \int_0^T (\alpha u(t), w\psi(t))_{L^2(\Gamma)}dt + (Y(0), U)\psi(0).$$

By comparison with (25),

$$(Y(0) - Y_0, U)\psi(0) = 0.$$

Now we choose ψ with $\psi(0) = 1$; thus

$$(Y(0) - Y_0, U) = 0. \quad \forall U \in V \times H_1(\Omega),$$

and (3) follows.

Uniqueness:

i) We first note that $b(U, U, \cdot)$ belongs to $L^2(0, T; V^* \times H^1(\Omega)^*)$, see Lemma 6.1, which implies that Y' also belongs to $L^2(0, T; V^* \times H^1(\Omega)^*)$.

This enables us to apply [15, Lemma 1.2 in Ch. III], which claims that Y is almost everywhere equal to a continuous function. Thus

$$Y \in \mathcal{C}([0,T]; H \times L^2(\Omega)),$$

and (4) follows immediately. The same lemma asserts that for any function Y in $L^2(0, T; V \times H^1(\Omega))$ which satisfies $Y' \in L^2(0, T; V^* \times (H^1(\Omega))^*)$, the equation

$$\frac{d}{dt}|Y(t)|^2 = 2\langle Y'(t), Y(t)\rangle \tag{27}$$

is valid, which will be used below.

ii) *Proof of uniqueness.* Let us assume that Y_1 and Y_2 are two solutions of (2)–(3), and let $Y = Y_1 - Y_2$. As shown before Y_1, Y_2, and thus Y are in $L^2(0,T; V^* \times (H^1(\Omega))^*)$. The difference $Y = Y_1 - Y_2$ satisfies

$$\frac{d}{dt}(Y,U) + a(Y,U) + (\gamma g \tau, v)_{L^2(\Omega)^2} + (\alpha \tau, \eta)_{L^2(\Gamma)} = b(y_2, Y_2, U) - b(y_1, Y_1, U)$$

$$\forall U \in V \times H^1(\Omega), \text{ and almost all } t \in (0,T), \tag{28}$$

$$Y(0) = 0. \tag{29}$$

Taking $U = Y(t)$ and using (27), we get

$$\frac{d}{dt}|Y(t)|^2 + 2\nu \|y(t)\|^2 + 2a \|\tau(t)\|^2 + 2\alpha |\tau(t)|^2_{L^2(\Gamma)} + 2\langle \gamma g \tau(t), y(t) \rangle$$
$$= 2b(y_2(t), Y_2(t), Y(t)) - 2b(y_1(t), Y_1(t), Y(t)).$$

Since $b(v, W, W) = 0 \quad \forall v \in V, W \in H_0^1(\Omega)^2 \times H^1(\Omega)$, the right-hand side is equal to

$$-2b(y(t), Y_2(t), Y(t)).$$

From Lemma 6.1 we deduce

$$|-2b(y(t), Y_2(t), Y(t))|$$

$$\leq c\,|y|^{\frac{1}{2}}_{L^2} |\nabla y|^{\frac{1}{2}}_{L^2} |y|^{\frac{1}{2}}_{L^2} |\nabla y|^{\frac{1}{2}}_{L^2} |\nabla y_2|_{L^2} + c\,|y|^{\frac{1}{2}}_{L^2} |\nabla y|^{\frac{1}{2}}_{L^2} |\tau|^{\frac{1}{2}}_{L^2} |\tau|^{\frac{1}{2}}_{H^1} |\tau_2|_{H^1}$$
$$\leq c\,|y|_{L^2} |\nabla y_2|_{L^2} |Y|_{V \times H^1(\Omega)} + c\,|y|_{L^2} |\tau_2|_{H^1} |Y|_{V \times H^1(\Omega)}$$
$$\leq c\,|Y|_{L^2} |Y_2|_{V \times H^1(\Omega)} |Y|_{V \times H^1(\Omega)}.$$

Using Young's inequality we estimate further

$$c|Y|_{L^2} |Y_2|_{V \times H^1(\Omega)} |Y|_{V \times H^1(\Omega)} \leq 2\min(\nu,a)|Y|^2_{V \times H^1(\Omega)} + \frac{1}{\epsilon}(c|Y|_{L^2} |Y_2|_{V \times H^1(\Omega)})^2.$$

The term

$$|2\langle \gamma g \tau(t), y(t) \rangle| \leq 2\gamma |g| |Y|^2_{L^2(\Omega)}$$

is also majorized. We can conclude

$$\frac{d}{dt}|Y(t)|^2 \leq (\frac{c^2}{\epsilon}|Y_2|^2_{V \times H^1(\Omega)} + 2\gamma |g|)|Y(t)|^2_{L^2(\Omega)} \quad \forall t \in [0,T],$$

so that

$$\frac{d}{dt}|Y(t)|^2 \leq c(\nu, a, \gamma g, Y_2(t), t)|Y(t)|^2_{L^2(\Omega)} \quad \forall t \in [0,T].$$

Integrating from 0 to s and using (29) gives

$$|Y(s)|^2 \leq \int_0^s c(\nu, a, \gamma g, Y_2(t), t)|Y(t)|^2 \, dt$$

Finally Gronwall's-Lemma implies $|Y(s)|^2 \leq 0 \ \forall s \in [0,T]$, which gives

$$Y_1 = Y_2,$$

so that the solution of (1) is unique. $\qquad \square$

References

[1] F. Abergel & R. Temam: *On Some Control Problem in Fluid Mechanics.* Theoretical and Computational Fluid Dynamics, 303–325, Springer, 1990.

[2] H. Choi, M. Hinze & K. Kunisch: *Instantaneous Control of Backward-Facing Step Flows.* Applied Numerical Mathematics 31, 133–158 (1999).

[3] P. Constantin & C. Foias: *Navier-Stokes Equations.* Chicago Lectures in Mathematics, The University of Chicago Press 1988.

[4] R. Glowinski: *Finite element methods for the numerical simulation of incompressible viscous flow; Introduction to the control of the Navier-Stokes equations.* Lectures in Applied Mathematics, 28, 1991.

[5] M.D. Gunzburger: *Perspectives in Flow Control and Optimization.* SIAM, 2003.

[6] M. Hinze: *Optimal and instantaneous control of the instationary Navier-Stokes equations.* Habilitationsschrift, 2000. Fachbereich Mathematik, Technische Universität Berlin.

[7] M. Hinze: *Instantaneous closed loop control of the Navier-Stokes system.* Siam J. Cont. Optim. 44, 564–583 (2005)

[8] M. Hinze & K. Kunisch: *Control strategies for fluid flows – optimal versus suboptimal control.* ENUMATH 97, Eds. H.G. Bock et al., World Scientific, Singapore, 351–358 (1997)

[9] M. Hinze & S. Volkwein: *Instantaneous control for the instationary Burgers equation – convergence analysis & numerical implementation.* Nonlinear Analysis T.M.A. 50, 1–26 (2002)

[10] M. Hinze & D. Wachsmuth: *Fast closed loop control of the Navier-Stokes system.* Modeling, Simulation and Optimization of Complex Processes, Eds. H.G. Bock et al., Springer, 189–201 (2004)

[11] K. Kunisch & X. Marduel: *Optimal control of non-isothermal viscoelastic fluid flow.* J. Non-Newtonian Fluid Mech. 88, No.3, 261–301 (2000).

[12] H.-C. Lee & O. Yu. Imanuvilov: *Analysis of Neumann Boundary optimal Control problems for the stationary Boussinesq equations including solid media.* SIAM J. Control Optim. Vol. 39, No.2, pp. 457–477 2000.

[13] X. Marduel & K. Kunisch: *Suboptimal Control of Transient Non-Isothermal Viscoelastic Fluid Flow.* Preprint, Special Research Center for Optimization and Control, University of Graz, (1999).

[14] U. Matthes: *Instantane Kontrolle der Boussinesq Approximation.* Diplomarbeit, 2003. Fachrichtung Mathematik, Technische Universität Dresden.

[15] R. Temam: *Navier-Stokes Equations, Theory and Numerical Analysis.* North-Holland 1984.

[16] D. Wachsmuth: *Numerische Analysis eines Verfahrens der Momentansteuerung.* Diplomarbeit, 2002. Fakultät für Mathematik, Technische Universität Chemnitz.

Michael Hinze and Ulrich Matthes, Institut für Numerische Mathematik
Technische Universität Dresden, D–01069 Dresden, Germany
e-mail: `hinze@math.tu-dresden.de`
e-mail: `matthes@math.tu-dresden.de`

International Series of Numerical Mathematics, Vol. 155, 175–192
© 2007 Birkhäuser Verlag Basel/Switzerland

Applications of Semi-smooth Newton Methods to Variational Inequalities

Kazufumi Ito and Karl Kunisch

Abstract. This paper discusses semi-smooth Newton methods for solving nonlinear non-smooth equations in Banach spaces. Such investigations are motivated by complementarity problems, variational inequalities and optimal control problems with control or state constraints, for example. The function $F(x)$ for which we desire to find a root is typically Lipschitz continuous but not C^1 regular. The primal-dual active set strategy for the optimization with the inequality constraints is formulated as a semi-smooth Newton method. Sufficient conditions for global convergence assuming diagonal dominance are established. Globalization strategies are also discussed assuming that the merit function $|F(x)|^2$ has appropriate descent directions.

1. Introduction

Examples which motivate our study include nonlinear variational inequalities of the form: find $x \in C$ such that

$$(f(x), y - x) \geq 0 \quad \text{for all} \quad y \in C, \tag{1.1}$$

where C is a closed convex set in a Hilbert space X and $f : X \to X$ is C^1. It can equivalently be written as

$$F(x) = x - Proj_C(x - f(x)) = 0, \tag{1.2}$$

where $Proj_C$ is the projection of X onto C. In particular, let Ω be a bounded domain in R^N and if C is a hypercube $\{x \mid \phi \leq x \leq \psi\}$ in $X = L^2(\Omega)$, with $\phi \leq \psi$ and the inequalities are defined pointwise, then (1.1) can be expressed as

$$F(x) = \mu - \max(0, \mu + x - \psi) - \min(0, \mu + x - \phi), \quad \mu = -f(x) \tag{1.3}$$

The first author is partially supported by the Army Research Office under DAAD19-02-1-039. and the second author is supported in part by the Fonds zur Förderung der wissenschaftlichen Forschung under SFB 03 "Optimierung und Kontrolle".

where $\mu \in X$ is the Lagrange multiplier. For example, consider a boundary control problem for the heat equation on a bounded open domain D in R^3;

$$\min_{u \in X} \frac{1}{2} \int_0^T \int_\Omega |y - \bar{y}|^2 \, dx \, dt + \frac{\alpha}{2} |u|_X^2 \tag{1.4}$$

subject to

$$\frac{\partial}{\partial t} y = \Delta y, \quad y(0, \cdot) = y_0 \text{ in } D$$

$$\tag{1.5}$$

$$\frac{\partial}{\partial \nu} y + y^3 = u \text{ on } (0, T) \times \partial D$$

with $X = L^2((0,T) \times \partial D)$, $\bar{y}$ and $y_0 \in L^2(\Omega)$. Then $f : X \to X$ is defined by

$$f(u) = \alpha \, u + p|_{\partial D}$$

where p satisfies the adjoint equation

$$\frac{\partial}{\partial t} p + \Delta p + y - \bar{y} = 0, \quad p(T, \cdot) = 0 \text{ in } D$$

$$\frac{\partial p}{\partial \nu} + 3y^2 \, p = 0 \text{ on } \partial D.$$

Note that F is a locally Lipschitz continuous functions but is not C^1, even if f is C^1.

If F is locally Lipschitz continuous on R^m, then according to Rademacher's theorem, F is differentiable almost everywhere. Let D_F denote the set of points at which F is differentiable and let $\partial_B F(x)$ be defined by

$$\partial_B F(x) = \left\{ J = \lim_{x_i \to x, \, x_i \in D_F} F'(x_i) \right\}. \tag{1.6}$$

We denote by $\partial F(x)$ the generalized derivative in the sense of Clarke, i.e.,

$$\partial F(x) = \text{the convex hull of } \partial_B F(x). \tag{1.7}$$

A generalized Newton iteration for solving the nonlinear equation $F(x) = 0$ is defined by

$$x^{k+1} = x^k - V_k^{-1} F(x^k), \text{ where } V_k \in \partial_B F(x^k). \tag{1.8}$$

In the finite-dimensional case a generalized Jacobian $V_k \in \partial_B F(x^k)$. Local convergence of $\{x^k\}$ to x^*, a solution of $F(x) = 0$, is based on the following concepts;

$$|F(x^* + h) - F(x^*) - V\,h| = o(|h|), \tag{1.9}$$

where $V = V(x^* + h) \in \partial_B F(x^* + h)$, for $x^* + h$ in a neighborhood of x^*. Thus, letting $h = x^k - x^*$ and $V^k = V(x^k)$ we have

$$|x^{k+1} - x^*| = |V_k^{-1}(F(x^k) - F(x^*) - V_k(x^k - x^*))| = o(|x^k - x^*|).$$

In the finite-dimensional case under appropriate assumptions (1.9) is equivalent to semi-smoothness of F at x^* [18, 17]. The notion of the semi-smooth was introduced originally by Miffin for functionals [15]. Convex functions, smooth functions and subsmooth functions are examples of semi-smooth functions in R^m. Rademacher's theorem does not hold in Banach spaces. The discussion on the generalized Newton's method suggests the following definition [6] in Banach spaces X, Z (we refer to [2, 14, 19] for similar concepts and discussions):

Definition 1.1.

(a) *Let $D \subset X$ be an open set. $F \colon D \subset X \to Z$ is called Newton differentiable at x, if there exists an open neighborhood $N(x) \subset D$ and mappings $G \colon N(x) \to \mathcal{L}(X, Z)$ such that*

$$\lim_{|h| \to 0} \frac{|F(x + h) - F(x) - G(x + h)h|_Z}{|h|_X} = 0. \tag{1.10}$$

The family $\{G(x) : x \in N(x)\}$ is called a N-derivative of F at x.

(b) *F is called semi-smooth at x, if it is Newton differentiable at x and*

$$\lim_{t \to 0^+} G(x + t h)h \quad \text{exists uniformly in } |h| = 1. \tag{1.11}$$

Now, we have a generalized Newton method in the Banach space X

$$x^{k+1} = x^k - V_k^{-1} F(x^k), \text{ where } V_k = G(x^k) \tag{1.12}$$

assuming G is a Newton derivative of F.

The outline of the paper is as follows. In Section 2 we discuss semi-smooth functions in infinite-dimensional spaces and convergence of the generalized Newton method (1.12). In Section 3 we formulate the primal-dual active set method for (1.3) as a semi-smooth Newton method and verify global convergence under diagonal dominant conditions. In Section 4 we discuss variational inequality problems in Hilbert spaces and the application of the semi-smooth Newton method for associated regularized problems. Globalization of the semi-smooth Newton method is addressed in Section 5.

2. Semi-smooth functions and local superlinear convergence

Let $D \subset X$ be an open set.
F is directionally differentiable at $x \in D$, if

$$\lim_{t \to 0^+} \frac{F(x + t h) - F(x)}{t} =: F'(x; h)$$

exists for all $h \in X$.
F is B (Bouligand)-differentiable at $x \in D$, if F is directionally differentiable at x and

$$\lim_{|h| \to 0} \frac{F(x + h) - F(x) - F'(x; h)}{|h|_X} = 0.$$

Lemma 2.1. *Suppose that $F : D \subset X \to Z$ is Newton differentiable at $x \in D$ with N-derivative G.*

(1) *F is directionally differentiable at x if and only if*

$$\lim_{t\to 0^+} G(x+t\,h)h \quad \text{exists for all } h \in X. \tag{2.1}$$

In this case $F'(x;h) = \lim_{t\to 0^+} G(x+th)h$ for all $h \in X$.

(2) *F is B-differentiable at x if and only if*

$$\lim_{t\to 0^+} G(x+t\,h)h \quad \text{exists uniformly in } |h| = 1. \tag{2.2}$$

Proof. (1) If (2.1) holds for $h \in X$ with $|h| = 1$, then

$$\lim_{t\to 0^+} \frac{F(x+th) - F(x)}{t} = \lim_{t\to 0^+} G(x+th)h.$$

Since $h \in X$ with $|h| = 1$ was arbitrary this implies that F is directionally differentiable at x and

$$F'(x;h) = \lim_{t\to 0^+} G(x+th)h.$$

Similarly, the converse holds.

(2) If F is directionally differentiable at x, then

$$\lim_{|h|\to 0} \frac{F(x+h) - F(x) - F'(x;h)}{|h|_X} = 0 \text{ if and only if}$$

$$\lim_{t\to 0^+} \frac{F(x+tv) - F(x)}{t} - F'(x;v) = 0 \text{ and the limit is uniform in } |v|_X = 1.$$

If F is B-differentiable at x then it is differentiable and from (1) we have

$$\lim_{t\to 0} \frac{F(x+tv) - F(x)}{t} = \lim_{t\to 0} G(x+tv)v.$$

The Bouligand property and the equivalence stated above imply that the $\lim_{t\to 0} G(x+tv)v$ exists uniformly in $|v| = 1$. The converse easily follows as well. $\qquad\square$

Corollary 2.1. *If F is semi-smooth at x, then F is B-differentiable at x and*

$$G(x+t\,h)h \to F'(x;h) \quad \text{uniformly in } |h| = 1$$

for arbitrary family G of N-derivative.

Example 2.1. *Let ψ is a Lipschitz, semi-smooth function at every $x \in \mathbb{R}$ and define the substitution function $F : L^q(\Omega) \to L^p(\Omega)$ by*

$$F(s) = \psi(x(s)) \quad a.e. \ s \in \Omega$$

where Ω is a bounded open domain in R^d. Let

$$G(x(s) + h(s)) = V(x(s) + h(s))$$

where $t \in R \to V(t)$ is a measurable selection such that $V(t) \in \partial\psi$. In fact, since ψ is semi-smooth,

$$|\psi(x+v) - \psi(x) - V(x+v)| = \epsilon(x, |v|)|v|, \quad x, v \in \mathbb{R}$$

with $\epsilon(x, t) \to 0$ as $t \to 0^+$. For $\delta > 0$ let

$$\Omega_\delta(h) = \{s \in \omega : |h(s)| \ge \delta\}$$

Then

$$|h|_q \ge \delta \, |\Omega_\delta|^{\frac{1}{q}}, \quad thus \quad \lim_{|h|_q \to 0} |\Omega_\delta| = 0 \quad for\ every\ \delta > 0.$$

Since ψ is Lipschitz there exists L such that $\epsilon(|h|) \le 2L$. Now,

$$|F(x + h) - F(x) - G(x + h)h|_p \le |\epsilon(|h|)|_r |h|_q$$

with $r = \frac{qp}{q-p}$, where

$$|\epsilon(|h|)|_r \le 2L \, |\Omega_\delta|^{\frac{1}{r}} + \int_\Omega |\epsilon(x(s), \delta)| \, ds$$

By the Lebesgue bounded convergence theorem, the second term of the right-hand side converges to 0 and therefore

$$\lim_{|h|_q \to 0} \frac{|F(x + h) - F(x) - G(x + h)h|_p}{|h|_q} \to 0.$$

Similarly, since

$$\lim_{|v| \to 0} \frac{|V(x + v)v - \psi'(x; v)|}{|v|} = 0,$$

F is semi-smooth and

$$F'(x; h) = \psi'(x(s); h(s)).$$

Example 2.2. Let $F : L^q(\Omega) \to L^p(\Omega)$ denote the pointwise max-operation $F(x) = \max(0, x)$ and define G by

$$G(x)(s) = \begin{cases} 0 & \text{if } x(s) < 0 \\ \delta & \text{if } x(s) = 0 \\ 1 & \text{if } x(s) > 0, \end{cases}$$

where $\delta \in \mathbb{R}$. It follows from Example 2.1 that F is semi-smooth from $L^q(\Omega)$ into $L^p(\Omega)$ provided that $1 \le p < q \le \infty$. If $p = q$, then it is easy to see that F is directionally differentiable at every $x \in L^p(\Omega)$. But F is not Newton-differentiable with G as N-derivative in general. For this purpose consider $x = -|s|$ on $\Omega = (-1, 1)$ and choose $h_n(s)$ as $\frac{1}{n}$ multiplied by the characteristic function of the interval $(-\frac{1}{n}, \frac{1}{n})$. Then $|h_n|_{L^p}^p = \dfrac{2}{n^{p+1}}$ and

$$\int_{-1}^{1} |F(x + h_n) - F(x) - G(x + h_n)h_n|^p \, ds = \int_{-\frac{1}{n}}^{\frac{1}{n}} |x(s)|^p = \frac{2}{p+1}(\frac{1}{n})^{p+1}.$$

Thus,

$$\lim_{n \to \infty} \frac{|F(x + h_n) - F(x) - G(x + h_n)h_n|_{L^p}}{|h_n|_{L^p}} = (\frac{1}{p+1})^{\frac{1}{p}} \ne 0,$$

and hence condition (1.10) is not satisfied at x for any $p \in [1, \infty)$. Similarly, for $p = \infty$ [6].

The following chain rule is proved in [6].

Lemma 2.2. *Suppose $H : D \subset X \to Y$ is continuously Fréchet differentiable at $x \in D$ and $\phi : Y \to Z$ is Newton differentiable at $H(x)$ with N-derivative G. Then $F = \phi(H)$ is Newton differentiable at x with N-derivative $G(H(x+h))H'(x+h) \in \mathcal{L}(X, Z)$ for h sufficiently small.*

We have the local superlinear convergence results for (1.12) [18, 2, 6]

Theorem 2.1. *Suppose that x^* is a solution to $F(x) = 0$ and that F is Newton differentiable at x^* with N-derivative G. If G is nonsingular for all $x \in N(x^*)$ and $\{\|G(x)^{-1}\| : x \in N(x^*)\}$ is bounded, then the Newton-iteration*

$$x^{k+1} = x^k - G(x^k)^{-1} F(x^k)$$

converges superlinearly to x^* provided that $|x^0 - x^*|$ is sufficiently small. If, moreover, F is Newton differentiable of order α at x^*, i.e., there exists a $\alpha > 0$ such that

$$\lim_{h \to 0} \frac{1}{|h|^{1+\alpha}} |F(x^* + h) - F(x^*) - G(x^* + h)h| = 0, \qquad (2.3)$$

then x^k converges to x^* with q-order $1 + \alpha$, i.e., we have $|x^{k+1} - x^*| = \mathcal{O}(|x^k - x^*|^{1+\alpha})$ as $k \to \infty$.

3. Primal-dual active set method and global convergence

The primal-dual active set method for (1.3) is defined by

Primal dual active set method

(1) Initialize x^0. Set $k = 0$.
(2) Set the index sets;

$$\mathcal{A}_k^+ = \{-f(x^k) + x^k - \psi > 0\}, \quad \mathcal{A}_k^- = \{-f(x^k) + x^k - \phi < 0\}, \quad \mathcal{I}_k = (\mathcal{A}_k^+ \cup \mathcal{A}_k^-)^c.$$

(3) Solve for x^{k+1}

$$f'(x^k)(x^{k+1} - x^k) + f(x^k) = 0 \quad \text{in } \mathcal{I}_k$$
$$x^{k+1} = \psi \text{ on } \mathcal{A}_k^+, \quad x^{k+1} = \phi \text{ on } \mathcal{A}_k^-,$$

(4) Stop, or set $k = k + 1$ and return to (2).

The primal-dual active set method for (1.3) is a specific semi-smooth Newton method [6] observing the fact that

$$G(x) = \begin{cases} 1 & \text{for } x > 0 \\ 0 & \text{for } x \le 0 \end{cases} \in \partial_B \max(0, x).$$

The primal-dual active set method is known to be extremely efficient for solving discretized variational inequalities and constrained optimal control problems [6]. Suppose $X = L^2(\Omega)$ and $f : X \to L^q(\Omega)$, $q > 2$ is continuously differentiable. It follows from Lemma 2.2 and Theorem 2.1 that the primal-dual active set method converges locally superlinearly. For the boundary control example in the introduction such a condition holds with $2 < q < 4$.

For the affine case $f(x) = Ax - b$ the primal-dual active set method (unilateral case) is equivalent to

(1) Initialize x^0, μ^0. Set $k = 0$.
(2) Set $\mathcal{I}_k = \{\mu^k + c(x^k - \psi) \leq 0\}, \quad \mathcal{A}_k = \{\mu^k + c(x^k - \psi) > 0\}$.
(3) Solve for (x^{k+1}, μ^{k+1})

$$Ax^{k+1} + \mu^{k+1} = b$$
$$x^{k+1} = \psi \text{ in } \mathcal{A}_k \quad \text{and} \quad \mu^{k+1} = 0 \text{ in } \mathcal{I}_k.$$

(4) Stop, or set $k = k + 1$ and return to (2).

Sufficient conditions for global convergence were established in [6] for the finite-dimensional case $X = \mathbb{R}^m$. The sufficient condition we discuss here is more general. It is related to diagonal dominance of A and will imply that

$$\mathcal{M}(x^{k+1}, \mu^{k+1}) = \max\left(\beta \int_\Omega |(x^{k+1} - \psi)^+| \, dx, \int_\Omega |(\mu^{k+1})^-| \, dx\right)$$

with $\beta > 0$ acts as a merit functional for the primal-dual algorithm. Here we set $\phi^+ = \max(\phi, 0)$ and $\phi^- = -\min(\phi, 0)$. Note that by step (3) of the algorithm we have

$$\mathcal{M}(x^{k+1}, \mu^{k+1}) = \max\left(\beta \int_{\mathcal{I}_k} |(x^{k+1} - \psi)^+| \, dx, \int_{\mathcal{A}_k} |(\mu^{k+1})^-| \, dx\right). \qquad (3.1)$$

The natural norm associated to this merit functional is the $L^1(\Omega)$-norm and consequently we assume that

$$A \in \mathcal{L}(L^1(\Omega)), \ a \in L^1(\Omega) \text{ and } \psi \in L^1(\Omega). \qquad (3.2)$$

The analysis of this section can also be used to obtain convergence in the $L^p(\Omega)$-norm for any $p \in (1, \infty)$, if the norms in the integrands of $\mathcal{M}$ are replaced by $|\cdot|^p$ norms and the $L^1(\Omega)$-norms below are replaced by $L^p(\Omega)$-norms as well.

We assume that there exist constants $\rho_i, i = 1, \ldots, 5$, such that for all partitions $\mathcal{A}$ and $\mathcal{I}$ of Ω and for all $\phi_\mathcal{A} \geq 0$ in $L^2(\mathcal{A})$ and $\phi_\mathcal{I} \geq 0$ in $L^2(\mathcal{I})$

$$|[A_\mathcal{I}^{-1} \phi_\mathcal{I}]^-| \leq \rho_1 |\phi_\mathcal{I}|$$
$$|[A_\mathcal{I}^{-1} A_{\mathcal{I}\mathcal{A}} \phi_\mathcal{A}]^+| \leq \rho_2 |\phi_\mathcal{A}| \qquad (3.3)$$

and

$$|[A_\mathcal{A} \phi_\mathcal{A}]^-| \leq \rho_3 |\phi_\mathcal{A}|$$
$$|[A_{\mathcal{A}\mathcal{I}} A_\mathcal{I}^{-1} \phi_\mathcal{I}]^-| \leq \rho_4 |\phi_\mathcal{I}| \qquad (3.4)$$
$$|[A_{\mathcal{A}\mathcal{I}} A_\mathcal{I}^{-1} A_{\mathcal{I}\mathcal{A}} \phi_\mathcal{A}]^+| \leq \rho_5 |\phi_\mathcal{A}|.$$

Here $|\cdot|$ denotes the $L^1(\Omega)$-norm. Assumption (3.3) requires in particular the existence of $A_\mathcal{I}^{-1}$. By a Schur-complement argument with respect to the sets $\mathcal{I}_k$ and $\mathcal{A}_k$ this implies existence of a solution to the linear systems in step (iii) of the algorithm for every k.

Theorem 3.1. *If* (3.2), (3.3), (3.4) *hold and* $\rho = \max(\beta\,\rho_1 + \rho_2, \frac{\rho_3}{\beta} + \rho_4 + \frac{\rho_5}{\beta}) < 1$, *then* $\mathcal{M}$ *is a merit function for the primal-dual algorithm of the reduced system and* $\lim_{k\to\infty}(x^k, \mu^k) = (x^*, \mu^*)$ *in* $L^1(\Omega) \times L^1(\Omega)$, *with* (x^*, μ^*) *a solution to* (1.3).

Proof. Let $\delta x = x^{k+1} - x^k$ and $\delta\mu = \mu^{k+1} - \mu^k$. Then,

$$A_{\mathcal{A}_k}\delta x_{\mathcal{A}_k} + A_{\mathcal{A}_k, \mathcal{I}_k}\delta x_{\mathcal{I}_k} + \delta\mu_{\mathcal{A}_k} = 0$$
$$A_{\mathcal{I}_k}\delta x_{\mathcal{I}_k} + A_{\mathcal{I}_k, \mathcal{A}_k}\delta x_{\mathcal{A}_k} - \mu^k_{\mathcal{I}_k} = 0. \tag{3.5}$$

For every $k \geq 1$ we have $(x^{k+1} - \psi)^+ \leq (x^{k+1} - x^k)^+$ on $\mathcal{I}_k$ and $(\mu^{k+1})^- = (\delta\mu)^-$ on $\mathcal{A}_k$. Therefore

$$\mathcal{M}(x^{k+1}, \mu^{k+1}) \leq \max(\beta \int_{\mathcal{I}_k} (\delta x_{\mathcal{I}_k})^+, \int_{\mathcal{A}_k} (\delta\mu_{\mathcal{A}_k})^-). \tag{3.6}$$

From (3.5) we deduce that

$$\delta x_{\mathcal{I}_k} = -A_{\mathcal{I}_k}^{-1}(-\mu^k_{\mathcal{I}_k}) + A_{\mathcal{I}_k}^{-1}A_{\mathcal{I}_k \mathcal{A}_k}(-\delta x_{\mathcal{A}_k}),$$

with $\mu^k_{\mathcal{I}_k} \leq 0$ and $\delta x_{\mathcal{A}_k} \leq 0$. By (3.3) therefore

$$|(\delta x_{\mathcal{I}_k})^+| \leq \rho_1 |\mu^k_{\mathcal{I}_k}| + \rho_2 |\delta x_{\mathcal{A}_k}| \tag{3.7}$$

$$= \rho_1 \int_{\mathcal{I}_k \cap \mathcal{A}_{k-1}} |(\mu^k_{\mathcal{I}_k})^-| + \rho_2 \int_{\mathcal{A}_k \cap \mathcal{I}_{k-1}} (x_k - \psi)^+ \leq (\rho_1 + \frac{\rho_2}{\beta})\mathcal{M}(x^k, \mu^k).$$

Similarly by (3.5),

$$\delta\mu_{\mathcal{A}_k} = A_{\mathcal{A}_k}(-\delta x_{\mathcal{A}_k}) + A_{\mathcal{A}_k \mathcal{I}_k}A_{\mathcal{I}_k}^{-1}(-\mu^k_{\mathcal{I}_k}) - A_{\mathcal{A}_k \mathcal{I}_k}A_{\mathcal{I}_k}^{-1}A_{\mathcal{I}_k \mathcal{A}_k}(-\delta x_{\mathcal{A}_k}).$$

Since $\delta x_{\mathcal{A}_k} \leq 0$ and $\mu^k_{\mathcal{I}_k} \leq 0$, we find by (3.4)

$$|(\delta\mu_{\mathcal{A}_k})^-| \leq \rho_3 |\delta x_{\mathcal{A}_k}| + \rho_4 |\mu^k_{\mathcal{I}_k}| + \rho_5 |\delta x_{\mathcal{A}_k}| \leq (\frac{\rho_3 + \rho_5}{\beta} + \rho_4)\mathcal{M}(x^k, \mu^k), \tag{3.8}$$

and therefore

$$\mathcal{M}(x^{k+1}, \mu^{k+1}) \leq \max(\beta\rho_1 + \rho_2, \frac{\rho_3 + \rho_5}{\beta} + \rho_4)\,\mathcal{M}(x^k, \mu^k) = \rho\,\mathcal{M}(x^k, \mu^k).$$

Thus, if $\rho < 1$, then $\mathcal{M}$ is a merit functional. Furthermore $\mathcal{M}(x^{k+1}, \mu^{k+1}) \leq \rho^k \mathcal{M}(x^1, \mu^1)$. Together with (3.7), (3.8) and (3.5) it follows that (x^k, μ^k) is a Cauchy sequence. Hence there exists (x^*, μ^*) such that $\lim_{k\to\infty}(x^k, \mu^k) = (x^*, \mu^*)$ and $Ax^* + \mu^* = a$, $\mu^*(x^* - \psi) = 0$ a.e. in Ω. Since $(x^k - \psi)^+ \to (x^* - \psi)^+$ as $k \to \infty$ and $\lim_{k\to\infty}\int_\Omega (x^{k+1} - \psi)^+ = 0$ it follows that $x^* \leq \psi$. Similarly one argues that $\mu^* \geq 0$. Thus (x^*, μ^*) is a solution to (1.3). $\qquad\square$

Similar results are established for the nonlinear case and bilateral case [13].

Remark 3.1. In the finite-dimensional case the integrals in the definition of $\mathcal{M}$ must be replaced by sums over the active/inactive index sets. If A is an M-matrix, then $\rho_1 = \rho_2 = 0$ and $\rho < 1$ if $\frac{\rho_3}{\beta} + \rho_4 + \frac{\rho_5}{\beta} < 1$. This is the case if A is diagonally dominant in the sense that $\rho_4 < 1$ and β is chosen sufficiently large. For such a matrix A the property $\rho < 1$ is stable under additive perturbation which are not necessarily M-matrices.

Remark 3.2. Consider the infinite-dimensional case with $A = \alpha I + K$, where $\alpha > 0$, $K \in \mathcal{L}(L^1(\Omega))$ and $K\phi \geq 0$ for all $\phi \geq 0$. This is the case for the operators in the boundary control of the heat equation in Introduction, as can be argued by using the maximum principle. Let $\|K\|$ denote the norm of $K \in \mathcal{L}(L^1(\Omega))$. For $\|K\| < \alpha$ and any $\mathcal{I} \subset \Omega$ we have $A_{\mathcal{I}}^{-1} = \frac{1}{\alpha} I_{\mathcal{I}} - \frac{1}{\alpha} K_{\mathcal{I}} A_{\mathcal{I}}^{-1}$ and hence $\rho_1 \leq \frac{\|K\|}{\alpha(\alpha - \|K\|)}$. Moreover $\rho_3 = 0$. The conditions involving ρ_2, ρ_4 and ρ_5 are satisfied with $\rho_2 = \frac{\|K\|}{\alpha - \|K\|}$, $\rho_4 = \frac{\|K\|^2}{\alpha(\alpha - \|K\|)}$ and $\rho_5 = \frac{\|K\|^2}{\alpha - \|K\|}$, and $\rho < 1$ if α is sufficiently large.

4. Elliptic and parabolic variational inequalities and regularization

We consider the case that $f(x) = Ax - g$ where the linear operator A of is not a continuous but only a closed operator in $X = L^2(\Omega)$. Such problem arise in the obstacle problems, the state constraint optimal control problem and in parabolic variational inequalities. We discuss the unilateral case $C = \{x \leq \psi\}$:

$$Ax + \mu = b, \quad \mu = \max(0, \mu + (x - \psi)). \tag{4.1}$$

Example 4.1. For the obstacle problem $A = -A_0$ where A_0 is a second order elliptic operator with $\operatorname{dom}(A) = H^2(\Omega) \cap H_0^1(\Omega)$. If $\partial\Omega$ and ψ are sufficiently regular, then the solution to (4.1) satisfies $(x, \mu) \in (H_0^1(\Omega) \cap H^2(\Omega)) \times L^2(\Omega)$.

Example 4.2. Consider the state constrained optimal control problem

$$\min_{u \in L^2(\Omega)} \quad \frac{1}{2}|y - \bar{y}|_{L^2(\Omega)}^2 + \frac{\beta}{2}|u|_{L^2(\Omega)}^2 \quad \text{subject to } Ey = u \text{ and } y \in C, \tag{4.2}$$

where E is a closed linear operator in $X = L^2(\Omega)$. We assume that E^{-1} exists and set $V = \operatorname{dom}(E)$ where V is endowed with the graph norm of E. The necessary and sufficient optimality condition for (4.2) is given by

$$\beta\,(Ey, E(v - y)) + (y, v - y) - (\bar{y}, v - y) \geq 0 \quad \text{for all } v \in V \cap C. \tag{4.3}$$

That is, $A = E^*E$ and $\mu \in V^*$ (not in $L^2(\Omega)$) in general.

Example 4.3. A class of parabolic variational inequalities is given by $A = \frac{d}{dt} - A_0$ and $X = L^2((0, T) \times \Omega)$. If $\partial\Omega$ and ψ are sufficiently regular, then $(x, \mu) \in (H^1(0, T; X) \cap L^2(0, T; H^2(\Omega) \cap C(0, T; H_0^1(\Omega)) \times X$ [11].

The primal-dual method can formally be applied to all these cases but due to the unboundedness of A the iterates μ^k are not necessarily in X. In order to remedy this difficulty we consider a one-parameter family of regularized problems based on smoothing of the complementarity condition by

$$\mu = \alpha \max(0, \mu + (x - \psi)), \quad \text{with } 0 < \alpha < 1. \tag{4.4}$$

This is a relaxation of the complementarity condition $\mu = \max(0, \mu + (y - \psi))$ with α as a continuation parameter. Note that an update for μ based on (4.4) results in $\mu \in X$. Equation (4.4) is equivalent to

$$\mu = \max(0, \frac{\alpha}{1 - \alpha}(x - \psi)), \tag{4.5}$$

with $\alpha/(1 - \alpha)$ ranging in $(0, \infty)$ for $\alpha \in (0, 1)$. We shall use a generalization of (4.4) and introduce an additional shift parameter $\bar{\mu} \in L^2(\Omega)$ in (4.5). Moreover we replace $\alpha/(1 - \alpha)$ by c and arrive at

$$\mu = \max(0, \bar{\mu} + c\,(x - \psi)), \quad \text{with } c \in (0, \infty). \tag{4.6}$$

This coincides with the generalized Yoshida-Morrey approximation for inequality constraints [8]. It results in a regularization of (4.1) given by

$$Ax + \mu = b$$
$$\mu = \max(0, \bar{\mu} + c\,(x - \psi)). \tag{4.7}$$

Semi-smooth Newton algorithm with regularization

   (i) Choose $\bar{\mu}, c, x_0$, set $k = 0$.
  (ii) Set $\mathcal{A}_k = \{s \colon (\bar{\mu} + c\,(x^k - \psi))(s) > 0\}$,
 (iii) Solve for $x^{k+1} \in X$:

$$Ax^{k+1} + \chi_{\mathcal{A}_k}(\bar{\mu} + c\,(x^{k+1} - \psi)) = b \tag{4.8}$$

 (iv) Stop or $k = k + 1$, go to (ii).

Assume that (4.8) has a unique solution and moreover that

$$|A^{-1}\phi|_{L^q(\Omega)} \le M\,|\phi|_{L^2(\Omega)}|, \quad \text{for some } q > 2.$$

It thus follows from Lemma 2.2 and Theorem 2.1 that If $|x_0 - x_c|_X$ is sufficiently small then x^k converges to x_c, the solution to (4.7) superlinearly in X and thus in $\mathrm{dom}\,(A)$.

Let x_c denotes the solution to (4.7). Suppose $\bar{\mu} \ge b - A\psi$ in distribution. In the obstacle problem and parabolic variational inequality it is shown in [9, 11] that for appropriately chosen $\bar{\mu}$, the regularizes solutions x_c are feasible, i.e., $x_c \le \psi$, as well as monotone with respect to c, i.e., we have

$$x_c \le x_{\hat{c}} \le x^*$$

and the bound

$$0 \le \mu_c = \max(0, \bar{\mu} + c\,(x_c - \psi)) \le \bar{\mu}$$

holds for all $0 < c < \hat{c}$ where x^* is the solution to (4.1). Moreover,

$$|x_c - x^*|_{L^\infty(\Omega)} \le \frac{|\bar{\mu}|_\infty}{c}. \tag{4.9}$$

It was also shown in [10, 11] that $x_c \le x^k \le x^{k+1}$ for the iterates of the semi-smooth Newton algorithms. Special properties also hold for the case $\bar{\mu} = 0$, see [10, 11].

Estimate (4.9) is particularly important since

(1) the free surface $S = \{x^* = \psi\}$ can be approximated by $S_c = \{\mu_c = 0\}$ with the rate $\dfrac{1}{c}$, and

(2) we can select c for the discretized problem to obtain the desired accuracy, say $c = O\left(\dfrac{1}{h^2}\right)$ with meshsize h for second order discretizations.

In [10] we use a homotopy method with respect $c \to \infty$ to accelerate the convergence of the algorithm to $x^* = \lim_{c \to \infty} x_c$. A path-following strategy for the choice of c is analyzed in [7].

5. Globalization

The globalization of the iteration (1.8) in R^m on the basis of the merit functional $\theta(x) = |F(x)|^2$ is achieved by the following

Algorithm Let β, $\gamma \in (0, 1)$ and $\sigma \in (0, \bar{\sigma})$. Choose $x^0 \in R^m$ and set $k = 0$. Given x^k with $F(x^k) \neq 0$. Then,

(i) if there exists a solution h^k to

$$V_k \, h^k = -F(x^k)$$

with $|h^k| \leq b|F(x^k)|$, and if further

$$|F(x^k + h^k)| < \gamma\,|F(x^k)|\,,$$

set $d^k = h^k$, $x^{k+1} = x^k + d^k$, $\alpha_k = 1$, and $m_k = 0$.

(ii) Otherwise choose $d^k = d(x^k)$ according to (A.2) and let $\alpha_k = \beta^{m_k}$, where m_k is the first positive integer m for which

$$\theta(x^k + \beta^m \, d^k) - \theta(x^k) \leq -\sigma \beta^m \, \theta(x^k).$$

Set $x^{k+1} = x^k + \alpha_k \, d^k$.

The following assumptions will be utilized:

- (A.1) $S = \{x \in R^m : |F(x)| \leq |F(x^0)|\}$ is bounded.
- (A.2) There exist $\bar{\sigma}$ and $b > 0$ such that for each $x \in S$ there exists $d = d(x) \in R^m$ satisfying

$$\theta'(x; d) \leq -\bar{\sigma}\theta(x) \quad \text{and} \quad |d| \leq b|F(x)|. \tag{5.1}$$

- (A.3) The following closure property holds: if $x_k \to \bar{x}$ and $d(x_k) \to \bar{d}$ with $x_k \in S$, then $\theta'(\bar{x}; \bar{d}) \leq -\bar{\sigma}\theta(\bar{x})$.
- (A.4) θ is subdifferentiability regular for all $x \in S$, i.e., $\theta^o(x; d) = \theta'(x; d)$ for all $d \in R^m$.

Here the Clarke generalized directional derivative $\theta^o(x; d)$ [1] of θ at x in the direction d is defined by

$$\theta^o(x; d) = \limsup_{y \to x,\, t \to 0^+} \frac{\theta(y + t\,d) - \theta(y)}{t}.$$

Convex, locally Lipschitz continuous functions, for example, are subdifferentiability regular [1]. Let us also point out that local Lipschitz continuity of F implies Fréchet-differentiability of θ at x^* if $F(x^*) = 0$. With reference to the closure property (A.3), note that it is not required that $\bar{d} = d(\bar{x})$.

Condition (A.2) is motivated by the work in [4] where it is shown to guarantee a nontrivial stepsize τ such that

$$\theta(x + \tau\,d) - \theta(x) \le -\sigma\tau\,\theta(x),$$

if $\sigma \in (0, \bar{\sigma})$ and $x \in S$. We show that combined with (A.1), (A.3) and (A.4) it also guarantees (subsequential) convergence of $\{x^k\}$ to a solution of $F(x) = 0$. Concerning conditions (A.2) and (A.3) we introduce the notion of quasi-directional derivative in Section 5.1. This notion will allow us to construct descent directions which satisfy these two conditions.

In [3], Chapter 8, in order to prove convergence, the following condition is assumed:

$$\limsup_{\tau_k \to 0^+} \frac{\theta(x^k + \tau_k\,d^k) - \theta(x^k) + \tau_k\,\theta(x^k)}{\tau_k} \le 0. \tag{5.2}$$

Alternatively, in [4] a generalized form of the condition

$$\lim_{k \to \infty} F(x^k)^T F'(x^k; d^k) \ge \limsup_{k \to \infty} \frac{\theta(x^k + \tau^k d^k) - \theta(x^k)}{\tau^k} \tag{5.3}$$

for any convergent sequence $\{(x^k, d^k, \tau^k)\}$, is used. In either of the two cases the directions d^k satisfy a condition like (A.2). Conditions (5.2), (5.3) resemble (A.3) and (A.4) but the latter are simpler to check and more transparent also for the case that F is C^1. Conditions (A.1)–(A.4) and the notion of quasi-directional derivative provide us with a rather axiomatic approach to globalization of the semi-smooth Newton method.

Theorem 5.1. *Suppose that $F : \mathbb{R}^m \to \mathbb{R}^m$ is locally Lipschitz and B-differentiable.*
(a) *Assume that (A.1)–(A.4) hold. Then the sequence $\{x^k\}$ generated by algorithm is bounded, it satisfies $|F(x^{k+1})| < |F(x^k)|$ for all $k \ge 0$, and each accumulation point x^* of $\{x^k\}$ satisfies $F(x^*) = 0$.*
(b) *If moreover for one such accumulation point*

$$|h| \le c\,|F'(x^*; h)| \quad \text{for all } h \in \mathbb{R}^m, \tag{5.4}$$

then the sequence x^k converges x^*.
(c) *If in addition to the above assumptions F is semi-smooth at x^* and all $V \in \partial_B F(x^*)$ are nonsingular, then x^k converges to x^* superlinearly.*

Proof. (a) First we prove that for each $x \in S$ such that $\theta(x) \neq 0$ and d satisfying $\theta'(x; d) \leq -\bar{\sigma}\theta(x)$, there exists a $\bar{\tau} > 0$ such that

$$\theta(x + \tau\, d) - \theta(x) \leq -\sigma\tau\, \theta(x) \text{ for all } \tau \in [0, \bar{\tau}].$$

If this is not the case, then there exists a sequence $\tau_n \to 0^+$ such that

$$\theta(x + \tau_n\, d) - \theta(x) > -\sigma\tau_n\, \theta(x).$$

Dividing both sides by τ_n and letting $n \to \infty$, we have by (5.1)

$$-\bar{\sigma}\theta(x) \geq \theta'(x; d) \geq -\sigma\, \theta(x).$$

Since $\sigma < \bar{\sigma}$, this shows $\theta(x) = 0$, which contradicts the assumption $\theta(x) \neq 0$. Hence for each level k at which $d^k = d(x^k)$ is chosen according to the second alternative in algorithm there exists $m^k < \infty$ and $\alpha_k > 0$ such that $|F(x^{k+1})| < |F(x^k)|$. By construction the iterates therefore satisfy $|F(x^{k+1})| < |F(x^k)|$ for each $k \geq 0$.

If $\limsup \alpha_k > 0$, then it is obvious that $\theta(x^k)$ monotonically converges to 0 and that each accumulation point x^* of $\{x^k\}$ satisfies $F(x^*) = 0$. If on the other hand $\limsup \alpha_k = 0$, then $\lim m_k \to \infty$. By the definition of m_k, for $\tau_k := \beta^{m_k - 1}$, we have $\tau_k \to 0$ and

$$\theta(x^k + \tau_k\, d^k) - \theta(x^k) > -\sigma\tau_k\, \theta(x^k). \tag{5.5}$$

By (A.1), (A.2) the sequence $\{(x^k, d^k)\}$ is bounded. Let $\{(x^k, d^k)\}_{k \in K}$ be any convergent subsequence with limit (x^*, d). Note that

$$\frac{\theta(x^k + \tau_k\, d^k) - \theta(x^k)}{\tau_k} = \frac{\theta(x^k + \tau_k\, d) - \theta(x^k)}{\tau_k} + \frac{\theta(x^k + \tau_k\, d^k) - \theta(x^k + \tau_k\, d)}{\tau_k},$$

where

$$\lim_{k \in K, k \to \infty} \frac{\theta(x^k + \tau_k\, d^k) - \theta(x^k + \tau_k\, d)}{\tau_k} \to 0,$$

since θ is locally Lipschitz continuous. Since $d^k = d(x^k)$ for all $k \in K$ it follows from (A.3) that $\theta'(x^*; d) \leq -\bar{\sigma}\, \theta(x^*)$. Then, from (5.5) and (A.4) we find

$$-\sigma\theta(x^*) \leq \limsup_{k \in K,\ k \to \infty} \frac{\theta(x^k + \tau_k\, d^k) - \theta(x^k)}{\tau_k} \leq \theta^\circ(x^*; d) = \theta'(x^*; d) \leq -\bar{\sigma}\, \theta(x^*). \tag{5.6}$$

It follows that $(\bar{\sigma} - \sigma)\, \theta(x^*) \leq 0$ and thus $\theta(x^*) = 0$.
For the proof of (b)–(c) we refer to [18, 12]. $\qquad\square$

Remark 5.1.
(1) We point out that the closure property (A.3) as well as the subdifferentiability regular property (A.4) are used in the proof of Theorem 5.1 only for the case that $\limsup_{k \to \infty} \alpha_k = 0$.
(2) Since $h \to F'(x^*; h)$ is positively homogeneous one can easily argue that (5.4) is equivalent to the condition that $F'(x^*; h) = 0$ implies that $h = 0$.
(3) If X is a Hilbert space, Theorem 5.1 holds assuming that the iterates (x^k, d^k) are sequentially compact.

5.1. Descent directions

We turn to a discussion of conditions (A.2) and (A.3) required for the descent directions d. For this purpose we introduce the following definition.

Definition 5.1. *Let* $F : \mathbb{R}^m \to \mathbb{R}^m$ *be directionally differentiable. Then* $G : S \times \mathbb{R}^m \to \mathbb{R}^m$ *is called a quasi-directional derivative of* F *on* $S \subset \mathbb{R}^m$ *if*

  (i) $(F(x), F'(x; d)) \le (F(x), G(x; d))$

  (ii) $G(x; td) = tG(x; d),$ *for all* $d \in \mathbb{R}^m, x \in S$ *and* $t \ge 0,$

  (iii) $(F(\bar{x}), G(\bar{x}; \bar{d})) \le \limsup_{x \to \bar{x},\, d \to \bar{d}}(F(x), G(x, d))$ *for all* $x \to \bar{x}$, $d \to \bar{d}$, *with* $x, \bar{x} \in S.$

Throughout the remainder of this section we assume that $F : \mathbb{R}^m \to \mathbb{R}^m$ is a Lipschitz continuous and directionally differentiable function and S refers to the set defined in (A.1).

(a) *Bouligand direction.* If there exists $\bar{b}$ such that

$$|h| \le \bar{b}\,|F'(x; h)| \text{ for all } x \in S,\ h \in \mathbb{R}^m \tag{5.7}$$

and if

$$F(x) + F'(x; d) = 0, \tag{5.8}$$

admits a solution d for each $x \in S$, then a first choice for the direction is given by the solution d to (5.8), see, e.g., [16]. By (5.7) we have $|d| \le \bar{b}|F(x)|$. Moreover

$$\theta'(x, d) = 2\,(F'(x; d), F(x)) = -2\,\theta(x),$$

and therefore the inequalities in (A.2) hold with $b = \bar{b}$ and $\bar{\sigma} = 2$. For this choice, however, (A.3) is not satisfied in general, see Section 5.2.

(b) *Generalized Bouligand direction.* As a second choice, see [16, 4], we assume that G is a quasi-directional derivative of F on S, that

$$|h| \le \bar{b}\,|G(x; h)| \text{ for all } x \in S,\ h \in \mathbb{R}^m \tag{5.9}$$

and

$$F(x) + G(x; d) = 0, \tag{5.10}$$

admits a solution d which is defined as the descent direction for each $x \in S$. One argues as for the first choice that the inequalities in (A.2) hold with $b = \bar{b}$ and $\bar{\sigma} = 2$. Moreover (A.3) is satisfied, since for any $(x, d) \to (\bar{x}, \bar{d})$ in $S \times \mathbb{R}$ with $d = d(x)$ we have

$$\theta'(\bar{x}, \bar{d}) \le 2(F(\bar{x}), G(\bar{x}; \bar{d})) \le 2 \limsup_{x \to \bar{x},\, d \to \bar{d}} (F(x), G(x; d))$$

$$= -2 \lim_{x \to \bar{x}} |F(x)|^2 = -2\theta(\bar{x}).$$

We refer to Section 5.2 for the construction of G for specific applications.

c) *Generalized gradient direction.* The following choice was discussed in [4]. Here d is chosen as the solution to

$$\min_{d} \quad J(x,d) = 2(F(x), G(x;d)) + \eta \, |d|^2, \qquad (5.11)$$

where $\eta > 0$ and $x \in S$. Assume that for some $L > 0$

$$\begin{cases} \quad h \to G(x;h) \text{ is continuous and} \\ \quad |G(x;h)| < L\,|h| \text{ for all } x \in S,\ h \in \mathbb{R}^m. \end{cases} \qquad (5.12)$$

Then, $d \to J(d)$ is coercive, bounded below, and continuous. Thus there exists an optimal solution d to (5.11) and we have the following lemma [4, 12].

Lemma 5.1. *Assume that $F : \mathbb{R}^m \to \mathbb{R}^m$ is Lipschitz continuous, directionally differentiable and that G is a quasi-directional derivative of S satisfying (5.12).*
(a) *If d is an optimal solution to (5.11), then*

$$(F(x), G(x;d)) = -\eta \, |d|^2.$$

(b) *If $d = 0$ is an optimal solution to (5.11), then $(F(x), G(x;h)) \geq 0$ for all $h \in \mathbb{R}^m$.*

By Lemma 5.1 the optimal value of the cost J in (5.11) is given by $-\eta|d|^2$. If this value is negative, then any solution to (5.11) provides a decay for θ since $\theta'(x;d) \leq -\eta|d|^2$. The optimal value of the cost is 0 if and only if $d = 0$ is the optimal solution. In this case Lemma 5.1 implies that x is a stationary point in the sense that $(F(x), G(x;h)) \geq 0$ for all $h \in \mathbb{R}^m$.

Let us now turn to the discussion of condition (A.2) for the direction given by the solution d to (5.11). We assume (5.9) and that (5.10) admits a solution for every $x \in S$. Since $J(0) = 0$, we have $2(F(x), G(x;d)) \leq -\eta \, |d|^2$ and therefore

$$\eta \, |d|^2 \leq -2(F(x), G(x;d)) \leq 2|F(x)|\,|G(x;d)| \leq 2L\,|d|\,|F(x)|.$$

Thus,

$$|d| \leq \frac{2L}{\eta} \, |F(x)|,$$

and the second condition in (A.2) holds. Turning to the first condition let $\hat{d}$ satisfy $F(x) + G(x;\hat{d}) = 0$. Then, using Lemma 5.1(a) and (5.9) we find at a solution d to (5.11)

$$\begin{aligned} J(x,d) = (F(x), G(x;d)) = -\eta \, |d|^2 &\leq 2\,(G(x;\hat{d}), F(x)) + \eta \, |\hat{d}|^2 \\ &\leq -2\,|F(x)|^2 + \eta \bar{b}^2 \, |F(x)|^2 = -(2 - \eta \bar{b}^2)\,\theta(x). \end{aligned} \qquad (5.13)$$

Since G is a quasi-directional derivative of F on S we have

$$\theta'(x;d) \leq 2(F(x), G(x;d)) \leq -2(2 - \eta \bar{b}^2)\,\theta(x)$$

and thus the direction d defined by (5.11) satisfies the first condition in (A.2) with $\bar{\sigma} = 2(2 - \eta \bar{b}^2)$, provided that $\eta < \frac{2}{\bar{b}^2}$.

To argue (A.3) let $x_k \to \bar{x}$, $d_k \to \bar{d}$, with $d_k = d(x_k)$, $x_k \in S$ and choose $\hat{d}_k$ such that $F(x_k) + G(x_k; \hat{d}_k) = 0$. Then

$$
\begin{aligned}
\frac{1}{2}\theta'(\bar{x}; \bar{d}) &= (F(\bar{x}), F'(\bar{x}, \bar{d})) \\
&\leq (F(\bar{x}), G(\bar{x}; \bar{d})) \\
&\leq \limsup_{k\to\infty}(F(x_k), G(x_k; d_k)) \\
&\leq \limsup_{k\to\infty}(2(F(x_k), G(x_k; d_k)) + \eta|d_k|^2) \\
&\leq \limsup_{k\to\infty}(2(F(x_k), G(x_k; \hat{d}_k)) + \eta|\hat{d}_k|^2) \\
&\leq -2|F(\bar{x})|^2 + \eta\,\bar{b}^2 \lim_{k\to\infty}|F(x_k)|^2 \\
&= -(2 - \eta\bar{b}^2)\theta(\bar{x}),
\end{aligned}
$$

and thus (A.3) holds if $\eta < \frac{2}{\bar{b}^2}$.

5.2. Box constraints

We return to the example with box constraints referred to below (1.1). In (1.3) we can eliminate μ by $\mu = -f(x)$ and work with the single variable x. This results in the equation

$$
F(x) = f(x) + \max(0, -f(x) + x - \psi) + \min(0, -f(x) + x - \phi) = 0.
$$

Define

$$
\begin{aligned}
&\mathcal{A}^+ = \{-f(x) + x - \psi > 0\} \quad \mathcal{A}^- = \{-f(x) + x - \phi < 0\} \\
&\mathcal{I}^1 = \{-f(x) + x - \psi = 0\}, \quad \mathcal{I}^2 = \{x - \psi < f(x) < x - \phi\} \quad \text{and} \\
&\mathcal{I}^3 = \{-f(x) + x - \phi = 0\},
\end{aligned}
$$

where $\mathcal{A}^+ = \{-f(x) + x - \psi > 0\}$ stands for $\{i : (-f(x) + x - \psi)_i > 0\}$, and analogously for the other sets. We obtain

$$
F'(x; d) = \begin{cases}
d & \text{on } \mathcal{A}^+ \cup \mathcal{A}^- \\
f'(x)d & \text{on } \mathcal{I}^2, \\
\max(f'(x)d, d) & \text{on } \mathcal{I}^1 \\
\min(f'(x)d, d) & \text{on } \mathcal{I}^3
\end{cases}
$$

and

$$
\begin{aligned}
\theta'(x; d) = {}&(x - \psi, d)_{\mathcal{A}^+} + (x - \phi, d)_{\mathcal{A}^-} + (f(x), f'(x)d)_{\mathcal{I}^2} \\
&+ (f(x), \max(f'(x)d, d))_{\mathcal{I}^1} + (f(x), \min(f'(x)d, d))_{\mathcal{I}^3}.
\end{aligned}
$$

Here 'on a set' means 'for all indices in the set', for example, $(F'(x;d))_i = (\min(f'(x)d,d))_i$ for $i \in \mathcal{I}^3$. The Bouligand direction (5.8) is given by

$$d + x - \psi = 0 \text{ on } \mathcal{A}^+,$$
$$d + x - \phi = 0 \text{ on } \mathcal{A}^-,$$
$$f'(x)d + f(x) = 0 \text{ on } \mathcal{I}^2, \tag{5.14}$$
$$\max(f'(x)d,d) + f(x) = 0 \text{ on } \mathcal{I}^1,$$
$$\min(f'(x)d,d) + f(x) = 0 \text{ on } \mathcal{I}^3.$$

It is shown in [12] that G defined by

$$G(x;d) = \begin{cases} d & \text{on } (\mathcal{A}^+ \cap \{f(x) \le 0\}) \cup (\mathcal{A}^- \cap \{f(x) \ge 0\}) \\ f'(x)d & \text{on } \mathcal{I}^2 \cap \{\phi \le x \le \psi\} \\ \max(f'(x)d,d) & \text{on } (\mathcal{A}^+ \cap \{f(x) > 0\}) \cup (\mathcal{I}^2 \cap \{x > \psi\}) \cup \mathcal{I}^1 \\ \min(f'(x)d,d) & \text{on } (\mathcal{A}^- \cap \{f(x) < 0\}) \cup (\mathcal{I}^2 \cap \{x < \phi\}) \cup \mathcal{I}^3 \end{cases}$$

is a quasi-directional derivative for F. The unilateral case was treated in [4].

References

[1] F.H. Clarke: *Optimization and Nonsmooth Analysis*, John Wiley and Sons, New York, 1983.

[2] X. Chen, Z. Nashed and L. Qi, Smoothing methods and semi-smooth methods for nondifferentiable operator equations, SIAM J. on Numerical Analysis, **38** (2000), pp. 1200–1216.

[3] F. Facchinei and J.-S. Pang: *Finite-dimensional variational inequalities and complementarity problems vol. II*, Springer-Verlag, 2003, Berlin.

[4] S.-H. Han, J.-S. Pang and N. Rangaray: Globally convergent Newton methods for nonsmooth equations, Mathematics of Operations Research, 17(1992), 586–607.

[5] A. Haraux: How to differentiate the projection on a closed convex set in Hilbert space. Some applications to variational inequalities, J. Math. Soc. Japan 29(1977), 615–631.

[6] M. Hintermüller, K. Ito and K. Kunisch: The primal-dual active set strategy as a semi-smooth Newton method, SIAM Journal on Optimization 13(2002), 865–888.

[7] M. Hintermüller and K. Kunisch: Path-following methods for a class of constrained minimization problems in function space, submitted.

[8] K. Ito and K. Kunisch, Augmented Lagrangian methods for nonsmooth convex optimization in Hilbert Spaces, Nonlinear Analysis, Theory, Methods and Applications **41** (2000), pp. 573–589.

[9] K. Ito and K. Kunisch, Optimal Control of elliptic variational inequalities, Applied Mathematics and Optimization **41** (2000), pp. 343–364.

[10] K. Ito and K. Kunisch, Semi-smooth Newton methods for variational inequalities of the first kind, M2AN Math. Model. Numer. Anal. 37 (2003), 41–62.

[11] K. Ito and K. Kunisch, Parabolic Variational Inequalities: The Lagrange Multiplier Approach, Journal de mathématiques pures et appliquées (2005), to appear.

[12] K. Ito and K. Kunisch, On the Semi-smooth Newton method and its globalization, Math. Programming, (2005), submitted.

[13] K. Ito and K. Kunisch, Convergence of the primal-dual active set strategy for diagonally dominant systems, SIAM J. Control and Optimization (2005), submitted.

[14] B. Kummer, Newton's method for nondifferentiable functions, in: J. Guddat et al., eds., *Mathematical Research, Advances in Optimization*, Akademie-Verlag Berlin, 1988, pp. 114–125.

[15] R. Mifflin: Semismooth and semiconvex functions in constrained optimization, SIAM J. Control and Optimization, 15(1977), 959–972.

[16] J.S. Pang: Newton's method for B-differentiable equations, Mathematics of Operations Research, 15(1990), 311–341.

[17] L. Qi and J. Sun: A nonsmooth version of Newton's method, Math. Programming 58(1993), 353–367.

[18] L. Qi: Convergence analysis of some algorithms for solving nonsmooth equations, Math. of Operations Research 18(1993), 227–244.

[19] M. Ulbrich, Semi-smooth Newton methods for operator equations in function spaces, 2000, to appear in SIAM J. on Optimization.

Kazufumi Ito
Center for Research in Scientific Computation
Department of Mathematics
North Carolina State University
Raleigh, NC 27695-8205, USA
e-mail: `kito@math.ncsu.edu`

Karl Kunisch
Institut für Mathematik
und wissenschaftliches Rechnen,
Universität Graz
Graz, Austria

International Series of Numerical Mathematics, Vol. 155, 193–215
© 2007 Birkhäuser Verlag Basel/Switzerland

Identification of Nonlinear Coefficients in Hyperbolic PDEs, with Application to Piezoelectricity

Barbara Kaltenbacher

Abstract. In this paper we consider the problem of determining parameters in nonlinear partial differential equations of hyperbolic type from boundary measurements. In order to investigate the qualitative behavior of this class of identification problems, we analyze the model problem of identifying c in the nonlinear wave equation $d_{tt} - (c(d_x)d_x)_x = 0$ and discuss stability and identifiability for this problem. Moreover, we derive applicability of these results to material parameter identification in piezoelectricity and provide numerical reconstruction results.

Mathematics Subject Classification (2000). Primary 35R30; Secondary 35L70.

Keywords. Parameter identification, nonlinear wave equation, piezoelectricity.

1. Introduction

Consider the model problem of identifying the function c in the nonlinear hyperbolic PDE

$$\rho d_{tt} - (c(d_x)d_x)_x = 0 \quad x \in (0, L), \quad t \in (0, T), \tag{1.1}$$

with boundary conditions

$$\begin{aligned} d(0, t) &= 0 \\ c(d_x(L, t))d_x(L, t) &= g(t) \end{aligned} \quad t \in (0, T), \tag{1.2}$$

and initial conditions

$$d(x, 0) = d_0(x), \quad d_t(x, 0) = d_1(x) \quad x \in (0, L), \tag{1.3}$$

for given $g : [0, T] \to \mathbb{R}$, $d_0 : [0, L] \to \mathbb{R}$, $d_1 : [0, L] \to \mathbb{R}$, from additional boundary measurements

$$y(t) = d(L, t). $$

Supported by the German Science Foundation DFG under grant Ka 1778/1.

This can, e.g., be seen as a model of a vibrating string of length L, with the elasticity coefficient c depending on the strain d_x. At the left boundary $x = 0$, the string is clamped, at the right boundary $x = L$, it is excited in longitudinal direction by a surface load (mechanical stress) g, and measurements of the displacement d are made.

Our motivation for studying (1.1) comes from the problem of determining material parameter curves for piezoelectric materials (see Section 4), in the piezoelectric PDEs

$$
\begin{aligned}
\rho\frac{\partial^2 \vec{d}}{\partial t^2} - \mathbf{B}^T\left(\mathbf{c}^E\mathbf{B}\vec{d} + \mathbf{e}^T\mathrm{grad}\phi\right) &= 0 \quad \text{in } \Omega \\
-\mathrm{div}\left(\mathbf{e}\mathbf{B}\vec{d} - \boldsymbol{\varepsilon}^S\mathrm{grad}\phi\right) &= 0 \quad \text{in } \Omega.
\end{aligned}
\tag{1.4}
$$

Here, $\vec{d}$ denotes the vector of mechanical displacements, ϕ the electric potential, ρ the mass density, and $\mathbf{B}$ a first order differential operator with respect to the space variables, that reflects the relation between displacements and strain (see, e.g., [17] and Section 4 below). The system (1.4) models the piezoelectric effect, i.e., a coupling between the electrical and the mechanical behavior of certain materials. The material tensors $\mathbf{c}^E$, $\boldsymbol{\varepsilon}^S$, and $\mathbf{e}$, appearing in (1.4) are the elasticity coefficients, the dielectric constants, and the piezoelectric coupling coefficients, respectively. When large excitations are applied, the material parameters will not be constants any more but depend on the field quantities, i.e., in (1.4), the entries of the material tensors $\mathbf{c}^E$, $\boldsymbol{\varepsilon}^S$, and $\mathbf{e}$ are functions of the amplitude of the electric field $|\vec{E}| = |\mathrm{grad}\phi|$ and/or the mechanical strain $|\vec{S}| = |\mathbf{B}\hat{d}|$. An interesting task is to reconstruct these parameter functions from overdetermined measurements at the boundary, e.g., voltage-current measurements at an electrode or displacement measurements at a surface point.

Parameter identification problems for nonlinear PDEs have been studied, e.g., in [6, 7, 8, 9, 21, 20, 30, 33]). However, in the situation considered here, identifiability is still an open problem, even in the one-dimensional model problem (1.1). Therefore it was our aim to do some investigations on the question of whether the parameter c in (1.1) can be uniquely determined from the given boundary data.

We just wish to mention two additional related applications, namely from electromagnetics and acoustics, respectively, where models containing such a kind of nonlinear wave equation play a role:

- the spatially one-dimensional case in Maxwell's equations

$$
\nabla \times \left(\frac{1}{\mu}\nabla \times \vec{A}\right) + \varepsilon\frac{\partial^2 \vec{A}}{\partial t^2} = \vec{\hat{J}}
\tag{1.5}
$$

for a magnetic vector potential $\vec{A}$ with $\vec{B} = \nabla \times \vec{A}$ the magnetic induction, $\vec{E} = -\frac{\partial \vec{A}}{\partial t}$ the electric field, ε the dielectric constant, $\vec{J}$ the impressed current density, and $\mu = \mu(|\vec{B}|)$ the magnetic permeability depending on the magnetic induction, see, e.g., [31].

- the acoustic wave equation

$$\Delta\psi = \frac{1}{c^2}\frac{\partial^2\psi}{\partial t^2}$$

for the acoustic velocity potential ψ with $\vec{v} = -\nabla\psi$ the acoustic particle velocity and c the speed of sound, see, e.g., [32]. Since the speed of sound can depend on the pressure p and we have the relation $p = \rho\frac{\partial\psi}{\partial t}$, with ρ the density, we arrive at an equation similar to (1.1) in the 1-d case, where the roles of time and space are interchanged.

The paper is organized as follows: In Section 2 we show well-posedness of the forward problem corresponding to our model problem. Section 3 contains considerations on stability and identifiability for this model problem. Application to the piezoelectric problem mentioned above is discussed in Section 4, where we also provide numerical reconstruction results. The proofs of the statements made in Sections 2 and 3 are given in Section 5, and we end with a short Section 6 on conclusions.

2. The forward model problem: Well-posedness

From a theoretical point of view, instead of determining the curve c, it is more convenient to consider the curve $\tilde{c} : \lambda \mapsto c(\lambda) \cdot \lambda$ as the searched for unknown.

In order to investigate the properties of our model problem in more detail, we consider the forward operator F mapping a parameter function $\tilde{c}$ into the measurement $d(L, \, . \,) =: y$ of a solution d to the PDE

$$\rho d_{tt} - (\tilde{c}(d_x))_x = 0 \quad x \in (0, L), \quad t \in (0, T), \tag{2.1}$$

with boundary conditions

$$\begin{aligned} d(0, t) &= 0 \\ \tilde{c}(d_x(L, t)) &= g(t) \end{aligned} \quad t \in (0, T), \tag{2.2}$$

and initial conditions

$$d(x, 0) = d_0(x), \quad d_t(x, 0) = d_1(x) \quad x \in (0, L). \tag{2.3}$$

The function d is supposed to be a classical solution, i.e., $d \in C^{2,2}([0, L] \times [0, T])$, as we will require it in our identifiability considerations in the next section.

Identifying $\tilde{c}$ now corresponds to solving the operator equation

$$F(\tilde{c}) = y \tag{2.4}$$

in appropriate function spaces to be specified below.

The derivative of F into some direction $\tilde{s}$ is formally given as

$$F'(\tilde{c})[\tilde{s}] = v(L, \, . \,) \tag{2.5}$$

where v is the solution of

$$\rho v_{tt} - (\tilde{c}'(d_x)v_x)_x - (\tilde{s}(d_x))_x = 0 \quad x \in (0, L), \quad t \in (0, T), \tag{2.6}$$

196   B. Kaltenbacher

with boundary conditions

$$v(0, t) = 0$$
$$\tilde{c}'(d_x(L, t))v_x(L, t) + \tilde{s}(d_x(L, t)) = 0 , \qquad t \in (0, T) , \tag{2.7}$$

$$v(x, 0) = 0 \quad v_t(x, 0) = 0 \quad x \in (0, L) , \tag{2.8}$$

and d solves (2.1), (2.2), (2.3).

In order to guarantee well-definedness and Fréchet differentiability of F, we assume that

$$g \in C^2(0, T) , \quad d_0 \in C^3(0, L), \quad d_1 \in C^2(0, L) \tag{2.9}$$

and that the compatibility conditions

$$(\tilde{c}^{-1} \circ g)(0) = d_0'(L) , \quad (\tilde{c}^{-1} \circ g)'(0) = d_1'(L) ,$$
$$\rho(\tilde{c}^{-1} \circ g)''(0) = (\tilde{c}(d_0'))''(L) \tag{2.10}$$

on the right-hand boundary, as well as

$$d_0(0) = d_1(0) = d_0''(0) = d_1''(0) = 0 \tag{2.11}$$

on the left-hand boundary hold. Thus, we choose its domain of definition as

$$D(F) = \{\tilde{c} \in X \mid \tilde{c}'(\lambda) \geq \gamma, \tilde{c}''(\lambda) \leq \bar{C} \; \forall \lambda \in [0, \Lambda] , \text{ and } (2.10) \text{ holds}\}, \tag{2.12}$$

for some constants $\gamma, \bar{C} > 0$. Here X denotes the linear function space

$$X = \{\tilde{s} \in C^3(0, \Lambda) \mid \tilde{s}(0) = 0\} \tag{2.13}$$

with $\Lambda > 0$. Note that in the applications we are interested in, the parameter curves are typically strictly monotonically increasing and smooth, so choosing the domain of definition $D(F)$ and the space X according to (2.12), (2.13) makes sense. The assumed smoothness of $\tilde{c}$ is also of importance for the efficient solution of initial-boundary value problems for (2.1) (or (1.1)), after $\tilde{c}$ has been determined. If $\tilde{c}$ is sufficiently smooth, then for this purpose Newton's method is applicable and quadratically convergent.

Nonlinear hyperbolic PDEs bear the possibility of blow up of solutions (cf., e.g., [1, 14, 29, 34] and Section 11.3.2. in [13] as well as the references therein). Thus also here, existence of a smooth solution d to (2.1) cannot be expected on an arbitrarily large time interval, so that we have to restrict T in our well-posedness result. Although existence theory for nonlinear systems of hyperbolic conservation laws (cf., e.g., [4], [27], [18], [35], to name just a few recent publications, as well as the references therein and in [13]) would provide us with results on weak solutions to our model PDE (2.1) on all of $\mathbb{R}$, we here prove a simple result customized to our special case with the given boundary conditions and in the smoothness class that we will require for our identifiability and stability considerations in Section 3.

Proposition 2.1. (*Well-posedness of the forward problem*)
Assume that T is sufficiently small, Λ is sufficiently large, (2.9) holds and $D(F)$ is defined according to (2.12).

Then, for any $\tilde{c} \in D(F)$ there exists a unique solution $d \in C^{3,2}([0,L] \times [0,T])$ of (2.1), (2.2), (2.3). Hence, the forward operator

$$
\begin{aligned}
F : D(F) \subseteq X \;\; &\to \;\; C^2(0,T) \\
\tilde{c} \;\; &\mapsto \;\; d(L,\cdot) \; \text{where } d \text{ solves (2.1), (2.2), (2.3)}
\end{aligned}
$$

is well defined. Moreover with $X' := X \cap C^4(0,\lambda)$, $D'(F) := D'(F) \cap C^4(0,\lambda)$,

$$
F : D'(F) \subseteq X' \to C^2(0,T)
$$

is continuously Fréchet differentiable with its derivative given by (2.5), (2.6), (2.7), (2.8).

Proof. See Section 5. $\qquad\qquad\qquad\qquad\qquad\qquad\qquad\qquad\qquad\qquad\square$

Remark 2.2. An analogous result can be derived with homogeneous Neumann instead of Dirichlet boundary conditions on the left-hand boundary $x = 0$, which corresponds to a stress free left-hand string tip. In the proof of well-posedness, (see Section 5) the boundary conditions on $u = d_x$ change to

$$
\begin{aligned}
u(0,t) &= 0 \\
\tilde{c}(u(L,t)) &= g(t)
\end{aligned}
\qquad t \in (0,T)\,,
$$

and d is obtained from u via

$$
d(x,t) := \int_0^x u(\xi,t)d\xi + \frac{1}{\rho} \int_0^t \int_0^\tau (\tilde{c}(u))_x(0,\sigma)\, d\sigma\, d\tau\,.
$$

Remark 2.3. Usually measurements only of zero order derivative values of $d(\cdot,L)$ (i.e., displacement measurements) can be expected to be available, hence a natural choice of the image space of F is $Y := L^\infty(0,T)$. Observe that by Proposition 2.1, the actual range of F is contained in $C^2(0,T)$ and therewith non-closed in the data space $Y = L^\infty(0,T)$. This smoothing property of the forward operator corresponds to ill-posedness of the inverse problem. More precisely, a lower bound for the degree of ill-posedness can be quantified by the difference in smoothness order of the function spaces C^2 and L^∞. Therewith, the problem of identifying $\tilde{c} \in X$ from $y \in Y$ has to be expected to be at least as ill-posed as twice numerical differentiation.

In some applications, it can be realistic to assume that we have measurements also of first order derivative values of $d(\cdot,L)$ (i.e., velocity measurements). Then the degree of ill-posedness is still at least as high as for one numerical differentiation.

3. The inverse model problem: Identifiability

Using different norms both in preimage and in image space, we aim at deriving a stability result with the implication that on a certain interval $[0,\overline{\lambda}] \subseteq [0,\Lambda]$ the parameter curve $\lambda \mapsto \tilde{c}(\lambda)$ can be uniquely determined from the given measurements.

The difference $F(\check{c}) - F(\tilde{c})$ can be written as the right-hand boundary values of v solving

$$\rho v_{tt} - (a\, v_x\, +\, \phi)_x = 0 \quad x \in (0, L), \quad t \in (0, T), \tag{3.1}$$

with boundary conditions

$$\begin{aligned} v(0, t) &= 0 \\ a(L, t)\, v_x(L, t)\, +\, \phi(L, t) &= 0 \end{aligned} \quad t \in (0, T), \tag{3.2}$$

and homogeneous initial conditions, where we denote

$$\begin{aligned} a(x, t) &= \int_0^1 \tilde{c}'(\check{d}_x(x, t) + \theta(d_x(x, t) - \check{d}_x(x, t)))d\theta, \\ \phi(x, t) &= \tilde{s}(d_x(x, t)), \\ \tilde{s} &= \check{c} - \tilde{c}, \end{aligned} \tag{3.3}$$

and $\check{d}$ solves (2.1), (2.2), (2.3), with $\tilde{c}$ replaced by $\check{c}$, i.e.,

$$F(\check{c}) - F(\tilde{c}) = v(L, \cdot). \tag{3.4}$$

Hence, our aim is to deduce an estimate of the form

$$\|\tilde{s}\|_{\tilde{X}} \leq C \|v(L, \cdot)\|_{\tilde{Y}}$$

with appropriate function spaces $\tilde{X}$ and $\tilde{Y}$, that by Remark 2.3 have to be different from X and Y as defined in the previous section.

For the sake of simplicity, we first of all restrict ourselves to the special case of constant $a(x, t) = \bar{a} \in \mathbb{R}$ so that we deal with the wave equation

$$\rho v_{tt} - \bar{a} v_{xx} = \phi_x, \tag{3.5}$$

with boundary conditions

$$\begin{aligned} v(0, t) &= 0 \\ \bar{a} v_x(L, t)\, +\, \phi(L, t) &= 0 \end{aligned} \quad t \in (0, T), \tag{3.6}$$

and homogeneous initial conditions

$$v(x, 0) = v_t(x, 0) = 0 \quad x \in (0, L). \tag{3.7}$$

The following result can be obtained by integrating along the characteristic lines of (3.5) and therewith deriving a Volterra integral equation of the first kind for the difference $\tilde{s}$ between parameter curves.

Proposition 3.1. *Let v be a solution to (3.5) with boundary conditions (3.6) and initial conditions (3.7). Here, we assume that ϕ is determined by (3.3) with $d \in C^{3,2}([0, L] \times [0, T])$ satisfying the boundary conditions (2.2) and the initial conditions (2.3) with (2.9), $g(0) = 0$ and g strictly monotonically increasing, $d_0' \equiv 0$, $\tilde{c} \in D(F)$, and $\tilde{s} \in C^2([0, \Lambda_1])$, for some $\Lambda_1 > 0$ such that $\{d_x(x, t) \mid (x, t) \in [0, L] \times [0, T]\} \subseteq [0, \Lambda_1]$.*
Additionally, assume that

$$\left| \pm \sqrt{\bar{a}/\rho}\, d_{xx}(x, t) + d_{xt}(x, t) \right| \geq \kappa \quad \forall (x, t) \in (0, L) \times (0, \bar{t}) \tag{3.8}$$

holds for some $\kappa > 0$, $0 < \bar{t} \leq T$.

Then, with

$$\overline{\lambda} = \tilde{c}^{-1}(g(\overline{t})) > 0 \tag{3.9}$$

the estimate

$$\|\tilde{s}\|_{L^2(0,\overline{\lambda})} \leq C \|v(L,\cdot)\|_{H^1(0,\overline{t})} \tag{3.10}$$

holds with some $C > 0$.

Proof. See Section 5. $\qquad\square$

Since the technique of reformulation as a Volterra integral equation via integration along the characteristic curves should in principle be applicable also in the general case (3.1), Proposition 3.1 gives a hint on how the difference between two curves $\tilde{s} = \check{c} - \tilde{c}$ can be estimated in terms of the difference between the measurements $v(L,\cdot) = F(\check{c}) - F(\tilde{c})$:

Conjecture 3.2. (*Stability and identifiability*)
Let the assumptions of Proposition 2.1 on well-posedness of the forward problem be satisfied, with

$$g(0) = 0\,, \quad g(t) \geq 0\,, \quad g'(t) \geq \mu > 0 \quad \forall t \in [0,\overline{t}]\,, \tag{3.11}$$

$$d_0'(x) = 0 \quad \forall x \in [0,L]\,, \tag{3.12}$$

for some $\mu > 0$. Let $\tilde{c} \in D(F)$, and d be the solution of (2.1), (2.2), (2.3). Additionally, assume that

$$\left| \left(\pm\sqrt{\tilde{c}'(d_x)/\rho}\ d_{xx} + d_{xt} \right)(x(t),t) \right| \geq \kappa \quad \forall t \in [0,\overline{t}]\,, \tag{3.13}$$

holds on an interval $[0,\overline{t}] \subseteq [0,T]$, with some constant $\kappa > 0$, for all characteristic curves $t \mapsto x(t)$ of (2.1), and that $\overline{t}$ and L are sufficiently small.

Then, the measurements $d(L,t)$, $t \in [0,\overline{t}]$ uniquely determine $\tilde{c}$ on the interval $[0,\overline{\lambda}]$ with

$$\overline{\lambda} = \tilde{c}^{-1}(g(\overline{t})) > 0 \tag{3.14}$$

and the estimate

$$\|\check{c} - \tilde{c}\|_{L^2(0,\overline{\lambda})} \leq C \|F(\check{c}) - F(\tilde{c})\|_{H^1(0,\overline{t})} \tag{3.15}$$

holds with some $C > 0$ for all $\check{c} \in D(F) \cap B_r(\tilde{c})$ where $B_r(\tilde{c})$ is a ball of sufficiently small radius r (with respect to the C^3 norm) around $\tilde{c}$.

Idea of proof. See Section 5.

Remark 3.3. The assumptions on d_x and g that result from (3.11), (3.12), and the compatibility conditions (2.10), (2.11) can be naturally fulfilled by starting with vanishing displacement, and (up to some time $\overline{t}$) strictly increasing nonnegative excitation.

Moreover, if $\left| \left(\pm\sqrt{\tilde{c}'(d_0')/\rho}\ d_0'' + d_1' \right)(x) \right|$ is uniformly bounded away from zero for all $x \in [0,L]$, then by continuity (3.13) will hold for some $\kappa > 0$ as long as $\overline{t}$ is sufficiently small.

Smallness of L is natural, e.g., in the context of the piezoelectric application in Section 4, where L is the (typically small) thickness of the probe, see Figure 2. However, smallness of $\bar{t}$ (as we also require it for existence in Proposition 2.1) imposes a restriction on the interval on which $\tilde{c}$ can be identified. Although one might think of choosing a fast increasing g for enlarging $\bar{\lambda}$ in (3.14), this does not really help since the smallness condition on $\bar{t}$ coming from the perturbation argument in the last part of the idea of proof of Conjecture 3.2 depends on how fast g and therewith a as defined in (3.3) changes with time.

Note that the basic idea of integrating along characteristics and formulating the identification problem as an integral equation comes from the book by Isakov [19], where uniqueness results for identification of spatially varying parameters in linear hyperbolic PDEs are proven.

4. Application to material nonlinearities in piezoelectricity

Recall the piezoelectric PDEs

$$
\begin{aligned}
\rho\frac{\partial^2 \vec{d}}{\partial t^2} - \mathbf{B}^T\left(\mathbf{c}^E\mathbf{B}\vec{d} + \mathbf{e}^T\mathrm{grad}\phi\right) &= 0 \quad \text{in } \Omega \\
-\mathrm{div}\left(\mathbf{e}\mathbf{B}\vec{d} - \boldsymbol{\varepsilon}^S\mathrm{grad}\phi\right) &= 0 \quad \text{in } \Omega,
\end{aligned}
\tag{4.1}
$$

where $\mathbf{B}$ is the transposed of the divergence DIV of a dyadic,

$$
\mathbf{B} = \begin{pmatrix}
\frac{\partial}{\partial x} & 0 & 0 \\
0 & \frac{\partial}{\partial y} & 0 \\
0 & 0 & \frac{\partial}{\partial z} \\
0 & \frac{\partial}{\partial z} & \frac{\partial}{\partial y} \\
\frac{\partial}{\partial z} & 0 & \frac{\partial}{\partial x} \\
\frac{\partial}{\partial y} & \frac{\partial}{\partial x} & 0
\end{pmatrix},
$$

and the material tensors $\mathbf{c}^E \in \mathbb{R}^6_6$, $\mathbf{e} \in \mathbb{R}^6_3$, $\boldsymbol{\varepsilon}^S \in \mathbb{R}^3_3$ have a structure depending on the material class under consideration and the direction of polarization. For instance, for materials in the 6mm crystal class polarized in 3-direction, they are of the form

$$
\mathbf{c}^E = \begin{pmatrix}
c_{11}^E & c_{12}^E & c_{13}^E & 0 & 0 & 0 \\
c_{12}^E & c_{11}^E & c_{13}^E & 0 & 0 & 0 \\
c_{13}^E & c_{13}^E & c_{33}^E & 0 & 0 & 0 \\
0 & 0 & 0 & \frac{1}{2}(c_{11}^E - c_{12}^E) & 0 & 0 \\
0 & 0 & 0 & 0 & c_{44}^E & 0 \\
0 & 0 & 0 & 0 & 0 & c_{44}^E
\end{pmatrix},
\tag{4.2}
$$

$$
\mathbf{e} = \begin{pmatrix}
0 & 0 & 0 & 0 & 0 & e_{15} \\
0 & 0 & 0 & 0 & e_{15} & 0 \\
e_{31} & e_{31} & e_{33} & 0 & 0 & 0
\end{pmatrix},
\tag{4.3}
$$

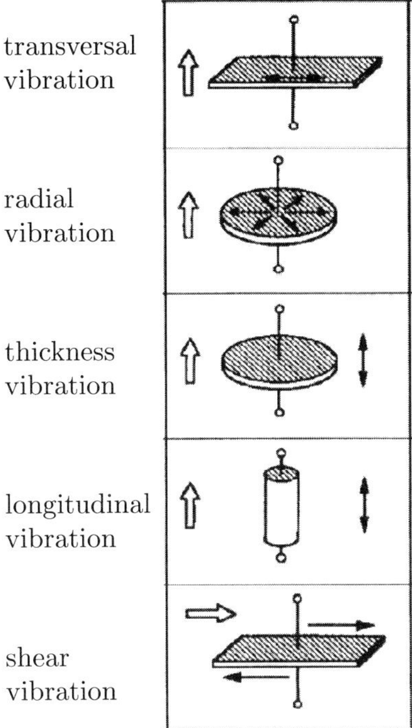

FIGURE 1. Test sample scheme according to the European Norm [12]

$$\boldsymbol{\varepsilon}^S = \begin{pmatrix} \varepsilon_{11}^S & 0 & 0 \\ 0 & \varepsilon_{11}^S & 0 \\ 0 & 0 & \varepsilon_{33}^S \end{pmatrix}, \tag{4.4}$$

(cf. [12], [17]).

By an appropriate experimental setup (cf. [12], [17]), the problem of determining the ten different scalar entries $c_{11}^E, c_{33}^E, c_{12}^E, c_{13}^E, c_{44}^E, e_{15}, e_{31}, e_{33}, \varepsilon_{11}^S, \varepsilon_{33}^S$ of the material tensors in (4.1) can be reduced to five basically spatially one-dimensional problems, corresponding to the five test samples in Figure 1. Moreover, by exposing the piezo-ceramic probe either to large mechanical pre-stressing or to high voltages, one can achieve predominance of dependence on *either* stress $|\mathbf{B}\vec{d}|$ *or* electric field $|\mathrm{grad}\phi|$.

Since we can collect two different sets of data for each test sample (e.g., for varying time, the curves of electrical charge and of mechanical displacement at an electrode[1]), we expect that we can identify two parameter curves per test sample, which by independence of the five sample experiments gives the full set of ten parameter curves in the tensors (4.2), (4.3), (4.4).

---

[1]that can be obtained from current and velocity measurements

B. Kaltenbacher

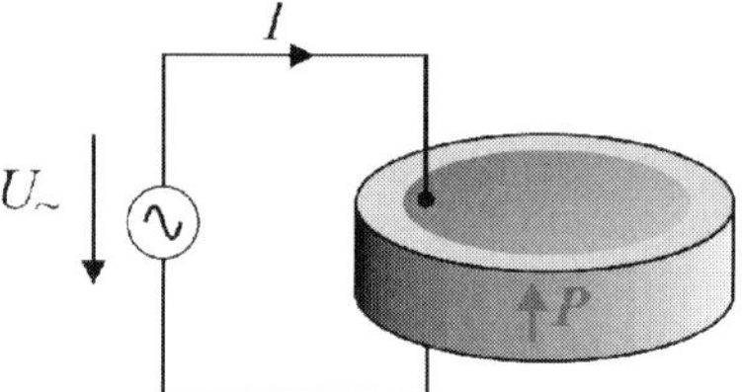

FIGURE 2. Schematic of the experimental setup: Piezoelectric disc polarized in thickness direction, electrodes placed on top and bottom of the sample.

To complete this uniqueness argument for the full parameter curve set by means of our identifiability Conjecture 3.2, it remains to get from the one-dimensional piezoelectric PDEs

$$
\begin{aligned}
\rho d_{tt} - (c^E d_x + e\phi_x)_x &= 0 \quad x \in (0, L), t \in [0, T] \\
-(e d_x - \varepsilon^S \phi_x)_x &= 0 \quad x \in (0, L), t \in [0, T],
\end{aligned}
\tag{4.5}
$$

(where in place of material tensors $\mathbf{c}^E, \mathbf{e}, \boldsymbol{\varepsilon}^S$ we only have scalars c^E, e, ε^S) to our model problem. In many cases of interest, this can be done by eliminating ϕ by means of the second line in (4.5):

As an example, consider a thin disc according to the third picture in Figure 1, so that the piezoelectric effect in thickness direction is predominant, see Figure 2. One of the electrodes is grounded, the other either loaded with a prescribed surface charge $q^L(t)$ or with an impressed voltage $\phi^L(t)$.

$$
\phi(0, t) = 0
$$
$$
\begin{cases}
\text{(i)} \ (e d_x - \varepsilon^S \phi_x)(L, t) = -\dfrac{q^L(t)}{A} \quad & \text{(charge excitation)} \\
\text{or} & \\
\text{(ii)} \ \phi(L, t) = \phi^L(t) \quad & \text{(voltage excitation)},
\end{cases}
\tag{4.6}
$$

where A is the surface area covered by the loaded electrode.

When exposing the probe to large mechanical pre-stressing $\sigma_0(t)$, $\sigma_L(t)$ on top and bottom, which corresponds to the boundary conditions

$$
\begin{aligned}
(c^E d_x + e\phi_x)(0, t) &= \sigma_0(t) \\
(c^E d_x + e\phi_x)(L, t) &= \sigma_L(t),
\end{aligned}
\tag{4.7}
$$

the elasticity coefficient c^E will exhibit a dependency on the strain d_x. Eliminating ϕ by resolving the second line of (4.5) together with the boundary conditions (4.6) for ϕ,

$$
\phi(x, t) = \frac{e}{\varepsilon^S}(d(x, t) - d(0, t)) + r(t) \cdot x
$$

with

$$\begin{cases} \text{(i) } r(t) = \frac{q^L(t)}{\varepsilon^S A} \\ \text{or} \\ \text{(ii) } r(t) = \frac{1}{L}\left(\phi^L(t) - \frac{e}{\varepsilon^S}(d(L,t) - d(0,t))\right), \end{cases}$$

and inserting this expression into the first line of (4.5) and the boundary conditions (4.7), we arrive at

$$d_{tt} - ((c(d_x))d_x)_x = 0$$

with boundary conditions

$$(c(d_x)d_x)(0,t) = \sigma_0(t) + e\, r(t)$$
$$(c(d_x)d_x)(L,t) = \sigma_L(t) + e\, r(t)$$

with

$$c(\lambda) := c^E(\lambda) + \frac{e^2}{\varepsilon^S}.$$

Assuming that e, ε^S are known (e.g., from experiments at low impressed currents/voltages and low boundary stresses, see, e.g., [12], [17]), and that we can measure the displacement difference $d(L,t) - d(0,t)$, we arrive at an identification problem for determining the curve $\lambda \mapsto c^E(\lambda)$, that is almost identical to the model problem (1.1), (1.2), (1.3).

4.1. Numerical reconstruction results

For a first implementation in a Matlab program, we concentrated on the spatially one-dimensional case (4.5). We assume that in (4.5), e and ε^S depend on the electric field $|\phi_x|$ and c^E is constant, (thus, we consider a situation that is in some sense complementary to the case of stress dependent c^E with constant e, ε^S as it was just shown to be dirctly reducible to our model problem). Therewith, in

$$\begin{aligned} \rho d_{tt} - \left(\mathbf{c}^E d_x + e(|\phi_x|)\phi_x\right)_x &= 0 \\ -\left(e(|\phi_x|)d_x - \varepsilon^S(|\phi_x|)\phi_x\right)_x &= 0 \end{aligned} \tag{4.8}$$

with the boundary conditions

$$\left(\mathbf{c}^E d_x + e(|\phi_x|)\phi_x\right)(0,t) = \left(\mathbf{c}^E d_x + e(|\phi_x|)\phi_x\right)(L,t) = 0$$
$$\phi(0,t) = 0 \qquad \left(e(|\phi_x|)d_x - \varepsilon^S(|\phi_x|)\phi_x\right) = -\frac{q^L(t)}{A}$$

(stress free surface, one grounded and one charge loaded electrode), the curves e and ε^S are to be identified. The given data are the mechanical displacement $d(L,\cdot)$ as well as the electric voltage $\phi(L,\cdot)$ and the electric charge q^L at the loaded electrode, the latter entering the boundary conditions.

With the forward operator $F : (e, \varepsilon^S) \to (d(L,\cdot), \phi(L,\cdot))$ mapping the parameter curves (e, ε^S) to the trace of the solution (d, ϕ) at $x = L$, we write this as a nonlinear operator equation

$$F(e, \varepsilon^S) = (d_L^{\mathrm{meas}}, \phi_L^{\mathrm{meas}})$$

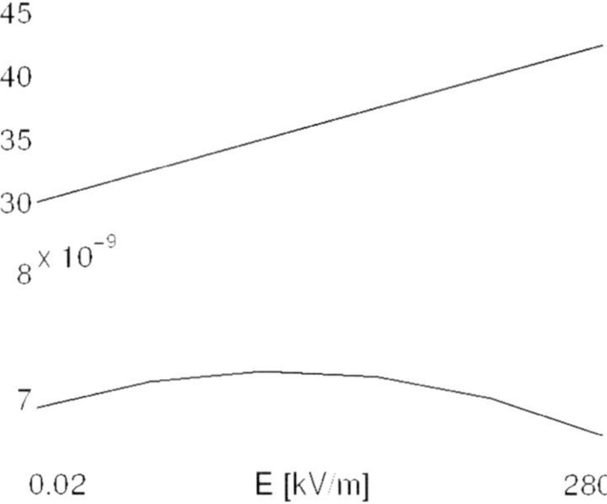

FIGURE 3. Exact curves for e and ε^S

that we approximately solve by applying Newton's method – note that the Jacobian can be expressed in terms of solutions to linearized versions of (4.8) – and discretization. The latter is here done by using a spline ansatz for the searched for curves and doing a space and time discretization by piecewise linear finite elements (i.e., due to the 1-d situation, equivalently, finite differences). Here the number of spline breakpoints is chosen relatively small, in order to regularize the inverse problem by coarse discretization (cf Remark 2.3, as well as Section 3.3. in [11], Chapter 3 in [26], and the references therein, as well as [23]). In contrast to that, a sufficiently large number of degrees of freedom in space and time is used for numerically solving (4.8), in order to keep the numerical approximation error in simulating measurements small, namely of the order of magnitude of the measurement noise level. Additionally, to avoid an inverse crime, we choose the discretization for generating synthetic data different from the one used in the inverse computations.

Figure 3 shows the exact curves taken from a paper by Anderssen et al. [2]. In Figures 4 and 5, we plot the results with exact data for our reconstruction of e, ε^S separately and simultaneously, respectively. The separate reconstructions are also shown for randomly perturbed data with a noise level of one per cent as typical in this context, see Figure 6. In all these graphics, the starting curves (constants) are marked by circles, the final ones by crosses, and the exact ones by a solid line.

The experiments show that the results are worse (and noise in the data is more critical) for e than for ε^S, which indicates that the measurements might be less sensitive with respect to the piezoelectric coupling coefficient.

However, we wish to remark that using a numerical reconstruction approach that is based on a formulation in frequency domain, much better results can be obtained, see [24].

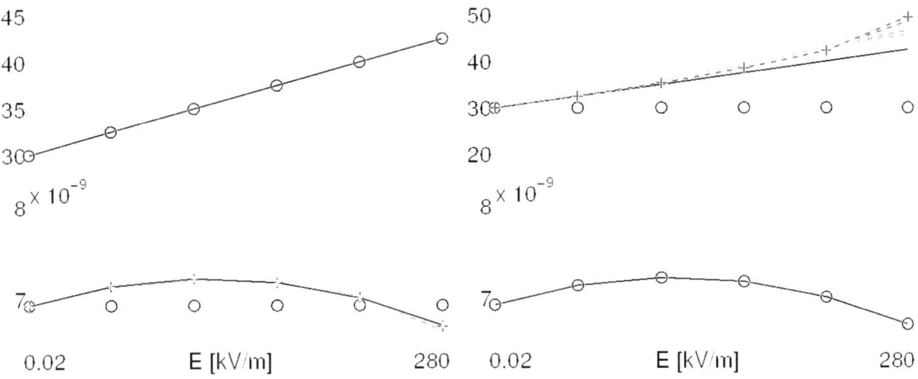

FIGURE 4. Separate reconstruction for e (left) and ε^S (right) from exact data

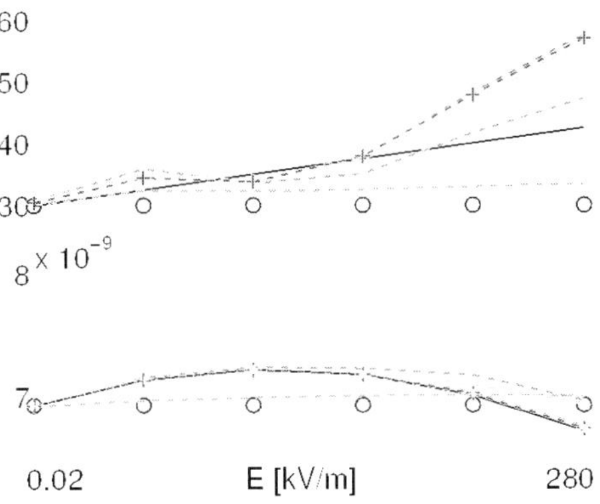

FIGURE 5. Simultaneous reconstruction for e and ε^S from exact data

5. Proofs of Propositions 2.1, 3.1, and idea of proof of Conjecture 3.2

Proof. (Proposition 2.1)

Consider $u := d_x$ as the searched for function in the PDE that is obtained by differentiating (2.1) with respect to x:

$$\rho u_{tt} - (\tilde{c}'(u)u_x)_x = 0 \quad x \in (0,L), \quad t \in (0,T).\tag{5.1}$$

We assume homogeneous Neumann boundary conditions on the left-hand boundary, and use the second line in (2.2) to get a condition on the right-hand boundary,

$$\begin{aligned}(\tilde{c}'(u)u_x)(0,t) &= 0\\ \tilde{c}(u(L,t)) &= g(t)\end{aligned} \quad t \in (0,T),$$

B. Kaltenbacher

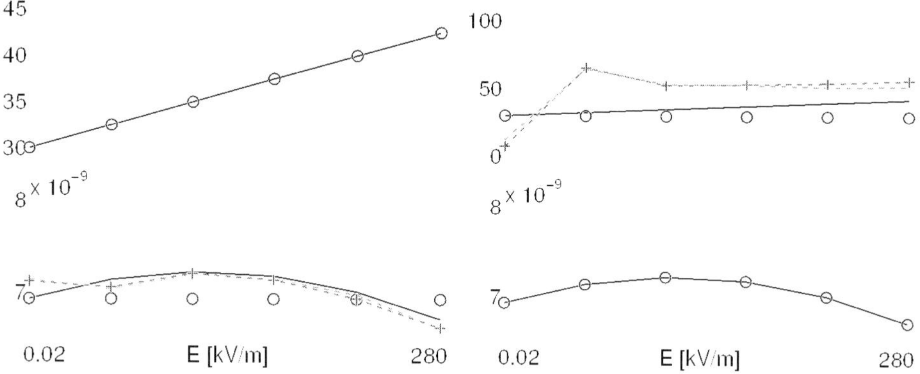

FIGURE 6. Separate reconstruction for e (left) and ε^S (right) from data contaminated with 1 per cent noise)

i.e.,

$$u_x(0,t) = 0$$
$$u(L,t) = \tilde{c}^{-1}(g(t)) \qquad t \in (0,T). \tag{5.2}$$

The initial conditions are

$$u(x,0) = d_{0x}(x), \quad u_t(x,0) = d_{1x}(x) \quad x \in (0,L). \tag{5.3}$$

To show existence of a solution to this nonlinear boundary value problem, we transfer the hyperbolic PDE (5.1) into standard form by a smooth transform of variables aligned to the characteristic curves given by $\dot{x}(t) = \pm\sqrt{(\tilde{c}'(u)(x(t),t)/\rho}$ (cf., e.g., Section 7.2.5 in [13] and Section 2-6. in [15]). Namely, we consider

$$U(\varphi(x,t) + \psi(x,t), \varphi(x,t) - \psi(x,t)) = u(x,t) \tag{5.4}$$

with φ, ψ solving

$$\sqrt{\rho}\varphi_t + \sqrt{\tilde{c}'(u)}\varphi_x = 0, \quad \sqrt{\rho}\psi_t - \sqrt{\tilde{c}'(u)}\psi_x = 0, \tag{5.5}$$

which leads to a PDE for U

$$U_{\beta\beta} - U_{\alpha\alpha} = \frac{1}{8}\left(\frac{a_x}{a} + \frac{a_t}{a^{3/2}}\right)\frac{1}{\psi_x}(U_\alpha + U_\beta) + \frac{1}{8}\left(\frac{a_x}{a} - \frac{a_t}{a^{3/2}}\right)\frac{1}{\varphi_x}(U_\alpha - U_\beta), \tag{5.6}$$

where $a := \tilde{c}'(u)/\rho$. To see that this transform of variables is in fact regular and C^2 smooth on some sufficiently small time interval $(0,\underline{t})$, provided $u \in C^2$, consider the Jacobi determinant

$$\det\begin{pmatrix} \varphi_x + \psi_x & \varphi_t + \psi_t \\ \varphi_x - \psi_x & \varphi_t - \psi_t \end{pmatrix} = \det\begin{pmatrix} \varphi_x + \psi_x & -\sqrt{\tilde{c}'(u)/\rho}(\varphi_x - \psi_x) \\ \varphi_x - \psi_x & -\sqrt{\tilde{c}'(u)/\rho}(\varphi_x + \psi_x) \end{pmatrix}$$
$$= -4\sqrt{\tilde{c}'(u)/\rho}\,\varphi_x\,\psi_x. \tag{5.7}$$

From the characteristic ODEs for φ

$$
\begin{array}{lll}
t_\tau(\tau, s) = 1 & t(0, s) = s & \Rightarrow t(\tau, s) = \tau + s \\
x_\tau(\tau, s) = \sqrt{\tilde{c}'(u(x(\tau, s), t(\tau, s))/\rho} & x(0, s) = 0 & \\
\varphi_\tau(\tau, s) = 0 & \varphi(0, s) = s & \Rightarrow \varphi(\tau, s) = s
\end{array}
\tag{5.8}
$$

for $\tau \geq 0$, $s = t(0, s) \geq 0$, it follows that

$$
1 = \varphi_s = \varphi_x \cdot x_s + \varphi_t \cdot t_s = \varphi_x(x_s - \sqrt{\tilde{c}'(u)/\rho}),
\tag{5.9}
$$

and therewith $\varphi_x \neq 0$. The same holds true for ψ, therefore the Jacobi determinant (5.7) does not vanish. On the other hand, (5.9) implies boundedness of $|\varphi_x|$ as long as

$$
x_s(\tau, s) \neq \sqrt{\tilde{c}'(u(x(\tau, s), \tau + s)/\rho}.
$$

The latter can be established for all t and therewith $\tau = t - s$ smaller than some $\underline{t} > 0$ that depends only on γ, $\bar{C}$, ρ, and the C^1-norm of u, by deriving an ODE for $\tilde{x} := x_s$ from (5.8):

$$
\tilde{x}_\tau(\tau, s) = \frac{\tilde{c}''(u(x(\tau, s), t(\tau, s)))(u_x(x(\tau, s), t(\tau, s))\tilde{x}(\tau, s) + u_t(x(\tau, s), t(\tau, s)))}{2\sqrt{\tilde{c}'(u(x(\tau, s), t(\tau, s)) \cdot \rho}},
$$

$$
\tilde{x}(0, s) = 0,
$$

which implies that

$$
|\tilde{x}_\tau(\tau, s)| \leq \frac{\bar{C}\|u\|_{C^1}}{2\sqrt{\gamma\rho}}(|\tilde{x}(\tau, s)| + 1),
$$

so

$$
|\tilde{x}(\tau, s)| \leq \exp\left(\frac{\bar{C}\|u\|_{C^1}}{2\sqrt{\gamma\rho}}\tau\right) - 1 < \sqrt{\gamma/\rho} \leq \sqrt{\tilde{c}'(u(x(\tau, s), \tau + s)/\rho}
$$

for all

$$
\tau \leq \underline{t} := \frac{2\sqrt{\gamma\rho}\ln(1 + \sqrt{\gamma/\rho})}{\bar{C}\|u\|_{C^1}}.
$$

Another differentiation of (5.9) with respect to s yields

$$
\begin{aligned}
0 &= \varphi_{ss} = \frac{d}{ds}(\varphi_x(x_s - \sqrt{\tilde{c}'(u)/\rho})) \\
&= (\varphi_{xx}x_s + \varphi_{xt}t_s)(x_s - \sqrt{\tilde{c}'(u)/\rho}) + \varphi_x(x_{ss} - (\sqrt{\tilde{c}'(u)/\rho})_s) \\
&= \varphi_{xx}(x_s - \sqrt{\tilde{c}'(u)/\rho})^2 - (\sqrt{\tilde{c}'(u)/\rho})_x\varphi_x(x_s - \sqrt{\tilde{c}'(u)/\rho}) \\
&\quad + \varphi_x(x_{ss} - (\sqrt{\tilde{c}'(u)/\rho})_s),
\end{aligned}
$$

so by reasoning similar to above and using our assumption $u \in C^2$, we arrive at boundedness also of φ_{xx} for $t \leq \underline{t}$. Analogously boundedness of all other first and second derivatives of φ and also of ψ can be verified. By the Inverse Function Theorem we therefore know that existence of a C^2 solution u to (5.1), (5.2), (5.3) follows from existence of a C^2 solution to (5.6) with transformed initial and boundary conditions. It therefore remains to show that the latter holds true, which we do by the Banach Fixed Point Theorem. Namely, we define the fixed point operator $\mathbf{T}$ mapping a C^2 function U to a solution $\mathbf{T}(U) := Z$ of

$$Z_{\beta\beta} - Z_{\alpha\alpha} = \frac{1}{8}\Big(\frac{a_x}{a} + \frac{a_t}{a^{3/2}}\Big)\frac{1}{\psi_x}(U_\alpha + U_\beta) + \frac{1}{8}\Big(\frac{a_x}{a} - \frac{a_t}{a^{3/2}}\Big)\frac{1}{\varphi_x}(U_\alpha - U_\beta)\,, \quad (5.10)$$

with the transformed initial and boundary conditions. Note that the right-hand side depends on U not only linearly via the first order derivative terms $U_\alpha + U_\beta$, $U_\alpha - U_\beta$ but also nonlinearly via $a = \tilde{c}'(U)/\rho$ and φ, ψ. Also note, that although the boundary and initial conditions (5.2), (5.3) are linear, they still depend on U since the transform of variables affects the curves in the α, β plane on which the initial and boundary conditions hold. The operator $\mathbf{T}$ is contractive, since Z is obtained by integrating the right-hand side in (5.10) over a – due to t small – short path along the characteristics $\alpha \pm \beta = \mathrm{const}$ and the right-hand side consists of the first order derivatives $U_\alpha + U_\beta$, $U_\alpha - U_\beta$, along with factors whose L^∞ norm can be bounded in terms of $\|U\|_{C^2}$, see the arguments following (5.9).

This implies existence of a solution $u \in C^2$ to the nonlinear initial-boundary value problem (5.1), (5.2), (5.3), and therewith, according to

$$d(x,t) := \int_0^x u(\xi,t)d\xi \qquad (5.11)$$

of a solution $d \in C^{3,2}$ to (1.1), (1.2), (1.3). Uniqueness of d follows from standard energy estimates, cf., e.g., Theorem 4 in Section 7.2 of [13], applied to the differential equation

$$\rho v_{tt} - \Big(\int_0^1 \tilde{c}'(d_x^2 + \theta(d_x^1 - d_x^2))d\theta\, v_x\Big)_x = 0$$

with homogeneous initial and boundary conditions, that the difference v between two different solutions d^1, d^2 has to satisfy.

Similar arguments can be used to show continuous dependence of the solution d in $C^{3,2}((0,L) \times (0,T))$ on the parameter $\tilde{c}$ in C^3.

To prove Fréchet differentiability, note that with $F(\tilde{c} + \tilde{s}) = \bar{d}(L, \cdot)$ where $\bar{d}$ solves (1.1), (1.2) with $\tilde{c}$ replaced by $\tilde{c} + \tilde{s}$, the function $w := \bar{d} - d - v$ solves

$$\rho w_{tt} - (\tilde{c}'(d_x)w_x)_x = f_x$$

with boundary conditions

$$w(0,t) = 0$$
$$\tilde{c}'(d_x(L,t))w_x(L,t) = f(L,t)\,,$$

where

$$f = \tilde{c}(\bar{d}_x) - \tilde{c}(d_x) - \tilde{c}'(d_x)(\bar{d}_x - d_x) \;+\; \tilde{s}(\bar{d}_x) - \tilde{s}(d_x)\,.$$

Therefore, again, the characteristic C^2 variable transform and a standard result for the wave equation imply

$$\|w\|_{C^2} \le C\|f_x\|_{C^1}$$
$$\le C(1 + \|d_x\|_{C^2}^2)\big(\|\tilde{c}\|_{C^4}\,\|\bar{d}_x - d_x\|_{C^2}^2 + \|\tilde{s}\|_{C^3}\,\|\bar{d}_x - d_x\|_{C^2}\big)\,.$$

By continuity of F, the right-hand side goes to zero like $o(\|\tilde{s}\|_{C^3})$, which proves Fréchet differentiability. Analogously, continuity of the Fréchet derivative can be shown by deriving a PDE describing the difference between two Fréchet derivative values. $\qquad\square$

Proof. (Proposition 3.1)
Without loss of generality, we set $\rho = L = \bar{a} = 1$. By integrating along the characteristic curves $x \pm t = \text{const}$, inserting the boundary conditions at the right-hand boundary

$$v_x(1,t) + \phi = 0 \,, \tag{5.12}$$

as well as the measurement difference (cf. (3.4))

$$m(t) := v(1,t) \,, \tag{5.13}$$

we arrive at the solution

$$
\begin{aligned}
v(x,t) = {} & \frac{1}{2}\left(m(1+t-x) + m(1+t-x-\min\{1+t-x, 2(1-x)\})\right) \\
& + \frac{1}{2}\int_{1+t-x-\min\{1+t-x,2(1-x)\}}^{1+t-x} \phi(1,\sigma)d\sigma \\
& - \int_0^{\min\{\frac{1}{2}(1+t-x),1-x\}} \int_0^{\nu} \phi_x(1-\tau, \tau+1+t-x-2\nu)\,d\tau d\nu \\
& - \int_{\min\{\frac{1}{2}(1+t-x),1-x\}}^{1-x} \int_{2\nu-(1+t-x)}^{\nu} \phi_x(1-\tau, \tau+1+t-x-2\nu)\,d\tau d\nu \,,
\end{aligned}
\tag{5.14}
$$

which additionally to (5.12) and (5.13) satisfies

$$(v_t + v_x)(x,0) = 0 \,. \tag{5.15}$$

From the initial conditions as well as the boundary conditions on the left-hand boundary

$$v(x,0) = 0, \quad v(0,t) = 0 \tag{5.16}$$

we obtain equations relating ϕ to the measurements, namely

$$
\begin{aligned}
0 = {} & v(x,0) \\
= {} & \frac{1}{2}(m(1-x) + m(0)) + \frac{1}{2}\int_0^{1-x} \phi(1,\sigma)d\sigma \\
& - \int_0^{\frac{1}{2}(1-x)} \int_0^{\nu} \phi_x(1-\tau, \tau+1-x-2\nu)\,d\tau d\nu \\
& - \int_{\frac{1}{2}(1-x)}^{1-x} \int_{2\nu-(1-x)}^{\nu} \phi_x(1-\tau, \tau+1-x-2\nu)\,d\tau d\nu \\
= {} & \frac{1}{2}\left(m(1-x) + m(0) + \int_0^{1-x} \phi(x+\sigma,\sigma)\,d\sigma\right) \,,
\end{aligned}
\tag{5.17}
$$

$$0 = v(0,t)$$

$$= \frac{1}{2}\left(m(1+t) + m(\max\{t-1,0\})\right) + \frac{1}{2}\int_{1+t-\min\{1+t,2\}}^{1+t} \phi(1,\sigma)d\sigma$$

$$- \int_0^{\min\{\frac{1}{2}(1+t),1\}} \int_0^{\nu} \phi_x(1-\tau,\tau+1+t-2\nu)\,d\tau d\nu$$

$$- \int_{\min\{\frac{1}{2}(1+t),1\}}^{1} \int_{2\nu-(1+t)}^{\nu} \phi_x(1-\tau,\tau+1+t-2\nu)\,d\tau d\nu$$

$$= \frac{1}{2}\left(m(1+t) + m(\max\{t-1,0\})\right) + \int_{\max\{t-1,0\}}^{t+1} \phi(|t-\sigma|,\sigma)\,d\sigma\right),$$

$$\tag{5.18}$$

where we have used the identities

$$\int_0^b \int_0^{\nu} \phi_x(1-\tau,\tau+1+\zeta-2\nu)\,d\tau d\nu$$

$$= \frac{1}{2}\int_{1+\zeta-2b}^{1+\zeta} \int_{\max\{\sigma-\zeta,\zeta-\sigma+2(1-b)\}}^{\min\{1,\zeta-\sigma+2\}=1} \phi_x(\xi,\sigma)\,d\xi\,d\sigma$$

$$= \frac{1}{2}\int_{1+\zeta-2b}^{1+\zeta} \left(\phi(1,\sigma) - \phi(\max\{\sigma-\zeta,\zeta-\sigma+2(1-b)\},\sigma)\right)\,d\sigma\,.$$

$$\int_{\frac{1}{2}(1+\zeta)}^{b} \int_{2\nu-(1+\zeta)}^{\nu} \phi_x(1-\tau,\tau+1+\zeta-2\nu)\,d\tau d\nu$$

$$= \frac{1}{2}\int_0^{1+\zeta} \int_{\max\{\sigma-\zeta,\zeta-\sigma+2(1-b)\}}^{\zeta-\sigma+2} \phi_x(\xi,\sigma)\,d\xi\,d\sigma$$

$$= \frac{1}{2}\int_0^{1+\zeta} \left(\phi(\zeta-\sigma+2,\sigma) - \phi(\max\{\sigma-\zeta,\zeta-\sigma+2(1-b)\},\sigma)\right)\,d\sigma\,,$$

with $\zeta = -x$ and $\zeta = t$.

Note that on the other hand (5.12), (5.13), (5.15), (5.16) imply that v as given in (5.14) solves the PDE (3.5) and satisfies the boundary and initial conditions (3.6) (3.7) (with $\rho = L = \bar{a} = 1$), and on the right-hand boundary coincides with m.

Thus, we obtain from (5.17) with $\zeta = -x \in [-1,0]$

$$-m(1+\zeta) = \int_0^{1+\zeta} \phi(|\sigma-\zeta|,\sigma)\,d\sigma\,, \tag{5.19}$$

where we have used the fact that $m(0) = v(1,0) = 0$. while (5.18) with $\zeta = t > 1$ implies

$$-m(1+\zeta) - m(-1+\zeta) = \int_{-1+\zeta}^{1+\zeta} \phi(|\sigma-\zeta|,\sigma)\,d\sigma\,. \tag{5.20}$$

Summing up (5.20) with ζ replaced by $\zeta - 2l$, $l = 0,\ldots,[\frac{1+\zeta}{2}]-1$ and adding (5.19) with ζ replaced by $\zeta - 2[\frac{1+\zeta}{2}]$, implies that (5.19) holds for all $\zeta \geq -1$. Inserting (3.3), and setting $t := \zeta + 1 \geq 0$, we can, due to (3.13), further transform the integrals with

$$\lambda := d_x(|\sigma - \zeta|, \sigma), \tag{5.21}$$
$$\tau := g^{-1}(\tilde{c}(\lambda))$$

to arrive at a Volterra type integral equation of the first kind for $\tilde{s} \circ \tilde{c}^{-1} \circ g$:

$$-m(t) = \int_{d_x(|t-1|,0)}^{d_x(1,t)} k(\lambda, t)\tilde{s}(\lambda)\, d\lambda$$
$$= \int_0^{\tilde{c}^{-1}(g(t))} k(\lambda, t)\tilde{s}(\lambda)\, d\lambda$$
$$= \int_0^t k(\tilde{c}^{-1}(g(\tau)), t)\frac{g'(\tau)}{\tilde{c}'(\tilde{c}^{-1}(g(\tau)))}\tilde{s}(\tilde{c}^{-1}(g(\tau)))\, d\tau \quad \forall t \in [0, \bar{t}], \tag{5.22}$$

where

$$k(\lambda, t) = \frac{1}{\operatorname{sign}(\sigma - t + 1)d_{xx}(|\sigma - t + 1|, \sigma) + d_{xt}(|\sigma - t + 1|, \sigma)}$$

with $\sigma = \sigma(\lambda, t - 1)$ according to the identity (5.21) (that due to (3.8) and the Implicit Function Theorem can be uniquely resolved with respect to σ). Now we apply the theory of Volterra integral equations (cf., e.g., Theorem 4.3 in [10]) using the fact that by our assumptions, the kernel $k(\tilde{c}^{-1}(g(\tau)), t)\frac{g'(\tau)}{\tilde{c}'(\tilde{c}^{-1}(g(\tau)))}$ is boundedly differentiable with respect to t and bounded away from zero on the diagonal set $\tau = t$, to conclude that

$$\|\tilde{s}\|_{L^2(0,\bar{\lambda})} \leq \frac{\|g\|_{C^1}}{\gamma}\|\tilde{s} \circ \tilde{c}^{-1} \circ g\|_{L^2(0,\bar{t})} \leq C\|m'\|_{L^2(0,\bar{t})},$$

i.e., (3.10), with $\bar{\lambda}$ according to (3.9). $\qquad\square$

Idea of proof. (Conjecture 3.2)
In the general situation of curved characteristics in (3.1), we expect that the same result as in Proposition 3.1 can be derived by using a characteristic transform of variables $\alpha = \varphi(x,t) + \psi(x,t)$, $\beta = \varphi(x,t) - \psi(x,t)$ where $\varphi_t + \sqrt{\frac{a}{\rho}}\varphi_x = 0$, $\psi_t - \sqrt{\frac{a}{\rho}}\psi_x = 0$, and defining $V(\alpha, \beta)$ by $V(\varphi(x,t) + \psi(x,t), \varphi(x,t) - \psi(x,t)) = v(x,t)$ as we did in the proof of Proposition 2.1.

Condition (3.13) implies that $|\frac{d}{dt}d_x(x(t), t)|$ is uniformly bounded away from zero for all characteristic curves $t \mapsto x(t)$ determined by $\dot{x}(t) = \sqrt{\tilde{c}'(d_x(x(t), t))/\rho}$ of (2.1), and therewith, by continuity arguments,

$$\frac{\kappa}{2} < |\frac{d}{dt}d_x(x(t), t)| = \left|\left(\pm\sqrt{a/\rho}\, d_{xx} + d_{xt}\right)(x(t), t)\right|, \tag{5.23}$$

also for the characteristic curves $t \mapsto x(t)$ of (3.1), as long as $\check{c}$ is sufficiently close to $\tilde{c}$. Due to (5.23), the integral transformation replacing (5.21) in the general case

$$\lambda := U(|\sigma - \beta + 1|, \sigma), \tag{5.24}$$

with $U(\alpha, \beta)$ defined (analogously to (5.4)) by

$$U(\varphi(x,t) + \psi(x,t), \varphi(x,t) - \psi(x,t)) = d_x(x,t)$$

is regular.

More precisely, we can derive the following:
With $\Phi(\varphi(x,t) + \psi(x,t), \varphi(x,t) - \psi(x,t)) = \phi(x,t)$, the PDE (3.1) becomes

$$V_{\beta\beta} - V_{\alpha\alpha} = \underbrace{\frac{1}{4a\psi_x}(\Phi_\alpha + \Phi_\beta)}_{=:\delta^+} + \underbrace{\frac{1}{4a\varphi_x}(\Phi_\alpha - \Phi_\beta)}_{=:\delta^-}$$

$$+ \underbrace{\frac{1}{8}\Big(\frac{a_x}{a} + \frac{a_t}{a^{3/2}}\Big)\frac{1}{\psi_x}(V_\alpha + V_\beta)}_{=:\gamma^+} + \underbrace{\frac{1}{8}\Big(\frac{a_x}{a} - \frac{a_t}{a^{3/2}}\Big)\frac{1}{\varphi_x}(V_\alpha - V_\beta)}_{=:\gamma^-}$$

$$=: r\,. \tag{5.25}$$

Since we can choose the characteristic variables α, β such that they are aligned to the original ones at $t = 0$ and $x = L$, i.e.,

$$\begin{aligned} \alpha(x,0) &= \tfrac{x}{L} & \alpha(L,t) &= 1 \\ \beta(x,0) &= 0 & \beta(L,t) &= t \end{aligned} \tag{5.26}$$

the solution formula (5.14) remains valid and we can make use of the initial values

$$V(\alpha, 0) = 0 \quad \alpha \in [0, 1]$$

(note that we do not use an analogon to the second equation in (5.16), since on the left-hand side $x = 0$, the characteristic transform of variables in general yields a curved boundary). In (5.14), we have to replace ϕ_x by the right-hand side r in (5.25), and ϕ by $\frac{L}{a}\Phi$, so that we arrive at

$$-m^\sharp(-1 + \zeta) = \int_0^{1+\zeta} \tfrac{L}{\bar{a}}\Phi(|\sigma - \zeta|, \sigma)\, d\sigma \tag{5.27}$$

in place of (5.19), where

$$\begin{aligned} m^\sharp(1 + \zeta) = {}& m(1 + \zeta) \\ & - \int_0^{1+\zeta} \Big(\tfrac{L}{\bar{a}} - \tfrac{L}{a}\Big)\, \Phi(|\sigma - \zeta|, \sigma)\, d\sigma \\ & -2 \int_0^{\frac{1}{2}(1+\zeta)} \int_0^\nu (r - \tfrac{L}{\bar{a}}\Phi_\alpha)(1 - \tau, \tau + 1 + \zeta - 2\nu)\, d\tau d\nu \\ & -2 \int_{\frac{1}{2}(1+\zeta)}^{1+\zeta} \int_{2\nu-(1+\zeta)}^\nu (r - \tfrac{L}{\bar{a}}\Phi_\alpha)(1 - \tau, \tau + 1 + \zeta - 2\nu)\, d\tau d\nu \end{aligned}$$

and $\bar{a}$ is a constant approximating a. Hence if it can be shown that

$$\|m^\sharp - m\|_{H^1} \leq C\|m\|_{H^1} + c\|\Phi\|_{L^2} \tag{5.28}$$

with c sufficiently small, and

$$\|\Phi\|_{L^2} = \|\phi\|_{L^2} = \sqrt{\int_0^L \int_0^{\bar{t}} \tilde{s}(d_x(x,t)\, dt\, dx} \leq C \|\tilde{s}\|_{L^2(0,\bar{\lambda})}, \tag{5.29}$$

then analogously to Proposition 3.1 with a perturbation argument one can conclude the assertion (3.15).

Indeed, from the fact that a is close (in $C^{2,2}$) to a constant in $[0,L] \times [0,\bar{t}]$ as long as L and $\bar{t}$ are sufficiently small, we can argue that the factors γ^+, γ^-, as well as $\delta_+ + \delta_- - \frac{L}{\bar{a}}$, $\delta_+ - \delta_-$ are small, since $\alpha_x \approx \frac{1}{L}$ and $\beta_x \approx 0$ by (5.26) Hence, it is clear that (5.28) with small c is likely to hold; however, it seems that this cannot be proven rigorously, at least not by standard results on hyperbolic PDEs. Moreover since for hyperbolic PDEs no maximum principles are available, it is not clear how (5.29) can be established. $\qquad\square$

6. Conclusions and remarks

In this paper we considered the problem of identifying a nonlinear coefficient c in a hyperbolic PDE from additional boundary measurements. We provided a well-based conjecture on stability and identifiability for the model case of a one-dimensional elastic string and demonstrated applicability to parameter identification in the piezoelectric PDEs.

For an alternative numerical solution approach to this identification problem, we refer to [24], where a method working in frequency domain is developed and numerical results are provided.

Future research work will be devoted to the study of the effect of damping, which plays a role in the applications we have in mind. As a matter of fact, damping makes the forward problem more stable and therewith can be expected to allow for more advanced assertions on the inverse problem. Here, we wish to refer to recent work by Lasiecka and coauthors, cf., e.g., [3].

Currently, we investigate nonlinearity in the sense of hysteresis, (cf., e.g., [5] and especially [28] for hysteresis in the context of the wave equation, as well as [16] for hysteresis identification), see [25].

Finally, we remark that the considerations of this paper might give hints on identifiability also for different identification problems for nonlinear PDEs of hyperbolic type.

Acknowledgment

The author wishes to thank Jean-Pierre Puel and Victor Isakov for interesting discussions and valuable comments.

References

[1] S. Alinhac, *Blowup for nonlinear hyperbolic equations*, Birkhäuser, Boston, 1995.

[2] B. Andersen, E. Ringgard, T. Bove, A. Albareda, and R. Pérez, *Performance of Piezoelectric Ceramic Multilayer Components Based on Hard and Soft PZT*, Proceedings of Actuator 2000, pages 419–422.

[3] V. Barbu, I. Lasiecka, M.A. Rammaha, *On nonlinear wave equations with degenerate damping and source terms* Trans. Amer. Math. Soc. 357 (2005), 2571–2611.

[4] A. Bressan, *Hyperbolic systems of conservation laws in one space dimension*, Proceedings of the International Congress of Mathematicians, Vol. I (Beijing, 2002), 159–178, Higher Ed. Press, Beijing, 2002.

[5] M. Brokate, J. Sprekels, *Hysteresis and Phase Transitions*, Springer, New York, 1996.

[6] J.R. Cannon, *Determination of the unknown coefficient $k(u)$ in the equation $\nabla k(u) \nabla u = 0$ from overspecified boundary data*, J. Math. Anal. Appl. 18 (1967), 112–114.

[7] J.R. Cannon and P. DuChateau, *An inverse problem for a nonlinear diffusion equation*, SIAM J. Appl. Math. 39 (1980), 272–289.

[8] P. DuChateau and W. Rundell, *Unicity in an inverse problem for an unknown reaction term in a reaction-diffusion equation*, J. Diff. Eq. 59 (1985), 155–164.

[9] H. Egger, H.W. Engl, and M.V. Klibanov, *Global uniqueness and Hölder stability for recovering a nonlinear source term in a parabolic equation*, Inverse Problems 21 (2005), 271–290.

[10] H.W. Engl, *Integralgleichungen*, Springer, Wien, 1997.

[11] H.W. Engl, M. Hanke, A. Neubauer, *Regularization of Inverse Problems*, Kluwer, Dordrecht, 1996.

[12] *Piezoelectric properties of ceramic materials and components*, European standard prEN 50324-1:1998, 50324-2:1998, 50324-3:2001, 1998–2001.

[13] L.C. Evans, *Partial Differential Equations*, AMS, 1998.

[14] R.T. Glassey, *Blow-up theorems for nonlinear wave equations*, Math. Z. 132 (1973), 183–203.

[15] R.B. Guenther, J.W. Lee *Partial Differential Equations of Mathematical Physics and Integral Equations*, Prentice-Hall, Eaglewood Cliffs, 1988.

[16] K.-H. Hoffmann, G.H. Meyer, *A least squares method for finding the Preisach Hysteresis operator from measurements*, Numer. Math. 55 (1989), 695–710.

[17] *IEEE Standard on Piezoelectricity*, ANSI/IEEE Std 176-1987, 1987.

[18] T. Iguchi and P.G. LeFloch, *Existence theory for hyperbolic systems of conservation laws with general flux functions*, Arch. Rational Mech. Anal. 168 (2003), 165–244.

[19] V. Isakov, *Inverse Problems for Partial Differential Equations*, Springer, New York, 1998.

[20] V. Isakov, *On uniqueness in inverse problems for semilinear parabolic equations*, Arch. Rat. Mech. Anal. 124 (1993), 1–12.

[21] V. Isakov and A.I. Nachman, *Global uniqueness for a two-dimensional semilinear elliptic inverse problem*, Trans. Amer. Mathem. Soc. 347 (1995), 3375–3390.

[22] H. Jaffe, D.A. Belincourt, *Piezoelectric transducer materials*, Proceedings of the IEEE 53 (1965), 1372–1386.

[23] B. Kaltenbacher, *Regularization by projection with a posteriori discretization level choice for linear and nonlinear ill-posed problems*, Inverse Problems 16 (2000), 1523–1539.

[24] B. Kaltenbacher, *Determination of material parameters in nonlinear hyperbolic PDEs via a multiharmonic formulation, used in piezoelectric material characterization*, to appear in Math. Mod. Meth. Appl. Sci (M3AS), 2006.

[25] B. Kaltenbacher, *Modelling and iterative identification of hysteresis via Preisach operators in PDEs* , LSE-report 04/05, University of Erlangen, 2005.

[26] A. Kirsch, *An Introduction to the Mathematical Theory of Inverse Problems*, Springer, New York, 1996.

[27] S. Klainerman and S. Selberg, *Bilinear Estimates and Applications to nonlinear wave equations*, Commun. Contemp. Math. 4 (2002), 223–295.

[28] P. Krejčí, *Global behaviour of solutions to the wave equation with hysteresis*, Adv. Math. Sci. Appl. 2 (1993), 1–23.

[29] H.A. Levine, *Instability and nonexistence of global solutions to nonlinear wave equations of the form $Pu_{tt} = -Au + F(u)$*, Trans. Amer. Math. Soc. 192 (1974), 1–21.

[30] A. Lorenzi and A. Lunardi, *An identification problem in the theory of heat conduction*, Differential and Integral Equations 3 (1990), 237–252.

[31] P. Monk, *A finite element method for approximating time-harmonic Maxwell equations*, Numer. Math. 63, 1992, 243–261.

[32] A.D. Pierce, *Acoustics: An Introduction to its Physical Principles and Applications*, Acoustical Society of America, Melville, 1989.

[33] M. Pilant and W. Rundell, *An inverse problem for a nonlinear elliptic differential equation*, SIAM J. Math. Anal. 18 (1987), 1801–1809.

[34] L.E. Payne, D.H. Sattinger, *Saddle points and instability of nonlinear hyperbolic equations*, Israel J. Math. 22 (1975), no. 3-4, 273–303.

[35] H. Smith and D. Tartaru, *Sharp local well-posedness results for the nonlinear wave equation*, Preprint.

Barbara Kaltenbacher
Institute for Numerical and Applied Mathematics
University of Göttingen
and Department of Sensor Technology
University of Erlangen, Germany
e-mail: barbara.kaltenbacher@lse.eei.uni-erlangen.de

International Series of Numerical Mathematics, Vol. 155, 217–248
© 2007 Birkhäuser Verlag Basel/Switzerland

An SQP Active Set Method for a Semilinear Optimal Control Problem with Nonlocal Radiation Interface Conditions

C. Meyer

Abstract. We consider a sequential quadratic programming (SQP) method for the solution of an optimal control problem governed by a semilinear elliptic equation with nonlocal interface conditions. These conditions arise from conductive-radiative heat transfer in non-convex domains. After stating first- and second-order optimality conditions, we introduce the SQP algorithm that uses an active set method to solve the linear quadratic subproblems arising in each step. The corresponding optimality systems are discretized by linear finite elements, using a partly exact summarized midpoint rule for the discretization of the nonlocal radiation interface conditions. The paper ends with some numerical results demonstrating the efficiency of the proposed method.

Mathematics Subject Classification (2000). 49M37, 65R20, 49K20.

Keywords. Optimal control, semilinear elliptic equations, nonlocal interface conditions, sequential quadratic programming, active set strategy.

1. Introduction

In this paper, we continue the work presented in [16] and [15] on an optimal control problem with nonlocal radiation interface conditions. While the contributions in [16] and [15] focus on theoretical aspects such as first- and second-order optimality conditions, the goal of this paper is the development of an efficient numerical algorithm including an accurate discretization of the nonlocal radiation interface conditions. The optimal control problem, discussed here, arises from the sublimation growth of semiconductor single crystals by the physical vapor transport (PVT) method. The semiconductor materials, produced with this method, such as silicon carbide (SiC) or aluminum nitrite (AlN), are used in numerous industrial applications, e.g., the production blue lasers. For the PVT method, polycrystalline

Supported by the DFG Research Center MATHEON "Mathematics for key technologies" in Berlin.

powder is placed under a low-pressure inert gas atmosphere at the bottom of a cavity inside a crucible. The crucible is heated up to 2000 till 3000 K by induction. Due to the high temperatures and the low pressure, the powder sublimates and crystallizes at a single-crystalline seed located at the cooled top of the cavity, such that the desired single crystal grows into the reaction chamber. See [13] for more details.

Here, we focus on the conductive-radiative heat transfer in the growth apparatus. Therefore, we consider a simplified setup of the growth apparatus, shown in Figure 1. Here, Ω_s denotes the domain of the solid graphite crucible, whereas

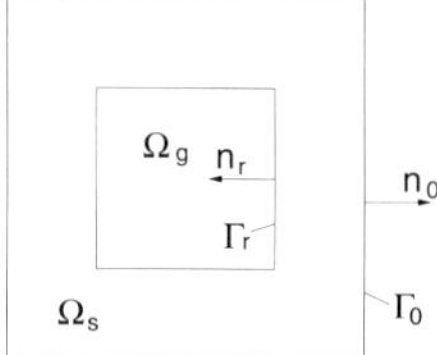

Figure 1: Exemplary domain for nonlocal radiative heat transfer.

Ω_g is the domain of gas phase inside. A very important determining factor for the crystal's quality and growth rate is the temperature gradient inside the gas phase [18]. Since we do not consider the electromagnetic induction, we will optimize the temperature gradient in the gas phase Ω_g by directly controlling the heat source u in Ω_s.

The temperature y inside the growth apparatus arises as the solution of the conductive-radiative heat transfer problem in the growth apparatus. This process is modelled by the following semilinear PDE (see [13] for details)

$$-\operatorname{div}(\kappa_s \nabla y) = u \qquad \text{in } \Omega_s$$
$$-\operatorname{div}(\kappa_g \nabla y) = 0 \qquad \text{in } \Omega_g$$
$$\kappa_g \left(\frac{\partial y}{\partial n_r}\right)_g - \kappa_s \left(\frac{\partial y}{\partial n_r}\right)_s = q_r \qquad \text{on } \Gamma_r \qquad (1.1)$$
$$\kappa_s \frac{\partial y}{\partial n_0} + \varepsilon\sigma\,|y|^3 y = \varepsilon\sigma\,y_0^4 \quad \text{on } \Gamma_0.$$

Here, n_0 is the outward unit normal on Γ_0, and n_r is the unit normal on Γ_r facing outward with respect to Ω_s (cf. Fig. 1). Furthermore, ε is the emissivity, σ the Boltzmann radiation constant, and κ_s, κ_g denote the thermal conductivities in Ω_s, Ω_g, respectively. By q_r we denote the additional radiative heat flux on Γ_r that is given by

$$q_r = (I - K)(I - (1-\varepsilon)K)^{-1}\varepsilon\,\sigma|y|^3 y := G\,\sigma|y|^3 y. \qquad (1.2)$$

The nonlocal operators K and G will be specified in Section 3. For an explicit description of the corresponding mathematical model, we refer to [20]. In addition to this semilinear state equation, we consider box constraints on the control. Thus, the optimal control problem, considered here, reads as follows:

$$(\text{P}) \quad \begin{cases} \text{minimize} & J(y,u) := \dfrac{1}{2} \int\limits_{\Omega_{\mathrm{g}}} |\nabla y - z|^2 \, dx + \dfrac{\nu}{2} \int\limits_{\Omega_{\mathrm{s}}} u^2 \, dx \\[2ex] \text{subject to} & (1.1) \\[1ex] \text{and} & u_a(x) \le u(x) \le u_b(x) \quad \text{a.e. in } \Omega_{\mathrm{s}}, \end{cases}$$

where z denotes the desired temperature gradient and $\nu > 0$ is a Tikhonov regularization parameter.

The nonlocal radiation on Γ_{r} represents the main characteristic of our problem, since the nonlinearity in the state equation (1.1) is in general not monotone due to the nonpositivity of G (see [20]). Therefore standard techniques for monotone nonlinearities cannot be applied. This complicates the analysis of the state equation (1.1), see Tiihonen [20], [21], and Laitinen and Tiihonen [14]. Furthermore, a consistent discretization of the nonlocal radiation operators K and G in a finite element framework is challenging, since the kernel of integral operator K exhibits a weak singularity (cf. [21]). Therefore, for the approximation of K and G, respectively, the critical part of K is integrated exactly which is possible due to the special shape of our computational domain (see Section 7 and Appendix A). Concerning the numerical treatment of K, our discretization slightly differs from an approach introduced by Tiihonen in [22], where the boundary integral over Γ_{r} is completely discretized using some quadrature rule. For a smoother class of domains than the one considered here, Tiihonen proved linear convergence of his finite element scheme in the H^1-norm (see [22]).

The numerical analysis of semilinear elliptic optimal control problems is well investigated in various articles. We only refer to Casas [6], Bonnans and Casas [5], Casas, Tröltzsch and Unger [7], or Bonnans [4], and the references therein. However, in case of problem (P), the standard arguments have to be modified, again due to the non monotony of the nonlinearity. This especially concerns first-order necessary conditions (see [16] and [15]). In Section 4, we will shortly sketch the arising difficulties and how to overcome them. To solve problem (P) numerically, we use an sequential quadratic programming (SQP) method. Moreover, the linear quadratic optimal control problems arising in each SQP step are solved by an active set method, see for instance [2] or [3]. Throughout the paper, we will refer to the overall method as the SQP active set algorithm. A similar method for the control of the Navier-Stokes equation is considered by Hintermüller and Hinze in a recent paper [11]. In the field of constrained optimal control of nonlinear PDEs, the SQP method is also investigated by Tröltzsch and Volkwein for the Burgers equation [25] and for a general class of semilinear PDEs in Goldberg and Tröltzsch [10], Tröltzsch [23], and Unger [26].

The paper is organized as follows: After stating the mathematical setting in Section 2, a summary of the main results concerning the semilinear state equation follows in Section 3. Sections 4 and 5 present first- and second-order optimality conditions for (P), whereas Sections 6 and 7 are devoted to the numerical treatment of (P). More precisely, the SQP active set algorithm is introduced in Section 6, and Section 7 is dedicated to the discretization of (P). The paper ends with some numerical examples in Section 8.

2. The mathematical setting

Throughout this paper, we assume the following conditions on the domain Ω and on the quantities and functions occurring in (P):

Assumption 1. We assume that $\Omega \subset \mathbb{R}^3$ is a bounded simply connected domain with Lipschitz boundary Γ_0. The boundary of the simply connected subdomain $\overline{\Omega}_{\mathrm{g}} \subset \Omega$, denoted by Γ_{r}, is assumed to be a closed Lipschitz surface that is piecewise $C^{1,\delta}$. Notice that the distance of Γ_{r} to Γ_0 is positive. Then, Ω_{s} is defined by $\Omega_{\mathrm{s}} = \Omega \backslash \overline{\Omega}_{\mathrm{g}}$.

The Boltzmann radiation constant is assumed to be positive, i.e., $\sigma \in \mathbb{R}^+$. For the thermal conductivity, we assume $\kappa \in L^\infty(\Omega)$ with

$$\kappa(x) = \left\{ \begin{array}{ll} \kappa_{\mathrm{s}}(x) & \text{in } \Omega_{\mathrm{s}} \\ \kappa_{\mathrm{g}}(x) & \text{in } \Omega_{\mathrm{g}} \end{array} \right.$$

and $\kappa(x) \geq \kappa_{\min} > 0$ a.e. on Ω. Furthermore, the emissivity $\varepsilon \in L^\infty(\Gamma_0 \cup \Gamma_{\mathrm{r}})$ is bounded by $1 \geq \varepsilon \geq \varepsilon_{\min} > 0$ a.e. on $\Gamma_0 \cup \Gamma_{\mathrm{r}}$.

Assumption 2. The desired temperature gradient z is given in $L^2(\Omega_{\mathrm{g}})^2$ and ν is a positive constant. For the box constraints, we assume $u_a, u_b \in L^\infty(\Omega_{\mathrm{s}})$ and $0 \leq u_a(x) < u_b(x)$ a.e. in Ω_{s}. The external temperature y_0 is a function in $L^{16}(\Gamma_0)$ and fulfills $y_0 \geq \vartheta$ a.e. on Γ_0 with a positive constant ϑ.

Moreover, we use the following notation:

Notation. We introduce the set of admissible controls by

$$U_{ad} := \{ u \in L^\infty(\Omega_{\mathrm{s}}) \mid u_a(x) \leq u(x) \leq u_b(x) \text{ a.e. in } \Omega_{\mathrm{s}} \}.$$

The identity operator in the respective function spaces is denoted by I. Moreover, τ_{r} is the trace operator on Γ_{r}, whereas τ_0 denotes the trace on Γ_0. Throughout this paper, c is a generic constant and φ denotes a generic function. Let W be a Banach space with its dual space W^*. Then, for $f \in W$ and $g \in W^*$, $\langle f, g \rangle$ denotes the associated pairing.

3. The semilinear state equations

In this section, some results of Laitinen and Tiihonen [14], Tiihonen [20], [21], and Meyer, Philip, and Tröltzsch [16] are recalled. First, we define the integral operator K.

Definition 3.1. The integral operator K is given by

$$(K\,y)(x) = \int_{\Gamma_r} \omega(x,z)\,y(z)\,ds_z,\tag{3.1}$$

where the kernel ω is defined by

$$\omega(x,z) = \begin{cases} \Xi(x,z)\,\dfrac{[n_r(z)\cdot(x-z)][n_r(x)\cdot(z-x)]}{2\|z-x\|^3} & ,\ \text{for } n=2 \\[3mm] \Xi(x,z)\,\dfrac{[n_r(z)\cdot(x-z)][n_r(x)\cdot(z-x)]}{\pi\|z-x\|^4} & ,\ \text{for } n=3. \end{cases}\tag{3.2}$$

In this definition, x, z denote two points on Γ_r, and $n_r(x)$ is the unit normal at x facing outward with respect to Ω_s (cf. Fig. 1). Moreover, Ξ is given by

$$\Xi(x,z) = \begin{cases} 0 & \text{if } \overline{xz}\cap\Omega_s \neq \emptyset, \\ 1 & \text{if } \overline{xz}\cap\Omega_s = \emptyset, \end{cases}\tag{3.3}$$

with $\overline{xz}$ denotes the line between x and z.

In [21], it is proven that $\omega(x,z)$ has a singularity at x of type $|x-z|^{-(1-\delta)}$ in the two-dimensional and $|x-z|^{-2(1-\delta)}$ in the three-dimensional case, which is, in both cases, integrable. Based on this result, Tiihonen and Laitinen proved in [14] some fundamental properties of K and G, in particular that G and G^* are a bounded linear operators from $L^p(\Gamma_r)$ to itself for all $1 \leq p \leq \infty$. Furthermore, using Brezis' existence theorem for pseudomonotone operators, Laitinen and Tiihonen derived in [14] that for every right-hand side $u \in H^1(\Omega_s)^*$ and $y_0 \in L^5(\Gamma_0)$, the state equation (1.1) admits unique solutions in the state space $V := \{v \in H^1(\Omega) \mid \tau_r\,v \in L^5(\Gamma_r),\, \tau_0\,v \in L^5(\Gamma_0)\}$. By standard truncation techniques, it is shown in [16] that, if the right-hand side is sufficiently regular, i.e., $u \in L^2(\Omega_s)$ and $y_0 \in L^{16}(\Gamma_0)$, solutions to (1.1) are bounded. Thus, we introduce the state space $V^\infty := H^1(\Omega)\cap L^\infty(\Omega)$. Notice that $y \in V^\infty$ implies $\tau_r y \in L^\infty(\Gamma_r)$ and $\tau_0 y \in L^\infty(\Gamma_0)$ (see [16, Remark 3.5]).

4. First-order necessary optimality conditions

Before we state first-order necessary conditions, let us first refer to the following theorem that covers the existence of an optimal solution for (P). It is proven in [16] by rather standard arguments.

Theorem 4.1. [16, Theorem 5.2] *Under the Assumptions 1 and 2, there exists an optimal control $\bar{u} \in L^\infty(\Omega_s)$ with associated state $\bar{y} \in V^\infty$.*

The key point in the proof of first-order necessary optimality conditions is to show the differentiability of the control-to-state operator $S : u \mapsto y$. In preparation of a corresponding theorem, we consider the following linear equation

$$\int_\Omega \kappa\nabla y\cdot\nabla v\,dx + 4\int_{\Gamma_0} \varepsilon\sigma\,|\bar{y}|^3 y\,v\,ds = \langle \varphi,\,v\rangle_{H^1(\Omega)^*,H^1(\Omega)} \quad \forall\,v \in H^1(\Omega)\tag{4.1}$$

222   C. Meyer

with a given $\varphi \in H^1(\Omega)^*$ and $\bar{y} \in V^\infty$ with $\bar{y} > 0$ a.e. in Ω. It is easy to verify that the bilinear form in (4.1) is bounded and coercive in $H^1(\Omega)$. Therefore, the Lax-Milgram lemma implies that (4.1) admits solutions in $H^1(\Omega)$ for every right-hand side in $\varphi \in H^1(\Omega)^*$. Thus, there exists a linear continuous operator $B_{\mathrm{d}} : H^1(\Omega)^* \to H^1(\Omega)$, mapping φ to y, such that the solution of (4.1) can be expressed as

$$y = B_{\mathrm{d}}\,\varphi. \tag{4.2}$$

Notice that the operator B_{d} depends on $\bar{y}$. However, to improve the readability, we simply write B_{d} instead of $B_{\mathrm{d}}(\bar{y})$ and proceed analogously with similar operators in all what follows. Next, we consider a slightly different equation:

$$\bar{a}[y, v] := \int_\Omega \kappa \nabla y \cdot \nabla v \, dx + 4 \int_{\Gamma_r} (G\,\sigma|\bar{y}|^3 y)v \, ds + 4 \int_{\Gamma_0} \varepsilon\sigma\,|\bar{y}|^3 y\,v \, ds$$
$$= \langle \varphi,\, v \rangle_{H^1(\Omega)^*, H^1(\Omega)} \quad \forall\, v \in H^1(\Omega). \tag{4.3}$$

Since G is not positive, the bilinear form $\bar{a}$ is in general not coercive. This is also confirmed by the following numerical example, where we evaluate $\int_{\Gamma_r} G(|\bar{y}|^3 v)\,v\,ds$ by numerical integration. The discrete scheme for the numerical integration is described later on in Section 7. This investigation is realized on the domain presented

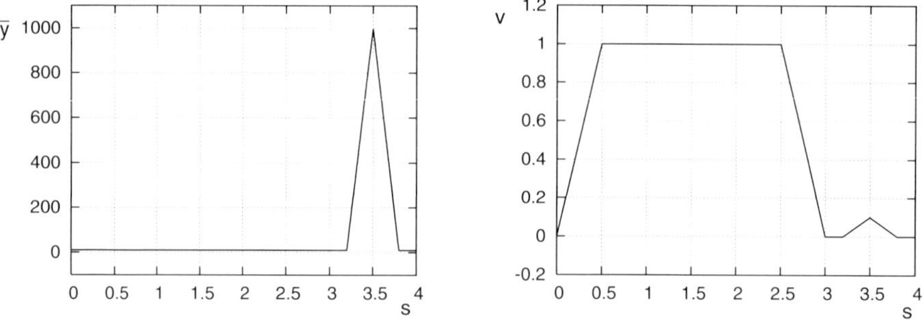

Figure 2: Example for non-coercivity.

in Figure 1, with side lengths 2 for the outer and 1 for the inner square, respectively. Furthermore, $\bar{y}$ and v are given by continuous piecewise linear functions that are shown in Figure 2. Here, s denotes the curve parameter associated to Γ_r. Table 1 shows the results of the numerical integration for different numbers of grid points n_r (cf. (7.1)).

As one can see, the results converge towards a negative number. This indicates that the exact value is negative, too, and hence, one cannot expect the bilinear form in (4.3) to be coercive. Thus, the Lax-Milgram lemma cannot be applied. However, (4.3) is equivalent to

$$y = B_{\mathrm{d}}\,\varphi - B_{\mathrm{r}}\,4\,G\sigma|\bar{y}|^3\tau_r y, \tag{4.4}$$

Table 1: Results of the numerical integration depending
on the mesh size for $\varepsilon = 0.8$.

n_{r}	2560	5120	7860	10240
$\int_{\Gamma_{\mathrm{r}}} G(\|\bar{y}\|^3 v)\, v\, ds$	-6.91145	-6.91152	-6.91153	-6.91153

where $B_{\mathrm{r}} : L^2(\Gamma_{\mathrm{r}}) \dashrightarrow H^1(\Omega)$ denotes the solution operator to (4.3) if φ can be expressed by

$$\langle \varphi,\, v \rangle_{H^1(\Omega)^*, H^1(\Omega)} = \int_{\Gamma_{\mathrm{r}}} f_{\mathrm{r}}\, v\, ds$$

with a function $f_{\mathrm{r}} \in L^2(\Gamma_{\mathrm{r}})$. Notice that it would be more appropriate to write $(G\sigma|\tau_{\mathrm{r}}\bar{y}|^3\, \tau_{\mathrm{r}}y)$ instead of $(G\sigma|\bar{y}|^3\tau_{\mathrm{r}}y)$ in this context. However, for the purpose of readability, in all what follows, we suppress the trace in connection with $\bar{y}$ since it represents a fixed reference state. Applying the trace operator to (4.4) yields

$$\tau_{\mathrm{r}}y + 4\,\tau_{\mathrm{r}}\, B_{\mathrm{r}}\, G\sigma|\bar{y}|^3\tau_{\mathrm{r}}y = \tau_{\mathrm{r}}B_{\mathrm{d}}\,\varphi. \tag{4.5}$$

To show the existence of solutions of this equation, we rely on the following assumption.

Assumption 3. $\lambda = -1$ is not an eigenvalue of

$$B(\bar{y})(\,\cdot\,) := 4\,\tau_{\mathrm{r}}\, B_{\mathrm{r}}\, G\sigma|\bar{y}|^3(\,\cdot\,), \tag{4.6}$$

with $B(\bar{y}) : L^2(\Gamma_{\mathrm{r}}) \to L^2(\Gamma_{\mathrm{r}})$.

Since $B_{\mathrm{r}} : L^2(\Gamma_{\mathrm{r}}) \to H^1(\Omega)$, we have that $\tau_{\mathrm{r}}\, B_{\mathrm{r}} : L^2(\Gamma_{\mathrm{r}}) \to H^{1/2}(\Gamma_{\mathrm{r}})$. Therefore, due to the compact embedding of $L^2(\Gamma_{\mathrm{r}})$ in $H^{1/2}(\Gamma_{\mathrm{r}})$, $B(\bar{y}) : L^2(\Gamma_{\mathrm{r}}) \to L^2(\Gamma_{\mathrm{r}})$ is a compact operator. Thus, thanks to Assumption 3, the theory of Fredholm operators ensures that $(I + B(\bar{y}))$ has a continuous inverse operator. Therefore, (4.5) admits a solution in $L^2(\Gamma_{\mathrm{r}})$, giving in turn the unique existence of solutions to (4.3). Moreover, by standard truncation techniques, it is proven in [16] that, if the inhomogeneity φ is sufficiently regular such that it can be expressed by

$$\langle \varphi,\, v \rangle_{H^1(\Omega)^*, H^1(\Omega)} = \int_{\Omega} f_{\Omega}\, v\, dx + \int_{\Gamma_{\mathrm{r}}} f_{\mathrm{r}}\, v\, ds + \int_{\Gamma_0} f_0\, v\, ds$$

with some functions $f_{\Omega} \in L^2(\Omega)$, $f_{\mathrm{r}} \in L^4(\Gamma_{\mathrm{r}})$, and $f_0 \in L^4(\Gamma_{\mathrm{r}})$, then the solution of (4.3) is bounded a.e. in Ω, i.e., $y \in V^\infty$. Based on these results, one shows the Fréchet differentiability of the control-to-state operator S by applying the implicit function theorem.

Theorem 4.2. [16, Theorem 7.1] *Under Assumptions 1–3, $S : L^2(\Omega_{\mathrm{s}}) \to V^\infty$ is twice continuously Fréchet-differentiable at $(\bar{y}, \bar{u})$. Its first derivative, denoted by*

$y = S'(\bar{u})h$, $h \in L^2(\Omega_{\mathrm{s}})$, *is given by*

$$-\mathrm{div}(\kappa_{\mathrm{s}}\,\nabla y) = h \qquad in \ \Omega_{\mathrm{s}}$$
$$-\mathrm{div}(\kappa_{\mathrm{g}}\,\nabla y) = 0 \qquad in \ \Omega_{\mathrm{g}}$$
$$\kappa_{\mathrm{s}}\left(\frac{\partial y}{\partial n_r}\right)_{\mathrm{s}} - \kappa_{\mathrm{g}}\left(\frac{\partial y}{\partial n_r}\right)_{\mathrm{g}} + 4\,G\,\sigma|\bar{y}|^3 y = 0 \qquad on \ \Gamma_{\mathrm{r}} \qquad (4.7)$$
$$\kappa_{\mathrm{s}}\frac{\partial y}{\partial n_0} + 4\,\varepsilon\sigma|\bar{y}|^3 y = 0 \qquad on \ \Gamma_0.$$

Moreover, the second derivative $w = S''(\bar{u})[h_1, h_2]$ *solves the equation*

$$-\mathrm{div}(\kappa_{\mathrm{s}}\,\nabla w) = 0 \qquad\qquad in \ \Omega_{\mathrm{s}}$$
$$-\mathrm{div}(\kappa_{\mathrm{g}}\,\nabla w) = 0 \qquad\qquad in \ \Omega_{\mathrm{g}}$$
$$\kappa_{\mathrm{s}}\left(\frac{\partial w}{\partial n_r}\right)_{\mathrm{s}} - \kappa_{\mathrm{g}}\left(\frac{\partial w}{\partial n_r}\right)_{\mathrm{g}} + 4\,G\,\sigma|\bar{y}|^3 w = -12\,G\,\sigma|\bar{y}|\bar{y}\,y_1 y_2 \qquad on \ \Gamma_{\mathrm{r}} \qquad (4.8)$$
$$\kappa_{\mathrm{s}}\frac{\partial w}{\partial n_0} + 4\,\varepsilon\sigma|\bar{y}|^3 w = -12\,\varepsilon\sigma|\bar{y}|\bar{y}\,y_1 y_2 \qquad on \ \Gamma_0$$

with $y_i = S'(\bar{u})h_i$, $i = 1, 2$.

Remark 4.3. Clearly, the implicit function theorem also gives that $S : L^2(\Omega_{\mathrm{s}}) \to V^\infty$ is twice continuously Fréchet differentiable in a neighborhood of $\bar{u}$.

Next we introduce the *adjoint equation* by

$$\int_\Omega \kappa\nabla p \cdot \nabla v\,dx + 4\int_{\Gamma_{\mathrm{r}}} \sigma|\bar{y}|^3 (G^*p)\,v\,ds + 4\int_{\Gamma_0} \varepsilon\sigma\,|\bar{y}|^3 p\,v\,ds$$
$$= \int_{\Omega_{\mathrm{g}}} (\nabla\bar{y} - z)\cdot\nabla v\,dx \quad \forall\,v \in H^1(\Omega). \qquad (4.9)$$

Due to $z \in L^2(\Omega_{\mathrm{g}})^2$ by Assumption 2 and $\bar{y} \in V^\infty$, the right-hand side clearly represents an element of $H^1(\Omega)^*$. Moreover, in [15] it is shown that the solution operator associated to (4.9) is equivalent to the adjoint of the solution operator corresponding to (4.3). Therefore, (4.9) admits a unique solution in $H^1(\Omega)$ provided that Assumption 3 holds true. For the derivation of first-order necessary optimality conditions to (P), we introduce the reduced objective functional by

$$j(u) := J(S(u), u) = \frac{1}{2}\|\nabla S(u) - z\|_{L^2(\Omega_{\mathrm{g}})}^2 + \frac{\nu}{2}\|u\|_{L^2(\Omega_{\mathrm{s}})}^2. \qquad (4.10)$$

Furthermore, we define the set of admissible controls by

$$U_{ad} := \{u \in L^2(\Omega) \,|\, u_a(x) \le u(x) \le u_b(x) \text{ a.e. in } \Omega_{\mathrm{s}}\}.$$

With Theorem 4.2 at hand, standard arguments give the following result.

Lemma 4.4. *Under Assumptions 1–3 , j is twice continuously Fréchet-differentiable from $L^2(\Omega_s)$ to $\mathbb{R}$ at $\bar{u}$. Its first derivative is given by*

$$j'(\bar{u})\,h = (\nu\,\bar{u} + p\,,\,h)_{L^2(\Omega_s)}, \tag{4.11}$$

where p solves the adjoint equation (4.9). For the second derivative, we obtain

$$j''(\bar{u})[h_1\,,\,h_2] = (\nabla y_1\,,\,\nabla y_2)_{L^2(\Omega_g)} + \nu(h_1\,,\,h_2)_{L^2(\Omega_s)}$$
$$- 12\left(\int_{\Gamma_r} G(\sigma\,|\bar{y}|\bar{y}\,y_1 y_2) p\,ds + \int_{\Gamma_0} \varepsilon\sigma\,|\bar{y}|\bar{y}\,y_1\,y_2\,p\,ds \right). \tag{4.12}$$

with $\bar{y} = S(\bar{u})$ and $y_i = S'(u)h_i$, $i = 1,2$.

According to the standard optimal control theory, an optimal solution $\bar{u}$ of (P) must satisfy the following variational inequality

$$j'(\bar{u})(u - \bar{u}) = (p + \nu\,\bar{u}\,,\,u - \bar{u})_{L^2(\Omega_g)} \geq 0 \quad \forall\,u \in U_{ad}. \tag{4.13}$$

A standard pointwise discussion of this inequality yields

$$\bar{u}(x) = \mathcal{P}_{ad}\left\{ -\frac{1}{\nu}\,p(x) \right\}, \tag{4.14}$$

where $\mathcal{P}_{ad}(x)$ denotes the pointwise projection operator on $[u_a(x), u_b(x)]$. In this way, we have derived the following theorem:

Theorem 4.5. *Suppose that Assumptions 1–3 are fulfilled and $\bar{u}$ is a locally optimal solution of (P) with associated state $\bar{y}$. Then there exists an adjoint state $p \in H^1(\Omega)$ such that the adjoint equation (4.9) and the condition (4.14) are satisfied.*

5. Second-order sufficient conditions

In this section, we follow the lines of [15], where the sufficiency of the second-order optimality conditions stated below is shown in detail. To obtain some flexibility, these conditions give local optimality in a L^s-neighborhood, where s is not necessarily equal to ∞, but can be chosen smaller. We introduce the *strongly active set* as follows:

Definition 5.1. Let $\tau > 0$ be given. Then the strongly active set A_τ is defined by

$$A_\tau := \{x \in \Omega\,|\,|p(x) + \nu\,\bar{u}(x)| \geq \tau\}.$$

Moreover, the corresponding *τ-critical cone* is defined in a standard way (cf. Dontchev et al. [9]).

Definition 5.2. The critical cone belonging to (P) is given by

$$C_\tau(\bar{u}) := \left\{ u \in L^2(\Omega) \;\middle|\; \begin{array}{ll} u(x) = 0 & \text{, a.e. in } A_\tau \\ u(x) \geq 0 & \text{, where } \bar{u}(x) = u_a(x) \text{ and } x \notin A_\tau \\ u(x) \leq 0 & \text{, where } \bar{u}(x) = u_b(x) \text{ and } x \notin A_\tau \end{array} \right\}. \tag{5.1}$$

In the following, let q be a real number with $4/3 \leq q \leq 2$. Then the second-order sufficient conditions for (P) are given by

$$\text{(SSC)} \quad \begin{cases} \text{Let } \delta > 0 \text{ and } \tau > 0 \text{ exist such that} \\ j''(\bar{u})\, u^2 \geq \delta \, \|u\|^2_{L^q(\Omega_{\mathrm{s}})} \quad \text{for all } u \in C_\tau(\bar{u}). \end{cases}$$

In [15], it is shown that condition (SSC) is indeed sufficient for local optimality. The corresponding rather technical proof is based on techniques introduced by Casas, Tröltzsch, and Unger in [7] and by Tröltzsch and Wachsmuth in [24]. It yields the following result.

Theorem 5.3. [15, Theorem 5.4] *Suppose that Assumptions 1–3 are fulfilled. Let $4/3 \leq q \leq 2$ be given. Define s by*

$$s := \begin{cases} q/(2-q) &, \quad \text{for } q < 2 \\ \infty &, \quad \text{for } q = 2. \end{cases} \tag{5.2}$$

Moreover, let $(\bar{y}, \bar{u})$ satisfy the first-order necessary optimality conditions for problem (P) and assume that condition (SSC) is fulfilled with some $\delta > 0$, $\tau > 0$. Then there exist $\bar{\varepsilon} > 0$ and $\bar{\sigma} > 0$ such that

$$j(u) \geq j(\bar{u}) + \bar{\sigma} \, \|u - \bar{u}\|^2_{L^q(\Omega_{\mathrm{s}})} \tag{5.3}$$

for all $u \in U_{ad}$ with $\|u - \bar{u}\|_{L^s(\Omega_{\mathrm{s}})} \leq \bar{\varepsilon}$.

Remark 5.4. Setting $q = 4/3$, we obtain $s = 2$, and hence Theorem 5.3 gives a $L^{4/3}$-quadratic growth condition in a L^2-neighborhood of $\bar{u}$. Choosing $q = 2$ and thus $s = \infty$, we obtain L^2-quadratic growth of j in a L^∞-neighborhood of $\bar{u}$.

6. SQP active set algorithm

Next, we present an infinite-dimensional algorithm to solve the semilinear elliptic problem (P). To keep the discussion concise, we rely on much more restrictive second-order conditions than (SSC). The stronger conditions are given by

$$\text{(S)} \quad \begin{cases} \text{Let } \delta > 0 \text{ exist such that} \\ j''(\bar{u})\, u^2 \geq \delta \, \|u\|^2_{L^2(\Omega_{\mathrm{s}})} \quad \text{for all } u \in L^2(\Omega_{\mathrm{s}}). \end{cases}$$

The SQP method is motivated by the following consideration: Let $u_n \in U_{ad}$, $n > 0$, be a given iterate, then an optimal descend direction v is given by a solution of

$$\min_{(u_n + v) \in U_{ad}} j(u_n + v).$$

With $v = u - u_n$, this is equivalent to the following optimization problem

$$\text{(P}_n\text{)} \qquad \min_{u \in U_{ad}} j\big(u_n + (u - u_n)\big).$$

Now, let us assume that u_n is located in a neighborhood of a stationary point $\bar{u}$ that satisfies Assumption 3, i.e., $\|u_n - \bar{u}\|_{L^2(\Omega_{\mathrm{s}})} \leq \rho$. As stated in Lemma 4.4, $j : L^2(\Omega_{\mathrm{s}}) \to \mathbb{R}$ is twice continuously Fréchet differentiable at $\bar{u}$. By means of the implicit function theorem, the same holds for u_n if ρ is sufficiently small (cf. Remark 4.3). Hence, we are allowed to use a Taylor expansion for j and approximate (P_n) by the following optimal control problem

$$(\mathrm{Q}_n) \qquad \min_{u \in U_{ad}} \left\{ j_n(u) := j'(u_n)(u - u_n) + \frac{1}{2} j''(u_n)(u - u_n)^2 \right\},$$

where we omit $j(u_n)$, since, as a constant, it does not influence the optimization. In view of (4.10), we continue with

$$j'(u_n)(u - u_n) = \big(\nabla S(u_n) - z\,,\ \nabla S'(u_n)(u - u_n)\big)_{L^2(\Omega_{\mathrm{g}})} + \nu\,(u_n\,,\ u - u_n)_{L^2(\Omega_{\mathrm{s}})}.$$

Next, we set $y_n = S(u_n)$, hence $y_n \in V^\infty$, and define $y := S'(u_n)(u - u_n) + y_n$. Thanks to the structure of S' (cf. (4.7)), it is clear that y solves the following linearized PDE

$$-\mathrm{div}(\kappa_{\mathrm{s}}\,\nabla y) = u \qquad\qquad \text{in } \Omega_{\mathrm{s}}$$

$$-\mathrm{div}(\kappa_{\mathrm{g}}\,\nabla y) = 0 \qquad\qquad \text{in } \Omega_{\mathrm{g}}$$

$$\kappa_{\mathrm{s}} \left(\frac{\partial v}{\partial n_r}\right)_{\mathrm{s}} - \kappa_{\mathrm{g}} \left(\frac{\partial v}{\partial n_r}\right)_{\mathrm{g}} + 4\,G\,\sigma |y_n|^3 y = 3\,G\,\sigma |y_n|^3 y_n \qquad \text{on } \Gamma_r \qquad (6.1)$$

$$\kappa_{\mathrm{s}} \frac{\partial y}{\partial n_0} + 4\,\varepsilon\sigma\,|y_n|^3 y = 3\,\varepsilon\sigma\,|y_n|^3 y_n + \varepsilon\sigma\,y_0^4 \quad \text{on } \Gamma_0.$$

Notice that, as mentioned above, the assumption $\|u_n - \bar{u}\|_{L^2(\Omega_{\mathrm{s}})} \le \rho$ ensures that S is Fréchet differentiable at u_n such that $S'(u_n) : L^2(\Omega_{\mathrm{s}}) \to V^\infty$ is well defined giving in turn the unique existence of solutions to (6.1). Formal integration by parts over Γ_0 and Γ_r then yields the associated variational formulation that is given by

$$a_n[y, v] := \int_\Omega \kappa \nabla y \cdot \nabla v\, dx + 4 \int_{\Gamma_r} (G\,\sigma |y_n|^3 y) v\, ds + 4 \int_{\Gamma_0} \varepsilon\sigma\,|y_n|^3 y\, v\, ds$$

$$= 3 \int_{\Gamma_r} (G\,\sigma |y_n|^3 y_n) v\, ds + 3 \int_{\Gamma_0} \varepsilon\sigma\,|y_n|^3 y_n\, v\, ds + \int_{\Omega_{\mathrm{s}}} u\, v\, dx + \int_{\Gamma_0} \varepsilon\sigma\,y_0^4\, ds \qquad (6.2)$$

for all $v \in H^1(\Omega)$. Using (4.12), j_n is transformed into

$$j_n(u) = (\nabla y_n - z\,,\ \nabla y - \nabla y_n)_{L^2(\Omega_{\mathrm{g}})} + \nu\,(u_n\,,\ u - u_n)_{L^2(\Omega_{\mathrm{s}})}$$

$$+ \frac{1}{2} \|\nabla y - \nabla y_n\|^2_{L^2(\Omega_{\mathrm{g}})} + \frac{\nu}{2} \|u - u_n\|^2_{L^2(\Omega_{\mathrm{s}})}$$

$$- 6 \int_{\Gamma_r} \left(G\sigma\,|y_n| y_n\,(y - y_n)^2\right) p_n\, ds - 6 \int_{\Gamma_0} \varepsilon\sigma\,|y_n| y_n\,(y - y_n)^2\, p_n\, ds$$

$$=: J_n(y, u),$$

where y and y_n are defined as above and p_n solves the following adjoint equation

$$a_n^*[p_n, v] := \int_\Omega \kappa \nabla p_n \cdot \nabla v\, dx + 4 \int_{\Gamma_r} \sigma |y_n|^3 (G^* p_n)\, v\, ds + 4 \int_{\Gamma_0} \varepsilon \sigma\, |y_n|^3 p_n\, v\, ds$$

$$= \int_{\Omega_g} (\nabla y_n - z) \cdot \nabla v\, dx \quad \forall v \in H^1(\Omega). \tag{6.3}$$

In [15, Lemma 6.1], it is shown that not only $S'(u)$ is well defined in a neighborhood of $\bar{u}$, but also the solution operator associated to (4.3) as a mapping from $H^1(\Omega)^*$ to $H^1(\Omega)$. Since the adjoint of this operator is equivalent to the solution operator of (6.3) and the inhomogeneity can clearly be identified with an element of $H^1(\Omega)^*$, there exists a unique $p_n \in H^1(\Omega)$ provided that Assumption 3 holds true and $\|u_n - \bar{u}\|_{L^2(\Omega_s)}$ is sufficiently small. Therefore, $J_n : V^\infty \times L^2(\Omega_s) \to \mathbb{R}$ is well defined and (Q_n) is equivalent to

$$(Q_n) \quad \begin{cases} \text{minimize } J_n(y, u) \\ \text{subject to (6.1)} \\ \text{and } u_a(x) \le u(x) \le u_b(x) \quad \text{a.e. in } \Omega_s. \end{cases}$$

Theorem 6.1. *Suppose that Assumptions 1–3, and condition (S) are fulfilled. Then, for every $u_n \in L^2(\Omega_s)$ satisfying $\|u_n - \bar{u}\|_{L^2(\Omega_s)} \le \rho$ with some sufficiently small $\rho > 0$, there exists a unique solution $\tilde{u}$ to (Q_n) with associated state $\tilde{y} = S'(u_n)(\tilde{u} - u_n) + y_n$.*

Proof. In the following, let $h \in L^2(\Omega_s)$ be an arbitrary direction. Condition (S) implies

$$\begin{aligned} j''(u_n) h^2 &= j''(\bar{u}) h^2 + \left(j''(u_n) - j''(\bar{u}) \right) h^2 \\ &\ge j''(\bar{u}) h^2 - c \|u_n - \bar{u}\|_{L^2(\Omega_s)} \|h\|_{L^2(\Omega_s)}^2 \\ &\ge (\delta - c\rho) \|h\|_{L^2(\Omega_s)}^2, \end{aligned}$$

since j is twice continuously Fréchet-differentiable from $L^2(\Omega_s)$ to $\mathbb{R}$. Thus, by choosing $\rho \le \delta/2c$, we obtain

$$j''(u_n) h^2 \ge \frac{\delta}{2} \|h\|_{L^2(\Omega_s)}^2. \tag{6.4}$$

In view of (4.11), we obtain for the first part of j_n

$$\begin{aligned} j'(u_n)(u - u_n) &= (\nabla y_n - z,\, \nabla y - \nabla y_n)_{L^2(\Omega_g)} + \nu\, (u_n,\, u - u_n)_{L^2(\Omega_s)} \\ &= (p_n + \nu\, u_n,\, u - u_n)_{L^2(\Omega_s)} \ge c, \end{aligned}$$

with a constant $c > -\infty$, since p_n is the solution of the adjoint equation (6.3) and $u_n, u \in U_{ad}$. Together with (6.4), this implies that j_n is bounded from below. The remaining part of the proof is along the lines of the standard theory for linear quadratic problems. First, we consider a sequence $\{u_m\}_{m=1}^\infty$ converging to the infimum of j_n. Due to $\nu/2 \|u - u_n\|_{L^2(\Omega_s)}^2$ within the objective functional j_n, we are allowed to select a weakly converging subsequence, w.l.o.g. $\{u_m\}$ itself. The corresponding weak limit is denoted by $\tilde{u}$. Since $S'(u_n)$ is linear and continuous, thus weakly continuous, we obtain weak convergence of the states in $H^1(\Omega)$ to a limit $\tilde{y}$ that satisfies the state equation (6.2) together with $\tilde{u}$. Moreover, the coercivity of $j''(u_n)$ by (6.4) and the linearity of the remaining part of j_n imply that j_n is convex on $L^2(\Omega)$. Therefore, j_n is weakly lower semicontinuous giving the optimality of $(\tilde{y}, \tilde{u})$. Furthermore, the convexity of j_n implies that $(\tilde{y}, \tilde{u})$ is the unique optimal solution. $\qquad\square$

Notice that, with second-order conditions of the form (SSC), one cannot derive an equation analogous to (6.4) that is essentially needed for the convexity and the boundedness of j_n. Therefore, to keep the discussion concise, the restrictive second-order conditions (S) are used here, that do not account for strongly active sets. However, it is possible to deal with strongly active sets in an SQP framework by considering second-order conditions of the form

$$\left\{ \begin{array}{l} Let\ \delta > 0\ and\ \tau > 0\ exist\ such\ that \\[2mm] \quad j''(\bar{u})\,u^2 \geq \delta\,\|u\|^2_{L^2(\Omega_{\mathrm s})} \quad for\ all\ u \in L^2(\Omega_{\mathrm s})\ with\ u(x) = 0\ on\ A_\tau. \end{array} \right.$$

In this case, the unique existence of solutions of the associated linearized problems analogous to (Q_n) is an immediate consequence of the convergence theory of the SQP method that follows from the theory of Newton's method for generalized equations. This is in detail discussed in Unger [26] for the elliptic case and in Goldberg and Tröltzsch [10] and Tröltzsch [23] for the parabolic case.

In a standard way, first-order necessary optimality conditions to (Q_n) are derived. The Lagrange formalism gives the following adjoint equation to (Q_n), here denoted in the variational formulation:

$$\begin{aligned} a_n^*[p, v] := & \int_\Omega \kappa \nabla p \cdot \nabla v\, dx + 4 \int_{\Gamma_{\mathrm r}} \sigma |y_n|^3 (G^* p)\, v\, ds + 4 \int_{\Gamma_0} \varepsilon\sigma\, |y_n|^3 p\, v\, ds \\[2mm] = & \int_{\Omega_{\mathrm g}} (\nabla y - z) \cdot \nabla v\, dx + 12 \int_{\Gamma_{\mathrm r}} \big(\sigma |y_n| y_n (G^* p_n)\big)(y_n - y)v\, ds \\[2mm] & + 12 \int_{\Gamma_0} (\varepsilon\sigma\, |y_n| y_n\, p_n)(y_n - y)v\, ds \quad \forall v \in H^1(\Omega), \end{aligned} \tag{6.5}$$

where $y \in H^1(\Omega)$ is the solution of (6.2). As it possesses the same bilinear form as (6.3) and the right-hand side is clearly an element of $H^1(\Omega)^*$ because of $y_n \in V^\infty$ and $\tilde{y}, p_n \in H^1(\Omega)$, the existence of solutions to (6.5) follows by the same arguments as in case of (6.3). Similar to the discussion in Section 4, a pointwise evaluation of the variational inequality to (Q_n) gives the well-known projection formula

$$\tilde{u}(x) = \mathcal{P}_{ad}\left\{ -\frac{1}{\nu}\, \tilde{p}(x) \right\}, \tag{6.6}$$

where $\tilde{p}$ denotes the solution of (6.5) with $y = \tilde{y}$. Problem (Q_n) is solved by a primal-dual active set strategy (see for instance [2] or [3]). To that end, we define

$$\lambda(x) := \tilde{p}(x) + \nu\,\tilde{u}(x)$$

and introduce the active and inactive sets by

$$
\begin{aligned}
\mathcal{A}_a &:= \{x \in \Omega_{\mathrm{s}} \mid \tilde{u}(x) - \lambda(x) < u_a(x)\} \\
\mathcal{A}_b &:= \{x \in \Omega_{\mathrm{s}} \mid \tilde{u}(x) - \lambda(x) > u_b(x)\} \\
\mathcal{I} &:= \Omega_{\mathrm{s}}\backslash\{\mathcal{A}_a \cup \mathcal{A}_b\}.
\end{aligned}
\tag{6.7}
$$

With (6.7) at hand, one shows by standard arguments that the optimality system, consisting of (6.2), (6.5), and (6.6), is equivalent to

$$
\left.
\begin{aligned}
a_n[\tilde{y}, v] &= 3\int_{\Gamma_{\mathrm{r}}} (G\,\sigma|y_n|^3 y_n)v\,ds + 3\int_{\Gamma_0} \varepsilon\sigma\,|y_n|^3 y_n\,v\,ds \\
&\quad + \int_{\Omega_{\mathrm{s}}} \tilde{u}\,v\,dx + \int_{\Gamma_0} \varepsilon\sigma\,y_0^4\,ds \quad \forall\,v \in H^1(\Omega) \\[2mm]
a_n^*[\tilde{p}, v] &= \int_{\Omega_{\mathrm{g}}} (\nabla\tilde{y} - z)\cdot\nabla v\,dx + 12\int_{\Gamma_{\mathrm{r}}} \big(\sigma|y_n|y_n(G^* p_n)\big)(y_n - \tilde{y})v\,ds \\
&\quad + 12\int_{\Gamma_0} (\varepsilon\sigma\,|y_n|y_n\,p_n)(y_n - \tilde{y})v\,ds \quad \forall\,v \in H^1(\Omega) \\[2mm]
\tilde{u}(x) &= u_a(x) \quad \text{a.e. in } \mathcal{A}_a \\
\tilde{u}(x) &= u_b(x) \quad \text{a.e. in } \mathcal{A}_b \\[1mm]
\nu\tilde{u}(x) &+ \tilde{p}(x) = 0 \quad \text{a.e. in } \mathcal{I},
\end{aligned}
\right\}
\tag{6.8}
$$

Altogether, the SQP method with primal-dual active set strategy proceeds as follows.

Algorithm 1.

1. Choose initial value u_0. Compute $y_0 = S(u_0)$ and p_0 as the solution of the adjoint equation (6.3). Set $n = 0$.
2. Compute inhomogeneities in (6.2) and (6.5) that only depend on y_n and p_n.
3. PRIMAL-DUAL ACTIVE SET STRATEGY:
   (a) Define initial sets $\mathcal{A}_a^{(0)} \subset \Omega_{\mathrm{s}}$ and $\mathcal{A}_b^{(0)} \subset \Omega_{\mathrm{s}}$ with $\mathcal{A}_a^{(0)} \cap \mathcal{A}_b^{(0)} = \emptyset$. Set $\mathcal{I}_h^{(0)} = \Omega_{\mathrm{s}}\backslash\{\mathcal{A}_b^{(0)} \cup \mathcal{A}_a^{(0)}\}$ and $i = 0$.
   (b) Find $\tilde{u}_i$, $\tilde{y}_i$, and $\tilde{p}_i$ by solving (6.8).

(c) Set $\lambda_i = \tilde{p}_i + \nu\,\tilde{u}_i$ and

$$\mathcal{A}_a^{(i+1)} := \{x \in \Omega_\mathrm{s} \mid \tilde{u}_i(x) - \lambda_i(x) < u_a(x)\}$$
$$\mathcal{A}_b^{(i+1)} := \{x \in \Omega_\mathrm{s} \mid \tilde{u}_i(x) - \lambda_i(x) > u_b(x)\}$$
$$\mathcal{I}^{(i+1)} := \Omega_\mathrm{s}\backslash\{\mathcal{A}_a^{i+1} \cup \mathcal{A}_b^{i+1}\}.$$

(d) If $\mathcal{A}_a^{(i+1)} = \mathcal{A}_a^{(i)}$ and $\mathcal{A}_b^{(i+1)} = \mathcal{A}_b^{(i)}$ then STOP, else:

Update $i = i + 1$ and goto (b).

4. Set $u_{n+1} = \tilde{u}_i$, $y_{n+1} = \tilde{y}_i$, and $p_{n+1} = \tilde{p}_i$.

5. if

$$\delta := \frac{1}{3}\left(\frac{\|u_{n+1} - u_n\|_{L^2(\Omega_\mathrm{s})}}{\|u_n\|_{L^2(\Omega_\mathrm{s})}} + \frac{\|y_{n+1} - y_n\|_{L^2(\Omega)}}{\|y_n\|_{L^2(\Omega)}} + \frac{\|p_{n+1} - p_n\|_{L^2(\Omega)}}{\|p_n\|_{L^2(\Omega)}} \right) \leq tol$$

then STOP, else: $n = n + 1$, goto 2.

In [26], [10], and [23], respectively, locally quadratic convergence of (y_n, u_n, p_n) to a locally optimal solution $(\bar{y}, \bar{u}, \bar{p})$ of (P) in $V^\infty \times L^\infty(\Omega_\mathrm{s}) \times H^1(\Omega)$ is shown. As already mentioned above, the underlying analysis is based on the theory of Newton's method for generalized equations. Moreover, since (Q_n) is a control constrained problem, the active set strategy can be interpreted as a semismooth Newton-method that is superlinearly converging (see, e.g., [12]). For a globalization of the SQP method, we use a projected gradient method with a line search according to the Armijo rule to find suitable initial values for the SQP method.

7. Discretization

The following section is devoted to the discretization of the linear PDEs in the optimality system for (Q_n), given by (6.8). To be more precise, we focus on the treatment of $b_{\mathrm{r},n}$, $b_{\mathrm{r},n}^*$, and $b_{0,n}$, i.e., the boundary integrals on Γ_r and Γ_0 in (6.2) and (6.5), since all other terms in both PDEs are discretized in a standard way using linear finite elements. The corresponding linear finite element ansatz functions are denoted by ϕ_i, $i = 1, \ldots, n_p$. The integral with the prescribed function z is approximated by third order Gauß quadrature.

The nonlinear integrals over Γ_r and Γ_0 are approximated by a summarized midpoint rule. The same technique is used by Atkinson and Chandler in [1] when they apply the midpoint rule to solve the radiosity equation in the two-dimensional case. Its advantage is that critical parts of the radiation integral operators can be evaluated exactly as demonstrated below. In the following, we explain the discretization of $b_{\mathrm{r},n}$ as an example for the integrals over Γ_r. The discretization of the integrals over Γ_0 is similar, but simpler, since they do not involve the nonlocal radiation operator. If we insert the ansatz for y, given by

$$y(x) \approx \sum_{j=0}^{n_p} \mathbf{y}_j\,\phi_j(x),$$

and the ansatz function ϕ_i as test function into $b_{\mathrm{r},n}$ in the bilinear form a_n in (6.2), we obtain the discrete version of boundary integral:

$$\int_{\Gamma_{\mathrm{r}}} \big(G\,d(x)\,y(x)\big)\phi_i(x)\,ds \approx \sum_{j=0}^{n_p} \int_{\Gamma_{\mathrm{r}}} \big(G\,d(x)\,\phi_j(x)\big)\phi_i(x)\,ds\,\mathbf{y}_j =: \sum_{j=0}^{n_p} \mathbb{B}_{ij}\,\mathbf{y}_j.$$

Here, d is defined by $d(x) := \sigma\,|y_n(x)|^3$. For the numerical integration of $\mathbb{B}_{ij}$, we divide the m intervals on Γ_{r} arising from the triangulation into n_{i} smaller intervals denoted by Γ_k, and, thus, Γ_{r} can be represented by

$$\Gamma_{\mathrm{r}} = \bigcup_{k=1}^{n_r} \Gamma_k \quad \text{with } n_r := n_{\mathrm{i}}\,m. \tag{7.1}$$

In all that follows, we denote the midpoint of the subinterval Γ_k by x_k. With that partition at hand, the midpoint rule reads as follows:

$$\mathbb{B}_{ij} = \int_{\Gamma_{\mathrm{r}}} \big(G\,d(x)\,\phi_j(x)\big)\phi_i(x)\,ds \approx \sum_{\Gamma_k \subset \Gamma_{\mathrm{r}}} \phi_i(x_k)\,|\Gamma_k|\,(G\,d\,\phi_j)(x_k). \tag{7.2}$$

According to the definition of G in (1.2), we obtain

$$(G\,d\,\phi_j)(x_k) = \big((I-K)\underbrace{(I-(1-\varepsilon)K)^{-1}\varepsilon\,d\,\phi_j}_{=:\,h}\big)(x_k), \tag{7.3}$$

where the integral operator K is defined by (3.1). Using again the summarized midpoint rule, K is approximated by

$$(K\,h)(x_k) \approx \sum_{\Gamma_l \subset \Gamma_{\mathrm{r}}} h(x_l) \int_{\Gamma_l} \omega(x_k,z)\,ds_z =: \sum_{\Gamma_l \subset \Gamma_{\mathrm{r}}} \mathbb{K}_{lk}\,h(x_l),$$

with ω as defined in (3.2). For our numerical investigation, we again choose the domain presented in Figure 1, with side lengths 2 for the outer and 1 for the inner square, respectively. Hence, according to (3.3), the convexity of Ω_{g} implies for the visibility factor $\Xi(x,z) \equiv 1$. As already mentioned, in Section 3, the kernel ω exhibits a singularity at $z = x_k$. In our case, this problem only occurs in the corners of Ω_{g}, since the definition of ω implies $\omega(x_k,z) = 0$ for all $z \in \Gamma_l$. Therefore, a numerical integration of $\int_{\Gamma_l} \omega(x_k,z)\,ds_z$ nearby the corners of Ω_{g} is critical. However, in our case, Ω_{g} is a convex polygon, and hence we can apply Lemma A.1 for the evaluation of $\mathbb{K}_{lk}$ (see Appendix A). Therefore, the integral $\mathbb{K}_{lk} = \int_{\Gamma_l} \omega(x_k,z)\,ds_z$ is exactly integrated in our case, and we obtain

$$\mathbb{K}_{lk} = \begin{cases} \dfrac{1}{2}\left(\dfrac{t_{\mathrm{r}}(x_k)\cdot(x_{l,1}-x_k)}{\|x_{l,1}-x_k\|} - \dfrac{t_{\mathrm{r}}(x_k)\cdot(x_{l,0}-x_k)}{\|x_{l,0}-x_k\|}\right) & \text{, if } x_k \notin \Gamma_l \\[2ex] 0 & \text{, if } x_k \in \Gamma_l. \end{cases}$$

Here, the unit tangential vector $t_{\mathrm{r}}(x_k)$ is defined as in (A.2). Moreover, $x_{l,0}$ and $x_{l,1}$ are the end points of the interval Γ_l ordered as required in Lemma A.1.

Lemma 7.1. *Assume that the emissivity is bounded from below by $\varepsilon_{\min} > 0$. Then the matrix $\mathbb{J} := \mathbb{I} - (1 - \varepsilon)\mathbb{K}$ is invertible.*

Proof. By construction we have that $\mathbb{K}_{ij} = 0$ if $x_j \in \Gamma_i$. Therefore, $\mathbb{K}_{ii} = 0$ and thus $\mathbb{J}_{ii} = 1$ hold true for all $1 \leq i \leq n_r$. Moreover, in case of a convex domain, the normal vector $n_{\mathrm{r}}(x)$ and the difference vector $z - x$ always point into the same half-space. Since the same holds for $n_{\mathrm{r}}(z)$ and $x - z$, $\omega(x, z) \geq 0$ follows for all $x, z \in \Gamma_{\mathrm{r}}$. Hence, we obtain $\mathbb{K}_{ij} \geq 0$ for all $1 \leq i, j \leq N_{\mathrm{r}}$. Together with $\varepsilon(x) \leq 1$ a.e. on Γ_{r} and $\mathbb{K}_{ii} = 0$, this yields

$$
\sum_{j \neq i} |\mathbb{J}_{ij}| = \sum_{j=1}^{N_{\mathrm{r}}} (1 - \varepsilon)\mathbb{K}_{ij}
$$
$$
\leq (1 - \varepsilon_{\min}) \sum_{j=1}^{N_{\mathrm{r}}} \int_{\Gamma_j} \omega(x_i, z) \, ds_z
$$
$$
= (1 - \varepsilon_{\min}) \int_{\Gamma_{\mathrm{r}}} \omega(x_i, z) \, ds_z.
$$

From [14, Lemma 1], it is known that $K\mathbf{1} = \mathbf{1}$ and hence $\int_{\Gamma_{\mathrm{r}}} \omega(x_i, z) \, ds_z = 1$. Therefore, the assumption on the emissivity implies

$$
\sum_{i \neq j} |\mathbb{J}_{ij}| \leq 1 - \varepsilon_{\min} < 1 = \mathbb{J}_{ii},
$$

which concludes the proof. $\qquad \square$

In view of $(I - (1 - \varepsilon)K)h = \varepsilon \sigma \, d \phi_j$, Lemma 7.1 allows us to continue with

$$
\mathbf{h} = (\mathbb{I} - (1 - \varepsilon)\mathbb{K})^{-1} \varepsilon \mathbb{D} \, \boldsymbol{\phi}_j \tag{7.4}
$$

where ϕ_j and $\mathbf{h}$ are vectors of the values of ϕ_j and h, respectively, at the midpoints x_k, i.e., $\boldsymbol{\phi}_j := \big(\phi_j(x_k)\big)_{k=1}^{n_r}$ and $\mathbf{h} = \big(h(x_k)\big)_{k=1}^{n_r}$. Moreover, $\mathbb{I}$ denotes the $n_r \times n_r$ identity matrix and $\mathbb{D}$ is defined by $\mathbb{D} := \mathrm{diag}\big(d(x_k)\big)_{k=1}^{n_r}$. The inverse of $\mathbb{I} - (1-\varepsilon)\mathbb{K}$ is calculated with the help of a Lapack LU decomposition. With (7.3) and (7.4) at hand, the nonlocal radiation operator is approximated by

$$
\big((G \, d\phi_j)(x_k)\big)_{k=1}^{n_r} \approx (\mathbb{I} - \mathbb{K})(\mathbb{I} - (1 - \varepsilon)\mathbb{K})^{-1} \varepsilon \mathbb{D} \, \boldsymbol{\phi}_j =: \mathbb{G} \mathbb{D} \boldsymbol{\phi}_j. \tag{7.5}
$$

Next, we introduce the matrix

$$
\Phi = (\boldsymbol{\phi}_1, \boldsymbol{\phi}_2, \ldots, \boldsymbol{\phi}_{n_p}).
$$

Notice that $\Phi \in \mathbb{R}^{n_r \times n_p}$, i.e., Φ is in general nonquadratic. Together with (7.2) and (7.5), this definition implies

$$
\mathbb{B} \approx \Phi^{\top} \mathbb{M}_r \mathbb{G} \mathbb{D} \Phi
$$

with $\mathbb{M}_r := \operatorname{diag}(|\Gamma_k|)_{k=1}^{n_r}$. Analogously, we obtain for $b^*_{r,n}$, i.e., the integral over Γ_r in the bilinear form a_n^* of the adjoint equation (6.5),

$$\int_{\Gamma_r} d(x)\,\phi_i(x)\,\big(G^*\,p(x)\big)\,ds = \int_{\Gamma_r} \big(G\,d(x)\,\phi_i(x)\big)p(x)\,ds \tag{7.6}$$

$$\approx \sum_{j=0}^{n_p}\int_{\Gamma_r} \big(G\,d(x)\,\phi_i(x)\big)\phi_j(x)\,ds\,\mathbf{p}_j = \sum_{j=0}^{n_p}\mathbb{B}_{ij}^{\top}\,\mathbf{p}_j \approx \sum_{j=0}^{n_p}(\Phi^{\top}\,\mathbb{D}\,\mathbb{G}^{\top}\,\mathbb{M}_r\,\Phi)_{ij}\,\mathbf{p}_j,$$

with $\mathbb{G}^{\top} = (\mathbb{I}-(1-\varepsilon)\mathbb{K})^{-\top}(\mathbb{I}-\mathbb{K}^{\top})$ and $\mathbb{D}$, $\mathbb{M}_r$, and Φ as defined above. On the other hand, we have

$$\int_{\Gamma_r} d(x)\,\phi_i(x)\big(G^*\,\phi_j(x)\big)\,ds \approx \sum_{\Gamma_k\subset\Gamma_r}\phi_i(x_k)\,d(x_k)\,|\Gamma_k|\,(G^*\,\phi_j)(x_k)$$

$$\approx (\Phi^{\top}\,\mathbb{D}\,\mathbb{M}_r\,\mathbb{G}^*\,\Phi)_{ij},$$

where $\mathbb{G}^*$ denotes a discrete version of G^*. Comparing this with (7.6), we choose

$$\mathbb{G}^* = \mathbb{M}_r^{-1}\mathbb{G}^{\top}\,\mathbb{M}_r,$$

and observe that $\mathbb{G}^*$ is the adjoint of $\mathbb{G}$ with respect to the weighted scalar product $\mathbf{v}^T\mathbb{M}_r\,\mathbf{w}$ as discretization of $\int_{\Gamma_r} v\,w\,ds$.

All other integrals over Γ_r in (6.2) and (6.5), have the same structure as the ones described above, except $\int_{\Gamma_r}(\sigma|y_n|y_n(G^*p_n))y\,v\,ds$. With $d = \sigma|y_n|y_n$, this is discretized by

$$\int_{\Gamma_r} \big(d(G^*p_n)\big)y\,\phi_i\,ds \approx \sum_{j=0}^{n_p}\int_{\Gamma_r}\big(d(G^*p_n)\big)\phi_j\,\phi_i\,ds\,\mathbf{y}_j$$

$$\approx \sum_{j=0}^{n_p}\mathbf{y}_j\sum_{\Gamma_k\subset\Gamma_r}\phi_i(x_k)\,\underbrace{d(x_k)|\Gamma_k|\,(G^*p_n)(x_k)}_{=:\,a(x_k)}\,\phi_j(x_k).$$

Similar to above, we obtain

$$a(x_k) \approx \big(\mathbb{D}\,\mathbb{M}_r\,\mathbb{G}^*\,\Phi\,\mathbf{p}_n\big)_k = \big(\mathbb{D}\,\mathbb{G}^{\top}\,\mathbb{M}_r\,\Phi\,\mathbf{p}_n\big)_k,$$

since p_n, as the current iterate of the SQP method, is also discretized by linear ansatz functions, and hence, the interpolation of p_n onto the intervals Γ_k, $k = 1,\dots,n_r$ on Γ_r is equivalent to $\Phi\,\mathbf{p}_n$.

With this discretization at hand, the optimality system (6.8) is approximated by a linear system of equation of the size $3\,n_p \times 3\,n_p$ for the unknowns $\tilde{u}$, $\tilde{y}$, and $\tilde{p}$. This system is solved numerically using the direct sparse LU factorization included in the UMFPACK library (see [8] and the references therein).

8. Numerical examples

The starting point of this section is an example that is already discussed in [16]. However, here, we present some additional results that highlight substantial characteristics of nonlocal radiative heat transfer in the context of optimal control. For the numerical investigations, we use the domain introduced above. Moreover, the material parameters are fixed at average values of the realistic distributions given in [17]. The specific values are given in Table 2.

Table 2: Material parameters for the numerical tests

$\kappa_g \left(\frac{W}{m\,K}\right)$	$\kappa_s \left(\frac{W}{m\,K}\right)$	ε	$\sigma \left(\frac{W}{m^2 K^4}\right)$
0.08	24.0	0.65	$5.6696 \cdot 10^{-8}$

Furthermore, the external temperature y_0 is assumed to be constant and equal to 293.0 K. Throughout the following numerical tests, the desired temperature gradient (in $\frac{K}{m}$) is given by $z \equiv (0, -20)^T$, and we took $u_a \equiv 125000$, and $u_b \equiv 750000$ for the control constraints (in $\frac{W}{m^3}$). Due to the comparatively large values of the control, one has to deal with rather small Tikhonov regularization parameters to control the influence of the cost term within the objective functional. Hence, we choose $\nu = 5 \cdot 10^{-10}$ for the first computation. However, later on, this is decreased to $\nu = 5 \cdot 10^{-11}$.

Before, we present the numerical results, let us shortly describe the used mesh. It consists of 98340 points and 196358 triangles. Due to the nonlocal radiation boundary condition, it is refined four times on the inner boundary Γ_r as shown in Figure 3, such that we obtain $m = 3049$ points on Γ_r. With $n_i = 4$, this results in $n_r = m\,n_i = 12196$ intervals for the numerical integration of the boundary integrals described in Section 7. Furthermore, since the corners of Γ_r are non convex with respect to the outer domain Ω_s (see Figure 1), the mesh is additionally refined eight times in each of these corners up to a mesh size of approximately 10^{-5} in a radius of 10^{-4} around the corners. This is illustrated in Figure 3, where each white framed box indicates the area that is show in the subsequent figure. To illustrate the convergence behavior of the SQP active set algorithm, several characteristic data are recorded during the iteration. First, the decrease of the objective functional is shown. Moreover, we present the residuals in the discrete versions of the state equation and the adjoint equation, respectively, and the error in the gradient equation in each iteration step. The residual in the discretized state equation is given by the following quantity:

$$r_y := \sum_{i=1}^{n_p} \left| a_h[y_n, \phi_i] - \int_{\Omega_s} u_n\, \phi_i\, dx - \int_{\Gamma_0} \varepsilon \sigma\, y_0^4\, \phi_i\, ds \right|, \qquad (8.1)$$

where, as before, the subscript n denotes the actual SQP iterate that is also discretized by linear finite element ansatz functions. Moreover, a_h is the discretization

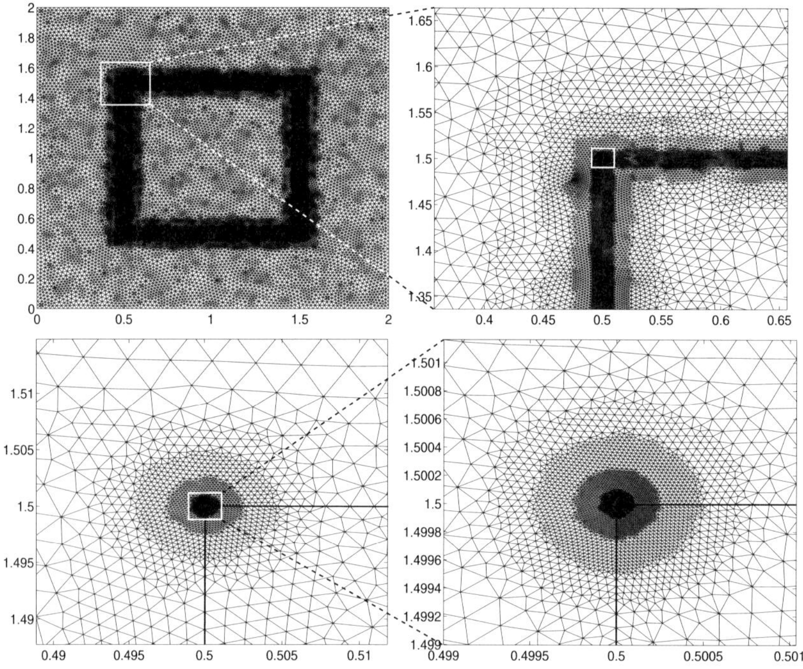

Figure 3: Mesh refinement on Γ_r and in the nonconvex corners.

of the left-hand side in the variational formulation of the state equation given by

$$\int_\Omega \kappa \nabla y_n \cdot \nabla v\, dx + \int_{\Gamma_r} (G\,\sigma |y_n|^3 y_n) v\, ds + \int_{\Gamma_0} \varepsilon \sigma\, |y_n|^3 y_n\, v\, ds.$$

Here, the integral over Γ_r is discretized as described in Section 7, i.e.,

$$\left(\int_{\Gamma_r} (G\,\sigma |y_n|^3 y_n)\phi_i\, ds \right)_{i=1}^{n_p} \approx \Phi^\top \mathbb{M}_r\, \mathbb{G}\, \mathbb{D}\, \Phi\, \mathbf{y}_n$$

with $\mathbb{D} = \operatorname{diag}\big(\sigma |y_n(x_k)|^3\big)_{k=1}^{n_r}$ and $y_n(x_k) = (\Phi\, \mathbf{y}_n)_k$. Notice that the last integral in (8.1) can be evaluated exactly in our case since y_0 is constant. In view of (4.9), the residual for the discrete adjoint equation is defined by

$$r_p := \sum_{i=1}^{n_p} \left| a_{n,h}^*[p_n, \phi_i] - \int_{\Omega_g} (\nabla y_n - z) \cdot \nabla \phi_i\, dx \right|,$$

where $a_{n,h}^*$ denotes the discretized version of the bilinear form of the adjoint equation in (4.9) which is clearly equal to a_n^* and is consequently discretized as described in Section 7. Since z is constant in our case, $\int_{\Omega_g} z \cdot \nabla \phi_i\, dx$ can also be computed exactly. The relative error in the gradient equation, i.e., the projection formula (4.14), is given by

$$r_{opt} := \frac{\left\| u_n - \mathcal{P}_{ad}\left\{ -\frac{1}{\nu}\, p_n \right\} \right\|_{L^2(\Omega_{\mathrm{s}})}}{\|u_n\|_{L^2(\Omega_{\mathrm{s}})}}.$$

Furthermore, at the end of each iteration, we compute three other quantities that can be interpreted as indicators for the error in the optimality system for (P) consisting of the state equation, the adjoint equation, and the gradient equation in form of (4.14). To that end, we introduce

$$e_y := \frac{\|\hat{y} - y_n\|_{L^2(\Omega)}}{\|y_n\|_{L^2(\Omega)}} \quad \text{with } \hat{y} = \sum_{i=1}^{N} \hat{\mathbf{y}}_i \phi_i(x) \text{ and } \hat{\mathbf{y}} := S_h(\mathbf{u}_n)$$

as an indicator for the relative error in the state equation. Here, $S_h : \mathbb{R}^{N_{\mathrm{s}}} \to \mathbb{R}^N$ denotes the discrete solution operator of the state equation. Moreover, N_{s} denotes the number of nodes in $\bar{\Omega}_{\mathrm{s}}$, and N_{g} is defined analogously. Within in our framework, S_h is numerically realized by a continuous Newton's method. It turns out that the linear PDE that has to be solved in each iteration step of Newton's method has the same structure as (6.1) and (6.2), respectively. Hence, it is discretized as described in Section 7. Due to the nonlocal radiation, the arising linear system of equations is not symmetric and solved with the help of the GMRES code included in the SPARSKIT library (cf. [19]). Here, an incomplete LU decomposition of the stiffness matrix was used for preconditioning. With $\hat{y}$ at hand, the indicator for the relative error in the adjoint equation is computed by

$$e_p := \frac{\|\hat{p} - p_n\|_{L^2(\Omega)}}{\|p_n\|_{L^2(\Omega)}} \quad \text{with } \hat{p} = \sum_{i=1}^{N} \hat{\mathbf{p}}_i \phi_i(x) \text{ and } \hat{\mathbf{p}} := S_h'(\hat{y})^* \, \hat{\mathbf{w}},$$

where $S_h'(\hat{y})^* : \mathbb{R}^{N_{\mathrm{g}}} \to \mathbb{R}^N$ denotes the discrete solution operator of the adjoint equation at $\hat{y}$. Moreover, $\hat{\mathbf{w}} \in \mathbb{R}^{N_{\mathrm{g}}}$ is defined by

$$\hat{\mathbf{w}}_i = \int_{\Omega_{\mathrm{g}}} (\nabla \hat{y} - z) \cdot \nabla \phi_i \, dx,$$

which can again be evaluated exactly, since $\hat{y}$ is discretized with linear finite elements and z is constant in our case. As already mentioned above, the bilinear form associated to the adjoint equation is discretized, as depicted in Section 7. The arising linear system of equations for $\hat{\mathbf{p}}$ is again solved with the SPARSKIT GMRES code. Finally, similarly to r_{opt}, we introduce the following quantity as an indicator for the relative error in the projection formula (4.14)

$$e_u = \frac{\|\hat{u} - u_n\|_{L^2(\Omega)}}{\|u_n\|_{L^2(\Omega)}} \quad \text{with } \hat{u} = \mathcal{P}_{ad}\left\{ -\frac{1}{\nu}\, \hat{p} \right\}.$$

Notice that we use the same numerical solvers for the semilinear state equation and the adjoint equation in the projected gradient method for the initial value search.

In the sections below, we present the result of a numerical test carried out with the setting mentioned above. Afterwards, the numerical solution is compared with two other tests, first without radiation on Γ_r and second with a lower Tikhonov parameter ν.

8.1. Example 1

For the given setting, we obtain the numerical solution that is shown in Figures 4–7. In the pictures, the numerical solutions are denoted by the subscript h. Figure 7 illustrates that there are significant differences of the optimal temperature distribution from the desired one. First the isotherms at the upper edge of Γ_r are not horizontal as required. In addition to that, with a value of about 17 K, the temperature difference between lower and upper edge of Γ_r is smaller than

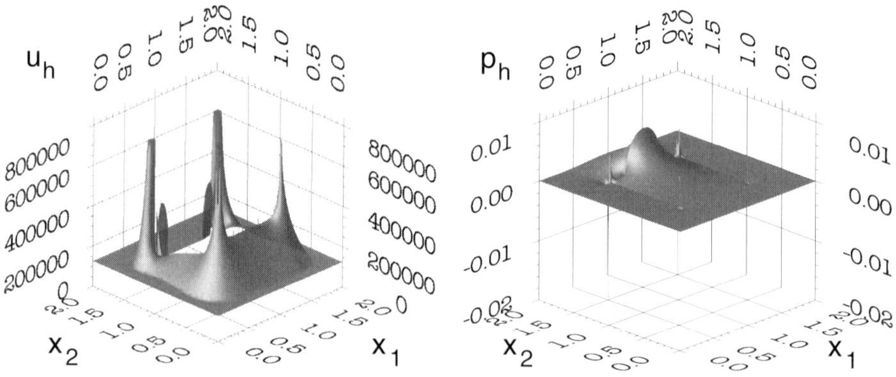

Figure 4: Control u_h.   Figure 5: Adjoint state p_h.

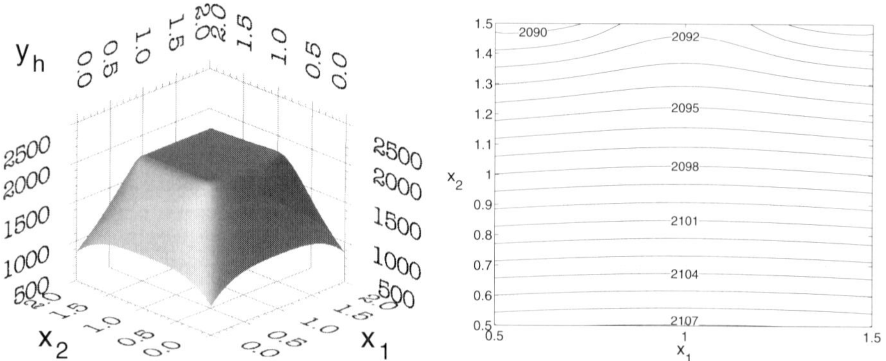

Figure 6: State y_h.   Figure 7: Isotherms in Ω_g.

the desired 20 K. Furthermore, the control possesses some peaks in the corners of Γ_r. As the computational mesh is refined in these corners several times, this does not seem to be a numerical effect. Table 3 illustrates the convergence behavior of the overall method, i.e., including the projected gradient method, by showing the decrease of the objective functional. Here, the rows above the double line refer to the projected gradient method, whereas the rows below correspond to the SQP active set algorithm. Notice that the initial value for the projected gradient method is given by $u_0 \equiv u_a + 0.3(u_b - u_a) = 312500$. The corresponding state is computed with Newton's method as described above. The initial value of the objective functional amounts 4.921569e+02. As one can see, the objective

Table 3: Convergence history for the first example.

it	$J(y,u)$	$1/2\|\nabla y - z\|^2_{L^2(\Omega_{\mathrm{g}})}$	$\nu/2\|u\|^2_{L^2(\Omega_{\mathrm{s}})}$
1	2.261592e+02	1.551524e+02	7.100687e+01
2	1.754913e+02	1.052975e+02	7.019378e+01
3	1.461291e+02	7.696015e+01	6.916894e+01
4	1.268671e+02	5.873924e+01	6.812786e+01
5	1.136532e+02	4.654510e+01	6.710807e+01
6	1.043283e+02	3.820065e+01	6.612768e+01
7	9.754344e+01	3.235032e+01	6.519312e+01
8	9.241677e+01	2.811423e+01	6.430253e+01
9	8.840054e+01	2.494841e+01	6.345213e+01
1	3.576389e+01	1.106707e+01	2.469682e+01
2	3.735623e+01	9.940953e+00	2.741527e+01
3	3.743465e+01	9.845256e+00	2.758939e+01
4	3.743563e+01	9.844839e+00	2.759079e+01

functional is heavily decreased after the first SQP iteration. Hence, as expected, the SQP method causes a speed up of the convergence rate. Table 4 shows the residuals, explained above, for this test case. Moreover, δ is the average difference between two iterates that was used for the stopping criterion (see Section 6), and #it$_{\mathrm{AS}}$ denote the number of active set iterations needed in the respective SQP step. The relative errors in the optimality system at the end of the iteration are listed in Table 5. As explained above, we use the projected gradient method for the globalization of the SQP method. To show the capability of this approach, a second computation with the same setting is made, except the initial value that is this time given by $u_0 \equiv u_a + 0.8(u_b - u_a) = 625000$. Table 6 illustrates that in both cases the same discrete optimum is achieved. Here, the subscript 0.3 refers to the solution with the smaller initial value, whereas 0.8 corresponds to the solution with the larger one.

Table 4: Residuals for the first example.

it_{SQP}	r_{opt}	r_y	r_p	δ	$\#\text{it}_{\text{AS}}$
1	1.589331e-04	5.482575e+04	4.737234e+00	5.220376e-01	9
2	1.414200e-04	6.502728e+02	2.171805e-01	1.737493e-01	9
3	6.143822e-05	1.328521e+00	1.284949e-03	1.263109e-02	2
4	1.034504e-16	1.706435e-04	1.361016e-04	8.339519e-05	1

Table 5: Errors for the first example.

e_u	e_y	e_p
1.871817e-05	3.475214e-11	8.845293e-05

Table 6: Comparison of solutions with different initial values.

$\dfrac{\|u_{0.3} - u_{0.8}\|_{L^2(\Omega_{\text{s}})}}{\|u_{0.3}\|_{L^2(\Omega_{\text{s}})}}$	$\dfrac{\|y_{0.3} - y_{0.8}\|_{L^2(\Omega_{\text{s}})}}{\|y_{0.3}\|_{L^2(\Omega)}}$	$\dfrac{\|p_{0.3} - p_{0.8}\|_{L^2(\Omega_{\text{s}})}}{\|p_{0.3}\|_{L^2(\Omega)}}$
3.487588e-05	6.218094e-06	1.069079e-04

8.2. Example 2

To demonstrate the influence of the nonlocal radiation on Γ_{r}, we now set $\varepsilon|_{\Gamma_{\text{r}}} = 0$. Hence, the interface condition on Γ_{r} is equivalent to

$$\kappa_{\text{g}} \left(\frac{\partial y}{\partial n_r} \right)_{\text{g}} - \kappa_{\text{s}} \left(\frac{\partial y}{\partial n_r} \right)_{\text{s}} = 0 \quad \text{on } \Gamma_{\text{r}}, \tag{8.2}$$

and all integrals over Γ_{r} in the variational formulations of the respective PDEs vanish. The physical meaning of (8.2) is the continuity of the normal heat flux on Γ_{r}. Notice that, on Γ_0, the emissivity is kept at 0.8. The corresponding numerical solution is plotted in Figures 8–11. First, we observe that the numerical solution differs significantly from the one for $\varepsilon = 0.8$ shown above. This indicates that it is indeed essential to account for radiation at this temperature level. In contrast to the results of Section 8.1, the difference between the optimal temperature distribution and the desired gradient is comparatively small. This is also confirmed by the Table 7, in particular by the third column, i.e., the values of $1/2\|\nabla y - z\|^2_{L^2(\Omega_{\text{g}})}$.

Notice that the difference in the values of the objective functional between the last projected gradient iteration and the first SQP step is even larger than in the first example. Moreover, we observe that the projected gradient method took 19 iterations to find an appropriate initial value. For lack of space, only the first and the last iteration are listed in Table 7. In addition to that, also the SQP algorithm

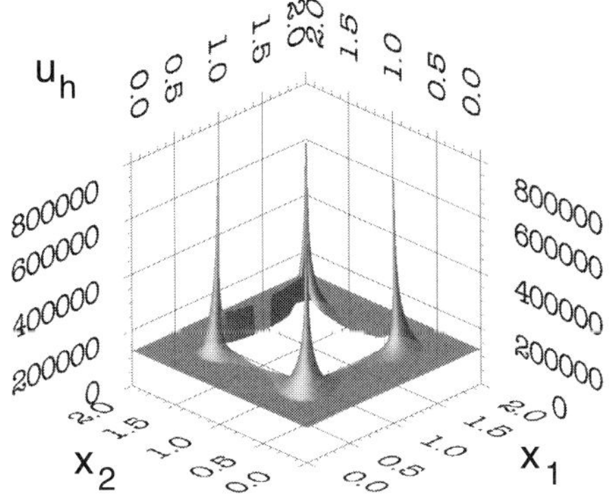

Figure 8: Control u_h for $\varepsilon|_{\Gamma_r} = 0$.

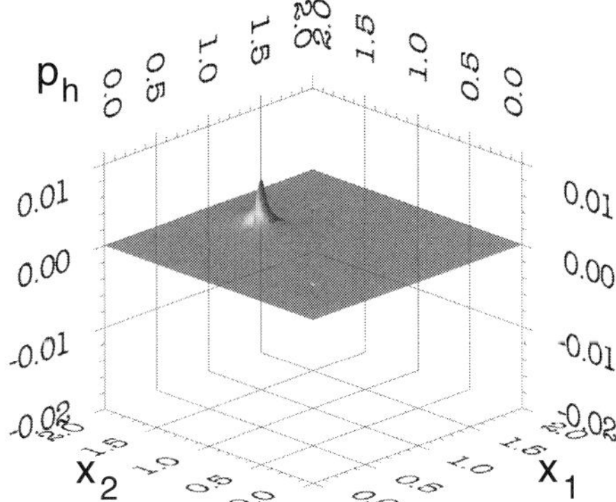

Figure 9: Adjoint state p_h for $\varepsilon|_{\Gamma_r} = 0$.

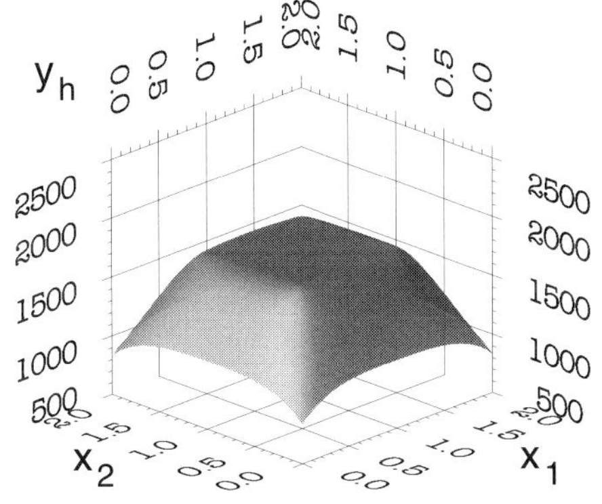

Figure 10: State y_h for $\varepsilon|_{\Gamma_r} = 0$.

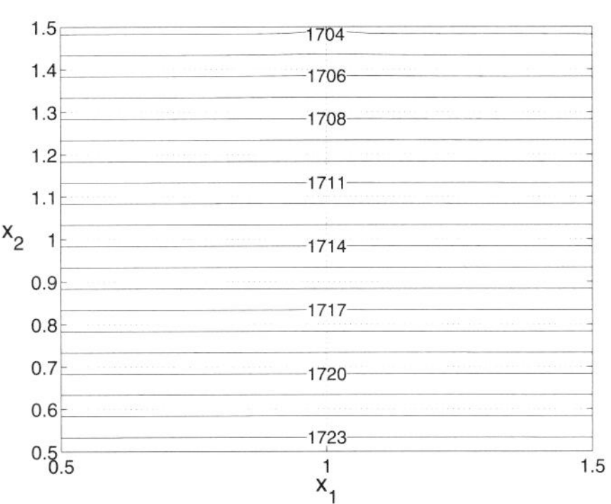

Figure 11: Isotherms in Ω_g for $\varepsilon|_{\Gamma_r} = 0$.

converges more slowly and in average more active set iterations are needed than in the first example, as Table 8 demonstrates. Table 9 shows the relative errors at the end of the iteration in this example. As one can see, the accuracy is similar to the first example (cf. Table 5). Beside the mentioned differences to the first example, the optimal control again exhibits the characteristics peaks in the corners of Γ_r. This observation indicates that this effect is not primary caused by the nonlocal radiation.

Table 7: Convergence history for the second example.

it	$J(y,u)$	$1/2\|\nabla y - z\|^2_{L^2(\Omega_g)}$	$\nu/2\|u\|^2_{L^2(\Omega_s)}$
1	1.490288e+04	1.481115e+04	9.173619e+01
19	1.020851e+03	9.251696e+02	9.568096e+01
1	9.823424e+01	2.771518e-02	9.820653e+01
2	2.115733e+01	5.352483e-03	2.115198e+01
3	1.416047e+01	1.341036e-02	1.414706e+01
4	1.418358e+01	2.119156e-02	1.416238e+01
5	1.418514e+01	2.144337e-02	1.416370e+01
6	1.418514e+01	2.144668e-02	1.416370e+01
7	1.418515e+01	2.144024e-02	1.416370e+01
8	1.418514e+01	2.144514e-02	1.416370e+01
9	1.418514e+01	2.144634e-02	1.416370e+01

Table 8: Residuals for the second example.

it_{SQP}	r_{opt}	r_y	r_p	δ	$\#it_{AS}$
1	6.312025e-08	4.492778e+03	1.115695e+00	5.678788e-01	45
2	9.292909e-06	9.878379e+04	2.483003e-01	5.419677e-01	13
3	1.460824e-05	1.979200e+04	3.898660e-02	9.871592e-01	14
4	1.523589e-08	4.001972e+02	1.430243e-03	3.732820e-01	7
5	4.097114e-07	1.972830e-01	6.144579e-05	4.653599e-03	2
6	8.876577e-17	3.419224e-05	8.094413e-05	1.310908e-04	1
7	3.058806e-16	6.033589e-05	8.884676e-05	4.754761e-04	1
8	3.620855e-17	2.886753e-05	1.144260e-05	1.656183e-04	1
9	9.094179e-17	2.397091e-05	3.672017e-05	4.463665e-05	1

Table 9: Errors for the second example.

e_u	e_y	e_p
3.059852e-05	2.572896e-11	2.144628e-04

8.3. Example 3

In the last example, the influence of the Tikhonov parameter ν is studied. To that end, we choose the values of the first example for all parameters and set $\nu = 10^{-10}$, and afterwards $\nu = 5 \cdot 10^{-11}$. Each time, the result of the computation with larger ν was used as initial value. The optimal values of the objective functional are shown

Table 10: Objective functional for different values of ν.

ν	$J(y, u)$	$1/2\|\nabla y - z\|^2_{L^2(\Omega_g)}$	$\nu/2\|u\|^2_{L^2(\Omega_s)}$	#it$_{SQP}$
5e-10	3.743563e+01	9.844839e+00	2.759079e+01	4
1e-10	1.286689e+01	5.682817e+00	7.184076e+00	4
5e-11	9.123665e+00	5.210145e+00	3.913520e+00	3

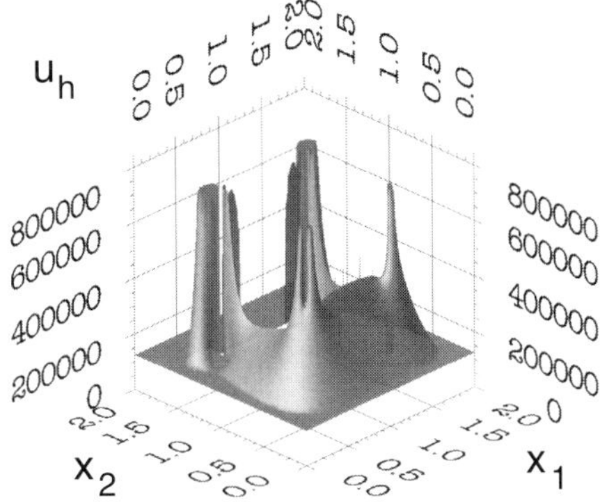

Figure 12: Control u_h for $\nu = 5 \cdot 10^{-11}$.

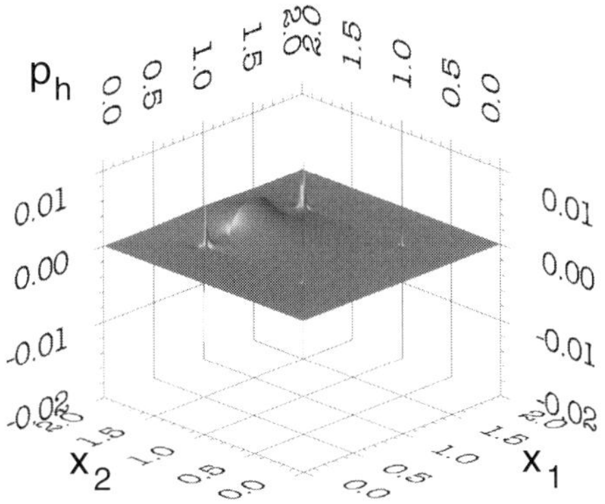

Figure 13: Adjoint state p_h for $\nu = 5 \cdot 10^{-11}$.

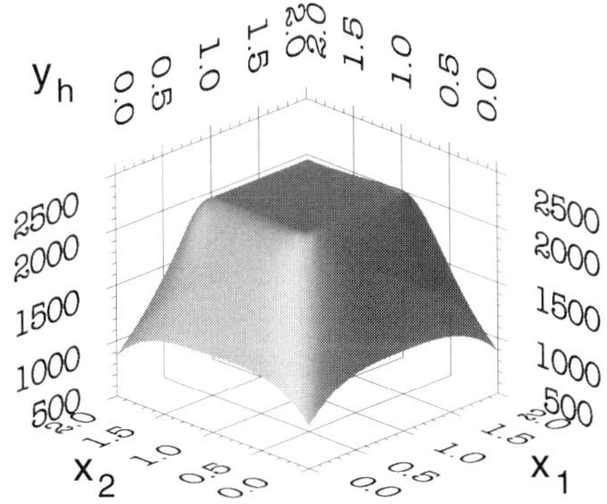

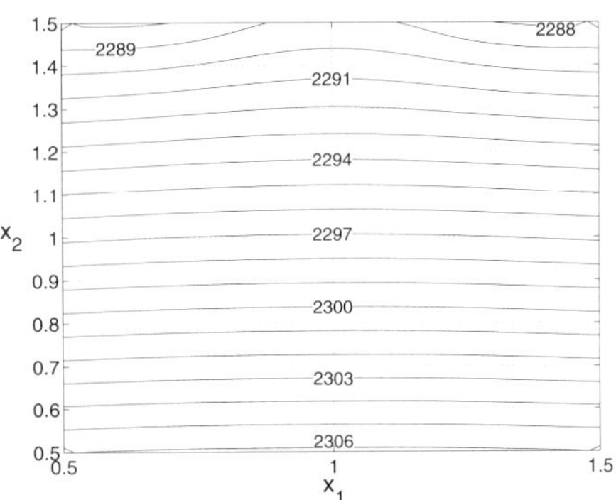

Figure 14: State y_h for $\nu = 5 \cdot 10^{-11}$.

Figure 15: Isotherms in Ω_g for $\nu = 5 \cdot 10^{-11}$.

in Table 10. Here, #it$_{SQP}$ denotes the number of SQP iterations. For $\nu = 5 \cdot 10^{-11}$, the value of J is not longer dominated by the Tikhonov part. However, as Figure

15 illustrates, the optimal state is still distinct from the desired one. Clearly, since the regularization parameter is smaller than in the first example, we expect the optimal control to be more irregular. As Figure 12 shows, this is indeed the case, especially in the corners of Γ_r. Table 11 again shows the relative errors.

Table 11: Errors for the third example.

e_u	e_y	e_p
2.715416e-04	3.960569e-12	7.569732e-05

Appendix A

In the following, we present a rather technical proof for the integration formula that is used to evaluate the matrix $\mathbb{K}$ arising from the discretization of the nonlocal radiation operator K (cf. Section 7).

Lemma A.1. *Assume that Ω_g is a polygon in $\mathbb{R}^2$ with boundary Γ_r and that the unit normal vector on Γ_r, denoted by n_r, is facing into the interior of Ω_g. Let a and b be two points located on the same edge of Γ_r and Γ_{ab} be defined by*

$$\Gamma_{ab} = \{z \in \mathbb{R}^2 \mid z = a + s(b-a) , \ 0 \le s \le 1\},$$

i.e., the line between a and b. Moreover, we assume that $\Xi(x,z) = 1$ for all $z \in \Gamma_{ab}$, and that a and b are ordered such that the orthogonal complement of $b-a$ given by

$$(b-a)^\perp := \begin{pmatrix} -(b_2 - a_2) \\ b_1 - a_1 \end{pmatrix} \tag{A.1}$$

is orientated in the direction of n_r, i.e., into the interior of Ω_g. Furthermore, let S denote the set of cornerpoints of Γ_r. Then, for every point $x \in \Gamma_r \backslash S$, the following equation holds true

$$\int_{\Gamma_{ab}} \omega(x,z)\, ds_z = \begin{cases} \dfrac{1}{2}\left(\dfrac{t_r(x) \cdot (b-x)}{\|b-x\|} - \dfrac{t_r(x) \cdot (a-x)}{\|a-x\|} \right) & , \ \text{if } x \notin \Gamma_{ab} \\[2mm] 0 & , \ \text{if } x \in \operatorname{int}\Gamma_{ab}, \end{cases}$$

where $t_r(x)$ is defined by

$$t_r(x) := n_r(x)^\perp = \begin{pmatrix} n_{r,2}(x) \\ -n_{r,1}(x) \end{pmatrix}. \tag{A.2}$$

Proof. Let us first consider the case $x \notin \Gamma_{ab}$. We start with the definition of ω in $\mathbb{R}^2$

$$\omega(x,z) = \frac{[n_r(z) \cdot (x-z)][n_r(x) \cdot (z-x)]}{2\|z-x\|^3} =: \frac{N}{2\,D}. \tag{A.3}$$

Notice that $x \notin S$ implies that $n_{\mathrm{r}}(x)$ is well defined. Due to $z \in \Gamma_{ab}$, it can be expressed by $z = a + s(b - a)$, $0 \leq s \leq 1$. Due to $n_{\mathrm{r}}(z) \cdot (b - a) = 0$, the numerator in (A.3) is equivalent to

$$N = \left[n_{\mathrm{r}}(z) \cdot (x - a) \right] \left[n_{\mathrm{r}}(x) \cdot (b - a) \right] s + \left[n_{\mathrm{r}}(z) \cdot (x - a) \right] \left[n_{\mathrm{r}}(x) \cdot (a - x) \right]$$

$$=: c_1 s + c_2.$$

For the denominator, we obtain

$$D = \left[(b - a)^2 s^2 + 2(a - x) \cdot (b - a) s + (x - a)^2 \right]^{3/2}$$

$$=: (k_1 s^2 + k_2 s + k_3)^{3/2}.$$

We continue with

$$\int_{\Gamma_{ab}} \omega(x, z) \, ds_z = \int_0^1 \frac{c_1 s + c_2}{2(k_1 s^2 + k_2 s + k_3)^{3/2}} \, \|b - a\| \, ds$$

$$= \frac{\|b - a\|}{2} \left(c_1 \int_0^1 \frac{s}{(k_1 s^2 + k_2 s + k_3)^{3/2}} \, ds \right.$$

$$\left. + c_2 \int_0^1 \frac{1}{(k_1 s^2 + k_2 s + k_3)^{3/2}} \, ds \right)$$

$$=: \frac{\|b - a\|}{2} (c_1 I_1 + c_2 I_2). \tag{A.4}$$

Integration via substitution yields for the first integral

$$I_1 = \left[-\frac{2 \, (k_2 s + 2 \, k_3)}{4 \vartheta \sqrt{k_1 s^2 + k_2 s + k_3}} \right]_0^1 = \frac{(x - a)^2}{\vartheta \, \|a - x\|} - \frac{(a - x) \cdot (b - x)}{\vartheta \, \|b - x\|}$$

where ϑ is defined by

$$\vartheta := (b - a)^2 (x - a)^2 - \left[(b - a) \cdot (a - x) \right]^2$$

By straightforward computation, it follows that

$$\vartheta = \left[(b_1 - a_1)(x_2 - a_2) - (b_2 - a_2)(x_1 - a_1) \right]^2 = \|b - a\|^2 \left[n_{\mathrm{r}}(z) \cdot (x - a) \right]^2,$$

since, by assumption, the unit normal on Γ_{ab} is equivalent to $n_{\mathrm{r}}(z) = \|b - a\|^{-1} (b - a)^{\perp}$. For the second integral, one finds

$$I_2 = \left[\frac{2 \, (2 \, k_1 s + k_2)}{4 \vartheta \sqrt{k_1 s^2 + k_2 s + k_3}} \right]_0^1 = -\frac{(b - a) \cdot (a - x)}{\vartheta \, \|a - x\|} + \frac{(b - a) \cdot (b - x)}{\vartheta \, \|b - x\|}$$

with ϑ as defined above. Hence, we obtain for the sum in (A.4)

$$c_1 I_1 + c_2 I_2 = \frac{1}{\|b - a\|^2 \, n_{\mathrm{r}}(z) \cdot (x - a)} \left\{ \frac{S_1}{\|b - x\|} - \frac{S_2}{\|a - x\|} \right\}. \tag{A.5}$$

with

$$S_1 := \big[n_{\mathrm{r}}(x) \cdot (a - x)\big]\big[(b - a) \cdot (b - x)\big] - \big[n_{\mathrm{r}}(x) \cdot (b - a)\big]\big[(a - x) \cdot (b - x)\big]$$

and

$$S_2 := \big[n_{\mathrm{r}}(x) \cdot (a - x)\big]\big[(b - a) \cdot (a - x)\big] - \big[n_{\mathrm{r}}(x) \cdot (b - a)\big](x - a)^2$$

Using the definition of t_{r} in (A.2), S_1 is transformed into

$$\begin{aligned}
S_1 &= \big[n_{\mathrm{r},2}(x)(b_1 - x_1) - n_{\mathrm{r},1}(x)(b_2 - x_2)\big] \\
&\quad \cdot \big[(a_2 - x_2)(b_1 - a_1) - (a_1 - x_1)(b_2 - a_2)\big] \qquad (\mathrm{A.6}) \\
&= \|b - a\| \big[t_{\mathrm{r}}(x) \cdot (b - x)\big]\big[n_{\mathrm{r}}(z) \cdot (x - a)\big].
\end{aligned}$$

Analogously, we find for S_2

$$S_2 = \|b - a\| \big[t_{\mathrm{r}}(x) \cdot (a - x)\big]\big[n_{\mathrm{r}}(z) \cdot (x - a)\big]. \qquad (\mathrm{A.7})$$

Inserting (A.7), (A.6), and (A.5) in (A.4) finally yields

$$\begin{aligned}
\int_{\Gamma_{ab}} \omega(x, z)\, ds_z &= \frac{\|b - a\|}{2}\,(c_1\, I_1 + c_2\, I_2) \\
&= \frac{1}{2}\left(\frac{t_{\mathrm{r}}(x) \cdot (b - x)}{\|b - x\|} - \frac{t_{\mathrm{r}}(x) \cdot (a - x)}{\|a - x\|}\right).
\end{aligned}$$

On the other hand, if $x \in \mathrm{int}\,\Gamma_{ab}$, then we have $n_{\mathrm{r}}(z) \cdot (z - x) = 0$ for all $z \in \Gamma_{ab}$, and hence, $\omega(x, z) = 0$ for every $z \in \Gamma_{ab}$ with $z \neq x$. Now, we partition the integral into one over a neighborhood $N(x)$ of x and one over $\Gamma_{ab} \backslash N(x)$. Due to $\omega(x, z) = 0$ if $z \neq x$, the latter one is clearly equal to zero. Moreover, when passing to the limit $|N(x)| \downarrow 0$, also the integral over $N(x)$ tends to zero, since $x \notin S$ and the singularity of ω is of order $1 - \delta$ on smooth parts of Γ_{r} (see Tiihonen [21]). This finally yields the assertion. $\qquad \square$

References

[1] K.E. ATKINSON AND G. CHANDLER, *The collocation method for solving the radiosity equation for unoccluded surfaces*, J. Int. Eqn. Appl., 10 (1998), pp. 253–290.

[2] M. BERGOUNIOUX, K. ITO, AND K. KUNISCH, *Primal-dual strategy for constrained optimal control problems*, SIAM J. Control and Optimization, 37 (1999), pp. 1176–1194.

[3] M. BERGOUNIOUX AND K. KUNISCH, *Primal-dual strategy for state-constrained optimal control problems*, Computational Optimization and Applications, 22(2002), pp. 193–224.

[4] J. BONNANS, *Second order analysis for control constrained optimal control problems of semilinear elliptic systems*, Appl. Math. Optimization, 38 (1998), pp. 303–325.

[5] J. BONNANS, E. CASAS, *Une principe de Pontryagine pour le contrôle des systèmes semilinéaires elliptiques*, J. Diff. Equations, 90 (1991), pp. 288–303.

[6] E. CASAS, *Pontryagin's principle for optimal control problems governed by semilinear elliptic equations*, International Series of Numerical Mathematics, 118 (1994), pp. 97–114.

[7] E. CASAS, F. TRÖLTZSCH, AND A. UNGER, *Second order sufficient optimality conditions for a nonlinear elliptic control problem* J. for Analysis and its Applications 15 (1996), pp. 687–707.

[8] T.A. DAVIS, *Algorithm 832: UMFPACK – an unsymmetric-pattern multifrontal method with a column pre-ordering strategy*, ACM Trans. Math. Software, 30 (2004), pp. 196–199.

[9] A.L. DONTCHEV, W.W. HAGER, A.B. POORE, AND B. YANG, *Optimality, stability, and convergence in optimal control*, Appl. Math. Optim., 31 (1995), pp. 297–326.

[10] H. GOLDBERG AND F. TRÖLTZSCH, *On a Lagrange-Newton method for a nonlinear parabolic boundary control problem*, Optimization Methods and Software, 8 (1998), pp. 225–247.

[11] H. HINTERMÜLLER AND M. HINZE, *A SQP-semi-smooth Newton-type algorithm applied to control of the instationary Navier-Stokes system subject to control constraints*, submitted to SIAM J. Optimization.

[12] M. HINTERMÜLLER, K. ITO, AND K. KUNISCH, *The primal-dual active set method as a semi-smooth Newton method*, to appear in SIAM J. Control Opt.

[13] O. KLEIN, P. PHILIP, AND J. SPREKELS, *Modeling and simulation of sublimation growth of SiC bulk single crystals*, Interfaces and Free Boundaries, 6 (2004), pp. 295–314.

[14] M. LAITINEN AND T. TIIHONEN, *Conductive-radiative heat transfer in grey materials*, Quart. Appl. Math., 59 (2001), pp. 737–768.

[15] C. MEYER, *Second-order Sufficient Optimality Conditions for a Semilinear Optimal Control Problem with Nonlocal Radiation Interface Conditions*, submitted to ESAIM: Control, Optimisation and Calculus of Variations.

[16] C. MEYER, P. PHILIP, AND F. TRÖLTZSCH, *Optimal control of a semilinear PDE with nonlocal radiation interface conditions*, submitted to SIAM J. Control Opt.

[17] P. PHILIP, *Transient Numerical Simulation of Sublimation Growth of SiC Bulk Single Crystals. Modeling, Finite Volume Method, Results*, PhD thesis, Department of Mathematics, Humboldt University of Berlin, Germany, 2003. Report No. 22, Weierstrass Institute for Applied Analysis and Stochastics, Berlin.

[18] H.-J. ROST, D. SICHE, J. DOLLE, W. EISERBECK, T. MÜLLER, D. SCHULZ, G. WAGNER, AND J. WOLLWEBER, *Influence of different growth parameters and related conditions on 6H-SiC crystals grown by the modified Lely method*, Mater. Sci. Eng. B, 61-62 (1999), pp. 68–72.

[19] Y. SAAD, *SPARSKIT and Sparse Examples*, Numer. Anal. Digest, 94 (1994).

[20] T. TIIHONEN, *A nonlocal problem arising from heat radiation on non-convex surfaces*, Eur. J. App. Math., 8 (1997), pp. 403–416.

[21] ——, *Stefan-Boltzmann radiation on non-convex surfaces*, Math. Meth. in Appl. Sci., 20 (1997), pp. 47–57.

[22] ——, *Finite element approximation of nonlocal heat radiation problems*, Math. Mod. and Meth. in Appl. Sci., 8 (1998), pp. 1071–1089.

[23] F. TRÖLTZSCH, *On the Lagrange-Newton-SQP method for the optimal control of semilinear parabolic equations*, SIAM J. Control Opt., 38 (1999), pp. 294–312.

[24] F. TRÖLTZSCH AND D. WACHSMUTH, *Second-order sufficient optimality conditions for the optimal control of Navier-Stokes equations*, accepted by ESAIM: Control, Optimisation and Calculus of Variations.

[25] F. TRÖLTZSCH AND S. VOLKWEIN, *The SQP-method for control constrained optimal control of the Burgers equation*, ESAIM: Control, Optimisation and Calculus of Variations, 6 (2001), pp. 649-674.

[26] A. UNGER, *Hinreichende Optimalitätsbedingungen 2. Ordnung und Konvergenz des SQP-Verfahrens für semilineare elliptische Randsteuerprobleme*, PhD thesis, Technical University Chemnitz, 1997.

Acknowledgment

The author is very grateful to P. Philip for several helpful discussions.

C. Meyer
Weierstrass Institute for
Applied Analysis and Stochastics
Mohrenstr. 39
D-10117 Berlin, Germany
e-mail: `meyer@wias-berlin.de`

International Series of Numerical Mathematics, Vol. 155, 249–267
© 2007 Birkhäuser Verlag Basel/Switzerland

Shape Optimization for Navier-Stokes Equations

Pavel I. Plotnikov and Jan Sokolowski

Abstract. The minimization of drag functional for the stationary, isothermal, compressible Navier-Stokes equations (N-S-E) in three spatial dimensions is considered. In order to establish the existence of an optimal shape the general result [26] on compactness of families of generalized solutions to N-S-E is applied. The family of generalized solutions to N-S-E is constructed over a family of admissible domains $\mathcal{U}_{ad}$. Any admissible domain $\Omega = B \setminus S$ contains an obstacle S, e.g., a wing profile. Compactness properties of the family of admissible domains are imposed. It turns out that we require the compactness of the family of admissible domains with respect to the Hausdorff metrics as well as in the sense of Kuratowski-Mosco. The analysis is performed for the range of adiabatic ratio $\gamma > 1$ in the pressure law $p(\rho) = \rho^\gamma$ and it is based on the technique proposed in [24] for the discretized N-S-E.

Mathematics Subject Classification (2000). Primary 35D05, 35Q30, 49Q10; Secondary 49Q20, 76N25.

Keywords. Compressible fluids, Navier-Stokes equations, Generalized solutions, Shape optimization, Drag minimization.

1. Introduction

From the point of view of applications, the shape optimization for N-S-E is a new and important issue. The main difficulty, in solution of such problems, is the lack of existence results for the nonlinear PDE's in the case of compressible fluids or gases. To be more precise, it seems that there are no general results, e.g., for the boundary value problems with the nonhomogeneous boundary conditions. Therefore, it is an interesting and difficult subject of current research. We refer the reader to monographs [17], [11], [19], [21] for the modern theory of nonlinear PDE's applied to N-S-E, which we apply in the paper for a specific problem. There

---

This work was completed during a stay of the first author at the Institute Elie Cartan of the University Henri Poincaré Nancy I.

is a real difference in complexity of analysis, as it can be seen in the paper, between linear and nonlinear problems from the point of view of shape optimization. Therefore, the framework introduced in the monographs [17], [11], [19], [21], is necessary for mathematical analysis of shape optimization of the drag functional $\mathbf{J}(\rho, \mathbf{u}, \Omega)$ for stationary, isothermal, compressible N-S-E. However, the precise results, we present here, are not included in the monographs, and are new to our best knowledge.

We are going to study the convergence of minimizing sequence $\{\rho_n, \mathbf{u}_n, \Omega_n\}_{n \geq 1}$ for the shape functional $J(\Omega) := \mathbf{J}(\rho, \mathbf{u}, \Omega)$ defined for generalized solutions $(\rho, \mathbf{u})$ of N-S-E. Our goal is to obtain an existence result for the class of shape optimization problems.

1.1. Shape optimization problems for N-S-E

In the present paper a class of shape optimization problems for stationary, isothermal, compressible N-S-E is considered. The model problem of shape optimization can be described as follows.

For a given domain $\Omega = B \setminus S$ find the velocity field $\mathbf{u}$, the density ρ, and the pressure $p = \rho^\gamma$ by solving the boundary value problem for nonlinear PDE's. The range $\gamma > 1$ of adiabatic ratio is required in the paper in view of potential applications.

Let $(\mathbf{u}, \rho)$ be a generalized solution to the boundary value problem posed in the geometrical domain $\Omega = B \setminus S$, where $B \subset \mathbb{R}^3$ is a fixed hold all domain and S is an obstacle.

$$-\nu \Delta \mathbf{u} - \xi \nabla \operatorname{div} \mathbf{u} + \rho \mathbf{u} \nabla \mathbf{u} + \nabla p(\rho) = \rho \mathbf{f} , \quad \operatorname{div}(\rho \mathbf{u}) = 0 ,$$

$$\mathbf{u} = 0 \text{ on } \partial S , \quad \mathbf{u} = \mathbf{U}^\infty \text{ on } \partial B ,$$

$$\rho = \rho^\infty \text{ on } \Sigma^+ = \{x \in \partial B : \mathbf{U}^\infty \cdot \mathbf{n}(x) < 0\} .$$

If there exist generalized solutions to the equations, next step is to minimize the integral cost functional $J(\Omega) := \mathbf{J}(\rho, \mathbf{u}, \Omega)$, with respect to the obstacle S within the family of admissible domains $\mathcal{U}_{ad}$, where

$$\mathbf{J}(\rho, \mathbf{u}, \Omega) = \int_\Omega (\Pi - \rho \mathbf{u} \otimes \mathbf{u} - p(\rho)\mathbf{I}) : \nabla \mathbf{u}^\infty dx + \int_\Omega (\mathbf{U}^\infty - \mathbf{u}^\infty) \cdot \mathbf{f}\rho dx .$$

We require that the set of admissible domains $\mathcal{U}_{ad}$ enjoys the following compactness conditions:

For any sequence of admissible domains $\{\Omega_n\}_{n \geq 1} \subset \mathcal{U}_{ad}$, there is a subsequence still denoted by $\{\Omega_n\}_{n \geq 1}$, $\Omega_n = B \setminus S_n \in \mathcal{U}_{ad}$, and a domain $\Omega \subset \mathcal{U}_{ad}$, such that

- Volumes of the obstacles are bounded from below

$$|S_n| = \operatorname{meas}(S_n) \geq \operatorname{Vol} ,$$

  where Vol is a given constant.
- The sequence $\{\Omega_n\}$ converges to Ω in the Hausdorff metrics.
- The sequence $\{\Omega_n\}$ converges to Ω in the sense of Kuratowski-Mosco.

The first condition is prescribed in order to assure that the optimal domains contains an obstacle, e.g., the wing profile. The second and the third conditions are sufficient for passage to the limit in nonlinear equations. Below, we provide more precise definitions of the required compactness properties. The main result of this work is the following existence theorem for the class of shape optimization problems.

Theorem 1.1. *Assume that the family of admissible shapes is compact in the sense described above, and that the set of generalized solutions to (1.1) for $\Omega = B \setminus S \in \mathcal{U}_{ad}$ is nonempty. Then there is at least one solution $\Omega^* = B \setminus S^* \in \mathcal{U}_{ad}$ for the shape optimization problem of the drag minimization:*

$$J(\Omega^*) := \mathbf{J}(\rho^*, \mathbf{u}^*, \Omega^*) \leq J(\Omega) := \mathbf{J}(\rho, \mathbf{u}, \Omega)$$

for all admissible domains $\Omega \in \mathcal{U}_{ad}$. The pair $(\rho^*, \mathbf{u}^*)$ is a generalized solution in Ω^*.

We obtain the existence result without any regularity assumptions on the boundaries of obstacles. In fact, we make use of a compactness result [26] obtained for the family of solutions to the N-S-E, so that the existence of optimal shapes is a simple corollary of the theorem on the compactness of the set of generalized solutions.

1.2. Mathematical analysis of shape optimization problems

Usually the mathematical analysis of shape optimization problems includes:

- The proof of existence of optimal shapes for a sufficiently large class of admissible domains.
- Derivation of necessary optimality conditions which characterize an optimal domain and can be used for numerical solution of the shape optimization problem.
- The convergence proof for numerical method which can be used to evaluate an optimal shape.

To our best knowledge, only the first point is studied in the literature for the compressible N-S-E and drag minimization. We refer the reader to [10] for the existence results for shape optimization problems in the case of evolution equations for the adiabatic ratio $\gamma > 3/2$, see also [9] for the related results. In the present paper the case of $\gamma > 1$ is considered in three spatial dimensions. The technique used, is introduced in [23], [24]. The optimality conditions for drag minimization are derived in [27].

We restrict ourselves to the first issue on the above list of problems to be solved, which is already quite difficult, since there are no existence results for the PDE's model itself in the range of parameters we are going to consider, i.e., for the adiabatic ratio $\gamma > 1$ in the law $p(\rho) = \rho^\gamma$ which gives the pressure p of the fluid in function of its density ρ in the isothermal regime. We also point out, that the existence of generalized solutions for N-S-E equations with nonhomogeneous boundary conditions for full range of parameters is an open problem. For some results in this direction, in two spatial dimensions and for the special geometry of flow region, we refer the reader to [14]–[16].

We show that, under appropriate assumptions, the convergence of the sequence of admissible domains implies the convergence of the associated sequence of shape functionals. To make our analysis representative we consider the drag functional, which is the standard choice in view of possible applications, e.g., in the shape optimization of a wing.

A related shape optimization problem is considered in [10] for evolution equations, and with the adiabatic ratio $\gamma > 3/2$. We consider the case of $\gamma > 1$ using the same technique as in [24]. The novelty of our results, in comparison with those of paper [24], is the PDE's model. We consider here the stationary problem. In [24] the discretized problem introduced in [17] is considered, such a problem depends on the parameter $\alpha > 0$ of time discretization of evolution equations [17], the limit case of $\alpha = 0$ becomes the stationary problem. It is clear, that the analysis of the stationary problem is more involved, compared to the analysis of more *regular* discretized problem.

1.3. Generalized solutions of N-S-E in three spatial dimensions

Suppose that compressible Newtonian fluid occupies the bounded region $\Omega \subset \mathbb{R}^3$. We will assume that $\Omega = B \setminus S$, where B is an open convex sufficiently large hold all containing inside a compact obstacle S. We could take, e.g., for B a ball of radius R, $B = \{x | |x| \le R\}$. We do not impose restrictions on the topology of the flow region. The cases of S with a finite number of connected components or $S = \emptyset$ are taken into consideration. The fluid density $\rho : \Omega \mapsto \mathbb{R}^+$ and the velocity field $\mathbf{u} : \Omega \mapsto \mathbb{R}^3$ are governed by the Navier-Stokes equations

$$-\nu \Delta \mathbf{u} - \xi \nabla \operatorname{div} \mathbf{u} + \rho \mathbf{u} \nabla \mathbf{u} + \nabla p(\rho) = \rho \mathbf{f} , \quad \operatorname{div}(\rho \mathbf{u}) = 0 ,$$

where ν, ξ are positive viscous coefficients and $\mathbf{f} : \Omega \mapsto \mathbb{R}^3$ is a given continuous vector field. We suppose that the flow is barotropic and $p(\rho) = \rho^\gamma$ with the adiabatic ratio $\gamma > 1$. If the viscous stress tensor is defined by the equality

$$\Pi = \nu(\nabla \mathbf{u} + \nabla \mathbf{u}^\top) + (\xi - \nu)\operatorname{div} \mathbf{u} \, \mathbf{I} , \tag{1.2}$$

then the governing equations can be written in the equivalent divergence form

$$\operatorname{div}(\rho \mathbf{u} \otimes \mathbf{u}) + \nabla p(\rho) - \rho \mathbf{f} = \operatorname{div} \Pi, \quad \operatorname{div}(\rho \mathbf{u}) = 0 \text{ in } \Omega . \tag{1.3a}$$

In view of possible applications, e.g., to the shape optimization problem of a wing, it is supposed that the velocity field satisfies the non-homogeneous boundary condition

$$\mathbf{u} = 0 \text{ on } \partial S , \quad \mathbf{u} = \mathbf{U}^\infty \text{ on } \partial B , \tag{1.3b}$$

and the density distribution is prescribed on the entrance set

$$\rho = \rho^\infty \text{ on } \Sigma^+ = \{x \in \partial B : \mathbf{U}^\infty \cdot \mathbf{n}(x) < 0\} . \tag{1.3c}$$

Here $\mathbf{n}$ is the outward unit normal vector to $\partial \Omega$. It is assumed that $\mathbf{U}^\infty \in \mathbb{R}^3$ is a given vector, and $\rho^\infty \in L_\infty(\Sigma^+)$ is a given non-negative constant.

Boundary condition (1.1) can be written in the form of the equality $\mathbf{u} = \mathbf{u}^\infty$ on $\partial\Omega$, where $\mathbf{u}^\infty(x)$ is a smooth function defined for any $x \in \mathbb{R}^3$, which vanishes in the vicinity of S and coincides with $\mathbf{U}^\infty$ in an open neighborhood of ∂B. The physical quantities which characterize the flow include the total energy $\mathbf{E}$, the volume rate of energy dissipation $\mathbf{D}$ and the drag $\mathbf{J}$, and are defined by

$$\mathbf{E} = \int_\Omega \left(\frac{1}{2}\rho|\mathbf{u}|^2 + \frac{\rho^\gamma}{\gamma - 1} \right) dx \ , \quad \mathbf{D} = \int_\Omega \left(\nu|\nabla\mathbf{u}|^2 + \xi|\mathrm{div}\,\mathbf{u}|^2 \right) dx \ , \tag{1.4}$$

$$\mathbf{J} = -\mathbf{U}^\infty \cdot \int_{\partial S} (\Pi - p(\rho)\mathbf{I}) \cdot \mathbf{n}\, dS \ .$$

The drag $\mathbf{J}$ accounts for the reaction of the surrounding fluid on the obstacle S. For our purposes, the formula for the drag can be written in the equivalent form

$$\mathbf{J}(\rho, \mathbf{u}, \Omega) = \int_\Omega (\Pi - \rho\mathbf{u} \otimes \mathbf{u} - p(\rho)\mathbf{I}) : \nabla\mathbf{u}^\infty dx + \int_\Omega (\mathbf{U}^\infty - \mathbf{u}^\infty) \cdot \mathbf{f}\rho dx \ . \tag{1.5}$$

We will consider the physically reasonable solutions to problems (1.3) and for which the density is non-negative and the total energies are bounded from above by some positive constant $\mathbf{E}$. In what follows we will denote by c various constants depending only on $\mathbf{E}$, data $\|\mathbf{f}\|_{C(B)}$, ρ^∞, $\|\mathbf{u}^\infty\|_{C^1(B)}$, material constants γ, ν, ξ, and the domain B.

In the paper the standard notation is used for the function spaces. The space $H^{1,r}(\Omega)$ is the Sobolev space of functions integrable along with the first order generalized derivatives in $L_r(\Omega)$ equipped with its natural norm. For $r = 2$ we use the notation $H^{1,2}(\Omega)$ rather than $H^1(\Omega)$; the notation $H_0^{1,r}(\Omega)$ stands for the closure of $C_0^\infty(\Omega)$ in the norm of $H^{1,r}(\Omega)$.

Definition 1.2. For given $\mathbf{U}^\infty \in \mathbb{R}^3$ and $\mathbf{f} \in C(\Omega)^3$ a generalized solution to problem (1.3) is the pair $(\rho, \mathbf{u})$, where $\rho \in L^\gamma(\Omega)$ is a non-negative function in Ω and $\mathbf{u} - \mathbf{u}^\infty \in H_0^{1,2}(\Omega)$, which satisfies the following conditions:

(a) The scalar function $\rho|\mathbf{u}|^2$ is integrable in Ω, i.e., the total energy $\mathbf{E}$ of the flow is finite. The mass density and the velocity field satisfy the energy inequality

$$\nu\|\nabla(\mathbf{u} - \mathbf{u}^\infty)\|_{L^2(\Omega)}^2 + \xi\|\mathrm{div}\,(\mathbf{u} - \mathbf{u}^\infty)\|_{L^2(\Omega)}^2 + \int_\Omega \rho\mathbf{u} \otimes \mathbf{u} : \nabla\mathbf{u}^\infty\, dx +$$

$$\int_\Omega p(\rho)\mathrm{div}\,\mathbf{u}^\infty\, dx - \int_\Omega \rho\mathbf{f} \cdot (\mathbf{u} - \mathbf{u}^\infty)\, dx + \frac{1}{\gamma - 1}\int_{\Gamma^+} (\rho^\infty)^\gamma(\mathbf{U}^\infty \cdot \mathbf{n})\, d\Gamma \leq 0 \tag{1.6}$$

(b) For all vector fields $\varphi \in C_0^1(\Omega)^3$,

$$\int_\Omega \left(\rho\mathbf{u} \otimes \mathbf{u} + p(\rho) \right) : \nabla\varphi\, dx + \int_\Omega \rho\mathbf{f} \cdot \varphi\, dx = \int_\Omega \Pi : \nabla\varphi\, dx. \tag{1.7a}$$

(c) The integral identity

$$\int_{\Omega} \Big(G(\rho)\mathbf{u} \cdot \nabla\psi + \big(G(\rho) - G'(\rho)\rho\big)\psi \mathrm{div}\,\mathbf{u} \Big)\, dx + \int_{\Sigma^+} \psi G(\rho^\infty)\mathbf{U}^\infty \cdot \mathbf{n} d\Sigma = 0,$$

$$(1.7\mathrm{b})$$

holds and any functions $\psi \in C^1(\Omega)$ vanishing on $\Sigma^- = \partial B \setminus \Sigma^+$, and any function $G \in C^1_{\mathrm{loc}}[0, \infty)$ with the properties

$$\limsup_{r \to \infty} |G(r)|/r < \infty, \quad [0, \infty) \ni r \mapsto G(r) - G'(r)r \in \mathbb{R} \text{ continuous and bounded.}$$

Condition (c) of the above definition means that we consider the renormalized weak solutions of the stationary problem, see [11] for a discussion on renormalized solutions. Such definition simplifies the further analysis without any loss of generality.

Remark 1.3. Denote by $\Sigma^- = \partial B \setminus \Sigma^+$ the *exit* part of the boundary ∂B. It follows from the definition of generalized solutions that the extensions of the density and of the velocity vector field onto the domain $\mathbb{R}^3 \setminus \Sigma^-$, given by the equalities

$$\rho(x) = 0 \text{ in } S, \quad \rho(x) = \rho^\infty \text{ in } \mathbb{R}^3 \setminus (B \cup \Sigma^-),$$

$$\mathbf{u} = 0 \text{ in } S, \quad \mathbf{u} = \mathbf{U}^\infty \text{ in } \mathbb{R}^3 \setminus (B \cup \Sigma^-),$$

$$(1.8)$$

satisfy the integral identity

$$\int_{\mathbb{R}^3} \Big(G(\rho)\mathbf{u} \cdot \nabla\psi + \big(G(\rho) - G'(\rho)\rho\big)\psi \mathrm{div}\,\mathbf{u} \Big)\, dx = 0 \qquad (1.9)$$

for any functions $\psi \in C^1_0(\mathbb{R}^3)$ vanishing near Σ^-.

1.4. Continuity of shape functionals

The cost functional for shape optimization problems is the drag $\mathbf{J}(\Omega, \mathbf{u}, \mu_\rho)$ defined by formula (1.5). In applications, the drag is usually minimized within the class of admissible shapes. To our best knowledge there are no results on the shape optimization problem in the framework of generalized solutions to stationary problems for the adiabatic ratio $\gamma > 1$, the case of evolution equations for adiabatic ratio $\gamma > 3/2$ is considered in [10].

The drag depends on the solution $(\rho, \mathbf{u})$ to problem (1.3), however such a solution, if it does exist, it is not in general unique. We point out, that the existence of solutions for the adiabatic ratio $\gamma \geq 1$ in three spatial dimensions is in general an open and difficult problem [18]. The case of discretized problems is considered in [24] for $\gamma \geq 1$, however no dependence of solutions on geometrical domains is considered in [24]. On the other hand, the case of drag minimization in two spatial dimensions is studied in [23]. Furthermore, the drag depends on an admissible shape of the obstacle S. The dependence of the drag on the admissible shapes is twofold, first, it depends directly on Ω since the integrals in (1.5) are defined over Ω, and it depends on the generalized solutions to N-S-E defined in Ω. The restrictions on the shapes of admissible obstacles S are defined in such a way that the set of admissible shapes and of the associated generalized solutions is compact. The precise conditions for admissible shapes are established below in the form of condition ($\mathfrak{N}$). In the present paper we do not derive the necessary optimality conditions for the problem of drag minimization, we prove only the compactness of the set of solutions over the set of admissible shapes. The compactness property leads to the existence of an optimal shape provided the set of generalized solutions is nonempty. The optimality conditions for the drag minimization problem are obtained in forthcoming paper [27] for a more restricted class of N-S-E.

We are now in a position to formulate the main result of the paper.

Suppose that a sequence of flow domains $\Omega_n = B \setminus S_n$ satisfies the following three conditions, we refer to as condition ($\mathfrak{N}$):

Condition ($\mathfrak{N}$)

- There is a compact $K_S \Subset B$ such that $\cup_n S_n \subset K_S$.
- If a compact set $K \Subset \Omega$, then $K \Subset \Omega_n$ for all large n.
- If $w_n \to w$ weakly in $H^{1,2}(B)$ and $w_n \in H_0^{1,2}(\Omega_n)$, then $w \in H_0^{1,2}(\Omega)$.

In this case we will write

$$\Omega_n \xrightarrow{\mathfrak{N}} \Omega \ . \tag{1.10}$$

It is easy to see that the convergence of the sequence $\{\Omega_n\}$ both in the Hausdorff metrics and in the sense of Kuratowski-Mosco implies all the three conditions listed in condition ($\mathfrak{N}$).

We deal with the system of PDE's which is not elliptic. Therefore, we need stronger conditions on the convergence of geometrical domains, compared to the classical Kuratowski-Mosco convergence, which is adapted to the elliptic case. In particular, our conditions are more restrictive when compared to the Kuratowski-Mosco convergence of domains Ω_n.

In order to prove Theorem 1.1 let us consider the minimizing sequence $\{\Omega_n\}$ for the shape functional $J(\Omega)$.

The main result of the paper is the existence of the limit for such a sequence, and it is based on the following compactness result for generalized solutions to N-S-E proved in [26].

Assume that there is a sequence $\{\Omega_n\}$ of domains which converges $\Omega_n \xrightarrow{\mathfrak{N}} \Omega$ and a sequence of generalized solutions $\{(\rho_n, \mathbf{u}_n)\}$ to compressible N-S-E

$$\operatorname{div}(\rho_n \mathbf{u}_n \otimes \mathbf{u}_n) + \nabla p(\rho_n) - \rho \mathbf{f} = \operatorname{div} \Pi, \quad \operatorname{div}(\rho_n \mathbf{u}_n) = 0 \ \text{in} \ \Omega_n \ , \tag{1.11a}$$

$$\mathbf{u}_n = 0 \ \text{on} \ \partial S_n \ , \quad \mathbf{u}_n = \mathbf{U}^\infty \ \text{on} \ \partial B \ , \tag{1.11b}$$

$$\rho_n = \rho^\infty \ \text{on} \ \Sigma^+ = \{x \in \partial B : \mathbf{U}^\infty \cdot \mathbf{n}(x) < 0\} \ . \tag{1.11c}$$

Suppose also that the total energies of the sequence $(\rho_n, \mathbf{u}_n)$ of generalized solutions to problem (1.11) are uniformly bounded by a constant c,

$$\mathbf{E}_n = \int_{\Omega_n} \left(\frac{1}{2}\rho|\mathbf{u}_n|^2 + \frac{1}{\gamma - 1}\rho_n^\gamma \right) dx \leq c \ . \tag{1.12}$$

Then there is a subsequence of the sequence $\{\rho_n, \mathbf{u}_n\}$, still denoted by $\{\rho_n, \mathbf{u}_n\}$, such that for any $r < \gamma$,

$$\rho_n \to \rho \ \text{in} \ L^r(B), \quad p(\rho_n) \to p(\rho) \ \text{in} \ L^1_{\mathrm{loc}}(\Omega), \tag{1.13}$$

$$\mathbf{u}_n \to \mathbf{u} \ \text{weakly in} \ H^{1,2}(B), \quad \mathbf{u} - \mathbf{u}^\infty \in H^{1,2}_0(\Omega) \ .$$

The pair of functions $(\rho, \mathbf{u})$ serves as a generalized solution to problem (1.3) for the limit domain Ω, furthermore, the shape functionals converge for $n \to \infty$,

$$\mathbf{J}(\rho_n, \mathbf{u}_n, \Omega_n) \to \mathbf{J}(\rho, \mathbf{u}, \Omega) \ . \tag{1.14}$$

The compactness result leads to the existence of optimal shapes provided that the set of generalized solutions is nonempty for the minimizing sequence of admissible shapes. The set of solutions is nonempty, in particular for sufficiently small data, since in such a case the existence of local solutions is proved at least for the homogeneous boundary conditions. We refer also to [14]-[16] for the existence results in two spatial dimensions with nonhomogeneous boundary conditions.

The convergences (1.13) and (1.14) are [proved in several steps, we refer the reader for complete analysis to [26]. Here, we provide the main lines of the proof which is quite complex. The case of two-dimensional spatial domains is treated in [23], for the adiabatic ratio $\gamma \geq 1$, where the existence of optimal shapes of obstacles is shown.

2. A priori estimates for generalized solutions

In this section we present the following theorem on local integrability of generalized solutions, which is interesting on its own.

Theorem 2.1. *Let $(\rho, \mathbf{u})$ be a generalized solution to problem (1.3) and Ω' be a subdomain of Ω with dist $(\Omega', \partial\Omega) > d > 0$. Then for $\kappa = 2(\gamma - 1)/(\gamma + 2) > 0$,*

$$\|\rho \mathbf{u}^2\|_{L^{1+\kappa}(\Omega')} \leq cd^{-1}, \quad \|\rho\|_{L^{\gamma(1+\kappa)}(\Omega')} \leq cN(\Omega') \ . \tag{2.1}$$

where the constant N only depends on Ω'.

The proof of Theorem 2.1 is based on following lemmas. The first lemma gives the estimate of the rate of energy dissipation in terms of the total energy of the fluid.

Lemma 2.2. *Under the assumptions of Theorem 2.1 the velocity vector field is bounded*

$$\|\mathbf{u}\|_{H^{1,2}(\Omega)} \leq c \ , \tag{2.2}$$

where the constant c only depends on the data of the boundary value problem.

Choose a domain Ω_0 with a smooth boundary so that

$$\Omega' \Subset \Omega_0 \Subset \Omega, \quad \text{dist} \ (\partial\Omega, \Omega_0) \geq d/3, \quad \text{dist} \ (\partial\Omega_0, \Omega')) \geq d/3 \ .$$

The second lemma shows that the Newtonian potential of the pressure is uniformly bounded on Ω_0.

Lemma 2.3. *Under the assumptions of Theorem 2.1,*

$$\operatorname*{ess\,sup}_{x\in\Omega_0} \int_\Omega \frac{\rho^\gamma(y)}{|x-y|}\,dy \leq cd^{-1}. \tag{2.3}$$

The third lemma shows that the energy is bounded.

Lemma 2.4. *Under the assumptions of Theorem 2.1,*

$$\int_{\Omega_0} \rho^\gamma |\mathbf{u}-\mathbf{u}^\infty|^2 dx \leq cd^{-1}. \tag{2.4}$$

3. Weak convergence

Since the notion of weak limits plays the crucial role in our analysis, we begin with short description of some basic facts concerning weak convergence and weak compactness. We refer the reader to, e.g., [28] for the proofs of basic results.

Let A be an arbitrary bounded, measurable subset of $\mathbb{R}^3$ and $1 < r \leq \infty$. Then for every bounded sequence $\{g_n\}_{n\geq 1} \subset L^r(A)$ there exist a subsequence, still denoted by $\{g_n\}$, and a function $g \in L^r(A)$, such that for $n \to \infty$,

$$\int_A g_n(x)h(x)dx \to \int_A g(x)h(x)dx \text{ for all } h \in L^{r/(r-1)}(A).$$

We say the sequence converges $g_n \to g$ weakly in $L^r(A)$ for $r < \infty$, and converges star-weakly in $L^\infty(A)$ in the limit case of $r = \infty$. In very special case of $r = 1$ it is known that the sequence of g_n contains a weakly convergent subsequence in $L^1(A)$, if and only if there is a continuous function $\Phi : \mathbb{R} \to \mathbb{R}^+$ such that

$$\lim_{s\to\infty} \Phi(s)/s = \infty \text{ and } \sup_{n\geq 1} \|\Phi(g_n)\|_{L^1(A)} < \infty.$$

If the sequence of g_n is only bounded in $L^1(A)$ and A is open, then after passing to a subsequence we can assume that g_n converges star-weakly to a bounded Radon measure measure μ_g, i.e.,

$$\lim_{n\to\infty} \int_A g_n(x)h(x)dx = \int_A h(x)d\mu_g(x) \text{ for all compactly supported } h \in C(A).$$

In the sequel, the linear space of compactly supported functions on a set A is denoted by $C_0(A)$, and its dual by $C_0(A)^*$.

The Ball's version [3] of the fundamental Tartar Theorem on Young measures gives a simple and effective representation of weak limits in the form of integrals over families of probabilities measures. The following lemma is a consequence of Ball's theorem.

Lemma 3.1. *Suppose that a sequence $\{g_n\}_{n\geq 1}$ is bounded in $L^r(A)$, $1 \leq r \leq \infty$, where A is an open, bounded subset of $\mathbb{R}^k$. Then we have the following characterizations of weak limits.*

(i) *There exists a subsequence, still denoted by $\{g_n\}_{n\geq 1}$, and a family of probability measures $\sigma_x \in C_0(\mathbb{R})^*$, $x \in A$, with a measurable distribution function $\Gamma(x,\lambda) := \sigma_x(-\infty, \lambda]$. The function $\lambda \mapsto \Gamma(x,\lambda)$ is monotone and continuous from the right, and admits the limits $1, 0$ for $\lambda \to \pm\infty$, respectively. Furthermore, for any continuous function $G : A \times \mathbb{R} \mapsto \mathbb{R}$ such that*

$$\lim_{|\lambda|\to\infty} \|G(\cdot,\lambda)\|_{C(A)}/|\lambda|^r = 0 \text{ for } r < \infty$$

$$\text{and } \sup_{|\lambda|} \|G(\cdot,\lambda)\|_{C(A)} < \infty \text{ for } r = \infty,$$

the sequence of $G(\cdot, g_n)$ converges weakly in $L^1(A)$ to a function

$$\overline{G}(x) = \int_{\mathbb{R}} G(x,\lambda)d_\lambda\Gamma(x,\lambda). \tag{3.1}$$

Moreover, the function

$$A \ni x \to \int_{\mathbb{R}} |\lambda|^r \, d_\lambda\Gamma(x,\lambda) \in \mathbb{R}$$

belongs to $L^1(A)$.

(ii) *If $G(x,\cdot)$ is convex and the sequence g_n converges weakly (star-weakly for $r = \infty$) to $g \in L^r(A)$, then $\overline{G}(x) \leq G(x,g(x))$. If the functions g_n satisfy the inequalities $g_n \leq M$ (resp. $g_n \geq m$), then $\Gamma(x,\lambda) = 1$ for $\lambda \geq M$ (resp. $\Gamma(x,\lambda) = 0$ for $\lambda < m$).*

(iii) *If $\Gamma(1 - \Gamma) = 0$ a.e. in A, then the sequence g_n converges to g in measure, and hence in $L^s(A)$ for positive $s < r$. Moreover, in this case $\Gamma(x,\lambda) = 0$ for $\lambda < g(x)$ and $\Gamma(x,\lambda) = 1$ for $\lambda \geq g(x)$*

Let us consider the sequence of generalized solutions $\{(\rho_n, \mathbf{u}_n)\}_{n\geq 1}$ to problem (1.3). Assume that the functions $(\rho_n, \mathbf{u}_n)$ are extended onto $\mathbb{R}^3$ by formulae (1.8), and fix an arbitrary bounded smooth domain D with $B \Subset D$. For such extended functions, by inequalities (1.12) and formulae (1.8), it follows that the sequence $(\rho_n, \mathbf{u}_n)$ contains a subsequence, still denoted by $(\rho_n, \mathbf{u}_n)$, such that

$$\rho_n \to \rho \text{ weakly in } L^\gamma(D), \quad \mathbf{u}_n \to \mathbf{u} \text{ weakly in } H^{1,2}(D),$$

$$\rho_n \to \rho \text{ weakly in } L^{(1+\kappa)\gamma}(\Omega') \text{ for all } \Omega' \Subset \Omega. \tag{3.2}$$

The behavior of the functions $p(\rho_n)$ is more complicated. Since they are uniformly bounded in $L^1(D)$, we can assume, after passing to a subsequence if necessary, that $p(\rho_n)$ converge weakly to some finite Borel measure μ_p on D. On the other hand, the sequence $\{p(\rho_n)\}$ is bounded in $L^{1+\kappa}(K)$ for any compact $K \Subset \Omega$. Using the diagonal process we obtain the existence of a subsequence which converges weakly on each compact $K \Subset \Omega$ to some function $\overline{p} \in L^{1+\kappa}_{\text{loc}}(\Omega)$. Since

$$\|\overline{p}\|_{L^1(K)} \leq \liminf_{n \to \infty} \|p(\rho_n)\|_{L^1(K)} \leq c$$

the function $\overline{p}$ is integrable over Ω. Next, we note that by construction the functions $p(\rho_n) = (\rho^\infty)^\gamma$ are bounded and independent of n on $D \setminus B$. From this we conclude that the extended function $\overline{p}(x), x \in D$, defined as follows

$$\overline{p}(x) = \overline{p}(x) \text{ in } \Omega, \quad \overline{p}(x) = (\rho^\infty)^\gamma \text{ in } D \setminus B,$$

belongs to the class $L^1(D \setminus S) \cap L^{1+\kappa}_{\text{loc}}(\Omega) \cap L^\infty(D \setminus B)$, which implies the equality

$$\int_D h(x) d\mu_p = \int_{D \setminus S} h(x)\overline{p}(x) dx + \int_{\partial B} h(x) d\mu_p + \int_S h(x) d\mu_p \text{ for all } h \in C_0(\mathbb{R}^3),$$

where the compact *obstacle* takes the form $S = B \setminus \Omega$. Applying Lemma 3.2 to the sequence of $g_n := \rho_n$ and to the sets $A = D, \Omega'$ leads to the following result on the representation of weak limits.

Lemma 3.2. *There exists a subsequence of the sequence $\{\rho_n, \mathbf{u}_n\}$, still denoted by $\{\rho_n, \mathbf{u}_n\}$, and a distribution function $\Gamma : D \times \mathbb{R} \mapsto [0, 1]$ such that*

(i) *$\Gamma(x, \lambda)$ meets all requirements of Lemma 3.1 and satisfies the equalities*

$$\Gamma(x, \lambda) = 0 \text{ for } \lambda < 0 \text{ a.e. in } D,$$

$$\Gamma(x, \lambda) = 0 \text{ for } \lambda < \rho^\infty(x), \quad \Gamma(x, \lambda) = 1 \text{ for } \lambda \geq \rho^\infty(x) \text{ a.e. in } D \setminus B.$$

(ii) *For any continuous function $G : D \times \mathbb{R}$ such that $\lim_{\rho \to \infty} \rho^{-\gamma} \|G(\cdot, \rho)\|_{C(D)} = 0$, the sequence $G(\cdot, \rho_n)$ converges weakly in $L^1(D)$ to the function,*

$$\overline{G}(x) = \int_{[0,\infty)} G(x, \lambda) \, d_\lambda \Gamma(x, \lambda) \text{ a.e. in } D. \tag{3.3}$$

In particular, the weak limit of ρ_n takes the form

$$\rho(x) = \int_{[0,\infty)} \lambda d\Gamma(x, \lambda) \equiv \int_{[0,\infty)} (1 - \Gamma(x, \lambda)) \, d\lambda \text{ a.e. in } D. \tag{3.4}$$

(iii) *The function $\overline{p}$ admits the representation*

$$\overline{p}(x) = \int_{[0,\infty)} \lambda^\gamma \, d_\lambda \Gamma(x, \lambda) \equiv \gamma \int_{[0,\infty)} \lambda^{\gamma-1}(1 - \Gamma(x, \lambda)) \, d\lambda \text{ a.e. in } D \setminus S.$$
$$\tag{3.5}$$

Since the embedding $H^{1,2}_0(D) \hookrightarrow L^r(D)$ is compact for $r < 6$, we can expect that $\rho_n \mathbf{u}_n$ converge weakly to $\rho \mathbf{u}$. The corresponding result is given by the following lemma.

Lemma 3.3. *For $\iota = (\gamma - 1)/(\gamma + 1) > 0$ and for any $\Omega' \Subset \Omega$,*

$$\rho_n \mathbf{u}_n \to \rho \mathbf{u} \text{ converges weakly in } L^{1+\iota}(D)^3,$$

$$\rho_n \mathbf{u}_n \otimes \mathbf{u}_n \to \rho \mathbf{u} \otimes \mathbf{u} \text{ weakly in } L^{1+\kappa}(\Omega')^9.$$

4. The effective viscous flux

Following [17] we introduce the quantity

$$V(\rho, \mathbf{u}) = p(\rho) - (\xi + \nu)\operatorname{div}\mathbf{u},$$

which is called the effective viscous flux. As it was shown in [17, 6, 8] the effective viscous flux enjoys many remarkable properties. The most important is the multiplicative relation

$$\overline{\varphi(\rho)V} = \overline{\varphi(\rho)}\ \overline{V}$$

for weak limits, which was proved in [17] for all $\gamma > 3/2$. The simple proof of this result, based on the new version of compensated compactness principle, was given in papers [6, 8]. In our case, by Theorem 2.1, the critical estimate $\||\rho_n|\mathbf{u}_n|^2\|_{L^{(1+\kappa)}(\Omega')} \le c(\Omega')$ holds for every $\Omega' \Subset \Omega$, which leads to the following local version of the compensated compactness result from [8].

Lemma 4.1. *Let there be given functions* $h \in C_0^\infty(\Omega)$, *and* $\varphi \in C^\infty(\mathbb{R}^+)$. *Then*

$$\int_\Omega h(x)\overline{\varphi V(\rho, \mathbf{u})}dx = \int_\Omega h(x)\overline{\varphi}\ \overline{V}dx, \ \text{where } \overline{V} = \overline{p} - (2 + \nu)\operatorname{div}\mathbf{u}. \tag{4.1}$$

Corollary 4.2. *Assume that the function* $\lambda \mapsto \varphi(\lambda)$ *belongs to the class* $C^\infty(\mathbb{R})$ *and vanishes for sufficiently large* λ. *Let* $\overline{\varphi p} \in L^\infty(D)$ *be* L^∞-*star weak limit of the sequence* $\{\varphi(\rho_n)p(\rho_n)\}$, $\overline{\varphi \operatorname{div}\mathbf{u}} \in L^2(D)$ *be* L^2-*weak limit of the sequence* $\{\varphi(\rho_n)\operatorname{div}\mathbf{u}_n\}$. *Then*

$$\frac{1}{\xi + \nu}\overline{\varphi p} - \overline{\varphi\operatorname{div}\mathbf{u}} = \frac{1}{\xi + \nu}\overline{\varphi}\,\overline{p} - \overline{\varphi}\operatorname{div}\mathbf{u} \ \text{in } D \setminus S, \tag{4.2}$$

where $\overline{\varphi}$ *and* $\overline{p}$ *are given by Lemma 3.2.*

5. The oscillation defect measure

The notion of oscillation defect measure was introduced in [6] in order to justify the existence theory for isentropic flows with *small* values of the adiabatic ratio γ. Following [6, 11] the r-oscillation defect measure associated with the sequence $\{\rho_n\}_{n \le 1}$ is defined as follows

$$\mathbf{osc}_r[\rho_n \to \rho\,](K) := \sup_{k \ge 1} \limsup_{n \to \infty} \|T_k(\rho_n) - T_k(\rho)\|_{L^r(K)}^r,$$

where $T_k(z) = kT(z/k)$, $T(z)$ is a smooth concave function, which is equal to z for $z \le 1$ and is a constant for $z \ge 3$. The smoothness properties of T_k are not important and we can take the simplest form $T_k(z) = \min\{z, k\}$. Note that the total energy estimates provide the boundedness of γ-oscillation defect measure on the whole domain D. The unexpected result was obtained by E. Feireisl et al. in papers [6, 8], where it was shown that $(1 + \gamma)$-oscillation defect measure associated with the sequence $\{\rho_n\}$ is uniformly bounded on all compact subsets of Ω.

Note that in the shape optimization problem we can not replace the compact subsets $K \Subset \Omega$ by the domain Ω itself, since the oscillation defect measure is not any *regular* set additive function on the family of compact subsets of Ω, i.e., it is not any measure it the sense of measure theory. In order to bypass this difficulty we observe that the finiteness of the oscillation defect measure on compacts gives some additional information on the properties of the distribution function Γ. Our task is to extract this information and then to use in the proof of Theorem 1.1. In order to formulate the appropriate auxiliary result we define the function $\mathcal{T}_\vartheta(x)$ by the equality

$$\mathcal{T}_\vartheta(x) = \overline{\min\{\rho, \vartheta\}}(x) - \min\{\rho(x), \vartheta(x)\} \text{ for each } \vartheta \in C(\Omega).$$

Lemma 5.1. *Under the assumptions of Theorem 1.1 and Lemma 3.2, there is a constant c independent of ϑ and K such that the inequalities*

$$\|\mathcal{T}_\vartheta\|_{L^{1+\gamma}(K)}^{1+\gamma} \leq \lim_{n\to\infty} \int_\Omega |\min\{\rho_n(x), \vartheta(x)\} - \min\{\rho(x), \vartheta(x)\}|^{1+\gamma} dx \leq c \quad (5.1)$$

hold for all $\vartheta \in C(\Omega)$ and $K \Subset \Omega$. We point out, that the limit in (5.1) does exist by the choice of the sequence ρ_n.

We reformulate the result of Lemma 5.1 in terms of the distribution function Γ. Recall that the functions $\min\{\rho_n, \lambda\}$ are uniformly bounded in $\mathbb{R}^3$ and $\min\{\rho_n, \lambda\}\mathrm{div}\,\mathbf{u}_n$ converges weakly in $L^2(D)$ for all non-negative λ. Introduce the functions

$$\mathcal{V}_\lambda = \overline{\left(\min\{\rho, \lambda\}\mathrm{div}\,\mathbf{u}\right)} - \overline{\min\{\rho, \lambda\}}\mathrm{div}\,\mathbf{u} \in L^2(D), \quad (5.2)$$

$$\mathfrak{H}(x) = \int_{[0,\infty)} \Gamma(x, s)(1 - \Gamma(x, s))\, ds, \qquad \mathfrak{H} \in L^\gamma(D).$$

Lemma 5.2. *There is a constant c independent of λ such that*

$$\|\mathfrak{H}\|_{L^{1+\gamma}(D\setminus S)} + \sup_\lambda \|\mathcal{V}_\lambda\|_{L^1(D\setminus S)} \leq c. \quad (5.3)$$

6. Kinetic formulation of the mass balance equation

In this section we show that the distribution function $\Gamma(x, \lambda)$ of the Young measure, associated with a given sequence of solutions to problem (1.3), satisfies some integro-differential transport equation which is called the kinetic equation. This result is given by the following lemma. Fix an arbitrary function $\zeta(x, \lambda)$ satisfying the conditions

$$\zeta \in C_0^\infty(D \times \mathbb{R}), \qquad \mathrm{spt}\, \zeta \Subset D \setminus \left(\Sigma^- \cup S\right). \quad (6.1)$$

We use the notation ∂_λ for the partial derivatives with respect to the variable λ, e.g., $\partial_\lambda\zeta := \frac{\partial\zeta}{\partial\lambda}$. The absolutely continuous measure is denoted by $d_\lambda\zeta := \partial_\lambda\zeta d\lambda$.

Recall that the compact obstacle is of the form $S = B \setminus \Omega$ and that $\Sigma^- \subset \partial B$ is the *exit* set.

Lemma 6.1. *Suppose that all assumptions of Theorem 1.1 are satisfied and Γ is a distribution function of the Young measure associated with a given sequence $\{\rho_n\}$ of solutions to problem (1.3). Then*

$$\int\limits_{(D\setminus S)\times\mathbb{R}} \Gamma(x,\lambda)\nabla_{x,\lambda}\zeta \cdot \mathbf{w}\, d\lambda\, dx + \int\limits_{(D\setminus S)\times\mathbb{R}_\lambda} \lambda\mathcal{M}(x,\lambda)\, d_\lambda\zeta dx = 0 \ . \tag{6.2}$$

Here $\mathbf{w}$ is the solenoidal vector field of the form $\mathbf{w}(x,\lambda) = (\mathbf{u}(x), -\lambda\mathrm{div}\,\mathbf{u})$, and the function $\mathcal{M}$ is defined by the equalities

$$\begin{aligned}
\mathcal{M}(x,\lambda) &= -\frac{1}{\xi+\nu} \int\limits_{(-\infty,\lambda)} (s^\gamma - \overline{p})\, d_s\Gamma(x,s) \\
&= \frac{1}{\xi+\nu} \int\limits_{[\lambda,\infty)} (s^\gamma - \overline{p})\, d_s\Gamma(x,s),
\end{aligned} \tag{6.3}$$

in which the weak limit for the pressure $\overline{p}(x) = \int_{\mathbb{R}} \lambda d_\lambda\Gamma(x,\lambda)$ is defined in Lemma 3.2. Integral identity (6.2) is equivalent, in the sense of distributions, to the kinetic equation

$$\frac{\partial}{\partial\lambda}\big[\lambda\mathrm{div}\,\mathbf{u}(x)\Gamma(x,\lambda)\big] - \mathrm{div}\left(\Gamma(x,\lambda)\mathbf{u}(x)\right) - \frac{\partial}{\partial\lambda}[\lambda\mathcal{M}(x,\lambda)] = 0 \ \text{in}\ \mathcal{D}'(D\setminus S). \tag{6.4}$$

Remark 6.2. Since the kinetic equation is understood in the sense of distributions, the equation remains valid if we replace the intervals of integration $[0,\lambda)$ and $[\lambda,\infty)$ in formulae (6.3) by $[0,\lambda]$ and (λ,∞) respectively, which creates some discomfort. In order to avoid such ambiguity, we observe that (6.2) also holds true if we replace the function $\mathcal{M}$ by its invariant form

$$\mathfrak{M}(x,\lambda) := \frac{1}{2}\left(\lim_{s\to\lambda+0}\mathcal{M}(x,s) + \lim_{s\to\lambda-0}\mathcal{M}(x,s)\right). \tag{6.5}$$

The next lemma describes the basic properties of the function $\mathcal{M}(x,\lambda)$, which are important for the further analysis.

Lemma 6.3. *For a.e. $x \in D\setminus S$,*

(i) *$\mathcal{M}(x,\cdot)$ is non-negative and vanishes on $\mathbb{R}^-$. Moreover, if the Borel function $\mathfrak{M}(x,.)$ given by (6.5) vanishes σ_x-almost everywhere on the interval (ω,∞) with $\omega = \overline{p}(x)^{1/\gamma}$, then $\sigma_x = d_\lambda\Gamma(x,\cdot)$ is a Dirac measure and*

$$\Gamma(x,\lambda) = 0 \ \text{for}\ \lambda < \overline{p}(x)^{1/\gamma}, \quad \Gamma(x,\lambda) = 1 \ \text{for}\ \lambda \geq \overline{p}(x)^{1/\gamma}.$$

(ii) *For all $g \in C_0^\infty(0,\infty)$,*

$$\int_{\mathbb{R}_\lambda} g(\lambda)\mathcal{M}(x,\lambda)d\lambda = -\int\limits_{[0,\infty)} g'(\lambda)\mathcal{V}_\lambda(x)d\lambda \tag{6.6}$$

where $\mathcal{V}_\lambda$ is defined by (5.2).

7. Renormalization of the kinetic equation

The notion of a renormalized solution, introduced in pioneering paper [4], plays an important role in the theory of compressible N-S-E developed by P.L. Lions and E. Feireisl et al. Moreover, the kinetic equation itself is a result of the renormalization procedure. Formally we can renormalize equation (6.4) multiplying the both sides by a function $\Psi'(\Gamma)$, which leads to the transport equation for the function $\Psi(\Gamma)$, but the justification of this construction is a delicate question. The corresponding result is given by the following lemma. Set $\Psi(\Gamma) = \Gamma(1 - \Gamma)$.

Lemma 7.1. *For all functions $h \in C_0^\infty(D \setminus S)$ with spt $h \Subset D \setminus (S \cup \Sigma^-)$ and for all functions $\eta \in C^\infty(\mathbb{R})$ vanishing near $+\infty$, we have the integral identity*

$$\int\limits_{(D\setminus S)\times\mathbb{R}} \mathcal{F}(x,\lambda)dx\,d\lambda = 2 \int\limits_{(D\setminus S)} \left(\int\limits_{[0,\infty)} \eta(\lambda)\lambda\mathfrak{M}(x,\lambda)d_\lambda\Gamma(x,\lambda) \right) h(x)\,dx\ , \qquad (7.1)$$

where

$$\mathcal{F} \equiv \eta(\lambda)\Psi(\Gamma)\mathbf{u}(x)\nabla h(x) - \lambda h(x)\Psi(\Gamma)\eta'(\lambda)\mathrm{div}\,\mathbf{u}(x) + \lambda h(x)\Psi'(\Gamma)\mathcal{M}(x,\lambda)\eta'(\lambda)\ .$$

In other words, the function $\Psi(\Gamma)$ satisfies the transport equation

$$\mathrm{div}_{\lambda,x}\big(\Psi(\Gamma)\mathbf{u}\big) + \frac{\partial}{\partial\lambda}\big(\lambda\Psi'(\Gamma)\mathfrak{M}\big) - 2\lambda\mathfrak{M}\frac{\partial\Gamma}{\partial\lambda} = 0 \ in\ \mathcal{D}'(D \setminus (\Sigma^- \cup S)).$$

8. Convergence of shape functionals

We provide here, for the convenience of the reader, a sketch of the proof for relations (1.13) and (1.14). The complete arguments can be found in [26].

We point out, that Theorem 1.1 will be proved if we show that any sequence of generalized solutions to problem (1.3) satisfying hypotheses (1.10), (1.12) and of Lemmas 3.2, 3.3, converges almost everywhere on $D \setminus S$. In the light of Lemmas 3.2, 3.3, it suffices to verify the equality $\Psi(\Gamma) = 0$ in $(D \setminus S) \times \mathbb{R}$.

We begin with proving that renormalized integral identity (7.1) after substituting $h = 1$ turns into the integral inequality

$$\int\limits_{(D\setminus S)\times\mathbb{R}} \left\{\lambda\Psi'(\Gamma)\mathcal{M}\eta' - \lambda\Psi(\Gamma)\eta'\mathrm{div}\,\mathbf{u}\right\}dx\,d\lambda$$

$$\geq 2 \int\limits_{(D\setminus S)} \left(\int\limits_{[0,\infty)} \eta(\lambda)\lambda\mathfrak{M}(x,\lambda)d_\lambda\Gamma(x,\lambda) \right) dx, \tag{8.1}$$

which holds true for all non-negative functions $\eta \in C^\infty(\mathbb{R})$ vanishing near $+\infty$. The proof is based on the following approximation result which is shown by an application of the Hedberg approximation Theorem [12].

Lemma 8.1. *For each $k > 1$ there exist a function $\zeta_k \in C^\infty(D)$ and a constant c independent of k such that ζ_k vanishes in a vicinity of S and*

$$0 \leq \zeta_k \leq 1, \quad \text{meas } A_k + \int_D |\nabla \zeta_k \mathbf{u}|\, dx \leq 1/k, \tag{8.2}$$

where $A_k = \{x \in D \setminus S : \zeta_k(x) \leq 1 - 1/k\} \subset D \setminus S$.

Following [10] let us consider the sequence of functions $\chi_k(x) = \chi\big(k \operatorname{dist}(x, \Sigma^- \cup \partial D)\big)$ with an arbitrary, smooth, monotone function χ such that $\chi(z) = 0$ for $z \leq 1/2$ and $\chi(z) = 1$ for $z \geq 1$. Since $\Psi(\Gamma)(\cdot, \lambda)$ vanishes outside of $D \setminus B$, we have for all sufficiently large k,

$$\nabla \chi_k(\mathbf{u} - \mathbf{u}^\infty) \to 0 \text{ in } L^1(D \setminus S), \quad \Psi(\Gamma)(\cdot, \lambda)\nabla \chi_k \mathbf{u}^\infty \leq 0 \text{ in } D. \tag{8.3}$$

Set $h_k = \zeta_k \chi_k$ and note that

$$\nabla h_k \mathbf{u} \leq |\nabla \zeta_k \mathbf{u}| + |\nabla \chi_k(\mathbf{u} - \mathbf{u}^\infty)| + \zeta_k \nabla \chi_k \mathbf{u}^\infty.$$

Recalling the inequality $\eta \geq 0$ and relations (8.2), (8.3) we obtain

$$\limsup_{k \to \infty} \int_{(D \setminus S) \times \mathbb{R}} \Psi(\Gamma)\eta \nabla h_k \mathbf{u}\, dx\, d\lambda \leq 0. \tag{8.4}$$

Moreover, the functions h_k converge to 1 in measure on $D \setminus S$. The functions h_k are Lipschitz continuous in $\mathbb{R}^3$ and vanish in the vicinity of $\Sigma^- \cup S$ therefore, can be used as test functions in (7.1). Substitution of h_k into (7.1) followed by the limit passage $k \to \infty$ in the resulting integral identity, leads to desired inequality (8.1).

Next, we claim that the right-hand side of (8.1) equals to zero. To this end, choose an arbitrary non-negative function $\upsilon \in C^\infty(\mathbb{R})$ with $\operatorname{spt} \upsilon \subset (-1, 1)$ and $\int_\mathbb{R} \upsilon(\lambda)d\lambda = 1$. For fixed $t > 2$, set $\eta(\lambda) = \int_\lambda^\infty \upsilon(s - t)\, ds$. Since $\eta'(\lambda) = 0$ and $\eta(\lambda) = 1$ for $\lambda \leq t - 1$, $\eta'(\lambda) = 0$ for $\lambda \geq t + 1$, we can use (8.1) to obtain

$$2 \int_{(D \setminus S)} \left(\int_{[0,t-1)} \lambda \mathfrak{M} d_\lambda \Gamma \right) dx \leq -(t+1) \int_{(D \setminus S) \times \mathbb{R}} \big\{ \mathcal{M} + \Psi(\Gamma)|\operatorname{div} \mathbf{u}(x)| \big\} \eta'\, dx\, d\lambda. \tag{8.5}$$

Using identity (6.6) and the relation

$$- \int_{(D \setminus S) \times \mathbb{R}} \Psi(\Gamma)\, \eta'(\lambda)\, |\operatorname{div} \mathbf{u}|\, dx\, d\lambda$$

$$\equiv \int_{[0,\infty)} \eta'' \left\{ \int_{D \setminus S} \left[\int_{\Omega \times [0,\lambda)} \Psi(\Gamma(x, s))\, ds \right] |\operatorname{div} \mathbf{u}|\, dx \right\} d\lambda$$

we can rewrite inequality (8.5) in the form

$$2 \int_{(D \setminus S)} \left(\int_{[0,t-1)} \lambda \mathfrak{M} d_\lambda \Gamma \right) dx \leq (1 + t) \int_{[1,\infty)} \eta''(\lambda) \wp(\lambda) d\lambda, \tag{8.6}$$

where the function $\wp : [0, \infty) \mapsto \mathbb{R}$ is given by

$$\wp(\lambda) = \int\limits_{(D\setminus S)\times[0,\lambda)} \Psi(\Gamma(x,s))|\mathrm{div}\,\mathbf{u}(x)|\,dx\,ds + \int\limits_{\Omega} \mathcal{V}_\lambda(x)dx \ .$$

Since

$$\int\limits_{[0,\lambda)} \Psi(\Gamma(x,s))\,ds \leq \int\limits_{[0,\infty)} \Psi(\Gamma(x,s))\,ds = \mathfrak{H}(x) \ ,$$

Lemma 5.2 implies the boundedness of $\wp$ on $\mathbb{R}^+$,

$$|\wp(\lambda)| \leq c\|\mathbf{u}\|_{H^{1,2}(D\setminus S)}\|\mathfrak{H}\|_{L^2(D\setminus S)} + \|\mathcal{V}_\lambda\|_{L^1(D\setminus S)} \leq c \ .$$

Taking into account that $\eta''(\lambda) = \partial_t \upsilon(\lambda - t)$, inequality (8.6) can be rewritten in the form

$$2 \int\limits_{(D\setminus S)} \left(\int\limits_{[0,t-1]} \lambda\mathfrak{M}d\Gamma(x,\lambda) \right) dx \leq (1+t)\frac{d}{dt}(\upsilon * \wp)(t) \ . \tag{8.7}$$

Since the smooth function $(\upsilon*\wp)(t)$ is uniformly bounded on $\mathbb{R}^+$, there is a sequence $t_k \to \infty$ such that $\lim\limits_{k\to\infty} (t_k + 1)\frac{d}{dt}(\upsilon * \wp)(t_k) \leq 0$. Substitution of $t = t_k$ into (8.7) followed by the limit passage $k \to \infty$ in (8.7) leads to

$$\int\limits_{[0,t-1]} \lambda\mathfrak{M}d\Gamma(x,\lambda) = 0 \text{ for a.e. } x \in D \setminus S \ .$$

In other words, $\mathcal{M}(x,\cdot)$ vanishes σ_x-almost everywhere on $(0,\infty)$, which along with Lemma 6.3 implies the equality $\Gamma(1 - \Gamma) = 0$ a.e. in $(D \setminus S) \times \mathbb{R}$. Hence ρ_n converges a.e. in $D \setminus S$. Estimates (1.12) imply the strong convergence of the sequence ρ_n in $L^r(D \setminus S)$ for all $r < \gamma$. Since $(\rho_n, \mathbf{u}_n)$ satisfy all assumptions of Theorem 2.1, we can make use of estimate (2.1) from this theorem, which yields the strong convergence ρ_n in $L^r_{\mathrm{loc}}(\Omega)$ for $r < \gamma(1 + \kappa)$. In particular, the sequence $p(\rho_n)$ converges to $\bar{p} = p(\rho)$ in $L^1_{\mathrm{loc}}(\Omega)$. After substituting $(\rho_n, \mathbf{u}_n)$ into integral identities (1.7a), (1.7b), followed by the limit passage $n \to \infty$, by Lemma 3.3, we can conclude that the pair $(\rho, \mathbf{u})$ is a generalized solution to problem (1.3). Finally, since $\nabla\mathbf{u}^\infty$ is compactly supported in Ω, limit passage in (1.14) for the sequence of drag functionals follows by Lemma 3.3 and by the strong convergence of the sequence $p(\rho_n)$ in $L^1_{\mathrm{loc}}(\Omega)$.

9. Concluding remarks

We have presented a result on the existence of optimal obstacle for compressible N-S-E. The result is derived by an application of the theory of N-S-E developed in particular by Pl.L. Lions, and E. Feireisl. In order to apply our result to a specific shape optimization problem, it is necessary to check, if the set of solutions to N-S-E is nonempty. Since we consider the nonhomogeneous boundary conditions, such a verification is not always simple task. The first order necessary optimality conditions for a class of drag minimization problems are derived in [26]. There is no numerical results for such problems, due in particular to the fact that the numerical methods for compressible N-S-E are still under studies.

References

[1] D. ADAMS *On the existence of capacitary strong type estimates in R^n* Arkiv for Matematik **14**, (1976) 125–140.

[2] D. ADAMS AND L. HEDBERG *Function Spaces and Potential Theory* Springer-Verlag, Berlin etc 1995).

[3] J.M. BALL *A version of the fundamental theorem for Young measures* In: *PDEs and continuum Models of Phase Trans.*, Lecture Notes in Physics, **344**(1989) 241–259.

[4] R.J. DiPERNA, P.L. LIONS *Ordinary differential equations, transport theory and Sobolev spaces* Invent. Math. **48**,(1989) 511–547.

[5] EDWARDS R.E. *Functional analysis. Theory and applications* (Holt, Rinehart and Wilson, N.Y.-Toronto-London, 1965)

[6] E. FEIREISL *On compactness of solutions to the compressible isentropic Navier-Stokes equations when the density is not square integrable* Comment. Math. Univ. Carolina **42**,(2001) 83–98.

[7] E. FEIREISL, Š. MATUŠŮ-NEČASOVÁ, H. PETZELTOVÁ, I. STRAŠKRABA *On the motion of a viscous compressible fluid driven by a time-periodic external force* Arch. Rational Mech. Anal.**149**, (1999) 69–96.

[8] E. FEIREISL, A.H. NOVOTNÝ, H. PETZELTOVÁ *On the existence of globally defined weak solutions to the Navier-Stokes equations* J. of Math. Fluid Mech. **3**, (2001), 358–392.

[9] E. FEIREISL, A.H. NOVOTNÝ, H. PETZELTOVÁ *On the domain dependence of solutions to the compressible Navier-Stokes equations of a barotropic fluid* Math. Methods Appl. Sci. 25 (2002), no. 12, 1045–1073.

[10] E. FEIREISL *Shape optimisation in viscous compressible fluids* Appl. Math. Optim. 47(2003), 59–78.

[11] E. FEIREISL *Dynamics of Viscous Compressible Fluids* (Oxford University Press, Oxford 2004).

[12] L. HEDBERG *Spectral synthesis in Sobolev spaces, and uniqueness of solutions of the Dirichlet problem.* Acta Math. 147 (1981), no. 3-4, 237–264.

[13] B. KAWOHL, O. PIRONNEAU, L. TARTAR AND J. ZOLESIO *Optimal Shape Design* Lecture Notes in Math. 1740, *Springer-Verlag,* 2000.

[14] JAE RYONG KWEON, R.B. KELLOGG *Regularity of solutions to the Navier-Stokes system for compressible flows on a polygon.* SIAM J. Math. Anal. 35 (2004), no. 6, 1451–1485

[15] JAE RYONG KWEON, R.B. KELLOGG *Regularity of solutions to the Navier-Stokes equations for compressible barotropic flows on a polygon.* Arch. Ration. Mech. Anal. 163 (2002), no. 1, 35–64.

[16] JAE RYONG KWEON, R.B. KELLOGG *Compressible Stokes problem on nonconvex polygonal domains.* J. Differential Equations 176 (2001), no. 1, 290–314.

[17] P.L. LIONS *Mathematical topics in fluid dynamics, Vol. 2, Compressible models* (Oxford Science Publication, Oxford 1998).

[18] P.L. LIONS *On some challenging problems in nonlinear partial differential equations* in V. Arnold (ed.) et al., *Mathematics: Frontiers and perspectives* (American Mathematical Society, Providence 2000) 121–135.

[19] J. MALEK, J. NECAS, M. ROKYTA, M. RUZICKA *Weak and measure-valued solutions to evolutionary PDE* (Chapman and Hall, London 1996).

[20] A. NOVOTNÝ, M. PADULA *Existence and Uniqueness of Stationary solutions for viscous compressible heat conductive fluid with large potential and small non-potential external forces* Siberian Math. Journal, 34, 1993, 120–146

[21] A. NOVOTNÝ, I. STRAŠKRABA *Introduction to the mathematical theory of compressible flow* Oxford Lecture Series in Mathematics and its Applications, Vol. 27. Oxford University Press, Oxford, 2004.

[22] P. PEDREGAL *Parametrized measures and variational principles* Progress in Nonlinear Differential Equations and their Applications, **30** Birkhäuser Verlag, Basel, 1997.

[23] P.I. PLOTNIKOV, J. SOKOLOWSKI *On compactness, domain dependence and existence of steady state solutions to compressible isothermal Navier-Stokes equations* Journal of Mathematical Fluid Mechanics (electronic).

[24] P.I. PLOTNIKOV, J. SOKOLOWSKI *Concentrations of solutions to time-discretized compressible Navier-Stokes equations* Communications in Mathematical Physics (electronic).

[25] P.I. PLOTNIKOV, J. SOKOLOWSKI *Stationary Boundary Value Problems for Navier-Stokes Equations with Adiabatic Index $\nu < 3/2$.* Doklady Mathematics. Vol. 70, No. 1, 2004, 535–538, Translated from Doklady Akademii Nauk, Volume 397, $N.^0$, 1-6, 2004.

[26] P.I. PLOTNIKOV, J. SOKOLOWSKI *Domain dependence of solutions to compressible Navier-Stokes equations* to appear.

[27] P.I. PLOTNIKOV, J. SOKOLOWSKI *Shape sensitivity analysis for compressible Navier-Stokes equations* in preparation.

[28] K. YOSHIDA *Functional Analysis and Its Applications* New York: Springer-Verlag, 1971.

[29] W.P. ZIEMER *Weakly differentiable functions* (Springer-Verlag, New York etc. 1989)

Pavel I. Plotnikov
Lavryentyev Institute of Hydrodynamics
Siberian Division of Russian Academy of Sciences
Lavryentyev pr. 15, Novosibirsk 630090, Russia
e-mail: `plotnikov@hydro.nsc.ru`

Jan Sokolowski
Institut Elie Cartan, Laboratoire de Mathématiques
Université Henri Poincaré Nancy I
B.P. 239, F-54506 Vandœuvre lès Nancy Cedex, France
e-mail: `Jan.Sokolowski@iecn.u-nancy.fr`

International Series of Numerical Mathematics, Vol. 155, 269–291
© 2007 Birkhäuser Verlag Basel/Switzerland

A Family of Stabilization Problems for the Oseen Equations

Jean-Pierre Raymond

Abstract. The feedback stabilization of the Navier-Stokes equations around an unstable stationary solution is related to the feedback stabilization of the Oseen equations (the linearized Navier-Stokes equations about the unstable stationary solution). In this paper we investigate the regularizing properties of feedback operators corresponding to a family of optimal control problems for the Oseen equations.

1. Introduction

Let Ω be a bounded and connected domain in $\mathbb{R}^3$ with a regular boundary Γ, $\nu > 0$, and consider a couple $(\mathbf{w}, \chi)$ – a velocity field and a pressure – solution to the stationary Navier-Stokes equations in Ω:

$$-\nu\Delta\mathbf{w} + (\mathbf{w}\cdot\nabla)\mathbf{w} + \nabla\chi = \mathbf{f} \quad \text{and} \quad \operatorname{div}\mathbf{w} = 0 \text{ in } \Omega, \qquad \mathbf{w} = \mathbf{u}_s^\infty \quad \text{on } \Gamma.$$

We assume that $\mathbf{w}$ is regular and is an unstable solution of the instationary Navier-Stokes equations.

The local feedback boundary stabilization of the Navier-Stokes equations consists in finding a Dirichlet boundary control $\mathbf{u}$, in feedback form, localized in a part of the boundary Γ, so that the corresponding control system:

$$\frac{\partial\mathbf{y}}{\partial t} - \nu\Delta\mathbf{y} + (\mathbf{y}\cdot\nabla)\mathbf{w} + (\mathbf{w}\cdot\nabla)\mathbf{y} + (\mathbf{y}\cdot\nabla)\mathbf{y} + \nabla p = 0 \quad \text{in } Q_\infty,$$

$$\operatorname{div}\mathbf{y} = 0 \quad \text{in } Q_\infty, \quad \mathbf{y} = M\mathbf{u} \quad \text{on } \Sigma_\infty, \quad \mathbf{y}(0) = \mathbf{y}_0 \text{ in } \Omega, \tag{1}$$

be stable for initial values $\mathbf{y}_0$ small enough in an appropriate space $\mathbf{X}(\Omega)$. In this setting, $Q_\infty = \Omega \times (0, \infty)$, $\Sigma_\infty = \Gamma \times (0, \infty)$, $\mathbf{X}(\Omega)$ is a subspace of $\mathbf{V}_n^0(\Omega) = \{\mathbf{y} \in \mathbf{L}^2(\Omega) \mid \operatorname{div}\mathbf{y} = 0 \text{ in } \Omega, \ \mathbf{y}\cdot\mathbf{n} = 0 \text{ on } \Gamma\}$, $\mathbf{n}$ is the unit normal to Γ outward Ω, $\mathbf{y}_0 \in \mathbf{X}(\Omega)$, and the operator M is a restriction operator defined in Section 2.

The feedback stabilization of equations (1) is closely related to the feedback stabilization of the Oseen equations:

$$\frac{\partial \mathbf{y}}{\partial t} - \nu \Delta \mathbf{y} + (\mathbf{w} \cdot \nabla)\mathbf{y} + (\mathbf{y} \cdot \nabla)\mathbf{w} + \nabla p = 0, \quad \text{in } Q_\infty,$$

$$\text{div } \mathbf{y} = 0 \quad \text{in } Q_\infty, \quad \mathbf{y} = M\mathbf{u} \text{ on } \Sigma_\infty, \quad \mathbf{y}(0) = \mathbf{y}_0 \quad \text{in } \Omega. \tag{2}$$

One way to determine a feedback control law able to stabilize system (2) consists in solving an optimal control problem with an infinite time horizon. Once the functional of the control problem is defined the feedback law can be determined by calculating the solution to the corresponding algebraic Riccati equation (if it exists and if it admits a unique solution). Changing the functional, we change the feedback operator. These functionals are generally of the form

$$J(\mathbf{y}, \mathbf{u}) = \frac{1}{2} \int_0^\infty \int_\Omega |C\mathbf{y}|^2 \, dx dt + \frac{1}{2} \int_0^\infty \int_\Gamma |\mathbf{u}|^2 \, dx dt,$$

where C, the observation operator, may be a bounded or an unbounded operator in $\mathbf{V}^0(\Omega) = \left\{ \mathbf{y} \in \mathbf{L}^2(\Omega) \mid \text{div } \mathbf{y} = 0 \text{ in } \Omega, \int_\Gamma \mathbf{y} \cdot \mathbf{n} = 0 \text{ on } \Gamma \right\}$. When C is unbounded let us denote by $D(C)$ its domain in $\mathbf{V}^0(\Omega)$.

Let us consider the optimal control problem

$$(\mathcal{Q}) \qquad \inf \left\{ J(\mathbf{y}, \mathbf{u}) \mid (\mathbf{y}, \mathbf{u}) \text{ satisfies (2)}, \ \mathbf{u} \in L^2(0, \infty; \mathbf{V}^0(\Gamma)), \right.$$

$$\left. \mathbf{y} \in L^2(0, \infty; D(C)) \right\},$$

where

$$\mathbf{V}^0(\Gamma) = \left\{ \mathbf{y} \in \mathbf{L}^2(\Gamma) \mid \int_\Gamma \mathbf{y} \cdot \mathbf{n} = 0 \right\}.$$

In two dimension [15], we have determined a feedback law by choosing $C = I$. For such a choice, we have shown that the optimal solution $(\mathbf{y}_{\mathbf{y}_0}, \mathbf{u}_{\mathbf{y}_0})$ to $(\mathcal{Q})$ obeys a feedback formula of the form

$$\mathbf{u}_{\mathbf{y}_0}(t) = -R_A^{-1} M B^* \Pi P \mathbf{y}_{\mathbf{y}_0}(t),$$

where P is the so-called Helmholtz or Leray projector in $\mathbf{L}^2(\Omega)$ onto $\mathbf{V}_n^0(\Omega) = \left\{ \mathbf{y} \in \mathbf{L}^2(\Omega) \mid \text{div } \mathbf{y} = 0 \text{ in } \Omega, \ \mathbf{y} \cdot \mathbf{n} = 0 \text{ on } \Gamma \right\}$, Π is the solution to the algebraic Riccati equation of $(\mathcal{Q})$, B^* is the adjoint of the boundary control operator B corresponding to the nonhomogeneous boundary condition in equation (2) (see Section 2),

$$R_A = M D_A^* (I - P) D_A M + I,$$

where D_A is the Dirichlet operator of the Oseen equations, and D_A^* its adjoint (see Section 2). Applying the same boundary feedback law to the Navier-Stokes equations, we have proved that the corresponding dynamical system is stable for initial data small enough in $\mathbf{V}_n^{1/2-\varepsilon}(\Omega)$ if $0 < \varepsilon < 1/4$.

This result cannot be directly extended to the three-dimensional case. Indeed even if $\mathbf{y}_0$ is very regular, the regularity of the optimal solution to $(\mathcal{Q})$ is not sufficient to deal with the stabilization problem of the Navier-Stokes equations in three dimension. In the three-dimensional case, Barbu, Lasiecka and Triggiani [4] have obtained a stabilization result with controls in the class of functions $\mathbf{u}$ obeying $\mathbf{u}(t) \cdot \mathbf{n} = 0$, when the control is applied everywhere on the boundary of Ω and when Ω is simply connected. For that, they have chosen C as an unbounded operator such that $|C\mathbf{y}|_{\mathbf{V}_n^0(\Omega)}$ is a norm in $\mathbf{V}_n^0(\Omega)$ equivalent to the usual norm of the space $\mathbf{H}^{3/2+\varepsilon}(\Omega)$ for some $\varepsilon > 0$. The idea in [1, 4] is to choose an operator C, unbounded in $\mathbf{V}_n^0(\Omega)$, so that the norm $|C\mathbf{y}|_{\mathbf{V}_n^0(\Omega)}$ be strong enough to dominate the nonlinearity of the Navier-Stokes equations.

In [16] we follow a completely opposite direction. We have chosen an operator C which is bounded, and which is even a smoothing operator. A key point in the analysis in [16] consists in studying problem the following family of optimal control problems

$$(\mathcal{P}) \qquad \inf\left\{ I(\mathbf{y}, \mathbf{u}) \mid (\mathbf{y}, \mathbf{u}) \text{ satisfies (2)}, \ \mathbf{u} \in L^2(0, \infty; \mathbf{V}^0(\Gamma)) \right\},$$

where

$$I(\mathbf{y}, \mathbf{u}) = \frac{1}{2} \int_0^\infty \int_\Omega |(-A_0)^{-\alpha} P\mathbf{y}|^2 \, dx dt + \frac{1}{2} \int_0^\infty \int_\Gamma |R_A^{1/2} \mathbf{u}|^2 \, dx dt,$$

$(-A_0) = -\nu P \Delta$ is the Stokes operator with homogeneous Dirichlet boundary conditions, and $0 \leq \alpha \leq 1/2$. Observe that, due to the definition of the operator R_A, problem $(\mathcal{P})$ is equivalent to the following one

$$\inf\left\{ J(\mathbf{y}, \mathbf{u}) \mid (\mathbf{y}, \mathbf{u}) \text{ satisfies (2)}, \ \mathbf{u} \in L^2(0, \infty; \mathbf{V}^0(\Gamma)) \right\},$$

where

$$J(\mathbf{y}, \mathbf{u}) = \frac{1}{2} \int_0^\infty \int_\Omega |(-A_0)^{-\alpha} P\mathbf{y}|^2 \, dx dt + \frac{1}{2} \int_0^\infty \int_\Omega |(I - P)\mathbf{y}|^2 \, dx dt$$
$$+ \frac{1}{2} \int_0^\infty \int_\Gamma |\mathbf{u}|^2 \, dx dt.$$

This is not at all a standard functional since $P\mathbf{y}$ and $(I - P)\mathbf{y}$ are involved with different norms. The equality $I(\mathbf{y}, \mathbf{u}) = J(\mathbf{y}, \mathbf{u})$ when $(\mathbf{y}, \mathbf{u})$ obeys (2) is a consequence of rewriting (2) in the form (9) (see the end of Section 2). The objective of the present paper is to study the regularity of optimal solutions of $(\mathcal{P})$. These results will next be used in [16] to study the feedback stabilization problem of the Navier-Stokes equations in three dimension (see also [17]).

The main results of the paper are given in Corollary 13 and Corollary 14 where we state regularity results for the optimal solution of problem $(\mathcal{P})$, and regularizing properties of the solution Π to the algebraic Riccati equation of $(\mathcal{P})$.

2. Functional framework and preliminary results

2.1. Notation and assumptions

Let us introduce the following spaces: $H^s(\Omega; \mathbb{R}^3) = \mathbf{H}^s(\Omega)$, $L^2(\Omega; \mathbb{R}^3) = \mathbf{L}^2(\Omega)$, the same notation conventions will be used for trace spaces and for the spaces $H_0^s(\Omega; \mathbb{R}^3)$. We also introduce different spaces of free divergence functions and some corresponding trace spaces:

$$\mathbf{V}^s(\Omega) = \Big\{ \mathbf{y} \in \mathbf{H}^s(\Omega) \mid \operatorname{div} \mathbf{y} = 0 \ \text{ in } \Omega, \ \langle \mathbf{y} \cdot \mathbf{n}, 1 \rangle_{H^{-1/2}(\Gamma), H^{1/2}(\Gamma)} = 0 \Big\}, \ s \geq 0,$$

$$\mathbf{V}_n^s(\Omega) = \Big\{ \mathbf{y} \in \mathbf{H}^s(\Omega) \mid \operatorname{div} \mathbf{y} = 0 \ \text{ in } \Omega, \ \mathbf{y} \cdot \mathbf{n} = 0 \text{ on } \Gamma \Big\} \quad \text{for } s \geq 0,$$

$$\mathbf{V}_0^s(\Omega) = \Big\{ \mathbf{y} \in \mathbf{H}^s(\Omega) \mid \operatorname{div} \mathbf{y} = 0 \ \text{ in } \Omega, \ \mathbf{y} = 0 \text{ on } \Gamma \Big\} \quad \text{for } s > 1/2,$$

$$\mathbf{V}^s(\Gamma) = \Big\{ \mathbf{y} \in \mathbf{H}^s(\Gamma) \mid \langle \mathbf{y} \cdot \mathbf{n}, 1 \rangle_{H^{-1/2}(\Gamma), H^{1/2}(\Gamma)} = 0 \Big\} \quad \text{for } s \geq -1/2.$$

In the above setting $\mathbf{n}$ denotes the unit normal to Γ outward Ω. We shall use the following notation $Q_T = \Omega \times (0, T)$, $\Sigma_T = \Gamma \times (0, T)$, $Q_{\bar{t}, T} = \Omega \times (\bar{t}, T)$ and $\Sigma_{\bar{t}, T} = \Gamma \times (\bar{t}, T)$ for $\bar{t} > 0$, and $0 < T \leq \infty$. For spaces of time-dependent functions we set

$$\mathbf{V}^{s, \sigma}(Q_T) = H^\sigma(0, T; \mathbf{V}^0(\Omega)) \cap L^2(0, T; \mathbf{V}^s(\Omega)),$$

and

$$\mathbf{V}^{s, \sigma}(\Sigma_T) = H^\sigma(0, T; \mathbf{V}^0(\Gamma)) \cap L^2(0, T; \mathbf{V}^s(\Gamma)).$$

We assume that Ω is of class C^6 and $\mathbf{w} \in \mathbf{V}^5(\Omega)$. (In the case when $\alpha = 0$ the regularity results that we state in this paper are true if Ω is of class C^4 and $\mathbf{w} \in \mathbf{V}^3(\Omega)$, see [15].)

In order to find a control $\mathbf{u}$, supported in an open regular subset Γ_c of Γ, we introduce a weight function $m \in C^5(\Gamma)$ with values in $[0, 1]$, with support in Γ_c, equal to 1 in Γ_0, where Γ_0 is an open, non empty, and regular subset in Γ_c. Associated with this function m we introduce the operator $M \in \mathcal{L}(\mathbf{V}^0(\Gamma))$ defined by

$$M\mathbf{u}(x) = m(x)\mathbf{u}(x) - \frac{m}{\int_\Gamma m} \left(\int_\Gamma m\mathbf{u} \cdot \mathbf{n} \right) \mathbf{n}(x).$$

By this way, we can replace the condition $\operatorname{supp}(\mathbf{u}) \subset \Gamma_c$ by considering a boundary condition of the form

$$\mathbf{y} = M\mathbf{u} \quad \text{on} \quad \Sigma_\infty.$$

The main interest of this operator M is that if $\mathbf{u} \in L^2(0, \infty; H^s(\Gamma_c; \mathbb{R}^3)) \cap H^{s/2}(0, \infty; L^2(\Gamma_c; \mathbb{R}^3))$ for some $0 < s \leq 9/2$, and if $\tilde{\mathbf{u}}$ denotes the extension of $\mathbf{u}$ by zero to $\Sigma_\infty \setminus (\Gamma_c \times (0, \infty))$, then $M\tilde{\mathbf{u}}$ belongs to $L^2(0, \infty; H^s(\Gamma; \mathbb{R}^3)) \cap H^{s/2}(0, \infty; L^2(\Gamma; \mathbb{R}^3))$, which is not true for $\tilde{\mathbf{u}}$.

For all $\psi \in H^{1/2+\varepsilon'}(\Omega)$, with $\varepsilon' > 0$, we denote by $c(\psi)$ and $c(m\psi)$ the constants defined by

$$c(\psi) = \frac{1}{|\Gamma|} \int_\Gamma \quad \text{and} \quad c(m\psi) = \frac{1}{|\Gamma|} \int_\Gamma m\psi. \tag{3}$$

2.2. Properties of some operators

In the following we consider the linearized Navier-Stokes equation

$$\frac{\partial \mathbf{y}}{\partial t} - \nu \Delta \mathbf{y} + (\mathbf{w} \cdot \nabla)\mathbf{y} + (\mathbf{y} \cdot \nabla)\mathbf{w} + \nabla p = 0, \quad \text{in } Q_T,$$

$$\text{div } \mathbf{y} = 0 \quad \text{in } Q_T, \quad \mathbf{y} = M\mathbf{u} \text{ on } \Sigma_T, \quad \mathbf{y}(0) = \mathbf{y}_0 \quad \text{in } \Omega, \tag{4}$$

and the adjoint equation

$$-\frac{\partial \mathbf{\Phi}}{\partial t} - \nu \Delta \mathbf{\Phi} - (\mathbf{w} \cdot \nabla)\mathbf{\Phi} + (\nabla \mathbf{w})^T \mathbf{\Phi} + \nabla \psi = \mathbf{y}, \quad \text{in } Q_T,$$

$$\text{div } \mathbf{\Phi} = 0 \quad \text{in } Q_T, \quad \mathbf{\Phi} = 0 \text{ on } \Sigma_T, \quad \mathbf{\Phi}(T) = 0 \quad \text{in } \Omega, \tag{5}$$

where T is finite or infinite. To study these equations, we introduce the Stokes and the Oseen operators associated with equations (4) and (5). Let P be the orthogonal projector in $\mathbf{L}^2(\Omega)$ onto $\mathbf{V}_n^0(\Omega)$, and denote by $(A_0, D(A_0))$, $(A, D(A))$, and $(A^*, D(A^*))$ the unbounded operators in $\mathbf{V}_n^0(\Omega)$ defined by

$$D(A_0) = \mathbf{H}^2(\Omega) \cap \mathbf{V}_0^1(\Omega), \quad A_0 \mathbf{y} = \nu P \Delta \mathbf{y} \quad \text{for all } \mathbf{y} \in D(A_0),$$

$$D(A) = \mathbf{H}^2(\Omega) \cap \mathbf{V}_0^1(\Omega), \quad A\mathbf{y} = \nu P \Delta \mathbf{y} - P((\mathbf{w} \cdot \nabla)\mathbf{y}) - P((\mathbf{y} \cdot \nabla)\mathbf{w}),$$

$$D(A^*) = \mathbf{H}^2(\Omega) \cap \mathbf{V}_0^1(\Omega), \quad A^* \mathbf{y} = \nu P \Delta \mathbf{y} + P((\mathbf{w} \cdot \nabla)\mathbf{y}) - P((\nabla \mathbf{w})^T \mathbf{y}).$$

Throughout the following we denote by $\lambda_0 > 0$ an element in the resolvent set of A satisfying

$$\left((\lambda_0 I - A)\mathbf{y}, \mathbf{y}\right)_{\mathbf{V}_n^0(\Omega)} \geq \omega_0 |\mathbf{y}|^2_{\mathbf{V}_0^1(\Omega)} \quad \text{for all } \mathbf{y} \in D(A),$$

and

$$\left((\lambda_0 I - A^*)\mathbf{y}, \mathbf{y}\right)_{\mathbf{V}_n^0(\Omega)} \geq \omega_0 |\mathbf{y}|^2_{\mathbf{V}_0^1(\Omega)} \quad \text{for all } \mathbf{y} \in D(A^*), \tag{6}$$

for some $0 < \omega_0 < \nu$ (see, e.g., Lemma 24 where the proof is given for $\omega_0 = \nu/2$). The following theorem may easily be deduced from (6), see, e.g., [14, Lemma 4.1].

Theorem 1. *The unbounded operator* $(A - \lambda_0 I)$ *(respectively* $(A^* - \lambda_0 I)$*) with domain* $D(A - \lambda_0 I) = D(A)$ *(respectively* $D(A^* - \lambda_0 I) = D(A^*)$*) is the infinitesimal generator of a bounded analytic semigroup on* $\mathbf{V}_n^0(\Omega)$*. Moreover, for all* $0 \leq \beta \leq 1$*, we have*

$$D((\lambda_0 I - A)^\beta) = D((\lambda_0 I - A^*)^\beta) = D((\lambda_0 I - A_0)^\beta) = D((-A_0)^\beta).$$

Observe that the semigroups $(e^{t(A - \lambda_0 I)})_{t \geq 0}$ and $(e^{t(A^* - \lambda_0 I)})_{t \geq 0}$ are exponentially stable on $\mathbf{V}_n^0(\Omega)$ and that

$$\|e^{t(A - \lambda_0 I)}\|_{\mathcal{L}(\mathbf{V}_n^0(\Omega))} \leq e^{-\omega t} \quad \text{and} \quad \|e^{t(A^* - \lambda_0 I)}\|_{\mathcal{L}(\mathbf{V}_n^0(\Omega))} \leq e^{-\omega t},$$

for all $\omega < \omega_0$ (see [5, Chapter 1, Theorem 2.12]).

Let us introduce D_A and D_p, two Dirichlet operators associated with A, defined as follows. For $\mathbf{u} \in \mathbf{V}^0(\Gamma)$, set $D_A \mathbf{u} = \mathbf{y}$ and $D_p \mathbf{u} = q$, where $(\mathbf{y}, q)$ is the unique solution in $\mathbf{V}^{1/2}(\Omega) \times (H^{1/2}(\Omega)/\mathbb{R})'$ to the equation

$$\lambda_0 \mathbf{y} - \nu \Delta \mathbf{y} + (\mathbf{w} \cdot \nabla)\mathbf{y} + (\mathbf{y} \cdot \nabla)\mathbf{w} + \nabla q = 0 \quad \text{in } \Omega,$$

$$\text{div } \mathbf{y} = 0 \text{ in } \Omega, \quad \mathbf{y} = \mathbf{u} \quad \text{on } \Gamma.$$

Lemma 2. *The operator D_A is a bounded operator from $\mathbf{V}^0(\Gamma)$ into $\mathbf{V}^0(\Omega)$, moreover it satisfies*

$$\|D_A\mathbf{u}\|_{\mathbf{V}^{s+1/2}(\Omega)} \leq C(s)\|\mathbf{u}\|_{\mathbf{V}^s(\Omega)} \qquad \text{for all } 0 \leq s \leq 2.$$

The operator $D_A^* \in \mathcal{L}(\mathbf{V}^0(\Omega), \mathbf{V}^0(\Gamma))$, the adjoint operator of $D_A \in \mathcal{L}(\mathbf{V}^0(\Gamma), \mathbf{V}^0(\Omega))$, is defined by

$$D_A^*\mathbf{g} = -\nu\frac{\partial \mathbf{z}}{\partial \mathbf{n}} + \pi\mathbf{n} - c(\pi)\mathbf{n}, \tag{7}$$

where $(\mathbf{z}, \pi)$ is the solution of

$$\lambda_0\mathbf{z} - \nu\Delta\mathbf{z} - (\mathbf{w}\cdot\nabla)\mathbf{z} + (\nabla\mathbf{w})^T\mathbf{z} + \nabla\pi = \mathbf{g} \quad \text{and div } \mathbf{z} = 0 \text{ in } \Omega,$$
$$\mathbf{z} = 0 \quad \text{on } \Gamma, \tag{8}$$

and $c(\pi)$ is defined by (3).

The first part of the lemma is well known when $\mathbf{w} = 0$, see, e.g., [18]. Its adaptation to the case when $\mathbf{w} \neq 0$, together with the second part of the lemma is proved in [14, Corollary 7.1 and Lemma 7.4].

Notice that the solution of equation (8) obeys $\frac{\partial \mathbf{z}}{\partial \mathbf{n}} \cdot \mathbf{n} = 0$. This can be deduced by a straightforward calculation using the divergence condition (see, e.g., [4, Lemma 3.3.1]).

Let us define the operators $\gamma_\tau \in \mathcal{L}(\mathbf{V}^0(\Gamma))$ and $\gamma_n \in \mathcal{L}(\mathbf{V}^0(\Gamma))$ by

$$\gamma_\tau\mathbf{u} = \mathbf{u} - (\mathbf{u}\cdot\mathbf{n})\mathbf{n} \qquad \text{and} \qquad \gamma_n\mathbf{u} = (\mathbf{u}\cdot\mathbf{n})\mathbf{n} = \mathbf{u} - \gamma_\tau\mathbf{u} \qquad \text{for all } \mathbf{u} \in \mathbf{V}^0(\Gamma).$$

Introducing the spaces

$$\mathbf{V}^0_\tau(\Gamma) = \left\{\mathbf{u} \in \mathbf{V}^0(\Gamma) \mid \gamma_\tau\mathbf{u} = 0\right\} \qquad \text{and} \qquad \mathbf{V}^0_n(\Gamma) = \left\{\mathbf{u} \in \mathbf{V}^0(\Gamma) \mid \gamma_n\mathbf{u} = 0\right\},$$

we have $\mathbf{V}^0(\Gamma) = \mathbf{V}^0_\tau(\Gamma) \oplus \mathbf{V}^0_n(\Gamma)$.

Lemma 3. ([15, Lemma 2.3]) *The operator M obeys the following properties*

$$M = M^*, \qquad M\gamma_\tau = \gamma_\tau M = m\gamma_\tau, \quad \text{and} \quad M\gamma_n = \gamma_n M.$$

The operators γ_τ and γ_n satisfy:

$$\gamma_\tau = \gamma_\tau^*, \qquad \gamma_n = \gamma_n^* \quad \text{and} \quad (I - P)D_A = (I - P)D_A\gamma_n.$$

In the next lemma we study the properties of the operator R_A which plays a crucial role in optimality conditions of control problems that we consider.

Lemma 4. ([15, Lemma 2.4]) *The operator*

$$R_A = MD_A^*(I - P)D_A M + I$$

is an isomorphism from $\mathbf{V}^0(\Gamma)$ into itself. Moreover, for all $0 \leq s \leq 9/2$, its restriction to $\mathbf{V}^s(\Gamma)$ is an isomorphism from $\mathbf{V}^s(\Gamma)$ into itself. In addition R_A satisfies

$$R_A\gamma_n = \gamma_n R_A\gamma_n \qquad \text{and} \qquad R_A\gamma_\tau = \gamma_\tau R_A\gamma_\tau = \gamma_\tau.$$

The restriction of R_A to $\mathbf{V}_\tau^0(\Gamma)$ is an isomorphism from $\mathbf{V}_\tau^0(\Gamma)$ into itself, and we have

$$R_A^{-1}\mathbf{u} = (\gamma_n R_A \gamma_n)^{-1}\mathbf{u} = \gamma_n R_A^{-1}\gamma_n \mathbf{u} = \gamma_n R_A^{-1}\mathbf{u} \quad \text{for all } \mathbf{u} \in \mathbf{V}_\tau^0(\Gamma).$$

In ([15, Lemma 2.4]) the above result is only stated for $0 \le s \le 3/2$. Here the regularity of Ω, m, and $\mathbf{w}$ have been increased so that the operators M, D_A, and D_A^* have better regularity properties, and the restriction of R_A to $\mathbf{V}^s(\Gamma)$ is an automorphism for $0 \le s \le 9/2$.

We introduce the operators

$$B_n = (\lambda_0 I - A)PD_A\gamma_n, \qquad B_\tau = (\lambda_0 I - A)D_A\gamma_\tau, \qquad B = B_n + B_\tau.$$

Let us set

$$B_{n,\beta} = (\lambda_0 I - A)^{\beta-1}B_n = (\lambda_0 I - A)^\beta PD_A\gamma_n,$$

$$B_{\tau,\beta} = (\lambda_0 I - A)^{\beta-1}B_\tau = (\lambda_0 I - A)^\beta D_A\gamma_\tau, \quad \text{and} \quad B_\beta = B_{n,\beta} + B_{\tau,\beta}.$$

Theorem 5. ([15, Theorem 2.5]) *For all $\beta \in]0, \frac{1}{4}[$, $B_{n,\beta}$ and $B_{\tau,\beta}$ belong to $\mathcal{L}(\mathbf{V}^0(\Gamma), \mathbf{V}_n^0(\Omega))$.*

Proposition 6. ([15, Proposition 2.6]) *For all $\mathbf{\Phi} \in D(A^*)$, $B^*\mathbf{\Phi}$ belongs to $\mathbf{V}^{1/2}(\Gamma)$, we have*

$$B^*\mathbf{\Phi} = D_A^*(\lambda_0 I - A^*)\mathbf{\Phi}, \quad B_\tau^*\mathbf{\Phi} = \gamma_\tau D_A^*(\lambda_0 I - A^*)\mathbf{\Phi}, \quad B_n^*\mathbf{\Phi} = \gamma_n D_A^*(\lambda_0 I - A^*)\mathbf{\Phi},$$

and

$$B^*\mathbf{\Phi} = -\nu\frac{\partial\mathbf{\Phi}}{\partial\mathbf{n}} + \psi\mathbf{n} - c(\psi)\mathbf{n},$$

with

$$\nabla\psi = (I - P)\Big[\nu\Delta\mathbf{\Phi} + (\mathbf{w}\cdot\nabla)\mathbf{\Phi} - (\nabla\mathbf{w})^T\mathbf{\Phi}\Big],$$

and $c(\psi)$ is defined by (3). In particular if $\mathbf{\Phi} \in \mathbf{V}^s(\Omega) \cap \mathbf{V}_0^1(\Omega)$ with $s > 3/2$, the following estimate holds

$$|B^*\mathbf{\Phi}|_{\mathbf{V}^{s-3/2}(\Gamma)} \le C|\mathbf{\Phi}|_{\mathbf{V}^s(\Omega)\cap\mathbf{V}_0^1(\Omega)}.$$

In [15] we have shown that $\mathbf{y}$ is a solution of equation (4) in the sense of transposition if and only if $P\mathbf{y}$ and $(I - P)\mathbf{y}$ are the solutions of the system

$$P\mathbf{y}' = AP\mathbf{y} + BM\mathbf{u} \quad \text{in } (0, T), \qquad P\mathbf{y}(0) = \mathbf{y}_0,$$

$$(I - P)\mathbf{y} = (I - P)D_A M\mathbf{u} = (I - P)D_A\gamma_n M\mathbf{u} \quad \text{in } (0, T). \tag{9}$$

3. A regularizing feedback operator

Since equation (4) is written in the form (9), to study the control problem $(\mathcal{P})$ stated in the introduction, it is sufficient to study the following problem in which $\mathbf{y}$ plays the role of $P\mathbf{y}$:

$$(\mathcal{P}_{0,\mathbf{y}_0}) \qquad \inf\Big\{I(\mathbf{y}, \mathbf{u}) \mid (\mathbf{y}, \mathbf{u}) \text{ satisfies (10)}, \ \mathbf{u} \in \mathbf{V}^{0,0}(\Sigma_\infty)\Big\},$$

where

$$I(\mathbf{y}, \mathbf{u}) = \frac{1}{2} \int_0^\infty \int_\Omega |(-A_0)^{-\alpha}\mathbf{y}|^2 \, dxdt + \frac{1}{2} \int_0^\infty |R_A^{1/2}\mathbf{u}(t)|^2_{\mathbf{V}^0(\Gamma)} \, dt,$$

and

$$\mathbf{y}' = A\mathbf{y} + BM\mathbf{u} \quad \text{in } (0, \infty), \qquad \mathbf{y}(0) = \mathbf{y}_0. \tag{10}$$

Notice that, due to Lemma 4, we have:

$$|R_A^{1/2}\mathbf{u}(t)|^2_{\mathbf{V}^0(\Gamma)} = |\gamma_\tau\mathbf{u}(t)|^2_{\mathbf{V}^0(\Gamma)} + |R_A^{1/2}\gamma_n\mathbf{u}(t)|^2_{\mathbf{V}^0(\Gamma)}.$$

3.1. A finite time horizon control problem

We first study the following family of finite time horizon control problems

$$(\mathcal{P}^T_{s,\zeta}) \qquad \inf\left\{ I_T(s, \mathbf{y}, \mathbf{u}) \ \mid \ (\mathbf{y}, \mathbf{u}) \text{ satisfies (11)}, \ \mathbf{u} \in \mathbf{V}^{0,0}(\Sigma_{s,T}) \right\},$$

where

$$\mathbf{y}' = A\mathbf{y} + BM\mathbf{u} \quad \text{in } (s, T), \qquad \mathbf{y}(s) = \zeta, \tag{11}$$

and

$$I_T(s, \mathbf{y}, \mathbf{u}) = \frac{1}{2} \int_s^T \int_\Omega |(-A_0)^{-\alpha}\mathbf{y}|^2 + \frac{1}{2} \int_s^T \int_\Gamma |R_A^{1/2}\gamma_n\mathbf{u}|^2 + \frac{1}{2} \int_s^T \int_\Gamma |\gamma_\tau\mathbf{u}|^2.$$

Theorem 7. *For all $s \in [0, T]$ and all $\zeta \in \mathbf{V}_n^0(\Omega)$, problem $(\mathcal{P}^T_{s,\zeta})$ admits a unique solution $(\mathbf{y}^s_\zeta, \mathbf{u}^s_\zeta)$. The optimal control $\mathbf{u}^s_\zeta$ is characterized by*

$$\mathbf{u}^s_\zeta = -MB_\tau^*\mathbf{\Phi}^s_\zeta - R_A^{-1}MB_n^*\mathbf{\Phi}^s_\zeta \quad \text{in } (s, T), \tag{12}$$

where $\mathbf{\Phi}^s_\zeta$ is solution to the equation

$$-\mathbf{\Phi}' = A^*\mathbf{\Phi} + (-A_0)^{-2\alpha}\mathbf{y}^s_\zeta \quad \text{in } (s, T), \qquad \mathbf{\Phi}(T) = 0. \tag{13}$$

Conversely the system

$$\begin{aligned}
\mathbf{y}' &= A\mathbf{y} - B_\tau M^2 B_\tau^*\mathbf{\Phi} - B_n M R_A^{-1}MB_n^*\mathbf{\Phi} \quad \text{in } (s, T), \qquad \mathbf{y}(s) = \zeta, \\
-\mathbf{\Phi}' &= A^*\mathbf{\Phi} + (-A_0)^{-2\alpha}\mathbf{y} \quad \text{in } (s, T), \qquad \mathbf{\Phi}(T) = 0,
\end{aligned} \tag{14}$$

admits a unique solution $(\mathbf{y}^s_\zeta, \mathbf{\Phi}^s_\zeta)$ in $L^2(s, T; \mathbf{V}_n^0(\Omega)) \times \left(\mathbf{V}^{2,1}(Q_{s,T}) \cap L^2(s, T; \mathbf{V}_0^1(\Omega))\right)$, and $(\mathbf{y}^s_\zeta, -MB_\tau^*\mathbf{\Phi}^s_\zeta - R_A^{-1}MB_n^*\mathbf{\Phi}^s_\zeta) = (\mathbf{y}^s_\zeta, -R_A^{-1}MB^*\mathbf{\Phi}^s_\zeta)$ is the optimal solution to $(\mathcal{P}^T_{s,\zeta})$.

Proof. The proof is similar to that of [15, Theorem 3.1]. $\qquad\square$

As in [15, Lemma 3.1], we can show that

$$B^*\mathbf{\Phi}^s_\zeta(t) = -\nu\frac{\partial\mathbf{\Phi}^s_\zeta}{\partial\mathbf{n}}(t) + \psi^s_\zeta(t)\mathbf{n} - c(\psi^s_\zeta(t))\mathbf{n} \quad \text{for almost all } t \in (s, T), \tag{15}$$

where ψ_ζ^s, the pressure associated with $\boldsymbol{\Phi}_\zeta^s$, is related to $\boldsymbol{\Phi}_\zeta^s$ by the equation

$$-\frac{\partial \boldsymbol{\Phi}_\zeta^s}{\partial t} - \nu \Delta \boldsymbol{\Phi}_\zeta^s - (\mathbf{w} \cdot \nabla)\boldsymbol{\Phi}_\zeta^s + (\nabla \mathbf{w})^T \boldsymbol{\Phi}_\zeta^s + \nabla \psi_\zeta^s$$

$$= (-A_0)^{-2\alpha} \mathbf{y}_\zeta^s, \quad \text{in } Q_{s,T},$$

$$\operatorname{div} \boldsymbol{\Phi}_\zeta^s = 0 \quad \text{in } Q_{s,T}, \quad \boldsymbol{\Phi}_\zeta^s = 0 \text{ on } \Sigma_{s,T}, \quad \boldsymbol{\Phi}_\zeta^s(T) = 0 \quad \text{in } \Omega.$$

$$(16)$$

The pressure ψ_ζ^s is also defined by

$$\nabla \psi_\zeta^s = (I - P)\Big[\nu \Delta \boldsymbol{\Phi}_\zeta^s + (\mathbf{w} \cdot \nabla)\boldsymbol{\Phi}_\zeta^s - (\nabla \mathbf{w})^T \boldsymbol{\Phi}_\zeta^s\Big].$$

Therefore the optimal control $\mathbf{u}_\zeta^s$ is defined by

$$\mathbf{u}_\zeta^s = m\nu \frac{\partial \boldsymbol{\Phi}_\zeta^s}{\partial \mathbf{n}} - R_A^{-1} M\psi_\zeta^s \mathbf{n} = -MB_\tau^* \boldsymbol{\Phi}_\zeta^s - R_A^{-1} MB_n^* \boldsymbol{\Phi}_\zeta^s. \quad (17)$$

In the following theorem we improve the regularity result of the optimal solution.

Theorem 8. *If* $\zeta \in \mathbf{V}_n^0(\Omega)$, *the solution* $(\mathbf{y}_\zeta^s, \boldsymbol{\Phi}_\zeta^s)$ *to system* (14) *belongs to* $\mathbf{V}^{1,1/2}(Q_{s,T}) \times L^2(s,T; \mathbf{V}^{(3+4\alpha)\wedge\sigma}(\Omega) \cap \mathbf{V}_0^1(\Omega)) \cap H^{3/2}(s,T; \mathbf{V}_n^{4\alpha\wedge(\sigma-3)}(\Omega))$ *for all* $\sigma < 7/2$.

Proof. Since $B^* \boldsymbol{\Phi}_\zeta^s \in L^2(s,T; \mathbf{V}^0(\Gamma))$, due to Lemmas 16 and 17, $\mathbf{y}_\zeta^s$ belongs to $\mathbf{V}^{1/2-\varepsilon,1/4-\varepsilon/2}(Q_{s,T})$ for all $\varepsilon > 0$. From Lemmas 22 and 23 it follows that

$$\frac{\partial \boldsymbol{\Phi}_\zeta^s}{\partial \mathbf{n}} \in \mathbf{V}^{1-\varepsilon,1/2-\varepsilon/2}(\Sigma_{s,T})$$

and

$$\psi_\zeta^s \in L^2(s,T; H^{3/2-\varepsilon}(\Omega)) \cap H^{1/4-\varepsilon/2}(s,T; H^1(\Omega)),$$

for all $\varepsilon > 0$, where ψ_ζ^s is the pressure appearing in equation (16) (we have not yet used the additional regularity coming from $(-A_0)^{-2\alpha}$, we shall use it at the end of the proof). From this regularity result and from (15), we deduce that $B^* \boldsymbol{\Phi}_\zeta^s \in \mathbf{V}^{1/2-\varepsilon,1/4-\varepsilon/2}(\Sigma_{s,T})$ and $\mathbf{u}_\zeta^s \in \mathbf{V}^{1/2-\varepsilon,1/4-\varepsilon/2}(\Sigma_{s,T})$ for all $\varepsilon > 0$, where $\mathbf{u}_\zeta^s = -MB_\tau^* \boldsymbol{\Phi}_\zeta^s - R_A^{-1} MB_n^* \boldsymbol{\Phi}_\zeta^s$ is the optimal control of problems $(\mathcal{P}_{s,\zeta}^T)$. Still applying Lemma 17, we obtain $\mathbf{y}_\zeta^s \in \mathbf{V}^{1-\varepsilon,1/2-\varepsilon}(Q_{s,T})$ for all $\varepsilon > 0$, and repeating the above analysis for $B^* \boldsymbol{\Phi}_\zeta^s$, we can show that $\mathbf{u}_\zeta^s \in \mathbf{V}^{1-\varepsilon,1/2-\varepsilon/2}(\Sigma_{s,T})$ for all $\varepsilon > 0$. Still with Lemmas 16 and 17, we prove that $\mathbf{y}_\zeta^s$ belongs to $\mathbf{V}^{1,1/2}(Q_{s,T})$. Due to Lemma 22, $\boldsymbol{\Phi}_\zeta^s$ belongs to $L^2(s,T; \mathbf{V}^{(3+4\alpha)\wedge\sigma}(\Omega) \cap \mathbf{V}_0^1(\Omega)) \cap H^{3/2}(s,T; \mathbf{V}_n^{4\alpha\wedge(\sigma-3)}(\Omega))$ for all $\sigma < 7/2$. The proof is complete. $\qquad\square$

Corollary 9. *For all* $s \in [0,T]$ *and all* $\zeta \in \mathbf{V}_n^0(\Omega)$, *the unique solution* $(\mathbf{y}_\zeta^s, \mathbf{u}_\zeta^s)$ *to problem* $(\mathcal{P}_{s,\zeta}^T)$ *and the corresponding solution* $\boldsymbol{\Phi}_\zeta^s$ *to equation* (16) *obey*

$$I_T(s, \mathbf{y}_\zeta^s, \mathbf{u}_\zeta^s) = \frac{1}{2} \int_\Omega \boldsymbol{\Phi}_\zeta^s(s) \cdot \zeta.$$

Proof. The proof is similar to that of [15, Corollary 3.1]. $\qquad\square$

Let $\Pi(s)$ be the operator defined by

$$\Pi(s) \ : \ \zeta \longmapsto \boldsymbol{\Phi}_\zeta^s(s), \tag{18}$$

where $(\mathbf{y}_\zeta^s, \boldsymbol{\Phi}_\zeta^s)$ is the unique solution to system (14). From Theorem 8 it follows that $\Pi(s) \in \mathcal{L}(\mathbf{V}_n^0(\Omega), \mathbf{V}^2(\Omega) \cap \mathbf{V}_0^1(\Omega))$. Actually using Theorem 8 we can prove that the family of operators $(\Pi(s))_{s \in [0,T]}$ defined by (18) belongs to $C_s([0,T]; \mathcal{L}(\mathbf{V}_n^0(\Omega)))$ (the space of functions Π from $[0,T]$ into $\mathcal{L}(\mathbf{V}_n^0(\Omega))$ such that, for all $\mathbf{y} \in \mathbf{V}_n^0(\Omega)$, $\Pi(\cdot)\mathbf{y}$ is continuous from $[0,T]$ into $\mathbf{V}_n^0(\Omega)$). Next, using the optimality system (14) we can show that Π is the unique solution in $C_s([0,T]; \mathcal{L}(\mathbf{V}_n^0(\Omega)))$ to the Riccati equation

$$\begin{aligned}
&\Pi^*(t) = \Pi(t) \quad \text{and} \quad \Pi(t) \geq 0, \\
&|\Pi(t)\zeta|_{\mathbf{V}^2(\Omega) \cap \mathbf{V}_0^1(\Omega)} \leq C|\zeta|_{\mathbf{V}_n^0(\Omega)} \quad \text{for all } \zeta \in \mathbf{V}_n^0(\Omega), \ t \in [0,T], \\
&-\Pi'(t) = A^*\Pi(t) + \Pi(t)A - \Pi(t)B_\tau M^2 B_\tau^*\Pi(t) \\
&\qquad\qquad - \Pi(t)B_n M R_A^{-1} M B_n^*\Pi(t) + (-A_0)^{-2\alpha}, \\
&\Pi(T) = 0.
\end{aligned} \tag{19}$$

In (19), $\Pi^*(t) \in \mathcal{L}(\mathbf{V}_n^0(\Omega))$ is the adjoint of $\Pi(t) \in \mathcal{L}(\mathbf{V}_n^0(\Omega))$. From the definition of Π, from Theorem 7 and Corollary 9 we deduce the following theorem. We also refer to [12, Theorem 1.2.2.1] where the existence of a unique solution to equation (19) is established.

Theorem 10. *The solution* $(\mathbf{y}, \mathbf{u})$ *to problem* $(\mathcal{P}_{0,\mathbf{y}_0}^T)$ *belongs to* $C([0,T]; \mathbf{V}^0(\Omega)) \times C([0,T]; \mathbf{V}^0(\Gamma))$, *it obeys the feedback formula*

$$\mathbf{u}(t) = -\left(M B_\tau^* + R_A^{-1} M B_n^*\right)\Pi(t)P\mathbf{y}(t),$$

and the optimal cost is given by

$$J(\mathbf{y}, \mathbf{u}) = \frac{1}{2}\Big(\Pi(0)\mathbf{y}_0, \mathbf{y}_0\Big)_{\mathbf{V}_n^0(\Omega)}.$$

If we set $\widehat{\Pi}(t) = \Pi(T-t)$, then $\widehat{\Pi}$ is the unique solution in $C_s([0,T]; \mathcal{L}(\mathbf{V}_n^0(\Omega)))$ to the Riccati equation

$$\begin{aligned}
&\widehat{\Pi}^*(t) = \widehat{\Pi}(t) \quad \text{and} \quad \widehat{\Pi}(t) \geq 0, \\
&|\widehat{\Pi}(t)\zeta|_{\mathbf{V}^2(\Omega) \cap \mathbf{V}_0^1(\Omega)} \leq C|\zeta|_{\mathbf{V}_n^0(\Omega)} \quad \text{for all } \zeta \in \mathbf{V}_n^0(\Omega), \ t \in [0,T], \\
&\widehat{\Pi}'(t) = A^*\widehat{\Pi}(t) + \widehat{\Pi}(t)A - \widehat{\Pi}(t)B_\tau B_\tau^*\widehat{\Pi}(t) \\
&\qquad\qquad - \widehat{\Pi}(t)B_n R_A^{-1} B_n^*\widehat{\Pi}(t) + (-A_0)^{-2\alpha}, \\
&\widehat{\Pi}(0) = 0.
\end{aligned} \tag{20}$$

From the definition of $\widehat{\Pi}$ it follows that $\Pi(0) = \widehat{\Pi}(T)$.

3.2. An infinite time horizon control problem

In this section we study problem $(\mathcal{P}_{0,\mathbf{y}_0})$.

Theorem 11. *For all* $\mathbf{y}_0 \in \mathbf{V}_n^0(\Omega)$, *problem* $(\mathcal{P}_{0,\mathbf{y}_0})$ *admits a unique solution* $(\mathbf{y}_{\mathbf{y}_0}, \mathbf{u}_{\mathbf{y}_0})$. *There exists* $\Pi \in \mathcal{L}(\mathbf{V}_n^0(\Omega))$, *obeying* $\Pi = \Pi^*$, *such that the optimal cost is given by*

$$I(\mathbf{y}_{\mathbf{y}_0}, \mathbf{u}_{\mathbf{y}_0}) = \frac{1}{2}\Big(\Pi \mathbf{y}_0, \mathbf{y}_0\Big)_{\mathbf{V}_n^0(\Omega)}.$$

Proof. Let us first prove that there exist controls $\mathbf{u} \in L^2(0,\infty; \mathbf{V}^0(\Gamma))$ such that $I(\mathbf{y}_{\mathbf{u}}, \mathbf{u}) < \infty$, where $\mathbf{y}_{\mathbf{u}}$ is the solution of equation (10) corresponding to $\mathbf{u}$. For that we are going to use null controllability results stated in [7] for the Oseen equations with a distributed control, and an extension procedure. Let $\mathcal{O}$ be an open regular subset in $\mathbb{R}^3 \setminus \overline{\Omega}$ such that $\overline{\mathcal{O}} \cap \overline{\Omega} = \Gamma_1$, where Γ_1 is an open regular subset of Γ_0 (recall that $\Gamma_0 \subset \Gamma_c$ and that we look for controls $\mathbf{u}$ with support in Γ_c). We could extend equation (2) to $\mathrm{int}(\overline{\mathcal{O}} \cup \overline{\Omega})$, but this set is not necessarily regular. Thus to construct an extension in a regular set Ω_e, we proceed as follows. Let ω be an open set such that $\overline{\omega} \subset \mathcal{O}$, and let Ω_e be open set with a regular boundary satisfying

$$(\Omega \cup \omega) \subset \Omega_e \subset \mathrm{int}(\overline{\mathcal{O}} \cup \overline{\Omega}).$$

To extend equation (2) to Ω_e, we have to extend $\mathbf{y}_0$ and $\mathbf{w}$. Since $\mathbf{w}$ is regular, $\mathbf{u}_s^\infty|_{\Gamma_1}$ is also regular and it can be extended to $\partial\mathcal{O}$ in such a way that, if $\tilde{\mathbf{u}}$ denotes such an extension, then $\tilde{\mathbf{u}} \in \mathbf{V}^{9/2}(\Gamma)$. Let $\mathbf{v} \in \mathbf{V}^5(\Omega)$ be the solution of the Stokes equation:

$$-\nu\Delta\mathbf{v} + \nabla q = 0 \quad \text{and} \quad \mathrm{div}\,\mathbf{v} = 0 \text{ in } \mathcal{O}, \qquad \mathbf{v} = \tilde{\mathbf{u}} \quad \text{on } \partial\mathcal{O}.$$

We set

$$\mathbf{w}_e(x) = \left\{ \begin{array}{ll} \mathbf{w}(x) & \text{if} \quad x \in \Omega, \\ \mathbf{v}(x) & \text{if} \quad x \in \mathcal{O}. \end{array} \right.$$

Since the traces of $\mathbf{w}$ and $\mathbf{v}$ on Γ_1 are equal, it is clear that $\mathrm{div}\,\mathbf{w}_e = 0$ in $\mathrm{int}(\overline{\mathcal{O}} \cup \overline{\Omega})$, and therefore in Ω_e. Let $\mathbf{z}_0$ be the extension of $\mathbf{y}_0$ by zero to $\Omega_e \setminus \Omega$. Since $\mathbf{y}_0 \cdot \mathbf{n} = 0$ on Γ, it is clear that $\mathbf{z}_0$ belongs to $\mathbf{V}_n^0(\Omega_e)$. Now we consider the following control system:

$$\frac{\partial\mathbf{z}}{\partial t} - \nu\Delta\mathbf{z} + (\mathbf{w}_e \cdot \nabla)\mathbf{z} + (\mathbf{z} \cdot \nabla)\mathbf{w}_e + \nabla q = \chi_\omega\,\mathbf{g}, \quad \text{in } \Omega_e \times (0,T), \tag{21}$$

$$\mathrm{div}\,\mathbf{z} = 0 \ \text{ in } \Omega_e \times (0,T), \ \mathbf{z} = 0 \text{ on } \partial\Omega_e \times (0,T), \ \mathbf{z}(0) = \mathbf{z}_0 \ \text{ in } \Omega_e,$$

for a given $T > 0$, where χ_ω is the characteristic function of ω. Set $\Gamma_e = \Gamma \cap \Omega_e$. From [7] it follows that there exists $\hat{\mathbf{g}} \in L^2(0,T; \mathbf{L}^2(\omega))$ such that the solution $\hat{\mathbf{z}}$ to equation (21) satisfies $\hat{\mathbf{z}}(T) = 0$ in Ω_e. Now if we set

$$\hat{\mathbf{u}}(x,t) = \left\{ \begin{array}{ll} \hat{\mathbf{z}}(x,t) & \text{if} \quad (x,t) \in \Gamma_e \times (0,T), \\ 0 & \text{if} \quad (x,t) \in \Sigma_\infty \setminus (\Gamma_e \times (0,T)), \end{array} \right.$$

it is clear that $\mathbf{y}_{\hat{\mathbf{u}}}$, the solution to equation (2) corresponding to $\hat{\mathbf{u}}$, obeys $\mathbf{y}_{\hat{\mathbf{u}}}|_{\Omega \times (0,T)} = \mathbf{z}|_{\Omega \times (0,T)}$, and that $\mathbf{y}_{\hat{\mathbf{u}}}(t) = 0$ for $t > T$. Thus we have proved that there exists a control $\hat{\mathbf{u}} \in L^2(0, \infty; \mathbf{V}^0(\Gamma))$ such that $I(\mathbf{y}_{\hat{\mathbf{u}}}, \hat{\mathbf{u}}) < \infty$. The existence of a unique solution $(\mathbf{y}_{\mathbf{y}_0}, \mathbf{u}_{\mathbf{y}_0})$ to $(\mathcal{P}_{0,\mathbf{y}_0})$ follows from classical arguments.

For the end of the proof we adapt to the case $0 \le \alpha \le 1/2$ what is done in [15, Theorem 4.1] for $\alpha = 0$. The new point is that we need Lemmas 25 and 26 in the proof. From the dynamic programming principle, it follows that the mapping

$$T \longmapsto \left(\widehat{\Pi}(T)\mathbf{y}_0, \mathbf{y}_0\right)_{\mathbf{V}_n^0(\Omega)}$$

is nondecreasing, and we have

$$\frac{1}{2}\left(\widehat{\Pi}(T)\mathbf{y}_0, \mathbf{y}_0\right)_{\mathbf{V}_n^0(\Omega)} \le I(\mathbf{y}_{\mathbf{y}_0}, \mathbf{u}_{\mathbf{y}_0}) < \infty.$$

As in [6], or in [12], we can show that there exists an operator $\Pi \in \mathcal{L}(\mathbf{V}_n^0(\Omega))$ satisfying $\Pi = \Pi^* \ge 0$ and

$$\Pi\mathbf{y}_0 = \lim_{T \to \infty} \widehat{\Pi}(T)\mathbf{y}_0 \qquad \text{for all } \mathbf{y}_0 \in \mathbf{V}_n^0(\Omega).$$

Let us show that $I(\mathbf{y}_{\mathbf{y}_0}, \mathbf{u}_{\mathbf{y}_0}) = \frac{1}{2}\left(\Pi\mathbf{y}_0, \mathbf{y}_0\right)_{\mathbf{V}_n^0(\Omega)}$. Problem $(\mathcal{P}_{0,\mathbf{y}_0}^k)$ admits a unique solution $(\mathbf{y}_k, \mathbf{u}_k)$ characterized by

$$\mathbf{y}_k' = A\mathbf{y}_k + BM\mathbf{u}_k \quad \text{in } (0, k), \qquad \mathbf{y}_k(0) = \mathbf{y}_0,$$

$$-\boldsymbol{\Phi}_k' = A^*\boldsymbol{\Phi}_k + (-A_0)^{-2\alpha}\mathbf{y}_k \quad \text{in } (0, k), \qquad \boldsymbol{\Phi}_k(k) = 0, \tag{22}$$

$$\gamma_\tau \mathbf{u}_k = -MB_\tau^*\boldsymbol{\Phi}_k, \qquad \gamma_n \mathbf{u}_k = -R_A^{-1}MB_n^*\boldsymbol{\Phi}_k.$$

Convergence of $\mathbf{y}_k$ and $\mathbf{u}_k$. Denote by $\tilde{\mathbf{u}}_k$ the extension by zero of $\mathbf{u}_k$ to (k, ∞), and by $\tilde{\mathbf{y}}_k$ the extension by zero of $\mathbf{y}_k$ to (k, ∞). Since we have

$$\int_0^k \int_\Omega |(-A_0)^{-\alpha}\mathbf{y}_k|^2 \, dx dt + \int_0^k |R_A^{1/2}\mathbf{u}_k(t)|_{\mathbf{V}^0(\Gamma)}^2 \, dt$$

$$\le \int_0^\infty \int_\Omega |(-A_0)^{-\alpha}\mathbf{y}_{\mathbf{y}_0}|^2 \, dx dt + \int_0^\infty |R_A^{1/2}\mathbf{u}_{\mathbf{y}_0}(t)|_{\mathbf{V}^0(\Gamma)}^2 \, dt\,,$$

the sequences $(\tilde{\mathbf{y}}_k)_k$ and $(\tilde{\mathbf{u}}_k)_k$ are respectively bounded in $L^2(0, \infty; (D((-A_0)^\alpha))')$ and $L^2(0, \infty; \mathbf{V}^0(\Gamma))$. Thus there exist $\mathbf{y}_\infty \in L^2(0, \infty; (D((-A_0)^\alpha))')$ and $\mathbf{u}_\infty \in L^2(0, \infty; \mathbf{V}^0(\Gamma))$ such that

$$\tilde{\mathbf{u}}_k \rightharpoonup \mathbf{u}_\infty \quad \text{weakly in } L^2(0, \infty; \mathbf{V}^0(\Gamma)),$$

$$\tilde{\mathbf{y}}_k \rightharpoonup \mathbf{y}_\infty \quad \text{weakly in } L^2(0, \infty; (D((-A_0)^\alpha))').$$

By passing to the limit in the above inequality we obtain

$$\int_0^\infty \int_\Omega |(-A_0)^{-\alpha}\mathbf{y}_\infty|^2 \, dx dt + \int_0^\infty |R_A^{1/2}\mathbf{u}_\infty(t)|_{\mathbf{V}^0(\Gamma)}^2 \, dt$$

$$\le \int_0^\infty \int_\Omega |(-A_0)^{-\alpha}\mathbf{y}_{\mathbf{y}_0}|^2 \, dx dt + \int_0^\infty |R_A^{1/2}\mathbf{u}_{\mathbf{y}_0}(t)|_{\mathbf{V}^0(\Gamma)}^2 \, dt\,.$$

Rewriting the first equation in (22) in the form:

$$\mathbf{y}_k' = A\mathbf{y}_k - \lambda_0(-A_0)^{-\alpha}\mathbf{y}_k + \lambda_0(-A_0)^{-\alpha}\mathbf{y}_k + BM\mathbf{u}_k \quad \text{in } (0, k), \qquad \mathbf{y}_k(0) = \mathbf{y}_0,$$

and by passing to the limit in this equation, with Lemmas 25 and 26, we can show that

$$\mathbf{y}'_\infty = A\mathbf{y}_\infty + BM\mathbf{u}_\infty \quad \text{in } (0,\infty), \qquad \mathbf{y}_\infty(0) = \mathbf{y}_0.$$

Thus the pair $(\mathbf{y}_\infty, \mathbf{u}_\infty)$ is admissible for $(\mathcal{P}_{0,\mathbf{y}_0})$ and we have

$$(\mathbf{y}_\infty, \mathbf{u}_\infty) = (\mathbf{y}_{\mathbf{y}_0}, \mathbf{u}_{\mathbf{y}_0}),$$

because $I(\mathbf{y}_\infty, \mathbf{u}_\infty) \le I(\mathbf{y}_{\mathbf{y}_0}, \mathbf{u}_{\mathbf{y}_0})$. Therefore we can claim that

$$\tilde{\mathbf{u}}_k \to \mathbf{u}_{\mathbf{y}_0} \quad \text{in } L^2(0,\infty; \mathbf{V}^0(\Gamma)) \quad \text{and} \quad \tilde{\mathbf{y}}_k \to \mathbf{y}_{\mathbf{y}_0} \quad \text{in } L^2(0,\infty; (D((-A_0)^\alpha))').$$

From Lemmas 25 and 26, it follows that $(\mathbf{y}_k)_k$ tends to $\mathbf{y}_{\mathbf{y}_0}$ in $L^2(0,\infty; \mathbf{V}_n^0(\Omega))$. Since

$$I_k(0, \mathbf{y}_k, \mathbf{u}_k) = \frac{1}{2}\left(\widehat{\Pi}(k)\mathbf{y}_0, \mathbf{y}_0\right)_{\mathbf{V}_n^0(\Omega)},$$

by passing to the limit when k tends to infinity, we obtain

$$I(\mathbf{y}_{\mathbf{y}_0}, \mathbf{u}_{\mathbf{y}_0}) = \frac{1}{2}\left(\Pi\mathbf{y}_0, \mathbf{y}_0\right)_{\mathbf{V}_n^0(\Omega)}. \qquad \square$$

We denote by $\varphi(\mathbf{y}_0)$ the value function of problem $(\mathcal{P}_{0,\mathbf{y}_0})$, that is:

$$\varphi(\mathbf{y}_0) = I(\mathbf{y}_{\mathbf{y}_0}, \mathbf{u}_{\mathbf{y}_0}).$$

Lemma 12. *For every* $\mathbf{y}_0 \in \mathbf{V}_n^0(\Omega)$, *the system*

$$\mathbf{y}' = A\mathbf{y} - B_\tau M^2 B_\tau^* \mathbf{\Phi} - B_n M R_A^{-1} M B_n^* \mathbf{\Phi} \quad \text{in } (0,\infty), \qquad \mathbf{y}(0) = \mathbf{y}_0,$$

$$-\mathbf{\Phi}' = A^* \mathbf{\Phi} + (-A_0)^{-2\alpha}\mathbf{y} \quad \text{in } (0,\infty), \qquad \mathbf{\Phi}(\infty) = 0, \tag{23}$$

$$\mathbf{\Phi}(t) = \Pi\mathbf{y}(t) \quad \text{for all } t \in (0,\infty),$$

admits a unique solution in $L^2(0,\infty; \mathbf{V}_n^0(\Omega)) \times \mathbf{V}^{2,1}(Q_\infty)$. *This solution belongs to* $C_b(\mathbb{R}^+; \mathbf{V}_n^0(\Omega)) \cap \mathbf{V}^{1,1/2}(Q_\infty) \times \left(L^2(0,\infty; \mathbf{V}^{3+4\alpha}(\Omega)) \cap H^{3/2}(0,\infty; \mathbf{V}^{4\alpha}(\Omega))\right)$ *and it satisfies:*

$$\|\mathbf{y}\|_{C_b(\mathbb{R}^+; \mathbf{V}_n^0(\Omega))} + \|\mathbf{y}\|_{\mathbf{V}^{1,1/2}(Q_\infty)} + \|\mathbf{\Phi}\|_{L^2(0,\infty; \mathbf{V}^{3+4\alpha}(\Omega)) \cap H^{3/2}(0,\infty; \mathbf{V}^{4\alpha}(\Omega))}$$

$$\le C|\mathbf{y}_0|_{\mathbf{V}_n^0(\Omega)}.$$

The pair $(\mathbf{y}, -MB_\tau^* \mathbf{\Phi} - R_A^{-1} M B_n^* \mathbf{\Phi})$ *is the solution of* $(\mathcal{P}_{0,\mathbf{y}_0})$.

Proof. The above lemma is already stated in [14] in the case when $\alpha = 0$. The extension to the case when $0 \le \alpha \le 1/2$ is still obtained thanks to Lemmas 25 and 26. For notational simplicity the solution to $(\mathcal{P}_{0,\mathbf{y}_0})$ will now be denoted by $(\hat{\mathbf{y}}, \hat{\mathbf{u}})$, that is $(\hat{\mathbf{y}}, \hat{\mathbf{u}}) = (\mathbf{y}_{\mathbf{y}_0}, \mathbf{u}_{\mathbf{y}_0})$. We denote by $\varphi_k(0, \mathbf{y}_0)$ the value function of problem $(\mathcal{P}_{0,\mathbf{y}_0}^k)$ and by $\varphi_k(\bar{t}, \varsigma)$ the value function of problem $(\mathcal{P}_{\bar{t},\varsigma}^k)$.

Let $(\mathbf{y}_k, \mathbf{u}_k)$ be the solution of $(\mathcal{P}_{0,\mathbf{y}_0}^k)$ characterized by (22), and let $(\mathbf{y}_k^{\bar{t}}, \mathbf{u}_k^{\bar{t}})$ be the solution of $(\mathcal{P}_{\bar{t},\mathbf{y}_k(\bar{t})}^k)$. Denote by $\mathbf{\Phi}_k^{\bar{t}}$ the adjoint state corresponding to $(\mathbf{y}_k^{\bar{t}}, \mathbf{u}_k^{\bar{t}})$, and by $\mathbf{\Phi}_k$ the adjoint state corresponding to $(\mathbf{y}_k, \mathbf{u}_k)$. From the dynamic programming principle it follows that $(\mathbf{y}_k^{\bar{t}}, \mathbf{u}_k^{\bar{t}}, \mathbf{\Phi}_k^{\bar{t}})(t) = (\mathbf{y}_k, \mathbf{u}_k, \mathbf{\Phi}_k)(t)$ for all $t \in$

$(\bar{t}, k)$. Therefore we have $\boldsymbol{\Phi}_k^{\bar{t}}(\bar{t}) = \boldsymbol{\Phi}_k(\bar{t}) \in \partial_y \varphi_k(\bar{t}, \mathbf{y}_k(\bar{t}))$, that is $\boldsymbol{\Phi}_k(\bar{t}) = \widehat{\Pi}(k - \bar{t})\mathbf{y}_k(\bar{t})$.

In the proof of Theorem 11, denoting by $\tilde{\mathbf{y}}_k$ the extension by zero of $\mathbf{y}_k$ to (k, ∞), we have shown that $(\tilde{\mathbf{y}}_k)_k$ converges to $\hat{\mathbf{y}}$ in $L^2(0, \infty; \mathbf{V}_n^0(\Omega))$. Thus

$$|\boldsymbol{\Phi}_k(\bar{t})|_{\mathbf{V}_n^0(\Omega)} \leq \|\widehat{\Pi}(k - \bar{t})\| |\mathbf{y}_k(\bar{t})|_{\mathbf{V}_n^0(\Omega)} \leq C |\mathbf{y}_k(\bar{t})|_{\mathbf{V}_n^0(\Omega)},$$

and

$$\|\tilde{\boldsymbol{\Phi}}_k\|_{L^2(0,\infty;\mathbf{V}_n^0(\Omega))} \leq C \|\tilde{\mathbf{y}}_k\|_{L^2(0,\infty;\mathbf{V}_n^0(\Omega))},$$

where $\tilde{\boldsymbol{\Phi}}_k$ is the extension by zero of $\boldsymbol{\Phi}_k$ to (k, ∞). Therefore $(\tilde{\boldsymbol{\Phi}}_k)_k$ is also bounded in $L^2(0, \infty; \mathbf{V}_n^0(\Omega))$. Observe that $\tilde{\boldsymbol{\Phi}}_k$ is also the solution of the equation

$$-\tilde{\boldsymbol{\Phi}}_k' = (A^* - \lambda_0 I)\tilde{\boldsymbol{\Phi}}_k + (-A_0)^{-2\alpha}\tilde{\mathbf{y}}_k + \lambda_0 \tilde{\boldsymbol{\Phi}}_k, \qquad \tilde{\boldsymbol{\Phi}}_k(\infty) = 0.$$

Thus

$$\tilde{\boldsymbol{\Phi}}_k(t) = \int_t^\infty e^{(A^* - \lambda_0 I)(\tau - t)}((-A_0)^{-2\alpha}\tilde{\mathbf{y}}_k(\tau) + \lambda_0 \tilde{\boldsymbol{\Phi}}_k(\tau))\, d\tau \qquad \text{for all } t \geq 0.$$

From Young's inequality for convolutions it follows that $(\tilde{\boldsymbol{\Phi}}_k)_k$ is also bounded in $L^\infty(0, \infty; \mathbf{V}_n^0(\Omega))$. There then exists $\hat{\boldsymbol{\Phi}} \in L^\infty(0, \infty; \mathbf{V}_n^0(\Omega)) \cap L^2(0, \infty; \mathbf{V}_n^0(\Omega))$ such that, after extraction of a subsequence, we have

$$\tilde{\boldsymbol{\Phi}}_k \rightharpoonup \hat{\boldsymbol{\Phi}} \quad \text{weakly in } L^2(0, \infty; \mathbf{V}_n^0(\Omega)) \quad \text{and} \quad \text{weak-star in } L^\infty(0, \infty; \mathbf{V}_n^0(\Omega)),$$

and $\hat{\boldsymbol{\Phi}}$ obeys the equation

$$\hat{\boldsymbol{\Phi}}(t) = \int_t^\infty e^{(A^* - \lambda_0 I)(\tau - t)}((-A_0)^{-2\alpha}\hat{\mathbf{y}}(\tau) + \lambda_0 \hat{\boldsymbol{\Phi}}(\tau))\, d\tau \qquad \text{for all } t \geq 0.$$

Step 3. Regularity of $\hat{\boldsymbol{\Phi}}$. We have

$$-\hat{\boldsymbol{\Phi}}' = (A^* - \lambda_0 I)\hat{\boldsymbol{\Phi}} + \lambda_0 \hat{\boldsymbol{\Phi}} + (-A_0)^{-2\alpha}\hat{\mathbf{y}} \quad \text{in } (0, \infty), \qquad \hat{\boldsymbol{\Phi}}(\infty) = 0.$$

Since $\hat{\boldsymbol{\Phi}}$ belongs to $L^2(0, \infty; \mathbf{V}_n^0(\Omega))$, we deduce that $\hat{\boldsymbol{\Phi}}$ belongs to $L^2(0, \infty; \mathbf{V}^{2+4\alpha}(\Omega)) \cap H^1(0, \infty; \mathbf{V}^{4\alpha}(\Omega))$ (see Lemma 19). Moreover due to Lemma 21, the sequence $(\tilde{\mathbf{u}}_k)_k = (-MB_\tau^*\tilde{\boldsymbol{\Phi}}_k - R_A^{-1}MB_n^*\tilde{\boldsymbol{\Phi}}_k)_k$ converges weakly in $L^2(0, \infty; \mathbf{V}^0(\Gamma))$ to $-MB_\tau^*\hat{\boldsymbol{\Phi}} - R_A^{-1}MB_n^*\hat{\boldsymbol{\Phi}}$. Thus $\hat{\mathbf{u}} = -MB_\tau^*\hat{\boldsymbol{\Phi}} - R_A^{-1}MB_n^*\hat{\boldsymbol{\Phi}}$, and $\hat{\mathbf{y}}$ obeys the first equation in (23) corresponding to $\hat{\boldsymbol{\Phi}}$. Therefore we have proved that the first two equations of system (23) admits at least one solution in $L^2(0, \infty; \mathbf{V}_n^0(\Omega)) \times \mathbf{V}^{2,1}(Q_\infty)$.

Step 4. Let us show that if $(\mathbf{y}, \boldsymbol{\Phi}) \in L^2(0, \infty; \mathbf{V}_n^0(\Omega)) \times \mathbf{V}^{2,1}(Q_\infty)$ obeys the first two equations of system (23), then $\mathbf{y}$ belongs $C_b([0, \infty); \mathbf{V}_n^0(\Omega)) \cap \mathbf{V}^{1,1/2}(Q_\infty)$, $\boldsymbol{\Phi} \in L^2(0, \infty; \mathbf{V}^{3+4\alpha}(\Omega)) \cap H^{3/2}(0, \infty; \mathbf{V}^{4\alpha}(\Omega))$, and

$$\begin{aligned}
&\|\mathbf{y}\|_{C_b([0,\infty);\mathbf{V}_n^0(\Omega))} + \|\mathbf{y}\|_{\mathbf{V}^{1,1/2}(Q_\infty)} \\
&+ \|\boldsymbol{\Phi}\|_{L^2(0,\infty;\mathbf{V}^{3+4\alpha}(\Omega)) \cap H^{3/2}(0,\infty;\mathbf{V}^{4\alpha}(\Omega))} \\
&\leq C(|\mathbf{y}_0|_{\mathbf{V}_n^0(\Omega)} + \|\mathbf{y}\|_{L^2(0,\infty;\mathbf{V}_n^0(\Omega))} + \|\boldsymbol{\Phi}\|_{L^2(0,\infty;\mathbf{V}_n^0(\Omega))}).
\end{aligned} \tag{24}$$

To establish this result we rewrite the first two equations in system (23) as follows

$$\mathbf{y}' = (A - \lambda_0 I)\mathbf{y} - B_\tau M^2 B_\tau^* \mathbf{\Phi} - B_n M R_A^{-1} M B_n^* \mathbf{\Phi} + \lambda_0 \mathbf{y} \quad \text{in } (0, \infty),$$

$$\mathbf{y}(0) = \mathbf{y}_0, \tag{25}$$

$$-\mathbf{\Phi}' = (A^* - \lambda_0 I)\mathbf{\Phi} + (-A_0)^{-2\alpha}\mathbf{y} + \lambda_0 \mathbf{\Phi} \quad \text{in } (0, \infty), \qquad \mathbf{\Phi}(\infty) = 0.$$

Due to Lemma 21 we know that $B^*\mathbf{\Phi} \in L^2(0, \infty; \mathbf{V}^0(\Gamma))$. Applying Lemmas 16 and 17, we obtain:

$$\|\mathbf{y}\|_{\mathbf{V}^{1/2-\varepsilon',1/4-\varepsilon'/2}(Q_\infty)} \le C(|\mathbf{y}_0|_{\mathbf{V}_n^0(\Omega)} + \|\mathbf{y}\|_{L^2(0,\infty;\mathbf{V}_n^0(\Omega))} + \|\mathbf{\Phi}\|_{L^2(0,\infty;\mathbf{V}_n^0(\Omega))})$$

for all $\varepsilon' > 0$. Still from Lemma 21, we deduce that $B^*\mathbf{\Phi} \in \mathbf{V}^{1/2-\varepsilon',1/4-\varepsilon'/2}(\Sigma_\infty)$ for all $\varepsilon' > 0$. Applying successively Lemmas 16, 17 and Lemma 21 we can prove that $\mathbf{y}$ belongs to $\mathbf{V}^{1-\varepsilon',1/2-\varepsilon'/2}(Q_\infty)$ and $B^*\mathbf{\Phi}$ belongs to $\mathbf{V}^{1-\varepsilon',1/2-\varepsilon'/2}(\Sigma_\infty)$ for all $\varepsilon' > 0$. Another iteration gives $\mathbf{y} \in \mathbf{V}^{1,1/2}(Q_\infty) \cap C_b([0,\infty); \mathbf{V}_n^0(\Omega))$ (because $\mathbf{y}_0 \in \mathbf{V}_n^0(\Omega)$). From Lemma 19 we deduce that $\mathbf{\Phi} \in L^2(0,\infty; \mathbf{V}^{3+4\alpha}(\Omega)) \cap H^{3/2}(0,\infty; \mathbf{V}^{4\alpha}(\Omega))$, and the estimate (24) holds true.

Step 5. Let us prove that the pair $(\hat{\mathbf{y}}, \hat{\mathbf{\Phi}})$ obeys the third equation in (23). With Lemma 19 we can show that

$$\tilde{\mathbf{\Phi}}_k(t) \rightharpoonup \hat{\mathbf{\Phi}}(t) \quad \text{weakly in } \mathbf{V}_n^0(\Omega) \qquad \text{for all } t \ge 0.$$

Since

$$\mathbf{\Phi}_k(t) \in \partial_{\mathbf{y}}\varphi_k(t, \mathbf{y}_k(t)), \qquad \mathbf{\Phi}_k(t) \rightharpoonup \hat{\mathbf{\Phi}}(t) \text{ weakly in } \mathbf{V}_n^0(\Omega),$$

and

$$\varphi_k(t, \mathbf{y}_k(t)) \to \varphi(\hat{\mathbf{y}}(t)) \text{ as } k \to \infty,$$

we deduce that

$$\hat{\mathbf{\Phi}}(t) \in \partial\varphi(\hat{\mathbf{y}}(t)), \qquad \text{i.e., } \hat{\mathbf{\Phi}}(t) = \Pi\hat{\mathbf{y}}(t).$$

Thus we have shown that $(\hat{\mathbf{y}}, \hat{\mathbf{\Phi}})$ obeys the third equation in (23).

Step 6. Uniqueness. If $(\mathbf{y}, \mathbf{\Phi}) \in L^2(0, \infty; \mathbf{V}_n^0(\Omega)) \times \mathbf{V}^{2,1}(Q_\infty)$ is a solution to system (23), due to Step 4 it obeys (24), and we can show that

$$\int_0^k |(-A_0)^{-\alpha}\mathbf{y}(t)|^2_{\mathbf{V}_n^0(\Omega)} dt + \int_0^k |R_A^{-1/2} M B^* \mathbf{\Phi}(t)|^2_{\mathbf{V}^0(\Gamma)} dt$$

$$= \int_\Omega \mathbf{y}_0 \mathbf{\Phi}(0) - \int_\Omega \mathbf{y}(k)\mathbf{\Phi}(k).$$

Passing to the limit when k tends to infinity we obtain

$$\int_0^\infty |(-A_0)^{-\alpha}\mathbf{y}(t)|^2_{\mathbf{V}_n^0(\Omega)} dt + \int_0^\infty |R_A^{-1/2} M B^* \mathbf{\Phi}(t)|^2_{\mathbf{V}^0(\Gamma)} dt = \int_\Omega \mathbf{y}_0 \mathbf{\Phi}(0),$$

because $\mathbf{y}$ belongs to $C_b([0,\infty); \mathbf{V}_n^0(\Omega))$. Thus if $\mathbf{y}_0 = 0$ we have $\mathbf{y} = 0$. From the relation $\mathbf{\Phi} = \Pi\mathbf{y}$ we deduce that $\mathbf{\Phi} = 0$, and the uniqueness is established.

Step 7. Final estimate. We first notice that

$$\int_0^\infty |(-A_0)^{-\alpha}\hat{\mathbf{y}}(t)|^2_{\mathbf{V}_n^0(\Omega)}dt + \int_0^\infty |R_A^{1/2}\hat{\mathbf{u}}(t)|^2_{\mathbf{V}^0(\Gamma)}dt = \int_\Omega \mathbf{y}_0\Pi\mathbf{y}_0. \qquad (26)$$

Thus rewriting the first equation in (23) in the form

$$\hat{\mathbf{y}}' = (A - \lambda_0(-A_0)^{-\alpha})\hat{\mathbf{y}} + BM\hat{\mathbf{u}} + \lambda_0(-A_0)^{-\alpha}\hat{\mathbf{y}} \quad \text{in } (0,\infty), \qquad \hat{\mathbf{y}}(0) = \mathbf{y}_0,$$

from Lemmas 25 and 26, it follows that

$$\begin{aligned}
&\|\hat{\mathbf{y}}\|_{L^2(0,\infty;\mathbf{V}_n^0(\Omega))} \\
&\leq C\big(|\mathbf{y}_0|_{\mathbf{V}_n^0(\Omega)} + \|(-A_0)^{-\alpha}\hat{\mathbf{y}}\|_{L^2(0,\infty;\mathbf{V}_n^0(\Omega))} + \|\hat{\mathbf{u}}\|_{L^2(0,\infty;\mathbf{V}^0(\Gamma))}\big).
\end{aligned} \qquad (27)$$

From (26) and (27), we deduce

$$\|\hat{\mathbf{y}}\|_{L^2(0,\infty;\mathbf{V}_n^0(\Omega))} \leq C|\mathbf{y}_0|_{\mathbf{V}_n^0(\Omega)}.$$

Since $\mathbf{\Phi} = \Pi\mathbf{y}$, the estimate of the lemma follows from (24). $\qquad\square$

Corollary 13. *If* $\mathbf{y}_0 \in \mathbf{V}_n^{1/2-\varepsilon}(\Omega)$ *for some* $0 < \varepsilon \leq 1/2$, *then the solution* $(\mathbf{y}, \mathbf{\Phi})$ *of system (23) belongs to* $\mathbf{V}^{3/2-\varepsilon,3/4-\varepsilon/2}(Q_\infty) \times \big(L^2(0,\infty;\mathbf{V}^{7/2+4\alpha-\varepsilon}(\Omega)) \cap H^{7/4-\varepsilon/2}(0,\infty;\mathbf{V}^{4\alpha}(\Omega))\big)$, *and we have:*

$$\begin{aligned}
&\|\mathbf{y}\|_{\mathbf{V}^{3/2-\varepsilon,3/4-\varepsilon/2}(Q_\infty)} + \|\mathbf{\Phi}\|_{L^2(0,\infty;\mathbf{V}^{7/2+4\alpha-\varepsilon}(\Omega))\cap H^{7/4-\varepsilon/2}(0,\infty;\mathbf{V}^{4\alpha}(\Omega))} \\
&+\|B^*\mathbf{\Phi}\|_{L^2(0,\infty;\mathbf{V}^{2+4\alpha-\varepsilon}(\Gamma))\cap H^{3/4+2\alpha-\varepsilon/2}(0,\infty;\mathbf{V}^{1/2}(\Gamma))} \leq C|\mathbf{y}_0|_{\mathbf{V}_n^{1/2-\varepsilon}(\Omega)}.
\end{aligned}$$

Proof. This result is established in the case where $\alpha = 0$ in [15, Corollary 4.1]. It is sufficient to rewrite the proof and to use Lemmas 19 and 21 to obtain the desired estimate. $\qquad\square$

Corollary 14. *The operator* Π *is continuous from* $\mathbf{V}_n^0(\Omega)$ *into* $\mathbf{V}^{2+4\alpha}(\Omega) \cap \mathbf{V}_0^1(\Omega)$, *and from* $\mathbf{V}_n^{1/2-\varepsilon}(\Omega)$ *into* $\mathbf{V}^{5/2+4\alpha-\varepsilon}(\Omega) \cap \mathbf{V}_0^1(\Omega)$ *for all* $0 < \varepsilon \leq 1/2$.

The operator $B^*\Pi$ *is continuous from* $\mathbf{V}_n^0(\Omega)$ *to* $\mathbf{V}^{1/2+4\alpha}(\Gamma)$, *and from* $\mathbf{V}_n^{1/2-\varepsilon}(\Omega)$ *to* $\mathbf{V}^{1+4\alpha-\varepsilon}(\Gamma)$ *for all* $0 < \varepsilon \leq 1/2$.

Proof. Due to Lemma 12, if $\mathbf{y}_0 \in \mathbf{V}_n^0(\Omega)$, then $\mathbf{\Phi}$ belongs to $C([0,\infty);\mathbf{V}^{2+4\alpha}(\Omega)\cap \mathbf{V}_0^1(\Omega))$. In particular $\mathbf{\Phi}(0)$ belongs to $\mathbf{V}^{2+4\alpha}(\Omega) \cap \mathbf{V}_0^1(\Omega)$. This means that Π is continuous from $\mathbf{V}_n^0(\Omega)$ into $\mathbf{V}^{2+4\alpha}(\Omega) \cap \mathbf{V}_0^1(\Omega)$. From Corollary 13 we deduce that if $\mathbf{y}_0 \in \mathbf{V}_n^{1/2-\varepsilon}(\Omega)$ then $\mathbf{\Phi} \in C([0,\infty);\mathbf{V}^{5/2+4\alpha-\varepsilon}(\Omega) \cap \mathbf{V}_0^1(\Omega))$, and Π is continuous from $\mathbf{V}_n^{1/2-\varepsilon}(\Omega)$ into $\mathbf{V}^{5/2+2\alpha-\varepsilon}(\Omega) \cap \mathbf{V}_0^1(\Omega)$.

Since B^* is continuous from $\mathbf{V}^{2+4\alpha}(\Omega)\cap\mathbf{V}_0^1(\Omega)$ into $\mathbf{V}^{1/2+4\alpha}(\Gamma)$ (see Proposition 6), the operator $B^*\Pi$ belongs to $\mathcal{L}(\mathbf{V}_n^0(\Omega), \mathbf{V}^{1/2+4\alpha}(\Gamma))$, and $B^*\Pi$ also belongs to $\mathcal{L}(\mathbf{V}_n^{1/2-\varepsilon}(\Omega), \mathbf{V}^{1+4\alpha-\varepsilon}(\Gamma))$ for all $0 < \varepsilon \leq 1/2$. $\qquad\square$

Theorem 15. *The unbounded operator* $(A_\Pi, D(A_\Pi))$ *defined by:*

$$D(A_\Pi) = \left\{\mathbf{y} \in \mathbf{V}_n^0(\Omega) \mid (A - BMR_A^{-1}MB^*\Pi)\mathbf{y} \in \mathbf{V}_n^0(\Omega)\right\},$$

$$A_\Pi\mathbf{y} = (A - BMR_A^{-1}MB^*\Pi)\mathbf{y} \qquad \text{for all } \mathbf{y} \in D(A_\Pi),$$

is the infinitesimal generator of an exponentially stable semigroup on $\mathbf{V}_n^0(\Omega)$. (*In the writing* $(A - BMR_A^{-1}MB^*\Pi)\mathbf{y}$ *the operators* A *and* $BMR_A^{-1}MB^*\Pi$ *are considered as operators from* $\mathbf{V}_n^0(\Omega)$ *into* $(D(A_0))'$, *which is meaningful due to the regularizing properties for* $B^*\Pi$ *stated in Corollary 14.*)

The operator Π *is the unique weak solution to the algebraic Riccati equation*

$$\Pi^* = \Pi \in \mathcal{L}(\mathbf{V}_n^0(\Omega)) \quad and \quad \Pi \geq 0,$$

$$\text{for all } \mathbf{y} \in \mathbf{V}_n^0(\Omega), \ \Pi\mathbf{y} \in \mathbf{V}^{2+4\alpha}(\Omega) \cap \mathbf{V}_0^1(\Omega)$$

$$\text{and} \quad |\Pi\mathbf{y}|_{\mathbf{V}^{2+4\alpha}(\Omega)} \leq C|\mathbf{y}|_{\mathbf{V}_n^0(\Omega)},$$

$$A^*\Pi + \Pi A - \Pi B_\tau M^2 B_\tau^*\Pi - \Pi B_n MR_A^{-1}MB_n^*\Pi + (-A_0)^{-2\alpha} = 0.$$

(28)

Proof. See [15, Theorem 4.2] for a similar result in the case where $\alpha = 0$. Notice in particular that due, to Lemma 25, the pair $(A, (-A_0)^{-\alpha})$ is exponentially detectable, which provides the uniqueness of solution to equation (28). $\qquad\square$

4. Appendix

In this section we state some regularity results for the state and adjoint equations.

Lemma 16. ([15, Lemma 7.1]) *If* $\mathbf{y}_0 \in \mathbf{V}_n^{1/2-\varepsilon}(\Omega)$ *with* $0 < \varepsilon \leq 1/2$, *the weak solution to equation*

$$\mathbf{y}' = (A - \lambda_0 I)\mathbf{y} \quad in \ (0, \infty), \qquad \mathbf{y}(0) = \mathbf{y}_0,$$

obeys

$$\|\mathbf{y}\|_{C_b([0,\infty);\mathbf{V}_n^{1/2-\varepsilon}(\Omega))} + \|\mathbf{y}\|_{\mathbf{V}^{3/2-\varepsilon,3/4-\varepsilon/2}(Q_\infty)} \leq C|\mathbf{y}_0|_{\mathbf{V}_n^{1/2-\varepsilon}(\Omega)}.$$

Lemma 17. *If* $\mathbf{u}$ *belongs to* $\mathbf{V}^{s,s/2}(\Sigma_\infty)$ *with* $0 \leq s < 1$, *then the weak solution to equation*

$$\mathbf{y}' = (A - \lambda_0 I)\mathbf{y} + BM\mathbf{u} \quad in \ (0, \infty), \qquad \mathbf{y}(0) = 0,$$

obeys

$$\|\mathbf{y}\|_{\mathbf{V}^{1/2+s-\varepsilon,1/4+s/2-\varepsilon/2}(Q_\infty)} \leq C\|\mathbf{u}\|_{\mathbf{V}^{s,s/2}(\Sigma_\infty)} \quad \text{for all } 0 < \varepsilon \leq 1/2.$$

Proof. This result is already proved in [15, Lemma 7.3]. $\qquad\square$

Lemma 18. *For all* $\mathbf{y} \in \mathbf{V}_n^s(\Omega)$ *with* $2 \leq s \leq 4$, *the solution* $\mathbf{\Phi} \in \mathbf{V}^2(\Omega) \cap \mathbf{V}_0^1(\Omega)$ *to the stationary equation* $\lambda_0 \mathbf{\Phi} - A^*\mathbf{\Phi} = \mathbf{y}$ *obeys*

$$|\mathbf{\Phi}|_{\mathbf{V}^{s+2}(\Omega)} \leq C|\mathbf{y}|_{\mathbf{V}_n^s(\Omega)}.$$

Proof. We rewrite the equation in the form

$$\lambda_0 \mathbf{\Phi} - \nu P\Delta\mathbf{\Phi} = \mathbf{y} + P((\mathbf{w} \cdot \nabla)\mathbf{\Phi}) - P((\nabla\mathbf{w})^T\mathbf{\Phi}) \quad in \ \Omega, \quad \mathbf{\Phi} = 0 \quad on \ \Gamma.$$

Since $\mathbf{w} \in \mathbf{V}^5(\Omega)$ and $\mathbf{\Phi} \in \mathbf{V}^2(\Omega)$, then $P((\mathbf{w} \cdot \nabla)\mathbf{\Phi})$ and $P((\nabla\mathbf{w})^T\mathbf{\Phi})$ belong to $\mathbf{V}^2(\Omega)$, which gives an estimate of $\mathbf{\Phi}$ in $\mathbf{V}^4(\Omega)$ since $\mathbf{y} \in \mathbf{V}_n^s(\Omega)$ with $2 \leq s \leq 4$. Knowing that $\mathbf{\Phi} \in \mathbf{V}^4(\Omega)$, $P((\mathbf{w} \cdot \nabla)\mathbf{\Phi})$ and $P((\nabla\mathbf{w})^T\mathbf{\Phi})$ belong to $\mathbf{V}^4(\Omega)$, which provides an estimate of $\mathbf{\Phi}$ in $\mathbf{V}^{s+2}(\Omega)$. The proof is complete. $\qquad\square$

Lemma 19. *If the function* $\mathbf{y}$ *belongs to* $\mathbf{V}^{s,s/2}(Q_\infty) \cap L^2(0,\infty; \mathbf{V}_n^0(\Omega))$ *with* $0 \leq s \leq 2$, *then the solution* $\mathbf{\Phi}$ *to the equation*

$$-\mathbf{\Phi}' = (A^* - \lambda_0 I)\mathbf{\Phi} + (-A_0)^{-2\alpha}\mathbf{y} \quad \text{in } (0,\infty), \qquad \mathbf{\Phi}(\infty) = 0, \tag{29}$$

satisfies

$$\|\mathbf{\Phi}\|_{L^2(0,\infty;\mathbf{V}^{2+s+4\alpha}(\Omega))} + \|\mathbf{\Phi}\|_{H^{1+s/2}(0,\infty;\mathbf{V}^{4\alpha}(\Omega))} \leq C\|\mathbf{y}\|_{\mathbf{V}^{s,s/2}(Q_\infty)}. \tag{30}$$

Proof. (i) Estimate (30) is already proved in [15, Lemma 7.5] in the case when $\alpha = 0$. Let us establish this estimate for $s = 0$ and $0 \leq \alpha \leq 1/2$. It is clear that $(-A_0)^{-2\alpha}\mathbf{y}$ belongs to $L^2(0,\infty; \mathbf{V}_n^0(\Omega))$. Thus applying [15, Lemma 7.6], we first prove that $\mathbf{\Phi}$ belongs to $\mathbf{V}^{2,1}(Q_\infty)$ with the corresponding estimate. Observe that $\mathbf{\Phi} = \mathbf{\Phi}_1 + \mathbf{\Phi}_2$, where $\mathbf{\Phi}_1$ is the solution to

$$-\mathbf{\Phi}_1' = (A_0 - \lambda_0 I)\mathbf{\Phi}_1 + (-A_0)^{-2\alpha}\mathbf{y} \quad \text{in } (0,\infty), \qquad \mathbf{\Phi}_1(\infty) = 0,$$

and $\mathbf{\Phi}_2$ is the solution to

$$-\mathbf{\Phi}_2' = (A_0 - \lambda_0 I)\mathbf{\Phi}_2 + P((\mathbf{w} \cdot \nabla)\mathbf{\Phi}) - P((\nabla\mathbf{w})^T\mathbf{\Phi}) \quad \text{in } (0,\infty), \quad \mathbf{\Phi}_2(\infty) = 0.$$

To study the regularity of $\mathbf{\Phi}_1$, we set $\hat{\mathbf{\Phi}} = (-A_0)^{2\alpha}\mathbf{\Phi}_1$. Since

$$\mathbf{\Phi}_1(t) = \int_t^\infty e^{(\tau-t)(A_0-\lambda_0 I)}(-A_0)^{-2\alpha}\mathbf{y}(\tau)\, d\tau,$$

then $\hat{\mathbf{\Phi}}$ is defined by

$$\hat{\mathbf{\Phi}}(t) = \int_t^\infty e^{(\tau-t)(A_0-\lambda_0 I)}\mathbf{y}(\tau)\, d\tau.$$

Thus $\hat{\mathbf{\Phi}}$ belongs to $\mathbf{V}^{2,1}(Q_\infty)$, and $\mathbf{\Phi}_1$ belongs to $L^2(0,\infty; \mathbf{V}^{2+4\alpha}(\Omega)) \cap H^1(0,\infty; \mathbf{V}^{4\alpha}(\Omega))$ if $0 \leq \alpha \leq 1/2$.

To study the regularity of $\mathbf{\Phi}_2$, we observe that $\mathbf{\Phi} \in \mathbf{V}^{2,1}(Q_\infty)$ and $\mathbf{w} \in \mathbf{V}^5(\Omega)$. We can verify that $P((\mathbf{w} \cdot \nabla)\mathbf{\Phi}) - P((\nabla\mathbf{w})^T\mathbf{\Phi})$ belongs to $\mathbf{V}^{2,1}(Q_\infty)$. Thus applying [15, Lemma 7.6] we claim that $\mathbf{\Phi}_2$ belongs to $\mathbf{V}^{4,2}(Q_\infty)$, which proves estimate (30) for $s = 0$ if $0 \leq \alpha \leq 1/2$.

(ii) Let us first prove estimate (30) for $s = 2$. As in step (i) we can show that $\hat{\mathbf{\Phi}}$ belongs to $\mathbf{V}^{4,2}(Q_\infty)$, and $\mathbf{\Phi}_1$ belongs to $L^2(0,\infty; \mathbf{V}^{4+4\alpha}(\Omega)) \cap H^2(0,\infty; \mathbf{V}^{4\alpha}(\Omega))$. Moreover $\mathbf{\Phi}_2$ belongs to $\mathbf{V}^{4,2}(Q_\infty)$. Thus $\mathbf{\Phi} = \mathbf{\Phi}_1 + \mathbf{\Phi}_2$ belongs to $\mathbf{V}^{4,2}(Q_\infty)$. Since $\mathbf{w} \in \mathbf{V}^5(\Omega)$, we can verify that $P((\mathbf{w} \cdot \nabla)\mathbf{\Phi}) - P((\nabla\mathbf{w})^T\mathbf{\Phi})$ belongs to $\mathbf{V}^{4,2}(Q_\infty)$. Therefore $\mathbf{\Phi}_2$ belongs to $\mathbf{V}^{6,3}(Q_\infty)$, which proves estimate (30) for $s = 2$ and $\alpha \leq 1/2$.

Consequently estimate (30) is proved for $s = 2$. The intermediate result can be proved by interpolation. $\qquad\square$

Lemma 20. *Let* $\mathbf{\Phi}$ *be the solution to equation* (29) *and let* ψ *be the pressure associated with* $\mathbf{\Phi}$, *that is the function* ψ *satisfying*

$$-\frac{\partial \mathbf{\Phi}}{\partial t} - \nu\Delta\mathbf{\Phi} - (\mathbf{w}\cdot\nabla)\mathbf{\Phi} + (\nabla\mathbf{\Phi})^T\mathbf{w} + \lambda_0\mathbf{\Phi} + \nabla\psi$$

$$= (-A_0)^{-2\alpha}\mathbf{y} \quad \text{in } Q_\infty, \tag{31}$$

$$\operatorname{div}\mathbf{\Phi} = 0 \quad \text{in } Q_\infty, \quad \mathbf{\Phi} = 0 \quad \text{on } \Sigma_\infty, \quad \mathbf{\Phi}(\infty) = 0 \quad \text{in } \Omega.$$

If in (29) $\mathbf{y}$ *belongs to* $\mathbf{V}^{s,s/2}(Q_\infty)$ *with* $0 \le s < 2$, *then the function* ψ *belongs to* $L^2(0,\infty; H^{s+1+4\alpha}(\Omega)) \cap H^{s/2+2\alpha}(0,\infty; H^1(\Omega))$.

Proof. First observe that

$$\nabla\psi = (I - P)\Big((-A_0)^{-2\alpha}\mathbf{y} + \tfrac{\partial\mathbf{\Phi}}{\partial t} + \nu\Delta\mathbf{\Phi} + (\mathbf{w}\cdot\nabla)\mathbf{\Phi} - (\nabla\mathbf{\Phi})^T\mathbf{w}\Big)$$

$$= (I - P)\Big(\nu\Delta\mathbf{\Phi} + (\mathbf{w}\cdot\nabla)\mathbf{\Phi} - (\nabla\mathbf{\Phi})^T\mathbf{w}\Big).$$

Assume that $\mathbf{y}$ belongs to $L^2(0,\infty; \mathbf{V}_n^s(\Omega)) \cap H^{s/2}(0,\infty; \mathbf{V}_n^0(\Omega))$ with $0 \le s \le 2$. From Lemma 19 it follows that

$$\Big(\nu\Delta\mathbf{\Phi} + (\mathbf{w}\cdot\nabla)\mathbf{\Phi} - (\nabla\mathbf{\Phi})^T\mathbf{w}\Big) \in L^2(0,\infty; \mathbf{H}^{s+4\alpha}(\Omega)) \cap H^{s/2+2\alpha}(0,\infty; \mathbf{L}^2(\Omega)).$$

Thus $\nabla\psi$ belongs to $L^2(0,\infty; \mathbf{H}^{s+4\alpha}(\Omega)) \cap H^{s/2+2\alpha}(0,\infty; \mathbf{L}^2(\Omega))$, and the proof is complete. $\square$

Lemma 21. *Let* $\mathbf{\Phi} \in \mathbf{V}^{2,1}(Q_\infty)$ *be the solution to equation* (29), *and set*

$$\mathbf{u} = -(M\gamma_\tau + R_A^{-1}M\gamma_n)B^*\mathbf{\Phi}.$$

If in (29) *the function* $\mathbf{y}$ *belongs to* $L^2(0,\infty; \mathbf{V}_n^s(\Omega)) \cap H^{s/2}(0,\infty; \mathbf{V}_n^0(\Omega))$ *with* $0 \le s \le 2$, *then*

$$\|B^*\mathbf{\Phi}\|_{L^2(0,\infty;\mathbf{V}^{s+1/2+4\alpha}(\Gamma))\cap H^{s/2+2\alpha}(0,\infty;\mathbf{V}^{1/2}(\Gamma))}$$

$$+\|\mathbf{u}\|_{L^2(0,\infty;\mathbf{V}^{s+1/2+4\alpha}(\Gamma))\cap H^{s/2+2\alpha}(0,\infty;\mathbf{V}^{1/2}(\Gamma))} \le C\|\mathbf{y}\|_{\mathbf{V}^{s,s/2}(Q_\infty)}. \tag{32}$$

Proof. As in [15, Lemma 3.4] we can show that

$$\mathbf{u} = -(M\gamma_\tau + R_A^{-1}M\gamma_n)B^*\mathbf{\Phi} = -\nu m\frac{\partial\mathbf{\Phi}}{\partial\mathbf{n}} + R_A^{-1}M\big(\psi\,\mathbf{n}\big),$$

where ψ is the pressure associated with $\mathbf{\Phi}$.

Since $\mathbf{y}$ belongs to $L^2(0,\infty; \mathbf{V}_n^s(\Omega)) \cap H^{s/2}(0,\infty; \mathbf{V}_n^0(\Omega))$, from Lemma 19 we deduce that $\mathbf{\Phi}$ belongs to $L^2(0,\infty; \mathbf{V}^{s+2+4\alpha}(\Omega)) \cap H^{1+s/2}(0,\infty; \mathbf{V}^{4\alpha}(\Omega))$, and from Lemma 20 it follows that ψ belongs to $L^2(0,\infty; H^{s+1+4\alpha}(\Omega)) \cap H^{s/2+2\alpha}(0,\infty; H^1(\Omega))$. Thus $\mathbf{u}$ belongs to $L^2(0,\infty; \mathbf{V}^{s+1/2+4\alpha}(\Gamma)) \cap H^{s/2+2\alpha}(0,\infty; \mathbf{V}^{1/2}(\Gamma))$. $\square$

Lemma 22. *For all* $\mathbf{y} \in L^2(0,T; \mathbf{V}_n^0(\Omega))$, *the solution to the equation*

$$-\mathbf{\Phi}' = A^*\mathbf{\Phi} + (-A_0)^{-2\alpha}\mathbf{y} \quad \text{in } (0,T), \qquad \mathbf{\Phi}(T) = 0, \tag{33}$$

satisfies

$$\|\mathbf{\Phi}\|_{L^2(0,T;\mathbf{V}^{(2+4\alpha)\wedge\sigma}(\Omega))} + \|\mathbf{\Phi}\|_{H^1(0,T;\mathbf{V}^{4\alpha\wedge(\sigma-2)}(\Omega))} \le C\|\mathbf{y}\|_{L^2(0,T;\mathbf{V}_n^0(\Omega))} \tag{34}$$

for all $\sigma < 7/2$. If the function $\mathbf{y}$ belongs to $\mathbf{V}^{s,s/2}(Q_T)$ with $0 \le s < 3/2$, then the function $\mathbf{\Phi}$ belongs to $L^2(0,T;\mathbf{V}^{s+2+4\alpha}(\Omega)) \cap H^{s/2+1}(0,T;\mathbf{V}^{4\alpha}(\Omega))$, and the following estimate holds:

$$\|\mathbf{\Phi}\|_{L^2(0,T;\mathbf{V}^{(s+2+4\alpha)\wedge\sigma}(\Omega))} + \|\mathbf{\Phi}\|_{H^{s/2+1}(0,T;\mathbf{V}^{4\alpha\wedge(\sigma-2-s)}(\Omega))} \le C\|\mathbf{y}\|_{\mathbf{V}^{s,s/2}(Q_T)} \quad (35)$$

for all $\sigma < 7/2$.

Proof. We proceed as in the proof of Lemma 19.

(i) Let us establish estimate (34) for $0 \le \alpha \le 1/2$. It is clear that $(-A_0)^{-2\alpha}\mathbf{y}$ belongs to $L^2(0,\infty;\mathbf{V}_n^0(\Omega))$. Thus applying [15, Lemma 7.6], we first prove that $\mathbf{\Phi}$ belongs to $\mathbf{V}^{2,1}(Q_T)$ with the corresponding estimate. As in the proof of Lemma 19 we write $\mathbf{\Phi} = \mathbf{\Phi}_1 + \mathbf{\Phi}_2$, where $\mathbf{\Phi}_1$ is the solution to

$$-\mathbf{\Phi}_1' = A_0\mathbf{\Phi}_1 + (-A_0)^{-2\alpha}\mathbf{y} \quad \text{in } (0,T), \qquad \mathbf{\Phi}_1(T) = 0,$$

and $\mathbf{\Phi}_2$ is the solution to

$$-\mathbf{\Phi}_2' = A_0\mathbf{\Phi}_2 - P((\mathbf{w}\cdot\nabla)\mathbf{\Phi}) + P((\nabla\mathbf{w})^T\mathbf{\Phi}) \quad \text{in } (0,T), \qquad \mathbf{\Phi}_2(T) = 0.$$

Since

$$\mathbf{\Phi}_1(t) = \int_t^T e^{(\tau-t)A_0}(-A_0)^{-2\alpha}\mathbf{y}(\tau)\,d\tau,$$

then $\hat{\mathbf{\Phi}} = (-A_0)^{2\alpha}\mathbf{\Phi}_1$ is defined by

$$\hat{\mathbf{\Phi}}(t) = \int_t^T e^{(\tau-t)A_0}\mathbf{y}(\tau)\,d\tau.$$

Thus $\hat{\mathbf{\Phi}}$ belongs to $\mathbf{V}^{2,1}(Q_T)$, and $\mathbf{\Phi}_1$ belongs to $L^2(0,T;\mathbf{V}^{2+4\alpha}(\Omega)) \cap H^1(0,T; \mathbf{V}^{4\alpha}(\Omega))$ for all $0 \le \alpha \le 1/2$.

Let us study the regularity of $\mathbf{\Phi}_2$. Since $\mathbf{\Phi} \in \mathbf{V}^{2,1}(Q_T)$ and $\mathbf{w} \in \mathbf{V}^5(\Omega)$, we can verify that $P((\mathbf{w}\cdot\nabla)\mathbf{\Phi}) - P((\nabla\mathbf{w})^T\mathbf{\Phi})$ belongs to $\mathbf{V}^{2,1}(Q_T)$. Thus $\mathbf{\Phi}_2$ belongs to $\mathbf{V}^{\sigma,\sigma/2}(Q_T)$ for all $\sigma < 7/2$, which proves estimate (34) for $0 \le \alpha \le 1/2$.

(ii) To prove the second estimate for $\mathbf{\Phi}_1$ we proceed by interpolation. We assume that $\mathbf{y} \in \mathbf{V}^{2,1}(Q_T)$ and that $\mathbf{y}(T) \in \mathbf{V}_0^1(\Omega)$. As in step (i) we can show that $\hat{\mathbf{\Phi}}$ belongs to $\mathbf{V}^{4,2}(Q_T)$, and $\mathbf{\Phi}_1$ belongs to $L^2(0,T;\mathbf{V}^{4+4\alpha}(\Omega)) \cap H^2(0,T;\mathbf{V}^{4\alpha}(\Omega))$. If $\mathbf{y} \in \mathbf{V}^{s,s/2}(Q_T)$ with $0 \le s < 3/2$, by interpolation between the estimates proved for $s = 0$ and the one proved for $s = 2$, we obtain

$$\|\mathbf{\Phi}_1\|_{L^2(0,T;\mathbf{V}^{s+2+4\alpha}(\Omega))} + \|\mathbf{\Phi}_1\|_{H^{s/2+1}(0,T;\mathbf{V}^{4\alpha}(\Omega))} \le C\|\mathbf{y}\|_{\mathbf{V}^{s,s/2}(Q_T)}.$$

For $s < 3/2$ the regularity condition $y(T) \in \mathbf{V}_0^1(\Omega)$ is not needed.

Let us study the regularity of $\boldsymbol{\Phi}_2$. Since $\boldsymbol{\Phi} \in \mathbf{V}^{2,1}(Q_T)$ and $\mathbf{w} \in \mathbf{V}^5(\Omega)$, we can verify that $P((\mathbf{w} \cdot \nabla)\boldsymbol{\Phi}) - P((\nabla\mathbf{w})^T\boldsymbol{\Phi})$ belongs to $\mathbf{V}^{2,1}(Q_T)$. Therefore $\boldsymbol{\Phi}_2$ belong to $\mathbf{V}^{\sigma,\sigma/2}(Q_T)$ for all $\sigma < 7/2$, and $\boldsymbol{\Phi} = \boldsymbol{\Phi}_1 + \boldsymbol{\Phi}_2$ belongs to $\mathbf{V}^{\sigma,\sigma/2}(Q_T) + \left(L^2(0,T;\mathbf{V}^{s+2+4\alpha}(\Omega)) \cap H^{s/2+1}(0,T;\mathbf{V}^{4\alpha}(\Omega))\right)$ for all $\sigma < 7/2$, which proves estimate (35). $\qquad\square$

Lemma 23. *If in* (33) $\mathbf{y}$ *belongs to* $\mathbf{V}^{s,s/2}(Q_T)$ *with* $0 \le s < 3/2$, *then the function* ψ, *the pressure associated with* $\boldsymbol{\Phi}$, *belongs to* $L^2(0,T;H^{(s+1+4\alpha)\wedge(\sigma+1)}(\Omega)) \cap H^{(s/2+2\alpha)\wedge(\sigma/2-1)}(0,T;H^1(\Omega))$ *for all* $\sigma < 7/2$.

Proof. It is sufficient to adapt the proof of Lemma 20 and to use Lemma 22 to obtain the desired result. $\qquad\square$

Lemma 24. *There exists* $\lambda_0 > 0$ *such that*

$$(-A\mathbf{y},\mathbf{y})_{\mathbf{V}_n^0(\Omega)} + \lambda_0((-A_0)^{-\alpha}\mathbf{y},\mathbf{y})_{\mathbf{V}_n^0(\Omega)} \ge \frac{\nu}{2}|\mathbf{y}|^2_{\mathbf{V}_0^1(\Omega)}, \tag{36}$$

for all $\mathbf{y} \in \mathbf{V}_0^1(\Omega) \cap \mathbf{V}^2(\Omega)$, *and all* $0 \le \alpha \le 1/2$.

Proof. First observe that there exists a constant $C > 0$ independent of $0 \le \alpha \le 1/2$ such that

$$0 \le ((-A_0)^{-1/2}\mathbf{y},\mathbf{y})_{\mathbf{V}_n^0(\Omega)} \le C((-A_0)^{-\alpha}\mathbf{y},\mathbf{y})_{\mathbf{V}_n^0(\Omega)}$$

for all $\mathbf{y} \in \mathbf{V}_n^0(\Omega)$. Thus to prove inequality (36) it is sufficient to establish it for $\alpha = 1/2$.

Inequality (36) for $\alpha = 1/2$ is equivalent to

$$\frac{\nu}{2}\int_\Omega |\nabla\mathbf{y}|^2 - \frac{\nu}{2}\int_\Omega |\mathbf{y}|^2 + \int_\Omega (\mathbf{y} \cdot \nabla)\mathbf{w} \cdot \mathbf{y} + \lambda_0\int_\Omega (-A_0)^{-1/2}\mathbf{y} \cdot \mathbf{y} \ge 0.$$

Let us argue by contradiction. Assume that, for all $k > 0$, there exists $\mathbf{y}_k \in \mathbf{V}_0^1(\Omega) \cap \mathbf{V}^2(\Omega)$ such that

$$\frac{\nu}{2}\int_\Omega |\nabla\mathbf{y}_k|^2 - \frac{\nu}{2}\int_\Omega |\mathbf{y}_k|^2 + \int_\Omega (\mathbf{y}_k \cdot \nabla)\mathbf{w} \cdot \mathbf{y}_k + k\int_\Omega (-A_0)^{-1/2}\mathbf{y}_k \cdot \mathbf{y}_k < 0.$$

Since $\nabla\mathbf{w} \in (\mathbf{L}^\infty(\Omega))^3$, there exists $C > 0$ such that

$$\frac{\nu}{2}\int_\Omega |\nabla\mathbf{y}_k|^2 - (C + \frac{\nu}{2})\int_\Omega |\mathbf{y}_k|^2 + k\int_\Omega (-A_0)^{-1/2}\mathbf{y}_k \cdot \mathbf{y}_k < 0.$$

Setting $\mathbf{z}_k = \mathbf{y}_k/|\mathbf{y}_k|_{\mathbf{V}_n^0(\Omega)}$, we have

$$\frac{\nu}{2}\int_\Omega |\nabla\mathbf{z}_k|^2 + k\int_\Omega (-A_0)^{-1/2}\mathbf{z}_k \cdot \mathbf{z}_k < C + \frac{\nu}{2}.$$

From Poincaré's inequality it follows that the sequence $(\mathbf{z}_k)_k$ is bounded in $\mathbf{V}_0^1(\Omega)$. There then exist a subsequence, still indexed by k to simplify the notation, and $\mathbf{z}_\infty \in \mathbf{V}_0^1(\Omega)$, such that $(\mathbf{z}_k)_k$ converges to $\mathbf{z}_\infty$ for the weak topology of $\mathbf{V}_0^1(\Omega)$. But $(\mathbf{z}_k)_k$ converges to 0 in $(D((-A_0)^{1/4}))'$. Thus $\mathbf{z}_\infty = 0$. Since the imbedding from $\mathbf{V}_0^1(\Omega)$ into $\mathbf{V}_n^0(\Omega)$ is compact, the sequence $(\mathbf{z}_k)_k$ converges to 0 in $\mathbf{V}_n^0(\Omega)$. We obtain a contradiction with the identity $|\mathbf{z}_k|_{\mathbf{V}_n^0(\Omega)} = 1$, and the proof is complete. $\qquad\square$

Lemma 25. *The unbounded operator* $(A - \lambda_0(-A_0)^{-\alpha})$ *with domain* $D(A) = D(A_0)$ *in* $\mathbf{V}_n^0(\Omega)$ *is the infinitesimal generator of an exponentially stable analytic semigroup on* $\mathbf{V}_n^0(\Omega)$. *Moreover for all* $0 \le \beta \le 1$, $D((\lambda_0(-A_0)^{-\alpha} - A)^\beta) = D((-A_0)^\beta)$.

Proof. We already know that $(A, D(A))$ is the infinitesimal generator of an analytic semigroup on $\mathbf{V}_n^0(\Omega)$. Since $\lambda_0(-A_0)^{-\alpha}$ is a bounded operator in $\mathbf{V}_n^0(\Omega)$, $(A - \lambda_0(-A_0)^{-\alpha})$ with domain $D(A_0)$ is the infinitesimal generator of an analytic semigroup on $\mathbf{V}_n^0(\Omega)$, and $D((\lambda_0(-A_0)^{-\alpha} - A)^\beta) = D((-A_0)^\beta)$ for all $0 \le \beta \le 1$. The exponential stability follows from (36). $\qquad\square$

Lemma 26. *If* $\mathbf{u}$ *belongs to* $\mathbf{V}^{s,s/2}(\Sigma_\infty)$ *with* $0 \le s < 1$, *then the weak solution to the equation*

$$\mathbf{y}' = (A - \lambda_0(-A_0)^{-\alpha})\mathbf{y} + BM\mathbf{u} \quad \text{in } (0, \infty), \qquad \mathbf{y}(0) = 0,$$

obeys

$$\|\mathbf{y}\|_{\mathbf{V}^{1/2+s-\varepsilon, 1/4+s/2-\varepsilon/2}(Q_\infty)} \le C\|\mathbf{u}\|_{\mathbf{V}^{s,s/2}(\Sigma_\infty)} \quad \text{for all } \varepsilon > 0.$$

Proof. We can follow the lines of the proof given in [15, Lemma 7.3] if we are able to show that $(\lambda_0(-A_0)^{-\alpha} - A)^{1/4-\varepsilon/2}PD_A$ is bounded from $\mathbf{V}^0(\Gamma)$ into $\mathbf{V}_n^0(\Omega)$. We know that $(\lambda_0 I - A)^{1/4-\varepsilon/2}PD_A$ is bounded from $\mathbf{V}^0(\Gamma)$ into $\mathbf{V}_n^0(\Omega)$. Thus it is enough to prove that $(\lambda_0(-A_0)^{-\alpha} - A)^{1/4-\varepsilon/2}(\lambda_0 I - A)^{-1/4+\varepsilon/2}$ is bounded in $\mathbf{V}_n^0(\Omega)$. It is the case since $(\lambda_0 I - A)^{-1/4+\varepsilon/2}$ is an isomorphism from $\mathbf{V}_n^0(\Omega)$ into $\mathbf{V}_n^{1/2-\varepsilon}(\Omega)$, and $(\lambda_0(-A_0)^{-\alpha} - A)^{1/4-\varepsilon/2}$ is an isomorphism from $\mathbf{V}_n^{1/2-\varepsilon}(\Omega)$ into $\mathbf{V}_n^0(\Omega)$ (which is a consequence of Lemma 25). $\qquad\square$

References

[1] V. Barbu, *Feedback stabilization of the Navier-Stokes equations*, ESAIM COCV, **9** (2003), 197–206.

[2] V. Barbu, S.S. Sritharan, H^∞ – *control theory of fluid dynamics*, Proc. R. Soc. Lond. A **454** (1998), 3009–3033.

[3] V. Barbu, R. Triggiani, *Internal stabilization of Navier-Stokes equations with finite-dimensional controllers*, Indiana University Journal, **52**(5) (2004), 1443–1494.

[4] V. Barbu, I. Lasiecka, R. Triggiani, *Boundary stabilization of Navier-Stokes equations*, Memoirs of the A.M.S., 2005, to appear.

[5] A. Bensoussan, G. Da Prato, M.C. Delfour, S.K. Mitter, *Representation and Control of Infinite-Dimensional Systems*, Vol. 1, Birkhäuser, 1992.

[6] A. Bensoussan, G. Da Prato, M.C. Delfour, S.K. Mitter, *Representation and Control of Infinite-Dimensional Systems*, Vol. 2, Birkhäuser, 1993.

[7] E. Fernandez-Cara, S. Guerrero, O. Yu.Imanuvilov, J.-P. Puel, *Local exact controllability of the Navier-Stokes system*, J. Math. Pures Appl., Vol. 83 (2004), 1501–1542.

[8] A.V. Fursikov, M.D. Gunzburger, L.S. Hou, *Inhomogeneous boundary value problems for the three-dimensional evolutionary Navier-Stokes equations*, J. Math. Fluid Mech., **4** (2002), 45–75.

[9] A.V. Fursikov, *Stabilizability of two-dimensional Navier-Stokes equations with help of a boundary feedback control*, J. Math. Fluid Mech., 3 (2001), 259–301.

[10] A.V. Fursikov, *Stabilization for the 3D Navier-Stokes system by feedback boundary control*, Discrete and Cont. Dyn. Systems, 10 (2004), 289–314.

[11] I. Lasiecka, R. Triggiani, *The regulator problem for parabolic equations with Dirichlet boundary control, part 1*, Appl. Math. Optim., Vol. 16 (1987), 147–168.

[12] I. Lasiecka, R. Triggiani, *Control Theory for Partial Differential Equations*, Vol. 1, Cambridge University Press, 2000.

[13] I. Lasiecka, R. Triggiani, *Control Theory for Partial Differential Equations*, Vol. 2, Cambridge University Press, 2000.

[14] J.-P. Raymond, *Stokes and Navier-Stokes equations with nonhomogeneous boundary conditions*, 2005, submitted.

[15] J.-P. Raymond, *Boundary feedback stabilization of the two-dimensional Navier-Stokes equations*, 2005, to appear in SIAM J. Control and Optim..

[16] J.-P. Raymond, *Boundary feedback stabilization of the three-dimensional Navier-Stokes equations*, in preparation.

[17] J.-P. Raymond, *Feedback boundary stabilization of the Navier-Stokes equations*, in Control Systems: Theory, Numerics and Applications, Proceedings of Science, PoS(CSTNA2005)003, htpp://pos.sissa.it.

[18] R. Temam, Navier-Stokes equations, North-Holland, 1984.

Jean-Pierre Raymond
Université Paul Sabatier,
Laboratoire MIP, UMR CNRS 5640,
F-31062 Toulouse Cedex 9, France
e-mail: `raymond@mip.ups-tlse.fr`

International Series of Numerical Mathematics, Vol. 155, 293–309
© 2007 Birkhäuser Verlag Basel/Switzerland

Beyond Bilinear Controllability: Applications to Quantum Control

Gabriel Turinici

Abstract. Quantum control is traditionally expressed through bilinear models and their associated Lie algebra controllability criteria. But, the first order approximation are not always sufficient and higher order developments are used in recent works. Motivated by these applications, we give in this paper a criterion that applies to situations where the evolution operator is expressed as sum of possibly non-linear real functionals of the **same** control that multiplies some time independent (coupling) operators.

Mathematics Subject Classification (2000). Primary 34H05, 93B05; Secondary 35Q40.

Keywords. Controllability, bilinear controllability, quantum control, laser control.

Contents

1 Background on quantum control 294

2 Background on controllability criteria 294

    2.1 Infinite-dimensional bilinear control 296

    2.2 Finite-dimensional bilinear control 298

3 Criteria for non linear operators 301

4 Applications to quantum control 304

    4.1 Density matrix 304

    4.2 Wave function 306

References 307

This work was partially supported by INRIA-Rocquencourt and CERMICS-ENPC, Champs sur Marne, France. The author acknowledges an ACI-NIM grant from the Ministère de la Recherche, France.

1. Background on quantum control

Controlling the evolution of molecular systems at quantum level has been envisioned from the very beginnings of the laser technology. However, approaches based on designing laser pulses based on intuition alone did not succeed in general situations due to the very complex interactions that are at work between the laser and the molecules to be controlled, which results, e.g., in the redistribution of the incoming laser energy to the whole molecule. Even if this circumstance initially slowed down investigations in this area, the realization that this inconvenient can be recast and attacked with the tools of (optimal) control theory [18] greatly contributed to the first positive experimental results [2, 21, 33, 6, 5, 17, 20].

The regime that is relevant for this work is related to time scales of the order of the femtosecond (10^{-15}) up to picoseconds (10^{-12}) and the space scales from the size of one or two atoms to large polyatomic molecules.

Historically, the first applications that were envisioned were the manipulation of chemical bonds (e.g., selective dissociation) or isotopic separation. Although initially, only few atoms molecules were investigated (di-atomics) the experiments soon were designed to treat more complex situations [2] as selective bond dissociation in an organi-metalic complex $CpFe(CO)_2Cl$ (Cp is the cyclopentadienyl ion) by maximizing or minimizing the quotient of $CpFeCOCl^+$ ions obtained with respect to $FeCl^+$ ions.

Continuing this breakthrough, other poly-atomic molecules were considered in strong fields. For instance, in [21] the molecules are the acetone $(CH_3)_2CO$, the trifluoroacetone CH_3COCF_3 and the acetophenone $C_6H_5COCH_3$. Using tailored laser pulses it was shown possible to obtain CH_3CO from $(CH_3)_2CO$, CF_3 (or CH_3) from CH_3COCF_3 but also $C_6H_5CH_3$ (toluene) from $C_6H_5COCH_3$.

But the applications of laser control do not stop here. High Harmonic Generation) [7] is a technique that allows to obtain output lasers whose frequency is large integer multiples of the input pulses.

A different class of applications works in a different regime of shorter time scales and large intensity. This regime is additionally not compatible with the standard Born-Oppenheimer approximation and requires to consider both nucleari and electrons as quantum particles with entangled wave function [4].

In a different framework, the manipulation of quantum states of atoms and molecules allows to envision the construction of quantum computers [13, 27]

Finally, biologically related applications are also the object of ongoing research.

2. Background on controllability criteria

We start in this section to investigate the theoretical controllability results that are nowadays available for quantum systems. The evolution of the system will be described by the driving Schrödinger equation (we work here in atomic units, i.e., $\hbar = 1$)

$$i\frac{\partial}{\partial t}\Psi(t,x) = H(t)\Psi(t,x)$$
$$\Psi(t_0,x) = \Psi_0(x). \tag{2.1}$$

where $H(t)$ is the Hamiltonian of the system and $x \in \mathbb{R}^\gamma$ the set of internal degrees of freedom. We introduce the Hilbert space structure given by the scalar product

$$\langle f, g \rangle = \int_{\mathbb{R}^\gamma} \overline{f(x)}g(x)dx \tag{2.2}$$

where $\overline{a+ib} = a - ib$ the conjugate of a complex number.

We only consider in this paper situations when the Hamiltonian is autoadjoint $H(t)^\dagger = H(t)$; we denoted by $T^\dagger$ the adjoint of a operator T. The autoadjointness of H implies that the $L_x^2(\mathbb{R}^\gamma)$ norm of the evolving state is conserved. Indeed

$$\frac{d}{dt}\|\Psi(x,t)\|_{L_x^2(\mathbb{R}^\gamma)} = \frac{d}{dt}\langle \Psi(x,t), \Psi(x,t)\rangle$$
$$= \langle \frac{d}{dt}\Psi(x,t), \Psi(x,t)\rangle + \langle \Psi(x,t), \frac{d}{dt}\Psi(x,t)\rangle \tag{2.3}$$
$$= \langle \frac{H(t)}{i}\Psi(x,t), \Psi(x,t)\rangle + \langle \Psi(x,t), \frac{H(t)}{i}\Psi(x,t)\rangle = 0.$$

Thus

$$\|\Psi(x,t)\|_{L_x^2(\mathbb{R}^\gamma)} = \|\Psi_0\|_{L^2(\mathbb{R}^\gamma)}, \ \forall t > 0, \tag{2.4}$$

so the wave function $\Psi(t)$, evolves on the (complex) unit sphere

$$S = \left\{\psi \in L^2(\mathbb{R}^\gamma) \ : \ \|\psi\|_{L^2(\mathbb{R}^\gamma)} = 1\right\}.$$

When the system evolves freely under its own internal dynamics, i.e., when isolated molecules are considered, the free evolution Hamiltonian H_0 is introduced. This Hamiltonian is the sum of the kinetic part T and the potential operator $V(x)$: $H_0 = T + V(x)$. A prototypical example of T is the Laplace operator while for $V(x)$ one can encounter Coulomb potential or Lennard-Jones type dependence. We obtain the following evolution in the absence of external interaction:

$$i\frac{\partial}{\partial t}\Psi(t,x) = H_0\Psi(t,x)$$
$$\Psi(t_0,x) = \Psi_0(x). \tag{2.5}$$

But, when the free evolution of the system does not generate a satisfactory dynamical output, an external interaction is introduced to *control* it. An example of external control of paramount importance is a laser source of intensity $\epsilon(t) \in \mathbb{R}$, $t \geq 0$.

The purpose of control may be formulated as to drive the system from its initial state Ψ_0 to take a convenient dynamical path to a final state compatible with predefined requirements. The control is here the laser intensity $\epsilon(t)$. We will come back later with details on the laser field $\epsilon(t)$.

This laser will modify the Hamiltonian $H(t)$ of the system. A first order approximation can be considered by introducing a time-independent dipole moment operator $\mu(x)$ resulting in the dynamics:

$$i\frac{\partial}{\partial t}\Psi(t,x) = \left(H_0 + \epsilon(t)\mu\right)\Psi(t,x)$$
$$\Psi(t_0,x) = \Psi_0(x). \tag{2.6}$$

This is the so-called *bi-linear* framework (the control enters linearly multiplying the state), that is the object of much theoretical and numerical work in quantum control. We also review below some of the results that are available in this formulation. However, recently, higher order field dependence has been considered in different circumstances see, e.g., [14, 15] for details. In these situations the Hamiltonian $H(t)$ is developed further as :

$$H(t) = H_0 + \epsilon(t)\mu_1 + \epsilon^2(t)\mu_2 + \cdots + \epsilon^L(t)\mu_L. \tag{2.7}$$

The question that will be of interest to us in this work is the study of all possible final states for the quantum system. This question is important in order to understand the capabilities that a laboratory experiment will be able to provide and also, in a more general setting, to accompany the introduction of new experimental protocols.

More specifically, we will show how the criteria available for bilinear control can be extended to treat the Hamiltonian (2.7) where a **single** control amplitude $\epsilon(t)$ appears before different coupling operators $\mu_1, \ldots, \mu_L$.

Many of the questions regarding the properties of the quantum control procedures, such as controllability, optimal control definition, etc., ... need, in order to be defined, to specify the admissible control class, i.e., the set $\mathcal{U}$ where the control $\epsilon(t)$ is allowed to vary. Among the properties that can define this admissible set, some are related to the regularity of the time-dependence (L^2, H^1, ... etc.) or of the Fourier expression (sum of sinusoidal functions multiplied by an overall envelope, etc., ...) or to additional structure: e.g., piecewise continuous, piecewise constant, locally bounded ....

The choice of one or several conditions in the list above is motivated in practice by capability to reproduce that particular form or to inherent experimental restrictions (finite total laser energy/fluence, etc). As the laser technology is constantly evolving, the first class of constraints becomes less critical and thus it is realistic to consider very weak constraints on the control set, e.g., $\mathcal{U} = L^2(\mathbb{R}) \cap L^\infty_{loc}(\mathbb{R})$.

However, to treat even more general situations, we will consider in this work controls $\epsilon(t)$ that are piecewise constant, taking any value in a set V, which will remain fully general.

2.1. Infinite-dimensional bilinear control

When compared to the finite-dimensional control equations (see Section 2.2), controllability of the infinite-dimensional version of the bilinear Time Dependent

Schrödinger Equation is much less understood at this time. In fact, most of the progress obtained so far takes the form of negative results, in contradiction with the positive results available in finite-dimensional settings. However we see the absence of positive controllability results is rather a failure of today's control theory tools to provide insight into controllability rather than an actual restriction. We do believe that new tools and concepts will make positive results possible.

Let us write the solution of (2.6) in the following form:

$$\Psi(t) = e^{-iH_0 t}\Psi_0 - i\int_0^t \epsilon(s)e^{-iH_0(t-s)}\mu\Psi(s)ds. \tag{2.8}$$

This formulation (see [11] for details) is granted by the properties of the operator $\mu : H_0^1(\mathbb{R}^\gamma) \to H^{-1}(\mathbb{R}^\gamma)$ which is continuous when μ is bounded; we also recall that the control ϵ can be considered bounded in both L^∞ and L^2.

The application $\epsilon(t) \mapsto \Psi(x,t)$ possesses an important compacity property which is the key of the controllability results (we refer the reader interested in details to [3, 25]) :

Lemma 2.1. *Suppose that $\mu : X \to X$ is a bounded operator and that H_0 generates a C^0 semigroup of bounded linear operators on some Banach space X (e.g., $X = H_0^1(\mathbb{R}^\gamma)$). Denote for $T > 0$ and $\epsilon \in L^1([0,T])$ by $\Psi_\epsilon(x,t)$ the solution of (2.6) with control ϵ. Then $\epsilon \mapsto \Psi_\epsilon$ is a compact mapping in the sense that for any ϵ_n that converges weakly to ϵ in $L^1([0,T])$ Ψ_{ϵ_n} converges strongly in $C([0,T];X)$ to Ψ_ϵ.*

This compactness property allows to give *negative results* for general bilinear controllability settings as in [3] where they were applied to the wave and rod equations. Specific statements for quantum control have been latter derived (Thm. 1 from [30] ; see also [3, 29]) and can be stated as:

Theorem 2.2. *Let S be the complex unit sphere of $L^2(\mathbb{R}^\gamma)$. Let μ be a bounded operator from the Sobolev space X (e.g., $X = H_x^1(\mathbb{R}^\gamma)$) to itself and let H_0 generate a C^0 semigroup of bounded linear operators on X. Denote by $\Psi_\epsilon(x,t)$ the solution of (2.6). Then the set of attainable states from Ψ_0 defined by*

$$\mathcal{AS} = \cup_{T>0}\{\Psi_\epsilon(x,T); \epsilon(t) \in L^2([0,T])\} \tag{2.9}$$

is contained in a countable union of compact subsets of X. In particular its complement $S \cap X \setminus \mathcal{AS}$ with respect to $S \cap X$ is everywhere dense on $S \cap X$. The same holds true for the complement with respect to S.

In a different formulation, the theorem implies that for any $\Psi_0 \in X \cap S$, within any open set around an arbitrary point $\Psi \in X \cap S$ there exists a state unreachable from Ψ_0 with L^2 controls.

Remark 2.3. Note that the result does not give information on the closure of the set $\mathcal{AS}$. In particular it may well be that while $\mathcal{AS}$ still has dense complement its closure be the whole space X. This would be the so-called approximate controllability, i.e., the possibility to reach targets arbitrarily close to any given final state. Despite some attempts in the literature, at this time there is no answer (positive or negative) to this question. Among the ingredients that make this study difficult we can mention the possibility to use arbitrary large final time T, the necessity to treat the continuous spectrum of the operator H_0 and the intrinsically unbounded domain on which the problem is posed.

To complicate even more the landscape, situations exists where the results obtained in infinite and finite-dimensional representation are of different nature. We will illustrate with a classical result on the harmonic oscillator.

Lemma 2.4. *The infinite-dimensional harmonic oscillator $H_0 = -\frac{\partial^2}{\partial x^2} + x^2$, $\mu = x$ is not controllable. Moreover the set of all admissible states is a low-dimensional manifold of L^2.*

Proof. Let us begin by noting that the operators $-iH_0$ and $-i\mu$ form a Lie algebra of dimension 4. Indeed, let us compute the iterated commutators of $H_0 = -\frac{\partial^2}{\partial x^2} + x^2$ and $\mu = x$:

$$\left[i\left(-\frac{\partial^2}{\partial x^2} + x^2 \right), ix \right] = 2\frac{\partial}{\partial x} \tag{2.10}$$

$$\left[ix, \frac{\partial}{\partial x} \right] = -i \tag{2.11}$$

$$\left[i\left(-\frac{\partial^2}{\partial x^2} + x^2 \right), i\frac{\partial}{\partial x} \right] = -2ix. \tag{2.12}$$

Thus the dimension of the Lie algebra $= 4$ and as such the system cannot be controllable (all the states are on a low-dimensional manifold of L^2). We refer to [23] for recent contributions when the algebra of the operators H_0 and μ is finite-dimensional. $\square$

This result is to be contrasted with additional works that show that any (spectral) truncation of the harmonic oscillator is controllable (see [26] for details).

What can be deduced from the above result is that truncating an infinite-dimensional system is not always justified and care must be taken to check that the control obtained in the resulting finite-dimensional approximation remain a good control for the initial, infinite-dimensional system. Of course, this is not needed for situations which are inherently finite-dimensional quantum systems (e.g., spins).

2.2. Finite-dimensional bilinear control

Here our focus will be on finite-dimensional systems. We introduce an orthonormal basis $D = \{\psi_i(x); i = 1, \dots, N\}$ for a finite-dimensional space. An important example of such a space is the one spanned by the first N eigenstates of the internal Hamiltonian H_0. This example is also motivated in bi-linear settings by the "perturbation" argument that considers the control term $\epsilon(t)\mu$ as a first order development of $H(t)$ around H_0. Note however that no concept of "smallness" is introduced in the definition of admissible controls $\mathcal{U}$.

Denote by M the linear space that D generates, and let $H_{0;a,b} = \langle H_0\psi_a, \psi_b \rangle$ and $\mu_{\ell;a,b} = \langle \psi_a, \mu_\ell\psi_b \rangle$ be the expressions of the operators H_0 and μ_ℓ with respect to this basis, $\ell = 1, \ldots, L$. To keep notations simple we will still denote from now on by H_0 and μ_ℓ the resulting $N \times N$ symmetric matrices.

In the Galerkin approach, expressing the Schrödinger equation in the space M is equivalent to supposing $\Psi(x, t) = \sum_{i=1}^{N} \psi_i(x) c_i(t)$.

$$i \frac{dc(t; \epsilon; c_0)}{dt} = H_0 c(t; \epsilon; c_0) + \left[\epsilon(t)\mu_1 + \cdots + \epsilon^L(t)\mu_L\right] c(t; \epsilon; c_0)$$

$$c(t = 0; \epsilon; c_0) = c_0.$$

(2.13)

In the following, when no ambiguity prevents it, we will also simply denote $c = c(t; \epsilon; c_0)$. The finite-dimensional counterpart of the norm conservation property (2.4) reads:

$$\sum_{n=1}^{N} |c_n|^2 = 1,$$

i.e., the state c evolves on the unit sphere S_N of $\mathbb{C}^N$. The controllability can be formulated in this case as:

Definition 2.5. The system $(H_0, \mu_1, \ldots, \mu_L)$ is called (wave function) controllable, if for any two states $c_k \in S_N$, $k = 1, 2$ there exists a final time $T < \infty$ and control $\epsilon(t) \in L^2([0, T])$ such that the solution of Eqn. (2.13) starting from c_1 ends in c_2 at final time T: $c(T; \epsilon; c_1) = c_2$.

Although specific results for this setting exist [31, 32], a different alternative is to see (2.13) as a system posed on $U(N)$[1]. We introduce the evolution equation on $U(N)$:

$$\frac{dU(t; \epsilon)}{dt} = \left[H_0 + \epsilon(t)\mu_1 + \cdots + \epsilon^L(t)\mu_L\right] U(t; \epsilon)$$

$$U(t = 0; \epsilon) = Id.$$

(2.14)

Since H_0 and μ_ℓ are symmetric matrices, $U(t; \epsilon)$ will remain unitary for all $t \geq 0$. It is classical to remark then that the evolution of $c(t; \epsilon; c_0)$ can be obtained from the evolution of $U(t; \epsilon)$ by

$$c(t; \epsilon; c_0) = U(t; \epsilon)c_0.$$

In particular it follows that if the set of all attainable matrices $U(t; \epsilon)$ is at least $SU(N)$ then the system is controllable. This is almost a necessary condition for controllability, a notable exception being the circumstance when N is even: in this case, if the set of all attainable matrices contains $Sp(N/2)$ then controllability still holds. We refer to [10, 1] for more detailed information.

Let us just mention that different representations of the system include the density matrix formulation with time dependent density matrix operator $\rho(t)$

---

[1] $U(N)$ is the set of all $N \times N$ complex unitary matrices.

satisfying

$$i\frac{\partial}{\partial t}\rho(t;\epsilon;\rho_0) = [H_0 + \epsilon(t)\mu_1 + \cdots + \epsilon^L(t)\mu_L, \rho(t;\epsilon;\rho_0)]$$
$$\rho(t=0;\epsilon;\rho_0) = \rho_0. \tag{2.15}$$

Then one can show $\rho(t;\epsilon;\rho_0) = U(t;\epsilon)\rho_0 U^\dagger(t;\epsilon)$. Controllability in this case is the possibility to steer any initial mixed state ρ_0 to any other state ρ^f unitarily equivalent to it[2].

Note that the density matrix controllability is equivalent to requiring that the set of all matrices attainable from identity be at least $SU(N)$.

At a general level, the evolution equation (2.14) can be re-written as

$$\frac{dx(t;\epsilon;x_0)}{dt} = (A + \epsilon(t)B_1 + \cdots + \epsilon^L(t)B_L)x(t;\epsilon;x_0)$$
$$x(0) = x_0. \tag{2.16}$$

where $x(t;\epsilon;x_0)$ belongs to a Lie group G (see [16, 8, 9] for basic facts about the Lie groups) and $A, B_1, \ldots, B_K$ to its associated Lie algebra $L(G)$. The equation above is to be taken in the usual sense (using the exponential map) when, e.g., $\epsilon(t)$ is piecewise continuous/constant and in a weak sense (integral form) for general $\epsilon(t)$ (see, e.g., [3] for additional details). For the quantum control problem $A = -iH_0$ and $B_\ell = -i\mu_\ell$, $G = U(N)$.

Remark 2.6. Everything that will be said in this and following sections applies with trivial modifications to the situation of several laser fields. For notational convenience we will only give here the results for a unique laser field.

We will denote by $L_{A,B_1,\ldots,B_k} \subset L(G)$ the Lie algebra spanned by A, B_k, $k = 1, \ldots, K$ and by e the unity of G.

Let us now consider the set of all reachable states from an initial state y:

$$\mathcal{R}_\mathcal{U}^t(y) = \{x(t;\epsilon;y) \text{ solution of } (2.16) \ ; \epsilon \in \mathcal{U}\}. \tag{2.17}$$

It is immediate to see that

$$\mathcal{R}_\mathcal{U}^t(y) = \mathcal{R}_\mathcal{U}^t(e)y \tag{2.18}$$

and thus, describing the set $\mathcal{R}_\mathcal{U}^t(e)$ allows to completely describe all other reachable sets. When the final time is not specified, we will denote

$$\mathcal{R}_\mathcal{U}(y) = \cup_{t \geq 0}\mathcal{R}_\mathcal{U}^t(y). \tag{2.19}$$

The central question is to characterize $\mathcal{R}_\mathcal{U}(e)$. When the bi-linear setting is considered, i.e., $L = 1$ and we note $B = B_1$, we have the following result [19, 22]:

Theorem 2.7. *Consider the system* (2.16) *defined on a Lie group G with associated Lie algebra $L(G)$ containing A and B. If G is compact and the Lie algebra $L_{A,B}$ generated by A and B is the complete algebra $L(G)$: $L_{A,B} = L(G)$ then the set*

---

[2]An $N \times N$ matrix ρ_2 is said unitarily equivalent to an $N \times N$ matrix ρ_1 if there exists $M \in U(N)$ such that $\rho_2 = M\rho_0 M^\dagger$.

$\mathcal{R}_\mathcal{U}(e)$ *of all states from the identity is the Lie group G. Moreover, there exists* $0 < T < \infty$ *such that* $\mathcal{R}_\mathcal{U}^{T'}(e) = G$ *for all* $T' \geq T$.

This gives, when applied to quantum control [24]: ($L = 1$, $\mu = \mu_1$):

Theorem 2.8. *If the Lie algebra $L_{-iH_0,-i\mu}$ generated by $-iH_0$ and $-i\mu$ has dimension N^2 (as a vector space over the real numbers) then the system (2.14) is density matrix controllable. Furthermore, if both $-iH_0$ and $-i\mu$ are traceless then a sufficient condition for the density matrix (thus wave function) controllability of quantum system is that the Lie algebra $L_{-iH_0,-i\mu}$ has dimension $N^2 - 1$.*

Although the results above conveniently address the situation of a bi-linear setting, we are not aware of any similar results for the general quantum control situations (2.7). In particular, we know by the result above that, if $u_1, \ldots, u_L$ are *independent controls*, i.e.,

$$\frac{dx(t; \epsilon; x_0)}{dt} = (A + u_1(t)B_1 + \cdots + u_L(t)B_L)x(t; \epsilon; x_0)$$
$$x(0) = x_0, \tag{2.20}$$

an equivalent condition for the controllability of the above system on its compact Lie group G is that $A, B_1, \ldots, B_L$ generate the whole Lie algebra $L(G)$. But, there is no obvious way to say what will happen when the controls u_ℓ are not independent but related by the condition $u_\ell = \epsilon^\ell(t)$. This study is the purpose of the next section.

3. Criteria for non linear operators

In order to extend the controllability results above beyond bi-linear interaction Hamiltonians, we will introduce in this section a more general setting: we will rewrite the control equation (2.16) as

$$\frac{dx(t; \epsilon; x_0)}{dt} = (F_1(\epsilon(t))B_1 + \cdots + F_l(\epsilon(t))B_L)x(t; \epsilon; x_0) \tag{3.1}$$
$$x(0) = x_0, \tag{3.2}$$

where $F_k : V \to \mathbb{R}$ are real functionals. Note in particular that we do not impose **any** assumption on the regularity of the functionals F_k. Of course, one can recover the equation (2.16) by setting $F_k(x) = x^k$ and adding $F_0 = 1$.

In order to avoid trivialities, we will suppose in the following that

$$\text{the functionals } (F_k)_{k=1}^L \text{ are linearly independent.} \tag{3.3}$$

Otherwise one may just consider a subset of functionals that are linearly independent and adjust the matrices B_k accordingly. Of course, since we do not specify the set V that lists all the possible control values ϵ the hypothesis above needs to be understood in the following acception: the functionals F_k are said to be linearly dependent if there exist constants $\lambda_1, \ldots, \lambda_L \in \mathbb{R}$ such that $\sum_{j=1}^L \lambda_j F_j(v) = 0$ for all $v \in V$. Otherwise the functionals are said to be linearly independent.

In order to obtain the quantum controllability results, we begin in this section with a controllability criterion on compact Lie groups. These results build on classical references for bilinear controllability [19]. We give first a weak but intuitive form and then we state the fully general one.

Theorem 3.1. *Let* (3.1) *be a control system posed on a compact connected Lie group G, with linearly independent functionals $(F_k)_{k=1}^{L}$. Then if the Lie algebra generated by $B_1, \ldots, B_L$ is the full Lie algebra $L(G)$ of the group G, then the system is approximately controllable, i.e., for any $a, b \in G$, b is an accumulation point of the set of all states $x(t)$ attainable from $x(0) = a$ with admissible controls.*

Proof. Let us begin by noting that if F_k are independent then there exist values $e_j \in V$, $j = 1, \ldots, L$ such that the vectors $v(e_j) = (F_1(e_j), \ldots, F_L(e_j))$ are linearly independent. Suppose on the contrary that this is not true. Consider then a maximal set of vectors $v(E_1), \ldots, v(E_p)$ that are linearly independent. The matrix $(F_k(E_j))_{k=1;j=1}^{L;p}$ has rank precisely p and thus one can extract p functionals, denoted for notational convenience $F_1, \ldots, F_p$ such that $\operatorname{rank}(F_k(E_j))_{k=1;j=1}^{p;p} = p$. Take now some functional F_{p+1} not in this set. It follows that

$$\operatorname{rank}(F_k(E_j))_{k=1;j=1}^{p+1;p+1} = p \quad \text{and as such} \quad \det(F_k(E_j))_{k=1;j=1}^{p+1;p+1} = 0$$

for any $E_{p+1} \in V$. This determinant can be computed as:

$$\det(F_k(E_j))_{k=1;j=1}^{p+1;p+1} = \lambda_1 F_1(E_{p+1}) + \cdots + \lambda_{p+1} F_{p+1}(E_{p+1}) = 0. \tag{3.4}$$

Note that λ_k do not depend on E_{p+1} and that in particular

$$\lambda_{p+1} = \det(F_k(E_j))_{k=1;j=1}^{p;p} \neq 0.$$

Thus Eq. (3.4) implies that a linear combination with at least one non-null coefficient λ_{p+1} exists such that $\sum_{k=1}^{p+1} \lambda_k F_k(E) = 0$ for all $E \in V$. This is prevented by hypothesis.

We have thus proved the existence of $e_j \in V$, $j = 1, \ldots, L$ with the $v(e_j) = (F_1(e_j), \ldots, F_L(e_j))$ linearly independent. This means that $M_j = \sum_{k=1}^{L} F_k(e_j) B_k$ are also linearly independent and span the same linear space as B_k, $k = 1, \ldots, L$ and thus M_j span also the Lie algebra $L(G)$. Moreover, all states $\{e^{tM_j} x(0); t \in \mathbb{R}_+, j \leq L\}$ are attainable from $x(0)$ for the control system (3.1).

It is clear that to prove approximate controllability is sufficient to set $a = e$ the neutral element of the group G, i.e., we have to prove that the closure $\overline{\mathcal{R}_{\mathcal{U}}(e)}$ (with respect to the Lie group topology) of the reachable states from identity is the whole G. From the hypothesis and surjectivity of the exponential mapping this is equivalent to proving that

$$\{e^M; M \in L(G)\} \subset \overline{\mathcal{R}_{\mathcal{U}}(e)}.$$

We will begin by noting that $\overline{\mathcal{R}_{\mathcal{U}}(e)}$ is a group. Indeed, take two elements $x(t_1; \epsilon_1; e), x(t_2; \epsilon_2, e) \in \mathcal{R}_{\mathcal{U}}(e)$. Then, defining the control $\epsilon_{12} : [0, t_1 + t_2] \to \mathbb{R}$ by $\epsilon_{12}(t) = \epsilon_1(t)$ for all $0 \leq t \leq t_1$ and $\epsilon_{12}(t_1 + t) = \epsilon_2(t)$ for all $0 \leq t \leq t_2$ we obtain

$x(t_1+t_2; \epsilon; e) = x(t_2; \epsilon_2; e)x(t_1; \epsilon_1, e)$ and thus $x(t_2; \epsilon_2; e)x(t_1; \epsilon_1, e) \in \mathcal{R}_{\mathcal{U}}(e)$. Hence $\mathcal{R}_{\mathcal{U}}(e)$ is a semi-group which implies that $\overline{\mathcal{R}_{\mathcal{U}}(e)}$ is a semi-group too.

Let us now consider $a \in \overline{\mathcal{R}_{\mathcal{U}}(e)}$. Then $a^n \in \overline{\mathcal{R}_{\mathcal{U}}(e)}$ for any $n = 1, 2, \ldots$. Since $\overline{\mathcal{R}_{\mathcal{U}}(e)} \subset G$ which is a compact group, $\overline{\mathcal{R}_{\mathcal{U}}(e)}$ is compact at its turn. Then there exists a sequence, that we can take such that n_k with $n_k - n_{k-1} \geq 2$, with $a^{n_k} \to b \in G$. But then $\overline{\mathcal{R}_{\mathcal{U}}(e)} \ni a^{n_k - n_{k-1} - 1} \to bb^{-1}a^{-1}$ and thus $a^{-1} \in \overline{\mathcal{R}_{\mathcal{U}}(e)}$.

It is immediate to see that, since the solution for the control $\epsilon(t) \equiv e_j$ is $x(t; 0, c) = c^{tM_j}$ we have the inclusion $\{e^{tM_j}; t \geq 0, j \leq L\} \subset \overline{\mathcal{R}_{\mathcal{U}}(e)}$. Since $\overline{\mathcal{R}_{\mathcal{U}}(e)}$ is a group, we will also have $\{e^{tM_j}; t \in \mathbb{R}; j \leq L\} \subset \overline{\mathcal{R}_{\mathcal{U}}(e)}$. Consider now two matrices $X_1, X_2 \in L(G)$ such that

$$\{e^{tX_i}; t \geq 0\} \subset \overline{\mathcal{R}_{\mathcal{U}}(e)}, \quad i = 1, 2.$$

We invoke now the formula

$$e^{t[X_1, X_2]} = \lim_{n \to \infty} \left(e^{-tX_2/\sqrt{n}} e^{-tX_1/\sqrt{n}} e^{tX_2/\sqrt{n}} e^{tX_1/\sqrt{n}} \right)^n \tag{3.5}$$

to conclude that

$$\{e^{t[X_1, X_2]}; t \in \mathbb{R}\} \subset \overline{\mathcal{R}_{\mathcal{U}}(e)}.$$

Similarly, we use the formula $e^{t_1 X_1 + t_2 X_2} = \lim_{n \to \infty} \left(e^{t_1 X_1/n} e^{t_2 X_2/n} \right)^n$ to conclude that

$$\{e^{t_1 X_1 + t_2 X_2}; t_1, t_2 \in \mathbb{R}\} \subset \overline{\mathcal{R}_{\mathcal{U}}(e)}.$$

We have thus proved that the set $\{M \in L(G); e^{tM} \in \overline{\mathcal{R}_{\mathcal{U}}(e)}; \forall t \in \mathbb{R}\}$ contains M_j, $j = 1, \ldots, L$, is closed to commutation and is a real vector space. Thus it contains $L(G)$ hence the conclusion of the theorem. $\qquad \square$

The theorem above has the advantage to be both intuitive and self-contained. However it only gives approximate controllability results, which are not the strongest forms available. But, in order to obtain exact controllability more involved techniques are needed. In literature, similar situations are treated by making use of the Chow theorem [12] and of the bi-linear control techniques [19, 28]. The criterion can be stated as follows:

Theorem 3.2. *Let (3.1) be a control system posed on a compact connected Lie group G with linearly independent functionals $F_k : V \to \mathbb{R}$, $k = 1, \ldots, L$ and piecewise constant controls ϵ taking any value in some set V. Then a necessary and sufficient condition for the exact controllability is that the Lie algebra $L_{B_1, \ldots, B_L}$ generated by $B_1, \ldots, B_L$ be the full Lie algebra $L(G)$ of the group G.*

Proof. We recall (see also end of Section 2.2) that the set of attainable states is included in the set of attainable states for the system

$$\frac{dx(t; \epsilon; x_0)}{dt} = [u_1(t)B_1 + \cdots + u_L(t)B_L]x(t; \epsilon; x_0) \tag{3.6}$$

$$x(0) = e,$$

whose controllability is equivalent to "$L_{B_1, \ldots, B_L} = L(G)$". Thus $L_{B_1, \ldots, B_L} = L(G)$ is a necessary condition for controllability. To prove that is also sufficient, consider as in the proof of Theorem 3.1, the matrices $M_j = \sum_{k=1}^{L} F_k(e_j)B_k$, $j = 1,\ldots,L$ that generate the same Lie algebra $L_{B_1,\ldots,B_L}$. We recall that all states $\{e^{tM_j}; t \in \mathbb{R}_+, j \leq L\}$ and all finite products of such states are attainable from the identity e. We invoke now a technique present in the proof of Thm. 3.1 of [28]: for any $P \in \mathbb{N}$ and any multi-index $i = (i_1,\ldots,i_r) \in \{1,\ldots,L\}^r$ denote by $A(i,T)$ the attainable states with the sequence of operators i and total time less than T:

$$A(i,T) = \left\{ \prod_{\ell=1}^{r} e^{t_\ell M_{i_\ell}}; \sum_{\ell=1}^{r} |t_\ell| \leq P, t_1,\ldots,t_r \in \mathbb{R} \right\}.$$

We know by the Chow theorem that the union of the sets $A(i,T)$ is the whole Lie group G. Also, it is immediate that any $A(i,T)$ is image of a compact set thus compact. It follows by the Baire category theorem that $A(i,P)$ has non-empty interior at least for a couple (i,P). For such an $i = (i_1,\ldots,i_m)$ we introduce the mapping $F : \mathbb{R}^m \to G$ defined by $t = (t_1,\ldots,t_\ell) \mapsto F(t) = \prod_{\ell=1}^{r} e^{t_\ell M_{i_\ell}}$. This mapping is analytic and its image is has nonempty interior. By the Sard theorem its differential $dF(t)$ has full rank (i.e., equals the dimension of the tangent space TG of G) at least at some point t and thus in a neighborhood. But since $dF(t)$ depends analytically on t the set of points where the rank is full is dense in $\mathbb{R}^m$ and as such the rank is full for some t with all components strictly positive. Using a local inverse mapping theorem it follows that the image $F(\mathcal{T})$ has non-empty interior where $\mathcal{T}$ is an open subset of $\mathbb{R}_+^m$. But all points in $F(\mathcal{T})$ are realizable with admissible controls and thus the set of reachable points $\mathcal{R}_\mathcal{U}(e)$ contains an open subset D of G.

By the previous theorem, $\mathcal{R}_\mathcal{U}(e)$ is a subgroup, i.e., for any $y \in \mathcal{R}_\mathcal{U}(e)$ the set Dy is also reachable. Since in addition $\mathcal{R}_\mathcal{U}(e)$ is dense in G it follows that $\mathcal{R}_\mathcal{U}(e) = G$. $\qquad\square$

4. Applications to quantum control

The purpose of this section is to instantiate the results obtained previously to the specific situation of the quantum control. We will give two results, one for the density matrix formalism and the second for the wave function.

4.1. Density matrix

To consider the specific situation of the density matrix formalism, we use the results of the Section 3 for the Lie group $U(N)$. We obtain a first

Theorem 4.1. *Consider the system*

$$i\frac{\partial}{\partial t}\rho(t;\epsilon;\rho_0) = [H_0 + F_1(\epsilon(t))\mu_1 + \cdots + F_L(\epsilon(t))\mu_L, \rho(t;\epsilon;\rho_0)]$$
$$\rho(t=0;\epsilon;\rho_0) = \rho_0$$

$$(4.1)$$

and suppose that the family $\{1, F_1,\ldots,F_L\}$ is linearly independent.

Then, when at least one matrix $H_0, \mu_1, \ldots, \mu_L$ has nonzero trace, the equation (4.1) *is density matrix controllable if and only if the Lie algebra $L_{iH_0, i\mu_1, \ldots, i\mu_L}$ spanned by the matrices $iH_0, i\mu_1, \ldots, i\mu_L$ is the Lie algebra $u(N)$ of all skew-hermitian matrices or equivalently* $\dim_{\mathbb{R}} L_{iH_0, i\mu_1, \ldots, i\mu_L} = N^2$.

Otherwise, when all matrices $H_0, \mu_1, \ldots, \mu_L$ have zero trace, a necessary and sufficient condition for controllability is that $L_{iH_0, i\mu_1, \ldots, i\mu_L} = su(N)$ or equivalently $\dim_{\mathbb{R}} L_{iH_0, i\mu_1, \ldots, i\mu_L} = N^2 - 1$.

Proof. The first part of the conclusion follows from Theorem 3.2 for the Lie group $G = U(N)$.

When all matrices have zero trace one uses the same result for $G = SU(N)$ noting that if two matrices ρ_1 and ρ_2 are unitarily equivalent $\rho_2 = M\rho_1 M^\dagger$ then there exists $\gamma \in R$ with $M_{su} = Me^{i\gamma} \in SU(N)$ and $\rho_2 = M_{su}\rho_1 M_{su}^\dagger$. $\qquad\square$

An algorithmic verification of the above theorem can be devised as follows:

1. Test whether the functions $\{1, F_1, \ldots, F_L\}$ are linearly independent. If the answer is yes go to next step, otherwise keep only a subset $F_{i_1}, \ldots, F_{i_p}$ with $\{1, F_{i_1}, \ldots, F_{i_p}\}$ linearly independent and modify the $B_1, \ldots, B_L$ accordingly. For notational convenience we suppose all functionals are independent ($p = L$).

2. Construct the traceless matrices
$$\widetilde{H_0} = H_0 - \frac{Tr(H_0)}{N} Id, \quad \widetilde{\mu_1} = \mu_1 - \frac{Tr(\mu_1)}{N} Id, \ldots, \widetilde{\mu_L} = \mu_1 - \frac{Tr(\mu_L)}{N} Id.$$
Denote by $\mathcal{O} = \{i\widetilde{H_0}, i\widetilde{\mu_1}, \ldots, i\widetilde{\mu_L}\}$.

3. Write any element of $\mathcal{O}$ as a column vector and compute the rank $r = \text{rank}(\mathcal{O})$ over the real numbers.

4. Construct all commutators $\mathcal{C}$ of matrices in $\mathcal{O}^3$ and test whether $\text{rank}(\mathcal{O} \cup \mathcal{C}) = r$. If not, set $\mathcal{O} := \mathcal{O} \cup \mathcal{C}$ and return to previous step.

5. Test whether $r = N^2 - 1$. If yes the system is controllable, if not the controllability does not hold.

Even more precise results can be derived for the situation in Eqn. 2.15.

Theorem 4.2. *Consider the development of the interaction Hamiltonian $H = H_0 + \epsilon(t)\mu_1 + \cdots + \epsilon^L(t)\mu_L$ resulting in the following evolution equation*

$$i\frac{\partial}{\partial t}\rho(t; \epsilon; \rho_0) = [H_0 + \epsilon(t)\mu_1 + \cdots + \epsilon^L(t)\mu_L, \rho(t; \epsilon; \rho_0)]$$
$$\rho(t = 0; \epsilon; \rho_0) = \rho_0. \tag{4.2}$$

Then, when at least one matrix $H_0, \mu_1, \ldots, \mu_L$ has nonzero trace, the equation (4.2) *is density matrix controllable if and only if the Lie algebra $L_{iH_0, i\mu_1, \ldots, i\mu_L}$ spanned by the matrices $iH_0, i\mu_1, \ldots, i\mu_L$ is the Lie algebra $u(N)$ of all skew-hermitian matrices or equivalently* $\dim_{\mathbb{R}} L_{iH_0, i\mu_1, \ldots, i\mu_L} = N^2$.

---

[3]Some optimizations are possible at this point as only new commutators are generally needed to be computed. We do not enter into details here.

Otherwise, when all matrices $H_0, \mu_1, \ldots, \mu_L$ have zero trace, a necessary and sufficient condition for controllability is that $L_{iH_0, i\mu_1, \ldots, i\mu_L} = su(N)$ or equivalently $\dim_{\mathbb{R}} L_{iH_0, i\mu_1, \ldots, i\mu_L} = N^2 - 1$.

4.2. Wave function

To derive results for the wave function of the same nature as the two criterions above one has to analyse the transitive subsets of $U(N)$. We recall that a subset $A \subset U(N)$ is called transitive when for any two vectors a, b on the unit sphere of $\mathbb{C}^N$ there exists a matrix $X \in A$ with $b = Xa$. For the situation of quantum control, such a study is available in the literature [1]. To be able to state the corresponding result for this specific situation here, we introduce the centralizer $\mathcal{C}_G z$ of an element $z \in G$ which is defined as the set of all elements that commute with z:

$$\mathcal{C}_G z = \{x \in G : xz = zx\}.$$

We also define $P = i \cdot \mathrm{diag}(1, 0, \ldots, 0) \in U(N)$.

Theorem 4.3. *Consider the system*

$$\frac{dc(t; \epsilon; c_0)}{dt} = [H_0 + F_1(\epsilon(t))\mu_1 + \cdots + F_L(\epsilon(t))\mu_L]\, c(t; \epsilon; c_0)$$
$$c(t = 0; \epsilon; c_0) = c_0 \tag{4.3}$$

with $\|c_0\| = 1$. Suppose that the family $\{1, F_1, \ldots, F_L\}$ is linearly independent and denote by $L_{iH_0, i\mu_1, \ldots, i\mu_L}$ the Lie algebra spanned by the matrices $iH_0, i\mu_1, \ldots, i\mu_L$. Then the equation (4.3) is (wave function) controllable if and only if

$$\dim_{\mathbb{R}} L_{iH_0, i\mu_1, \ldots, i\mu_L} - \dim(L_{iH_0, i\mu_1, \ldots, i\mu_L} \cap \mathcal{C}_G P) = 2N - 2.$$

In particular a sufficient condition for controllability is that

$$\dim_{\mathbb{R}} L_{iH_0, i\mu_1, \ldots, i\mu_L} = N^2.$$

Proof. The proof follows from arguments in [1]. $\qquad\square$

The following procedure allows to implement the above criteria:

1. Test whether the functions $\{1, F_1, \ldots, F_L\}$ are linearly independent. If the answer is yes go to next step, otherwise keep only a subset $F_{i_1}, \ldots, F_{i_p}$ with $\{1, F_{i_1}, \ldots, F_{i_p}\}$ linearly independent and modify the $B_1, \ldots, B_L$ accordingly. For notational convenience we suppose all functionals are independent, i.e., $p = L$.
2. Denote $\mathcal{O} = \{iH_0, i\mu_1, \ldots, i\mu_L\}$.
3. Write any element of $\mathcal{O}$ as a column vector and compute the rank $r = \mathrm{rank}(\mathcal{O})$ over the real numbers.
4. Construct all commutators $\mathcal{C}$ of matrices in $\mathcal{O}^4$ and test whether $\mathrm{rank}(\mathcal{O} \cup \mathcal{C}) = r$. If not, set $\mathcal{O} := \mathcal{O} \cup \mathcal{C}$ and return to previous step.

---

[4] Here again, optimizations are possible.

5. Extract from $\mathcal{O}$ the matrices that commute with P and compute the rank d of this ensemble over $\mathbb{R}$. Test whether $r - d = 2N - 2$. If yes the system is controllable, if not the controllability does not hold.

We also obtain

Theorem 4.4. *Consider the development of the interaction Hamiltonian $H = H_0 + \epsilon(t)\mu_1 + \cdots + \epsilon^L(t)\mu_L$ resulting in the following evolution equation*

$$i\frac{dc(t; \epsilon; c_0)}{dt} = \left[H_0 + \epsilon(t)\mu_1 + \cdots + \epsilon^L(t)\mu_L \right] c(t; \epsilon; c_0)$$
$$c(t = 0; \epsilon; c_0) = c_0. \tag{4.4}$$

with $\|c_0\| = 1$. Denote by $L_{iH_0, i\mu_1, \ldots, i\mu_L}$ the Lie algebra spanned by the matrices $iH_0, i\mu_1, \ldots, i\mu_L$.

Then the equation (4.4) is (wave function) controllable if and only if

$$\dim_{\mathbb{R}} L_{iH_0, i\mu_1, \ldots, i\mu_L} - \dim(L_{iH_0, i\mu_1, \ldots, i\mu_L} \cap \mathcal{C}_G P) = 2N - 2.$$

In particular a sufficient condition for controllability is that

$$\dim_{\mathbb{R}} L_{iH_0, i\mu_1, \ldots, i\mu_L} = N^2.$$

Remark 4.5. All the above results basically state that controllability with linearly independent functionals of a single control ϵ is true whenever the same equation, but with completely independent controls, is controllable.

References

[1] Francesca Albertini and Domenico D'Alessandro. Notions of controllability for bilinear multilevel quantum systems. *IEEE Trans. Automat. Control*, 48(8):1399–1403, 2003.

[2] A. Assion, T. Baumert, M. Bergt, T. Brixner, B. Kiefer, V. Seyfried, M. Strehle, and G. Gerber. Control of chemical reactions by feedback-optimized phase-shaped femtosecond laser pulses. *Science*, 282:919–922, 1998.

[3] J.M. Ball, J.E. Marsden, and M. Slemrod. Controllability for distributed bilinear systems. *SIAM J.Control and Optimization*, 20(4):575–597, 1982.

[4] André D. Bandrauk and H.-Z. Lu. Numerical methods for molecular time-dependent schrödinger equations – bridging the perturbative to nonperturbative regime. In Ph.G. Ciarlet, editor, *Computational Chemistry, Special Volume (C. Le Bris Editor) of Handbook of Numerical Analysis, vol X*, pages 803–832. Elsevier Science B.V., 2003.

[5] C.J. Bardeen, V.V. Yakovlev, J.A. Squier, and K.R. Wilson. Quantum control of population transfer in green fluorescent protein by using chirped femtosecond pulses. *J. Am. Chem. Soc.*, 120:13023–13027, 1998.

[6] C.J. Bardeen, V.V. Yakovlev, K.R. Wilson, S.D. Carpenter, P.M. Weber, and W.S. Warren. Feedback quantum control of molecular electronic population transfer. *Chem. Phys. Lett.*, 280:151–158, 1997.

[7] R. Bartels, S. Backus, E. Zeek, L. Misoguti, G. Vdovin, I.P. Christov, M.M. Murnane, and H.C. Kapteyn. Shaped-pulse optimization of coherent emission of high-harmonic soft X-rays. *Nature*, 406:164–166, 2000.

[8] Nicolas Bourbaki. *Elements of mathematics. Lie groups and Lie algebras. Chapters 1–3*. Berlin: Springer, 1998.

[9] Nicolas Bourbaki. *Elements of mathematics. Lie groups and Lie algebras. Chapters 4–6*. Berlin: Springer, 2002.

[10] R.W. Brockett. Lie theory and control systems defined on spheres. *SIAM J. Appl. Math.*, 25:213–225, 1973. Lie algebras: applications and computational methods (Conf., Drexel Univ., Philadelphia, Pa., 1972).

[11] Thierry Cazenave. *Semilinear Schrödinger equations*, volume 10 of *Courant Lecture Notes in Mathematics*. New York University Courant Institute of Mathematical Sciences, New York, 2003.

[12] W.-L. Chow and B. L. van der Waerden. "Zur algebraischen Geometrie ix.". *Math. Ann.*, 113:692–704, 1937.

[13] D. Deutsch. Quantum theory, the Church-Turing principle and the universal quantum computer. *Proc. R. Soc. Lond. A*, 400:97–117, 1985.

[14] C.M. Dion, A.D. Bandrauk, O. Atabek, A. Keller, H. Umeda, and Y. Fujimura. Two-frequency IR laser orientation of polar molecules. numerical simulations for HCN. *Chem. Phys.Lett*, 302:215–223, 1999.

[15] C.M. Dion, A. Keller, O. Atabek, and A.D. Bandrauk. Laser-induced alignment dynamics of HCN: Roles of the permanent dipole moment and the polarizability. *Phys. Rev. A*, 59(2):1382, 1999.

[16] V.V. Gorbatsevich, A.L. Onishchik, and E.B. Vinberg. *Foundations of Lie Theory and Lie Transformation Groups*. Berlin Springer, 1997.

[17] T. Hornung, R. Meier, and M. Motzkus. Optimal control of molecular states in a learning loop with a parameterization in frequency and time domain. *Chem. Phys. Lett.*, 326:445–453, 2000.

[18] R.S. Judson and H. Rabitz. Teaching lasers to control molecules. *Phys. Rev. Lett*, 68:1500, 1992.

[19] V. Jurdevic and H. Sussmann. Control systems on Lie groups. *Journal of Differential Equations*, 12:313–329, 1972.

[20] J. Kunde, B. Baumann, S. Arlt, F. Morier-Genoud, U. Siegner, and U. Keller. Adaptive feedback control of ultrafast semiconductor nonlinearities. *Appl. Phys. Lett.*, 77:924, 2000.

[21] R.J. Levis, G.M. Menkir, and H. Rabitz. Selective bond dissociation and rearrangement with optimally tailored, strong-field laser pulses. *Science*, 292:709–713, 2001.

[22] C. Lobry. Controllability of nonlinear systems on compact manifolds. *SIAM J. Control*, 12:1–4, 1974.

[23] Mazyar Mirrahimi and Pierre Rouchon. Controllability of quantum harmonic oscillators. *IEEE Trans. Automat. Control*, 49(5):745–747, 2004.

[24] V. Ramakrishna, M. Salapaka, M. Dahleh, H. Rabitz, and A. Pierce. Controllability of molecular systems. *Phys. Rev. A*, 51 (2):960–966, 1995.

[25] J. Salomon. *Numerical analysis of simulations in quantum control*. PhD thesis, Paris VI University, 2005.

[26] S.G. Schirmer, H. Fu, and A.I. Solomon. Complete controllability of quantum systems. *Phys. Rev. A*, 63:063410, 2001.

[27] P.W. Shor. Algorithms for quantum computation: Discrete logarithms and factoring. In S. Goldwasser, editor, *Proceedings of the 35th Annual Symposium on the Foundations of Computer Science*, pages 124–134, Los Alamitos, CA, 1994. IEEE Computer Society.

[28] H.J. Sussmann. Controllability of nonlinear systems. *Journal of Differential Equations*, 12:95–116, 1972.

[29] G. Turinici. Controllable quantities for bilinear quantum systems. In *Proceedings of the 39th IEEE Conference on Decision and Control, Sydney, Australia,*, volume 2, pages 1364–1369, December 2000.

[30] G. Turinici. On the controllability of bilinear quantum systems. In M. Defranceschi and C. Le Bris, editors, *Mathematical models and methods for ab initio Quantum Chemistry*, volume 74 of *Lecture Notes in Chemistry*, pages 75–92. Springer, 2000.

[31] Gabriel Turinici and Herschel Rabitz. Quantum wave function controllability. *Chem. Phys.*, 267:1–9, 2001.

[32] Gabriel Turinici and Herschel Rabitz. Wavefunction controllability in quantum systems. *J. Phys.A.*, 36:2565–2576, 2003.

[33] T.C. Weinacht, J. Ahn, and P.H. Bucksbaum. Controlling the shape of a quantum wavefunction. *Nature*, 397:233–235, 1999.

Gabriel Turinici
CEREMADE
Université Paris Dauphine
Place du Maréchal De Lattre De Tassigny
F-75775 Paris Cedex 16, France
e-mail: `Gabriel.Turinici@dauphine.fr`

International Series of Numerical Mathematics, Vol. 155, 311–328
© 2007 Birkhäuser Verlag Basel/Switzerland

Optimal Control Problems with Convex Control Constraints

Daniel Wachsmuth

Abstract. We investigate optimal control problems with vector-valued controls. As model problem serve the optimal distributed control of the instationary Navier-Stokes equations. We study pointwise convex control constraints, which is a constraint of the form $u(x,t) \in U(x,t)$ that has to hold on the domain Q. Here, U is an set-valued mapping that is assumed to be measurable with convex and closed images. We establish first-order necessary as well as second-order sufficient optimality conditions. And we prove regularity results for locally optimal controls.

Mathematics Subject Classification (2000). Primary: 49M05,26E25; Secondary: 49K20.

Keywords. Optimal control, convex control constraints, set-valued mappings, regularity of optimal controls, second-order sufficient optimality condition, Navier-Stokes equations.

1. Introduction

In fluid dynamics the control can be brought into the system by blowing or suction on the boundary. Then the control is a velocity, which is a directed quantity, hence it is a vector in $\mathbb{R}^2$ respectively $\mathbb{R}^3$. That is, the optimal control problem is to find a vector-valued function $u \in L^p((0,T) \times \Omega)^n$. Distributed control can be realized for instance as a force induced by an outer magnet field in a conducting fluid, see, e.g., Kunisch and Griesse [14]. There, the control u is a function of class $L^2(Q)^2 = L^2(Q; \mathbb{R}^2)$. This illustrates that the control is a directed quantity: it consists of a direction and an absolute value. Or in other words, the control u at a point (x,t) is a vector in $\mathbb{R}^2$.

The optimization has to take into account that one is not able to realize arbitrarily large controls. To this end, control constraints are introduced. If the control $u(x,t)$ is only a scalar variable such as heating or cooling then there is only one choice of a convex pointwise control constraint: the so-called box constraints

$$u_a(x,t) \le u(x,t) \le u_b(x,t). \tag{1a}$$

For the analysis of optimal control of non-stationary Navier-Stokes equations using this particular type of control constraints, we refer to Hinze and Hintermüller [16], Roubíček and Tröltzsch [23], Tröltzsch and Wachsmuth [25], and Wachsmuth [27]. But these box constraints are not the only choice for vector-valued controls. For instance, if one wants to bound the $\mathbb{R}^2$-norm of the control, one gets a nonlinear constraint

$$|u(x,t)| = \sqrt{u_1(x,t)^2 + u_2(x,t)^2} \le \rho(x,t). \tag{1b}$$

What happens if the control is not allowed to act in all possible directions but only in directions of a segment with an angle less than π? Using polar coordinates $u_r(x,t)$ and $u_\phi(x,t)$ for the control vector $u(x,t)$, this can be formulated as

$$0 \le u_r(x,t) \le \psi(u_\phi(x,t), x, t), \tag{1c}$$

where the function ψ models the shape of the set of allowed control actions.

Here, we will use another – and more natural – representation of the constraints. Let us denote by U the set of admissible control vectors. Then we can write the control constraints (1a)–(1c) as an inclusion

$$u(x,t) \in U.$$

The advantage of this approach is that the analysis is based on rather elementary say geometrical arguments, hence there is no need of any constraint qualification. We will impose assumptions on U that allow to apply the common theory of existence and optimality condition: non-emptyness, convexity, and closedness, but no boundedness or further regularity of the boundary. We have to admit that the assumption of convexity gives some inherent regularity, the boundary of convex sets is locally Lipschitz. However, even in the convex case, there can be very irregular situations: one can construct convex sets in $\mathbb{R}^2$ with countably many corners, which lie dense on the boundary, see [10].

The formulation of the control constraint as an inclusion has a further benefit: the set of admissible control vectors can vary over time and space by simply writing

$$u(x,t) \in U(x,t),$$

without causing any additional problems. The main difficulty appears already in the non-varying case, see the discussion in Section 7.1 below.

Optimal control problems with such control constraints are rarely investigated in literature. Second-order necessary conditions for problems with the control constraint $u(\xi) \in U(\xi)$ were proven by Páles and Zeidan [20] involving second-order admissible variations. Second-order necessary as well as sufficient conditions were established in Bonnans [5], Bonnans and Shapiro [8], and Dunn [12]. However, the set of admissible controls has to be polygonal and independent of ξ, i.e., $U(\xi) \equiv U$. This results were extended by Bonnans and Zidani [9] to the case of finitely many convex contraints $g_i(u(\xi)) = 0$, $i = 1, \dots, l$. As already mentioned, we will follow another approach and treat the control constraint as an inclusion

$u(x, t) \in U(x, t)$. State constraints of the form $y(x, t) \in C$ are considered in the recent research paper by Griesse and de los Reyes [13].

As a model problem serves the optimal distributed control of the instationary Navier-Stokes equations in two dimensions. We emphasize that the restriction to two dimensions, i.e., $u \in L^2(Q)^2$, is only due to the limitation of the analysis of instationary Navier-Stokes equations. As long as there exists an applicable theory of a state equation in $\mathbb{R}^n$, all results regarding convex control constraints are ready for an extension to the n-dimensional case.

To be more specific, we want to minimize the following quadratic objective functional:

$$J(y, u) = \frac{\alpha_T}{2} \int_\Omega |y(x, T) - y_T(x)|^2 \, \mathrm{d}x + \frac{\alpha_Q}{2} \int_Q |y(x, t) - y_Q(x, t)|^2 \, \mathrm{d}x \, \mathrm{d}t$$

$$+ \frac{\alpha_R}{2} \int_Q |\operatorname{curl} y(x, t)|^2 \, \mathrm{d}x \, \mathrm{d}t + \frac{\gamma}{2} \int_Q |u(x, t)|^2 \, \mathrm{d}x \, \mathrm{d}t \quad (2)$$

subject to the instationary Navier-Stokes equations

$$\begin{aligned}
y_t - \nu \Delta y + (y \cdot \nabla)y + \nabla p &= u & \text{in } Q, \\
\operatorname{div} y &= 0 & \text{in } Q, \\
y(0) &= y_0 & \text{in } \Omega,
\end{aligned} \quad (3)$$

and to the control constraints $u \in U_{ad}$ with set of admissible controls defined by

$$U_{ad} = \{u \in L^2(Q)^2 : u(x, t) \in U(x, t) \text{ a.e. on } Q\}. \quad (4)$$

Here, Ω is a bounded domain in $\mathbb{R}^2$, Q denotes the time-space cylinder $Q := \Omega \times (0, T)$. Let us underline the fact that for $(x, t) \in Q$ the control $u(x, t)$ is a vector in $\mathbb{R}^2$.

The conditions imposed on the various ingredients of the optimal control problem are specified in Sections 2.1 and 4.1, see assumptions (A) and (AU).

For the optimal control of the non-stationary Navier-Stokes equations there are several articles about existence of solution and necessary optimality conditions, for instance Abergel and Temam [1], Gunzburger and Manservisi [15]. Sufficient optimality conditions and second-order optimization methods were investigated by Hinze [17], Hinze and Kunisch [18], Ulbrich [26], and Tröltzsch and Wachsmuth [25]. However, in these articles only the box constraints (1a) or even no control constraints are considered.

The plan of the article is as follows. At first we introduce some notation and results concerning the state equation in Section 2. Set-valued mappings are the subject of Section 3. The exact statement of our model problem can be found in Section 4 together with first-order necessary optimality conditions in Section 5. We prove regularity results for locally optimal controls in Section 6. Finally, we discuss sufficient optimality conditions and stability of optimal controls in Sections 7 and 8 respectively.

2. Notations and preliminary results

At first, we introduce some notations and results that we will need later on. To begin with, we define the spaces of solenoidal or divergence-free functions

$$H := \{v \in L^2(\Omega)^2 : \operatorname{div} v = 0\}, \qquad V := \{v \in H_0^1(\Omega)^2 : \operatorname{div} v = 0\}.$$

These spaces are Hilbert spaces with scalar products $(\cdot, \cdot)_H$ and $(\cdot, \cdot)_V$ respectively. The dual of V with respect to the scalar product of H we denote by V' with the duality pairing $\langle \cdot, \cdot \rangle_{V',V}$.

We will work with the standard spaces of abstract functions from $[0, T]$ to a real Banach space X, $L^p(0, T; X)$, endowed with its natural norm,

$$\|y\|_{L^p(X)} := \|y\|_{L^p(0,T;X)} = \left(\int_0^T |y(t)|_X^p \, dt \right)^{1/p} \qquad 1 \le p < \infty,$$

$$\|y\|_{L^\infty(X)} := \operatorname{ess\,sup}_{t \in (0,T)} |y(t)|_X.$$

In the sequel, we will identify the spaces $L^p(0, T; L^p(\Omega)^2)$ and $L^p(Q)^2$ for $1 < p < \infty$, and denote their norm by $\|u\|_p := |u|_{L^p(Q)^2}$. The usual $L^2(Q)^2$-scalar product we denote by $(\cdot, \cdot)_Q$ to avoid ambiguity.

In all what follows, $\| \cdot \|$ stands for norms of abstract functions, while $| \cdot |$ denotes norms of "stationary" spaces like H and V.

To deal with the time derivative in (3), we introduce the common spaces of functions y whose time derivatives y_t exist as abstract functions,

$$W^\alpha(0, T; V) := \{y \in L^2(0, T; V) : \ y_t \in L^\alpha(0, T; V')\}, \quad W(0, T) := W^2(0, T; V),$$

where $1 \le \alpha \le 2$. Endowed with the norm

$$\|y\|_{W^\alpha(0,T;V)} := \|y\|_{L^2(V)} + \|y_t\|_{L^\alpha(V')},$$

these spaces are Banach spaces. Every function of $W(0, T)$ is, up to changes on sets of zero measure, equivalent to a function of $C([0, T], H)$, and the imbedding $W(0, T) \hookrightarrow C([0, T], H)$ is continuous, cf. [2, 19].

2.1. The state equation

Before we start with the discussion of the state equation, we specify the requirements for the various ingredients describing the optimal control problem. In the sequel, we assume that the following conditions are satisfied:

$$(\mathbf{A}) \begin{cases} \text{1. } \Omega \text{ has Lipschitz boundary } \Gamma := \partial\Omega, \\ \text{2. } y_0, \, y_T \in H, \, y_Q \in L^2(Q)^2, \\ \text{3. } \alpha_T, \, \alpha_Q, \, \alpha_R \ge 0, \\ \text{4. } \gamma, \, \nu > 0. \end{cases}$$

The assumptions on the set-valued mapping U are given in the next section. Now, we will briefly summarize known facts about the solvability of the instationary

Navier-Stokes equations (3). First, we define the trilinear form $b : V \times V \times V \mapsto \mathbb{R}$ by

$$b(u, v, w) = ((u \cdot \nabla)v, w)_2 = \int_\Omega \sum_{i,j=1}^{2} u_i \frac{\partial v_j}{\partial x_i}\, w_j \ \mathrm{d}x.$$

Its time integral is denoted by b_Q,

$$b_Q(y, v, w) = \int_0^T b(y(t), v(t), w(t))\, \mathrm{d}t.$$

To specify the problem setting, we introduce a linear operator $A : L^2(0, T; V) \mapsto L^2(0, T; V')$ by

$$\int_0^T \langle (Ay)(t),\, v(t) \rangle_{V',V}\, \mathrm{d}t := \int_0^T (y(t), v(t))_V\, \mathrm{d}t,$$

and a nonlinear operator B by

$$\int_0^T \big\langle (B(y))(t),\, v(t) \big\rangle_{V',V}\, \mathrm{d}t := \int_0^T b(y(t), y(t), v(t))\, \mathrm{d}t.$$

For instance, the operator B is continuous and twice Fréchet-differentiable as operator from $W(0, T)$ to $L^2(0, T; V')$.

Now, we concretize the notation of weak solutions for the instationary Navier-Stokes equations (3) in the Hilbert space setting.

Definition 2.1 (Weak solution). *Let $f \in L^2(0, T; V')$ and $y_0 \in H$ be given. A function $y \in L^2(0, T; V)$ with $y_t \in L^2(0, T; V')$ is called weak solution of (3) if*

$$\begin{aligned} y_t + \nu Ay + B(y) &= f, \\ y(0) &= y_0. \end{aligned} \tag{5}$$

Results concerning the solvability of (5) are standard, cf. [24] for proofs and further details.

Theorem 2.2 (Existence and uniqueness of solutions). *For every source term $f \in L^2(0, T; V')$ and initial value $y_0 \in H$, the equation (5) has a unique solution $y \in W(0, T)$. Moreover, the mapping $(y_0, f) \mapsto y$ is locally Lipschitz continuous from $H \times L^2(0, T; V')$ to $W(0, T)$.*

It is well known that the control-to-state mapping is Fréchet-differentiable. The first derivative can be computed as the solution of a linearized equation, cf. [15, 17, 18].

Remark 2.3 (Linearized state equation). *We consider the linearized equation*

$$\begin{aligned} y_t + \nu Ay + B'(\bar{y})y &= f, \\ y(0) &= y_0, \end{aligned} \tag{6}$$

for a given state $\bar{y}$, which is usually the solution of the nonlinear system (5). Following the lines of Temam, existence and uniqueness of a weak solution y in the space $W(0, T)$ was proven for instance in [18, Prop. 2.4]. See also the discussion in [15].

3. Set-valued functions

Before we begin with the formulation of the optimal control problem with inclusion constraints, we will provide some background material. Here, we will specify the notation and assumptions for the admissible set $U(\cdot)$. It is itself a mapping from the control domain Q to the set of subsets of $\mathbb{R}^2$, it is a so-called *set-valued mapping*. We will use the notation $U : Q \rightsquigarrow \mathbb{R}^2$.

The optimal control problem is the minimization of the objective functional subject to the state equations and to the control constraint

$$u(x,t) \in U(x,t). \tag{7}$$

The controls are taken from the space $L^2(Q)^2$, so it is natural to require the fulfillment of (7) for (only) almost all $(x,t) \in Q$. And we have to impose at least some measurability conditions on the mapping U. In the sequel, we will work with measurable set-valued mappings. For an excellent – and for our purposes complete – introduction we refer to the textbook by Aubin and Frankowska [4].

Definition 3.1. *A set-valued mapping $F : Q \rightsquigarrow X$ with closed images is called measurable, if the inverse of each open set is measurable. In other words, for every open subset $\mathcal{O} \subset X$ the inverse image*

$$F^{-1}(\mathcal{O}) = \{\omega \in Q : F(\omega) \cap \mathcal{O} \neq \emptyset\}$$

has to be measurable.

Observe, that for a single-valued function f the definition of measurability coincides with the definition of measurability for the set-valued function $\tilde{f}$ given by

$$\tilde{f}(\omega) = \{f(\omega)\}.$$

However, this definition does not imply the existence of a measurable selection, which is a single-valued function f satisfying $f(x,t) \in U(x,t)$ almost everywhere on Q. The existence is guaranteed under additional assumptions on U.

Theorem 3.2. [4, Th. 8.1.4] *Let $F : Q \rightsquigarrow \mathbb{R}^2$ be a set-valued mapping with nonempty closed images. Then the following two statements are equivalent:*

1. *F is measurable*
2. *There exists a sequence of measurable selections $\{f_n\}_{n=1}^{\infty}$ of F such that for all $(x,t) \in Q$ it holds*

$$F(x,t) = \overline{\bigcup_{n \geq 1} \{f_n(x,t)\}}.$$

The theorem gives not only the existence of a measurable selection but also a tool to prove measurability of set-valued mappings based on countable approximations.

It is well known that every optimal control is the projection of its associated state on the admissible set. Such a characterization is also valid in the set-valued constraint case. But as a first step, we have to make sure that the pointwise projection on the set-valued mapping U preserves measurability.

Theorem 3.3. *[4, Cor. 8.2.13] Let $F : Q \rightsquigarrow \mathbb{R}^2$ be a set-valued measurable mapping with closed, non-empty, and convex images, and $f : Q \mapsto \mathbb{R}^2$ a measurable (single-valued) mapping. Then the projection*

$$g(x,t) = \mathrm{Proj}_{F(x,t)}(f(x,t))$$

is a single-valued measurable function too.

4. The optimal control problem

Here, we will investigate the optimal control problem with the control constraint (7). At first, we have to specify the assumptions to ensure existence of solutions.

4.1. Set of admissible controls

In this section, we want to investigate the convex control constraint, which has to hold pointwise

$$u(x,t) \in U(x,t) \text{ a.e. on } Q.$$

We recall the definition of the set of admissible controls U_{ad},

$$U_{ad} = \{u \in L^2(Q)^2 : u(x,t) \in U(x,t) \text{ a.e. on } Q\}.$$

Once and for all, we specify the requirements for the function U, which defines the control constraints.

(AU)
$\begin{cases}
\textit{The set-valued function } U : Q \rightsquigarrow \mathbb{R}^2 \textit{ satisfies:} \\
\quad 1. \ U \textit{ is a measurable set-valued function.} \\
\quad 2. \ \textit{The images of } U \textit{ are non-empty, closed, and convex a.e. on} \\
\qquad Q. \textit{ That is, the sets } U(x,t) \textit{ are non-empty, closed and convex} \\
\qquad \textit{for almost all } (x,t) \in Q. \\
\quad 3. \ \textit{There exists a function } f_U \in L^2(Q)^2 \textit{ with } f_U(x,t) \in U(x,t) \\
\qquad \textit{a.e. on } Q.
\end{cases}$

Please note, we did not impose any conditions on the sets $U(x,t)$ that are beyond convexity such as boundedness or regularity of the boundaries $\partial U(x,t)$. Assumptions (i) and (ii) guarantee that there exists a measurable selection of U, i.e., a measurable single-valued function f_M with $f_M(x,t) \in U(x,t)$ a.e. on Q. However, no measurable selection needs to be square-integrable as the following example shows.

Example 4.1. *Set $U(t) = [t^{-1/2}, 1 + t^{-1/2}]$, $0 < t \leq 1$. Assumptions (i) and (ii) are fulfilled. But every function f with $f(t) \in U(t)$ for almost all $0 < t \leq 1$ cannot be in $L^2(0,1)$, since the function $g(t) = t^{-1/2}$ is not square integrable on $[0,1]$.*

The existence of a square integrable, admissible function is then ensured by the third assumption. This implies that the set of admissible control is non-empty.

Corollary 4.2. *The set of admissible controls U_{ad} defined by*

$$U_{ad} = \{u \in L^2(Q)^2 : u(x,t) \in U(x,t) \text{ a.e. on } Q\}$$

is non-empty, convex and closed in $L^2(Q)^2$.

The assumption (AU) is as general as the analysis of the second-order condition allows it. In the case that the set-valued function U is a constant function, i.e., $U(x,t) \equiv U_0$, we can give a simpler characterization.

Corollary 4.3. *Let the set-valued function U be a constant function, i.e., $U(x,t) = U_0$ a.e. on Q for some $U_0 \subset \mathbb{R}^2$. Then the assumption (AU) is fulfilled if the set U_0 is non-empty, closed, and convex.*

Assuming (AU) we can derive another interesting result. Conditon (iii) allows us to prove that the pointwise projection on U_{ad} of a L^2-function is itself a L^2-function.

Corollary 4.4. *Let be given a function $u \in L^2(Q)^2$. Then the function v defined pointwise a.e. by*

$$v(x,t) = \text{Proj}_{U(x,t)}(u(x,t))$$

is also in $L^2(Q)^2$. Further, if for some $p \geq 2$ the functions u and f_U are in $L^p(Q)^2$, then the projection v is in $L^p(Q)^2$ as well.

Proof. By assumption (AU), the set-valued function U is measurable with closed and convex images, and u is a measurable single-valued function. Then by Theorem 3.3 the function v is measurable as well. By Lipschitz continuity of the pointwise projection, it holds

$$|v(x,t) - f_U(x,t)| = |\text{Proj}_{U(x,t)}(u(x,t)) - \text{Proj}_{U(x,t)}(f_U(x,t))|$$
$$\leq |u(x,t) - f_U(x,t)|$$

almost everywhere on Q. Thus, squaring and integrating gives

$$\|v - f_U\|_2^2 \leq \|u - f_U\|_2^2 < \infty,$$

which implies $v \in L^2(Q)^2$. If in addition, u and f_U are in $L^p(Q)^2$ for some $p > 2$, then we can prove analogously that the projection is also in L^p, i.e., $v \in L^p(Q)^2$. $\square$

4.2. Existence of optimal controls

Before we can think about existence of solution, we have to specify which problem we want to solve. We will assume that conditions (A) of Section 2.1 are satisfied. Moreover, we assume that $U(\cdot)$ fulfills the pre-requisite (AU). So we end up with the following optimization problem

$$\min J(y,u) \tag{8a}$$

subject to the state equation

$$y_t + \nu Ay + B(y) = u \quad \text{in } L^2(0,T;V'), \tag{8b}$$
$$y(0) = y_0 \quad \text{in } H, \tag{8c}$$

and the control constraint

$$u \in U_{ad}, \tag{8d}$$

where U_{ad} is given by (7).

Under the assumptions above, the optimal control problem (8) is solvable. We recall that in Section 2.1 the regularization parameter γ is supposed to be greater than zero. One can prove existence even with $\gamma = 0$ under the additional condition of boundedness of U_{ad} in L^2.

Theorem 4.5. *The optimal control problem admits a – global optimal – solution $\bar{u} \in U_{ad}$ with associated state $\bar{y} \in W(0, T)$.*

5. First-order necessary conditions

The necessary optimality conditions for the optimal control problem discussed in the present chapter differ slightly from the conditions that can be found in the literature, see, e.g., [25]. However, we will repeat the exact statement for convenience of the reader.

Theorem 5.1 (Necessary condition). *Let $\bar{u}$ be locally optimal in $L^2(Q)^2$ with associated state $\bar{y} = y(\bar{u})$. Then there exists a unique Lagrange multiplier $\bar{\lambda} \in W^{4/3}(0, T; V)$, which is the weak solution of the adjoint equation*

$$-\bar{\lambda}_t + \nu A\bar{\lambda} + B'(\bar{y})^*\bar{\lambda} = \alpha_Q(\bar{y} - y_Q) + \alpha_R \operatorname{curl}^*\operatorname{curl}\bar{y}$$
$$\bar{\lambda}(T) = \alpha_T(\bar{y}(T) - y_T). \tag{9}$$

Moreover, the variational inequality

$$(\gamma\bar{u} + \bar{\lambda}, u - \bar{u})_Q \geq 0 \quad \forall u \in U_{ad} \tag{10}$$

is satisfied.

Similar as in the box-constrained case, we can reformulate the variational inequality (10). The projection representation of the optimal control is now realized using the admissible sets $U(\cdot)$

$$\bar{u}(x, t) = \operatorname{Proj}_{U(x,t)}\left(-\frac{1}{\gamma}\bar{\lambda}(x, t)\right) \qquad \text{a.e. on } Q. \tag{11}$$

Here, it will be a little bit more difficult to prove regularity results for the optimal control using the regularity of the adjoint state. The projection formula is used in connection with Lipschitz stability of optimal controls [16, 23, 27]. The ideas there cannot be transferred to the case of set-valued constraints, see the discussion in Section 8 below.

Necessary optimality conditions of second order for optimal control problems with set-valued constraints were developed in [20]. It involves the use of the concept of second-order tangent, see, e.g., [11].

6. Regularity of optimal controls

Let us comment on the regularity of a locally optimal control $\bar{u}$. By (11), it inherits some regularity from the associated adjoint state $\bar{\lambda}$. Here, we will show, how the regularities $\bar{\lambda} \in L^p(Q)^2$ respectively $\bar{\lambda} \in C(\bar{Q})^2$ can be carried over to the control $\bar{u}$. However, it is not clear whether and how it is possible to prove $\bar{u} \in W^{1,p}(Q)^2$ if $\bar{\lambda} \in W^{1,p}(Q)^2$, and what assumptions on U are needed.

6.1. Optimal controls in L^p

Corollary 4.4 gives a hint, how we can prove the regularity $\bar{u} \in L^p(Q)^2$ provided $\bar{\lambda} \in L^p(Q)^2$ holds. We have to assume only the existence of an admissible L^p-function.

Theorem 6.1. *Let $\bar{u}$ be a locally optimal control of the optimal control problem (8) with associated adjoint state $\lambda \in L^p(Q)^2$, $p \leq \infty$. If there is an admissible function $f_p \in L^p(Q)^2 \cap U_{ad}$ for $p \leq \infty$ then the optimal control $\bar{u}$ is in that $L^p(Q)^2$, too.*

Proof. The proof follows immediately from the projection representation (11) and Corollary 4.4. $\qquad\qquad\qquad\qquad\qquad\qquad\qquad\qquad\qquad\qquad\qquad\qquad$ $\square$

We will complete this short section with the following corollary, which states the precise regularity assumptions on the problem data, such that the pre-requisites of the previous theorem are fulfilled, see also [27].

Corollary 6.2. *Let be given $y_0, y_T \in V$, $y_Q \in L^2(Q)^2$. Let the set-valued mapping U satisfy the assumption (AU). Further, we assume the existence of an admissible L^p-function $f_p \in L^p(Q)^2 \cap U_{ad}$ for $2 \leq p < \infty$.*
Then every locally optimal control of problem (8) is in $L^p(Q)^2$, $2 \leq p < \infty$.

The method of proof applied here does not work to obtain continuity of an optimal control. This is investigated in the next section.

6.2. Continuity of optimal controls

Now, we are going to prove continuity of an locally optimal control. We will rely in our considerations again on the projection formula (11), which says that the optimal control is the pointwise projection of a continuous function on the admissible sets. Hence, this admissible sets $U(x,t)$ vary over space and time. Here, we have to impose some continuity assumptions on the set-valued mapping U.

There are two equivalent characterizations of continuous *single*-valued functions:

1. the image of a converging sequence is also a converging sequence,
2. the preimages of open sets are open sets.

In the set-valued case, however both definitions of a continuous function are no longer equivalent. They define two independent kinds of semicontinuity.

Definition 6.3. *A set-valued mapping $F : D \subset X \rightsquigarrow Y$ is called* lower semicontinuous, *if for all $x \in D$, $y \in F(x)$, and any sequence $\{x_n\} \subset D$ converging to x there is a sequence of elements $y_n \in F(x_n)$ converging to y.*

Definition 6.4. *A set-valued mapping $F : D \subset X \rightsquigarrow Y$ is called* upper semicontinuous, *if for all $x \in D$ and all open sets $O \supset F(x)$ there exists $\delta = \delta(O)$ such that $F(x') \subset O$ for all x' with $|x - x'| \leq \delta$.*

Both definition are not equivalent and are independent. There are set-valued mappings, which are lower semicontinuous but not upper and vice-versa. It is natural to define a continuous mapping to have both semicontinuous properties.

Definition 6.5. *A set-valued mapping $F : D \subset X \rightsquigarrow Y$ is called* continuous, *if U is both lower and upper semicontinuous.*

The assumption (AU) on the set-valued mapping U contains the condition that $U(x,t)$ is non-empty, closed and convex almost everywhere on Q. Do these properties of the images hold everywhere provided U is continuous? At first, we want to show the improvement from 'non-empty almost everywhere' to 'non-empty everywhere' in the continuous case.

Lemma 6.6. *Let U fulfill (AU). Further let $U : \bar{Q} \rightsquigarrow \mathbb{R}^2$ be upper semicontinuous. Then $U(x,t)$ is non-empty for all $(x,t) \in \bar{Q}$.*

Proof. We will prove it by contradiction. Let $\xi = (x,t) \in \bar{Q}$ such that $U(\xi)$ is empty. Then we take sequences of points $\xi_n = (x_n, t_n) \in Q$ with $U(\xi_n) \neq \emptyset$ converging to ξ and $u_n \in U(\xi_n)$.

At first, we consider the case that the sequence $\{u_n\}$ admits a cluster point $\tilde{u}$. Then there is a subsequence $\{u_{n_k}\}$ converging to $\tilde{u}$. Let us define an open set by $O = \mathbb{R}^2 \setminus \overline{B_\rho(\tilde{u})}$ for some $\rho > 0$. Then, we have $\emptyset = U(\xi) \subset O$ and $U(\xi_{n_k}) \not\subset O$ for all k large enough, which is a contradiction to upper semicontinuity.

Now, let the sequence $\{u_n\}$ have no cluster point. Here, we take an arbitrary $\hat{u}$. Then for $\varepsilon > 0$ there is n_ε such that $|\hat{u} - u_n| > \varepsilon$ for all $n > n_\varepsilon$, otherwise there would exist a cluster point in $B_\varepsilon(\hat{u})$. Let us set $O = B_\varepsilon(\hat{u})$. By construction, we have $\emptyset = U(\xi) \subset O$ and $U(\xi_n) \not\subset O$ for $n > n_\varepsilon$, which is a contradiction to upper semicontinuity.

And the proof is complete. $\qquad\square$

Unfortunately, the property 'closedness of the images' cannot be transferred from 'almost everywhere' to 'everywhere' for continuous U as the following counterexample shows.

Example 6.7. *Define $F : [0,1] \rightsquigarrow \mathbb{R}$ by*

$$F(t) = \begin{cases} (0,1] & \text{if } t = 0 \\ [t, 1+t] & \text{otherwise.} \end{cases}$$

Clearly, F is lower semicontinuous. It is also upper semicontinuous: every open set that contains $F(0)$ contains also $F(\varepsilon)$ for sufficiently small ε. Hence F is continuous. It has closed images almost everywhere but not everywhere.

Now, let us prove a lemma, which will help us later on.

Lemma 6.8. *Let U fulfill* (AU)*. In addition, we assume that U is upper semicontinuous on $\bar{Q}$ with closed images $U(x,t)$ for all $(x,t) \in \bar{Q}$. Then for given sequences (x_n, t_n) converging to $(x,t) \in \bar{Q}$ and $y_n \in U(x_n, t_n)$ converging to y the limit y lies in $U(x,t)$, $y \in U(x,t)$.*

Proof. We will use again the notation $\xi = (x,t)$ and $\xi_n = (x_n, t_n)$. Let us assume $y \notin U(\xi)$. Set $\varepsilon = \mathrm{dist}(y, U(\xi))$, which is positive since $U(\xi)$ is closed. Then there exists N such that for all $n > N$ it holds $y_n \notin U(\xi)$ and $\mathrm{dist}(y_n, U(\xi)) \geq \frac{2}{3}\varepsilon$. Now, we construct an open set by $O := \{v : \mathrm{dist}(v, U(\xi)) < \frac{1}{3}\varepsilon\}$. It implies $y_n \notin O$ and $U(\xi_n) \not\subset O$ for $n \geq N$. This yields a contradiction to upper semicontinuity, since we have $O \supset U(\xi)$. Hence it holds $y \in U(\xi)$. $\qquad\square$

Furthermore, it turns out that the assumption of closed images is essential to prove the convexity of the images of U.

Lemma 6.9. *Let U fulfill* (AU)*. In addition, let U be continuous on $\bar{Q}$ with closed images $U(x,t)$ for all $(x,t) \in \bar{Q}$. Then $U(x,t)$ is convex for all $(x,t) \in \bar{Q}$.*

Proof. Let $\xi = (x,t) \in \bar{Q}$ be given with $y_1, y_2 \in U(\xi)$, $\lambda \in (0,1)$. We have to show that $\lambda y_1 + (1-\lambda)y_2$ is in $U(\xi)$. We take a sequence of points $\xi_n = (x_n, t_n) \in Q$, for which $U(\xi_n)$ is non-empty and convex, converging to ξ.

By lower semicontinuity there exists sequences of points $y_1^n, y_2^n \in U(\xi_n)$ converging to y_1 respectively y_2. The points $y^n := \lambda y_1^n + (1-\lambda)y_2^n$ are in $U(\xi_n)$ and converge to $y := \lambda y_1 + (1-\lambda)y_2$ for $n \to \infty$. The previous Lemma 6.8 implies that the limit $y = \lambda y_1 + (1-\lambda)y_2$ is in $U(\xi)$. Hence $U(\xi)$ is convex. $\qquad\square$

Assuming the continuity of the set-valued mapping U and the adjoint state λ we can prove continuity of an optimal control.

Theorem 6.10. *Let U satisfies the assumption* (AU)*. Furthermore, let $U : \bar{Q} \rightsquigarrow \mathbb{R}^n$ be continuous with closed images everywhere. Suppose $\bar{u}$ satisfies the first-order necessary optimality conditions together with the state $\bar{y}$ and adjoint $\bar{\lambda}$. If the adjoint is continuous, $\bar{\lambda} \in C(\bar{Q})$, so is the control as well, $\bar{u} \in C(\bar{Q})$.*

Proof. We will show that the projection

$$\mathrm{Proj}_{U(x,t)}\left(-\frac{1}{\gamma}\bar{\lambda}(x,t)\right) = \bar{u}(x,t)$$

results in a continuous function. We abbreviate $v(x,t) := -\bar{\lambda}(x,t)/\gamma$, which is a continuous function by assumption.

Let $\xi = (x,t) \in \bar{Q}$ be given. Take a sequence $\xi_n = (x_n, t_n) \in Q$ that converges to ξ. We have to show the convergence $\bar{u}(\xi_n) \to \bar{u}(\xi)$. We will give the proof in several steps.

Step 1: $U(x,t)$ is non-empty, closed and convex everywhere on $\bar{Q}$. This follows by the preceding Lemmata 6.6 and 6.9.

Step 2: U_{ad} contains a continuous function. Define the function $m : \bar{Q} \to \mathbb{R}^n$ as

$$m(x,t) = \arg\min\left\{|v| : v \in U(x,t)\right\},$$

which gives the elements of $U(x,t)$ with the smallest norm. It is called the minimal selection of U. Since $U(x,t)$ is non-empty, closed and convex, the function m is well defined. By [3, Chapt. 3, Sect. 1, Prop. 23, p. 120], the minimal selection m is continuous.

Step 3: Boundedness of $\{\bar{u}(\xi_n)\}$. Using Lipschitz continuity of the projection, we can estimate

$$|\bar{u}(\xi_n) - m(\xi_n)| = |\operatorname{Proj}_{U(\xi_n)}(v(\xi_n)) - \operatorname{Proj}_{U(\xi_n)}(m(\xi_n))|$$
$$\leq |v(\xi_n) - m(\xi_n)| \leq \|v - m\|_{C(\bar{Q})} < \infty,$$

which proves boundedness of the set $\{\bar{u}(\xi_n)\}$.

Step 4: Every accumulation point of $\{\bar{u}(\xi_n)\}$ is in $U(\xi)$. Since $\{\bar{u}(\xi_n)\}$ is bounded in $\mathbb{R}^n$, we can select a subsequence $\{\bar{u}(\xi_{n'})\}$ converging to some element $\tilde{u}$. By Lemma 6.8, we find that $\tilde{u}$ is in $U(\xi)$.

Step 5: There is exactly one accumulation point of $\{\bar{u}(\xi)\}$. Take an arbitrary element $z \in U(\xi)$. By lower semicontinuity, there is a sequence of elements $z_{n'} \in F(\xi_{n'})$ converging to $z \in U(\xi)$. Since $u(\xi_{n'}) = \operatorname{Proj}_{U(\xi_{n'})} v(\xi_{n'})$, we find

$$\big(u(\xi_{n'}) - v(\xi_{n'}), \, z_{n'} - u(\xi_{n'})\big) \geq 0 \quad \forall n'.$$

Hence

$$\big(u(\xi_{n'}) - v(\xi_{n'}), \, z_{n'} - z\big) + \big(u(\xi_{n'}) - v(\xi_{n'}), \, z - u(\xi_{n'})\big) \geq 0 \quad \forall n'.$$

Letting $n' \to \infty$, we find

$$\big(\tilde{u} - v(\xi), \, z - \tilde{u}\big) \geq 0.$$

Since $z \in U(x)$ was arbitrary, it holds

$$\tilde{u} = \operatorname{Proj}_{U(\xi)} v(\xi)$$

for every accumulation point of $\{\bar{u}(\xi_n)\}$. The projection is unique hence the set $\{u(\xi_n)\}$ has exactly one accumulation point.

Conclusion. By the previous step, we find the convergence $\bar{u}(\xi_n) \to \bar{u}(\xi)$. Hence the prove is complete, and $\bar{u}$ is a continuous function on $\bar{Q}$. $\qquad\square$

For an exact statement, which regularity of the data is sufficient for $\lambda \in C(\bar{Q})^2$, we refer to [27].

Remark 6.11. *The projection formula (11) remains true if one replaces $U(x,t)$ by its closure $\bar{U}(x,t)$, provided $U(\cdot)$ is closed almost everywhere on Q. Furthermore, one can show that for continuous $U : Q \rightsquigarrow \mathbb{R}^2$ the closure $\bar{U} : Q \rightsquigarrow \mathbb{R}^2$ is also a continuous set-valued mapping. In this way, we can construct a continuous representation of a locally optimal control $\bar{u}$ without the assumption of closedness of the images of U.*

7. Second-order sufficient optimality conditions

7.1. Normal directions

Before we start with the formulation of the sufficient optimality conditions, let us recall some notations of convex set theory. Let be given a convex set C. Then $\mathcal{N}_C(u)$ and $\mathcal{T}_C(u)$ are the normal and tangent cones of C at some point u. The space of normal directions is written $N_C(u) = \operatorname{span} \mathcal{N}_C(u)$ with its orthogonal complement $T_C(u)$.

Now, we want to use these notations with $C = U_{ad}$. Let be given an admissible control $u \in U_{ad}$. It is well known, that the sets $\mathcal{N}_{U_{ad}}(u)$, $\mathcal{T}_{U_{ad}}(u)$, $N_{U_{ad}}(u)$, and $T_{U_{ad}}(u)$ admit a pointwise representation as U_{ad} itself, cf. [4, 22]. For instance, for $u \in L^2(Q)^2$ the set $\mathcal{N}_{U_{ad}}(u)$ is given by

$$\mathcal{N}_{U_{ad}}(u) = \left\{ v \in L^2(Q)^2 : v(x,t) \in \mathcal{N}_{U(x,t)}(u(x,t)) \text{ a.e. on } Q \right\}.$$

In a while, we will need the projection of a test function w on the space of normal directions and its complement. We will denote the resulting functions by w_N and w_T respectively. They are defined pointwise by

$$w_N(x,t) = \operatorname{Proj}_{\left[N_{U(x,t)}(u(x,t)) \right]}(w(x,t)) \tag{12}$$

and

$$w_T(x,t) = \operatorname{Proj}_{\left[T_{U(x,t)}(u(x,t)) \right]}(w(x,t)). \tag{13}$$

It is not easy to prove that the functions w_N and w_T are measurable. At this point the method of Dunn [12] requires that the admissible set U is polyhedric and independent of (x,t). However, these restrictions can be overcome using the results for set-valued mappings. Let us sketch the method of the measurability proof. Behind the projections there are the following mappings:

1. $Q \ni (x,t) \mapsto \mathcal{N}_{U(x,t)}(u(x,t)) =: \mathcal{N}(x,t)$
2. $Q \ni (x,t) \mapsto \operatorname{span}\{\mathcal{N}(x,t)\} = \operatorname{span}\left\{ \mathcal{N}_{U(x,t)}(u(x,t)) \right\} =: N(x,t)$
3. $Q \ni (x,t) \mapsto \operatorname{Proj}_{N(x,t)}(w(x,t)) =: w_N(x,t)$.

Here, one can see, what happens if $U(x,t)$ is constant over Q: the mapping $\mathcal{N}$ is even in this case a set-valued mapping, which is not constant. Even the dimension of $\mathcal{N}(x,t)$ varies. So we would not have any advantage if we assume constant admissible sets $U(x,t) = U_0$.

Now, the measurability can be proven as follows. The mapping $(x,t) \rightsquigarrow \mathcal{T}_{U(x,t)}(u(x,t))$ is measurable if U has closed and convex images, cf. [4, Cor. 8.5.2]. The normal cone $\mathcal{N}$ is then the dual of $\mathcal{T}$, and one can prove that the dual cone operation does preserve measurability. The same can be done for the linear hull N. Here, the proof is based on the countability argument already stated in Theorem 3.2.

Now, the projection of measurable function on measurable set-valued mappings – here N – results in a measurable function, see Theorem 3.3. Altogether, both of the projections (12) and (13) results in measurable function. Let us remark that it is very difficult to prove this by hand: here one has to go step by step from regular convex sets and constant U to more irregular ones.

As a second point here, let us define some more notations in connection to convex sets. The relative interior of a convex set is defined by

$$\operatorname{ri} C = \{x \in \operatorname{aff} C : \exists \varepsilon > 0, B_\varepsilon(x) \cap \operatorname{aff} C \subset C\},$$

its complement in C is called the relative boundary

$$\operatorname{rb} C = C \setminus \operatorname{ri} C.$$

The distance of a point $u \in \mathbb{R}^n$ to a set $C \subset \mathbb{R}^n$ is defined by

$$\operatorname{dist}(u, C) = \inf_{x \in C} |u - x|.$$

7.2. Sufficiency

Let us come back to optimization with convex constraints. To motivate the following, we will investigate the finite-dimensional problem $\min_{x \in C} f(x)$ with $C \subset \mathbb{R}^n$ first. It is known that a local minimizer x of the function f over a convex set C fulfills

$$-\nabla f(x) \in \mathcal{N}_C(x).$$

A sufficient condition is then given by

$$-\nabla f(x) \in \operatorname{ri} \mathcal{N}_C(x) \tag{14}$$

and

$$f''(x)[y, y] > 0 \qquad \forall y \in T_C(x). \tag{15}$$

It consists of a first-order part: strict complementarity and a second-order part: coercivity. Now we want to adapt this formulation to the optimal control problem considered here. Condition (14) would become

$$-(\gamma \bar{u}(x,t) + \bar{\lambda}(x,t)) \in \operatorname{ri} \mathcal{N}_{U(x,t)}(\bar{u}(x,t)) \qquad \text{a.e. on } Q. \tag{16}$$

However, this is not enough for optimal control problems, we need the satisfaction of this condition in a uniform sense. We have to assume not only that $-(\gamma \bar{u}(x,t) + \bar{\lambda}(x,t))$ lies in the relative interior of the normal cone, we need moreover that $-(\gamma \bar{u}(x,t) + \bar{\lambda}(x,t))$ has a positive distance to the relative boundary of $\mathcal{N}_{U(x,t)}(\bar{u}(x,t))$. But this cannot be assumed for all (x,t): if $\bar{u}(x,t)$ is in the interior of the admissible set $U(x,t)$ then the normal cone consists only of the origin and has no relative interior. Therefore, we introduce the set of strongly active constraints as the set of points, where this condition is fulfilled,

$$Q_\varepsilon = \left\{(x,t) \in Q : \operatorname{dist}\left(-(\gamma \bar{u}(x,t) + \bar{\lambda}(x,t)), \operatorname{rb} \mathcal{N}_{U(x,t)}(\bar{u}(x,t))\right) > \varepsilon\right\}. \tag{17}$$

That is, we assume that (16) is only fulfilled on a subset of the domain Q. Consequently, we have to require the coercivity assumption for more directions than included in $T_{U_{ad}}(\bar{u})$. Furthermore, the inequality > 0 in (15) has to be replaced by a norm-square, since the proof in finite dimensions that '> 0' suffices is tied to compactness of the unit sphere, which does not hold in the infinite-dimensional case.

Altogether, we require that the following is fulfilled. We assume that the reference pair $(\bar{y}, \bar{u})$ satisfies the coercivity assumption on $\mathcal{L}''(\bar{y}, \bar{u}, \bar{\lambda})$, in the sequel called second-order sufficient condition:

$$
\textbf{(SSC)} \quad
\begin{cases}
\textit{There exist } \varepsilon > 0 \textit{ and } \delta > 0 \textit{ such that} & \\[4pt]
\qquad \mathcal{L}''(\bar{y}, \bar{u}, \bar{\lambda})[(z, h)]^2 \geq \delta \, \|h\|_2^2 & \text{(18a)} \\[4pt]
\textit{holds for all pairs } (z, h) \in W(0,T) \times L^2(Q)^2 \textit{ with} & \\[4pt]
\qquad h \in \mathcal{T}_{U_{ad}}(\bar{u}), \ h_N = 0 \textit{ on } Q_\varepsilon, & \text{(18b)} \\[4pt]
\textit{and } z \in W(0,T) \textit{ being the weak solution of the linearized equation} & \\[4pt]
\qquad \begin{aligned} z_t + Az + B'(\bar{y})z &= h \\ z(0) &= 0. \end{aligned} & \text{(18c)}
\end{cases}
$$

In (18b) h_N denotes the pointwise projection of h on the subspaces $N \subset \mathbb{R}^2$, compare (12). We required in (SSC) the coercivity of $\mathcal{L}''$ for more test functions than in (15). The set $\mathcal{T}_{U_{ad}}(\bar{u})$, which was user there, is only a subset of $\mathcal{T}_{U_{ad}}(\bar{u})$. However, the space of test functions in (SSC) can be reformulated as: $h \in \mathcal{T}_{U_{ad}}(\bar{u})$ with $h(x,t) \in \mathcal{T}_{U(x,t)}(\bar{u}(x,t))$ on Q_ε. We can use test functions with values in the spaces T only on the strongly active set, due to the strong complementarity, which holds there. On the rest of the domain, the values of the test function has to lie in the tangent cones $\mathcal{T}$.

Now, the next theorem states the sufficiency of (SSC).

Theorem 7.1. *Let $(\bar{y}, \bar{u})$ be admissible for the optimal control problem and suppose that $(\bar{y}, \bar{u})$ fulfills the first order necessary optimality conditions with associated adjoint state $\bar{\lambda}$. Assume further that (SSC) is satisfied at $(\bar{y}, \bar{u})$. Then there exist $\alpha > 0$ and $\rho > 0$ such that*

$$
J(y, u) \geq J(\bar{y}, \bar{u}) + \alpha \, \|u - \bar{u}\|_2^2
$$

holds for all admissible pairs (y, u) with $\|u - \bar{u}\|_\infty \leq \rho$.

The proof can be found in [28].

There are a number of sufficient second-order optimality conditions for finite-dimensional optimization problems with convex constraints, see for instance [7, 8, 21]. They all use the second-order tangent sets, and it is not clear how those results are related to the condition presented here. Also, the extension of the finite-dimensional results to optimal control problems is not a trivial exercise and requires further research.

Remark 7.2. *Let us comment on the definition of the strongly active set in (17) if U is formed by box constraints. Since this particular constraint is formed by two independent inequalities, one can refine the definition of strongly active sets, see [25], to contain more points than the active set introduced here.*

8. Stability of optimal controls

Usually, the fulfillment of a second-order sufficient condition implies stability of locally optimal controls under small perturbations. This is demonstrated in a great variety of articles for optimal control problems with box constraints. However, in the case of general convex control constraints the sufficient condition (SSC) is too weak to get stability of optimal controls. This is due to the fact, that tangent variations of the control are not necessarily admissible directions, which is an essential ingredient in the proofs for the box constrained case.

In finite-dimensional optimization, there are a few publications concerning stability of solutions to convex constrained optimization problems, see [6, 8]. They use the assumption of second-order regular sets. The extension of that conditions to the infinite-dimensional case considered here is not obvious, since the proofs argue by contradiction and rely on the finite-dimensionality, i.e., on compactness of the unit sphere. That means, one has to use methods which differ from the indirect methods of [6] as well as from the direct proofs from, e.g., [16, 23, 25].

Obviously, if we assume coercivity of $\mathcal{L}''$ for all test directions, we can prove such stability results. Since this would be only a technical exercise, we do not proceed in this direction.

References

[1] F. Abergel and R. Temam. On some control problems in fluid mechanics. *Theoret. Comput. Fluid Dynam.*, 1:303–325, 1990.

[2] R.A. Adams. *Sobolev spaces*. Academic Press, San Diego, 1978.

[3] J.-P. Aubin and I. Ekeland. *Applied Nonlinear Analysis*. Wiley, New York, 1984.

[4] J.-P. Aubin and H. Frankowska. *Set-valued analysis*. Birkhäuser, Boston, 1990.

[5] J.F. Bonnans. Second-order analysis for control constrained optimal control problems of semilinear elliptic equations. *Appl. Math. Optim.*, 38:303–325, 1998.

[6] J.F. Bonnans, R. Cominetti, and A. Shapiro. Sensitivity analysis of optimization problems under second order regular constraints. *Mathematics of Operations Research*, 23(4):806–831, 1998.

[7] J.F. Bonnans, R. Cominetti, and A. Shapiro. Second order optimality conditions based on parabolic second order tangent sets. *SIAM J. Optim.*, 9(2):466–492, 1999.

[8] J.F. Bonnans and A. Shapiro. *Perturbation analysis of optimization problems*. Springer, New York, 2000.

[9] J.F. Bonnans and H. Zidani. Optimal control problems with partially polyhedric constraints. *SIAM J. Control Optim.*, 37:1726–1741, 1999.

[10] T. Bonnesen and W. Fenchel. *Theorie der konvexen Körper*. Springer, Berlin, 1934.

[11] R. Cominetto. Metric regularity, tangent sets, and second-order optimality conditions. *Appl. Math. Opt.*, 21:265–287, 1990.

[12] J.C. Dunn. Second-order optimality conditions in sets of L^∞ functions with range in a polyhedron. *SIAM J. Control Optim.*, 33(5):1603–1635, 1995.

[13] R. Griesse and J.C. de los Reyes. State-constrained optimal control of the stationary Navier-Stokes equations. submitted, 2005.

[14] R. Griesse and K. Kunisch. A practical optimal control approach to the stationary MHD system in velocity-current formulation. RICAM Report 2005-02, 2005.

[15] M.D. Gunzburger and S. Manservisi. The velocity tracking problem for Navier-Stokes flows with bounded distributed controls. *SIAM J. Control Optim.*, 37:1913–1945, 1999.

[16] M. Hintermüller and M. Hinze. A SQP-semi-smooth Newton-type algorithm applied to control of the instationary Navier-Stokes system subject to control constraints. *SIAM J. Optim.*, 16(4):1177–1200, 2006.

[17] M. Hinze. *Optimal and instantaneous control of the instationary Navier-Stokes equations.* Habilitation, TU Berlin, 2002.

[18] M. Hinze and K. Kunisch. Second-order methods for optimal control of time-dependent fluid flow. *SIAM J. Control Optim.*, 40:925–946, 2001.

[19] J.L. Lions and E. Magenes. *Non-homogeneous boundary value problems and applications*, volume I. Springer, Berlin, 1972.

[20] Zs. Páles and V. Zeidan. Optimum problems with measurable set-valued constraints. *SIAM J. Optim.*, 11:426–443, 2000.

[21] J.-P. Penot. Second-order conditions for optimization problems with constraints. *SIAM J. Optim.*, 37:303–318, 1998.

[22] R.T. Rockafellar. *Conjugate duality and optimization.* SIAM, Philadelphia, 1974.

[23] T. Roubíček and F. Tröltzsch. Lipschitz stability of optimal controls for the steady-state Navier-Stokes equations. *Control and Cybernetics*, 32(3):683–705, 2002.

[24] R. Temam. *Navier-Stokes equations.* North Holland, Amsterdam, 1979.

[25] F. Tröltzsch and D. Wachsmuth. Second-order sufficient optimality conditions for the optimal control of Navier-Stokes equations. *ESAIM: COCV*, 12:93–119, 2006.

[26] M. Ulbrich. Constrained optimal control of Navier-Stokes flow by semismooth Newton methods. *Systems & Control Letters*, 48:297–311, 2003.

[27] D. Wachsmuth. Regularity and stability of optimal controls of instationary Navier-Stokes equations. *Control and Cybernetics*, 34:387–410, 2005.

[28] D. Wachsmuth. Sufficient second-order optimality conditions for convex control constraints. *J. Math. Anal. App.*, 2006. To appear.

Daniel Wachsmuth
Institut für Mathematik
Technische Universität Berlin
Str. des 17. Juni 136
D-10623 Berlin, Germany
e-mail: `wachsmut@math.tu-berlin.de`

International Series of Numerical Mathematics, Vol. 155, 329–382
© 2007 Birkhäuser Verlag Basel/Switzerland

Control of Moving Domains, Shape Stabilization and Variational Tube Formulations

Jean-Paul Zolésio

Abstract. This paper deals with the control of a moving dynamical domain in which a non cylindrical dynamical boundary value problem is considered. We consider weak Eulerian evolution of domains through the convection of a measurable set by (non necessarily smooth) vector field V. We introduce the concept of tubes by "product space" and we show a closure result leading to existence results for a variational shape principle. We illustrate this by new results: heat equation and wave equation in moving domains with various boundary conditions and also the geodesic characterisation for two Eulerian shape metrics leading to the Euler equation through the transverse field considerations. We consider the non linear Hamilton-Jacobi like equation associated with level set parametrization of the moving domain and give new existence result of possible topological change in finite time in the solution.

1. The use of BV perimeter in shape optimization

In order to derive existence results in classical shape optimization (see for example [9]) I introduced after 1984 ([38], [39]) the concept of a functional regularized with the perimeter:

$$J_\sigma(\Omega) = J(\Omega) + \sigma \, P_D(\Omega) \tag{1.1}$$

This has been in the context of large water wave modelling (non shallow water free boundary) in which the "small" parameter σ turns to be the surface tension. That result was emphasized in [36] after having been presented to a large audience, then in [37] (that paper was kept two years before being accepted without changes for publication).

For dynamical modelling (artery [11], fluid structure interaction [26], ...) the concept of a tube (ζ, V) and a tube functional to be extremized with respect to the tube was introduced in [40], [13], ... in the following form:

$$J(\zeta, V) = j(\zeta, V) + \sigma \int_0^\tau P_D(\Omega_t(V))\, dt. \tag{1.2}$$

Following that idea we consider in this paper new control functionals for non cylindrical heat equation, non cylindrical wave equations and shape metrics. Many results are completely new: the optimality conditions and existence results for both heat and wave problems as well as the existence of solution to the level set equation with possible topological change in finite time. The Euler equation for the shape geodesic generalizes in some sense the variational formulation for the incompressible Euler equation. In deriving the new geodesic conditions for the new shape metrics $\bar{\delta}(\Omega_1, \Omega_2)$ and $d_E(\Omega_1, \Omega_2)$ we introduce new technical results such as the expression for the boundary shape derivative v'_{Γ_t} and the new weak form for the transverse vector field evolution equation. Also the structure of the adjoint field Λ is clarified when the right-hand side is a shape gradient like measure $\gamma_t^*(g\,\vec{n}_t)$, then $\Lambda =! \nabla \lambda \nabla \chi_{\Omega_t}$. The cubic energy expression for the wave equation with homogeneous Dirichlet condition derived in 1984 is also generalized for the first time to the "co-normal Neumann" condition $\frac{\partial y}{\partial t} y + \frac{\partial y}{\partial \nu_t^*} y = 0$ on Γ_t.

2. Shape evolution

Let be given a bounded "universe" D in R^N with Lipschitzian continuous boundary and consider the set $L^1(D, \{0, 1\})$ of characteristic functions $\zeta \in L^1(D)$ such that $\zeta^2 = \zeta$. We shall consider the family of measurable subsets $\Omega \subset D$ such that $\zeta = \chi_\Omega$ (that family is then defined up to subsets with zero measure in D). The time evolution of Ω is described with the help of vector fields. The time interval being denoted by $I = [0, \tau]$,

$$p > 1, \ V \in E := \{V \in L^p(I \times D, R^N), \ \ s.t. \ \mathrm{div} V \in L^p(I \times D), \ \langle V, n_{\partial D}\rangle = 0\}.$$

The subspace

$$E^{\mathrm{lip}} := E \cap L^1(0, \tau, W^{1,\infty}(D, R^N))$$

plays an essential role: for any measurable subset $\Omega_0 \subset D$ and any $V \in E^{\mathrm{lip}}$ there exists a unique solution to the convection problem:

$$\mathbf{C} := L^1(0, \tau, L^1(D, \{0, 1\})) \cap C^0([0, \tau], L^1(D))$$

$$\zeta \in \mathbf{C}, \ \frac{\partial}{\partial t}\zeta + \nabla_x \zeta . V = 0, \ \zeta(0) = \chi_{\Omega_0}. \tag{2.1}$$

Indeed it exists the flow mapping $T_t(V)$ so that $\zeta(t,.) = \chi_{\Omega_0} o(T_t(V))^{-1}$. We denote $\Omega_t(V) := T_t(V)(\Omega_0)$ and shall refer to $Q_V := \cup_{0 < t < \tau} \{t\} \times \Omega_t(V)$ as being a classical tube (roughly speaking the regularity of the moving boundary being "controlled" by the smoothness of V and $\partial\Omega_0$). We consider the family of subsets with finite perimeter in D:

$$\mathbf{P}_D = \{\, \omega \subset D, \ \chi_\Omega \in BV(D)\,\}, \ \mathbf{H} = \mathbf{C} \cap L^1(I, BV(D)). \tag{2.2}$$

A main result for our topic is the following tubes closures:

Theorem 2.1. *Let $p > 1$, $\Omega_0 \subset D$ and a sequence $(\zeta_n, V_n) \in \mathbf{H} \times E$ verifying (2.1) and such that there exists a positive constant $M > 0$ with*

$$||V_n||_E + \int_0^\tau ||\nabla_x \zeta_n||_{M^1(D,R^N)}\, dt \ \leq \ M. \tag{2.3}$$

Then there exists a weakly converging subsequence in $L^\infty(]0,\tau[\times D) \times E$. Any limiting element (ζ, V) belongs to $\mathbf{H} \times E$, verifies (2.1) with the bound (2.3). Moreover the convergence is strong in the $L^1(0,\tau, L^1(D))$ norm and notice that we get the following continuity on the limiting element: $\zeta \in \mathbf{H}$.

Proof. As $\zeta \in L^\infty$ and $\operatorname{div}V$, $V \in L^p(I \times D)$ we have $\zeta V \in L^p$ so that

$$\frac{\partial}{\partial t}\zeta = -\nabla\zeta.V = -\operatorname{div}(\zeta V) + \zeta \operatorname{div}V \ \in L^p(0, \tau, W^{-1,1}(D)).$$

The conclusion follows from the "parabolic Helly compacity results" which is included in [13], [31], [33], (see also the book [26] for a $p = 2$ version). It states that if a sequence ζ_k remains bounded in $L^1(0, \tau, BV(D))$ with $\frac{\partial}{\partial t}\zeta_k$ bounded in $L^p(0, \tau, W^{-1,1}(D))$, then there exists a subsequence converging strongly in $L^1(0, \tau, L^1(D))$.

As we shall see in Section 10 the distribution space $W^{-1,1}(D)$ can be replaced by the Banach space of $M^1(D)$ of bounded measures over the bounded domain D.

We consider the weak closure of the family of classical tubes:

$$\mathbf{T}_{\Omega_0} := \{\, (\zeta, V) \in \mathbf{H} \times E \ \ s.t. \ \ \exists M > 0, \ (\zeta_n, V_n) \in \mathbf{H} \times E^{\mathrm{lip}}$$

$$\text{weakly converges to } (\zeta, V) \text{ with}: \ \int_0^\tau P_D(\, T_t(V_n)(\Omega_0)\,)\, dt \leq M \,\}.$$

Corollary 2.2. *The set $\mathbf{T}_{\Omega_0}$ is weakly closed in $\mathbf{H} \times E$.*

In Section 10 we shall use $M^1(D)$ and derive a similar closure result with $p = 1$ but, in order to recover the continuity $\zeta \in C^0(0, \tau, L^1(D))$, we shall also need $p > 1$. The weak convection (2.1) is studied in [33]. It is interesting to notice that when the initial condition is a smooth enough function Φ_0 then the solution convects the level sets of Φ_0. Also from [10], [1] we know that the oriented distance function $b_{\Omega_t(V)}$ itself is solution to equation (2.1) with speed vector field $V(t, p_t(x))$, where $p_t = I_d - b_{\Omega_t(V)}\, \nabla b_{\Omega_t(V)}$ is the projection onto the boundary $\partial\Omega_t(V)$. The convection (2.1) generalises to boundary measures as follows:

Let $\Gamma_t(V) = \partial\Omega_t(V)$ with $V \in L^1(0, \tau, W_0^{1,\infty}(D, R^N))$ and let Γ_0 be a smooth manifold. Then we consider the element

$$\gamma_t := -\nabla\chi_{\Omega_t(V)}.\nabla b_{\Omega_t(V)}$$

which is the usual boundary layer measure:

$$\langle \gamma_t, \Psi\rangle_{H^{-1}(D)\times H_0^1(D)} = \int_{\Gamma_t(V)} \Psi(x)\, d\Gamma_t(x). \tag{2.4}$$

That measure solves the following evolution

$$\gamma(0) = \Gamma_0 := \partial\Omega_0$$

$$\frac{\partial}{\partial t}\gamma(t) + \nabla\gamma(t).V(t) + \langle DV(t).\nabla b_{\Omega_t(V)}, \nabla b_{\Omega_t(V)}\rangle\gamma(t) = 0. \tag{2.5}$$

That results enables us to develop a *tube variational analysis* for the minimization over $\mathbf{T}_{\Omega_0}$ of functionals in the following form

$$j(\zeta,V) = \int_0^\tau \int_{\Omega_t(V)} F(t, y_{Q_V}, \nabla y_{Q_V})dx dt + \sigma \int_0^\tau P_D(\Omega_t(V))\,dt,$$

where y_{Q_V} stands for the solution of some boundary value problem associated with the tube Q_V. We shall propose two examples concerning the heat equation and the wave equation in that non cylindrical evolution domain Q_V. We shall choose adequate boundary conditions associated with that moving boundary. The equation (2.5) should permit to handle the convergence of boundary integrals and is under consideration in forecoming papers. This will permit to enlarge the present study to functionals J in the form

$$J(\zeta,V) = j(\zeta,V) + \int_0^\tau \int_{\partial\Omega_t(V)} f(t, y_{Q_V}, \nabla y_{Q_V})\,d\Gamma_t\,dt$$

$$= j(\zeta,V) + \int_0^\tau \langle\gamma(t), f(t,y,\nabla y)\rangle\,dt.$$

A main point in that study is that for smooth enough tubes in a minimizing sequence, say $V_n \in E \cap L^1(0,\tau, W_0^{1,\infty}(D, R^N))$ the tubes Q_{V_n} is smooth enough so that some classical analysis will furnish the existence and may be uniqueness for the solution y_{Q_n} to the boundary value problem under concern. The point is that any such analysis fails for non smooth limiting tube in $\mathbf{T}_{\Omega_0}$. We shall propose specific choice of function $F(t, y, \nabla y)$ so that the minimization of (ζ, V) will "create" the existence. We deal here with simple linear equation and in a forcoming work we shall extend to Navier-Stokes 3D equation.

Indeed that analysis would be efficient with open tubes. There are several obvious ways for dealing with open tubes. For example introducing mollifier on the considered vector fields. We propose here to replace the BV perimeter by the *density perimeter* that we recall now.

2.1. Boundedness of the density perimeter

2.1.1. Density perimeter. Following [3], [4], we consider for any closed set A in D the density perimeter associated to any $\gamma > 0$ by the following.

$$P_\gamma(A) = \sup_{\epsilon\in(0,\gamma)} \left[\frac{\text{meas}(A^\epsilon)}{2\,\epsilon}\right], \tag{2.6}$$

where A^ϵ is the dilation $A^\epsilon = \cup_{x\in A}B(x,\epsilon)$. We recall some main properties:
  – The mapping $\Omega \to P_\gamma(\partial\Omega)$ is lower-semi continuous in the H^c-topology.
  – The property $P_\gamma(\partial\Omega) < \infty$ implies that $\text{meas}(\partial\Omega) = 0$ and $\Omega \setminus \partial\Omega$ is open in D.

If $P_\gamma(\partial\Omega_n) \leq m$ and Ω_n converges in the H^c-topology to some open subset $\Omega \subset D$, then the convergence holds in the $L^2(D)$-topology.

The "parabolic" situation: whenever $V \in C^\infty(I \times \bar{D})$, the mapping $t \to P_\gamma(\partial\Omega_t)$ is not continuous. So that mapping cannot be an element of $H^1(0,\tau)$. For any smooth vector field, $V \in C^0([0,\tau], W_0^{1,\infty}(D, R^N))$, we consider,

$$\Theta_\gamma(V, \Omega_0) = \mathrm{Min}\left\{\int_0^\tau \left(\frac{\partial}{\partial t}\mu\right)^2 dt \,|\, \mu \in \mathbf{M}_\gamma(V, \Omega_0)\right\}, \qquad (2.7)$$

where

$$\mathbf{M}_\gamma(V, \Omega_0) = \{\mu \in H^1(0,\tau),\ P_\gamma(\partial\Omega_t(V)) \leq \mu(t)\ a.e.t,$$
$$\mu(0) \leq (1+\gamma)P_\gamma(\partial\Omega_0)\}.$$

In general this set is non empty. When that set is empty we put $\Theta_\gamma(V, \Omega_0) = +\infty$. Notice that even when the mapping $p = (t \longrightarrow P_\gamma(\Omega_t(V)))$ is an element of $H^1(0,\tau)$ (then $p \in \mathbf{M}_\gamma(V, \Omega_0)$ }), we may have: $\Theta(V, \Omega_0) < ||p'||^2_{L^2(0,\tau)}$ as the minimizer will escape to possible variation of the function p.

Proposition 2.3. *Let $V \in C^0([0,\tau], W_0^{1,\infty}(D, R^N)), \mathrm{div}\, V = 0$, we have:*

$$P_\gamma(\partial\Omega_t(V)) \leq 2P_\gamma(\partial\Omega_0) + \sqrt{\tau}\ \Theta(V, \Omega_0)^{1/2}. \qquad (2.8)$$

Moreover if $V_n \in C^0([0,\tau], W_0^{1,\infty}(D, R^N))$, verifies $V_n \longrightarrow V$ in $L^2((0,\tau)\times D, R^N)$ and the uniform boundedness: $\exists M > 0,\ \Theta(V_n, \Omega_0) \leq M$ Then

$$\Theta(V, \Omega_0) \leq \liminf \Theta(V_n, \Omega_0).$$

An alternative approach is to consider

$$\tilde{\Theta}_\gamma(V, \Omega_0) := ||p_\gamma||_{BV(0,\tau)},\ \ p_\gamma(t) := P_\gamma(\partial\Omega_t(V))$$

We would derive the same kind of estimates.

3. Heat equation with insulated boundary

We consider the non cylindrical situation: the boundary Σ is insulated or adiabatic. As the domains move it is not the usual Neumann boundary condition but the one described below.

Non cylindrical evolution problems, such as Navier-Stokes equation for moving boundaries in a fluid (see [5]) is a challenging optimal control issue. In the case of linear problems we deal with easier situations. Nevertheless a difficult issue is that we need to handle such problem with non smooth geometry. The study of non cylindrical heat equation is an old story. Far from being exhaustive here let us quote the works by P. Acquistapace [21], and recently in [23]. In these works the boundary of the moving domain should be smooth enough. The obvious technique was based on the transport into a cylindrical problem which, in terms of an abstract setting, leads to a dynamical system with a non autonomous operator with a moving domain. Here we revisit that analysis in the scope of the optimal control of the moving domain Ω_t. As classically in shape analysis, the control parameter will be the speed vector field $V(t,x)$ whose flow mapping $T_t(V)$ builds the non cylindrical evolution domain $Q_V = \cup_{0<t<\tau} \{t\} \times \Omega_t$. Let $V \in C^0([0,\tau], C^1(D, R^N))$ with $V.n = 0$ on ∂D, the moving domain is $\Omega_t := T_t(V)(\Omega_0)$ and its characteristic function is $\zeta = \zeta_0 o T_t(V)^{-1}$. We consider the unique solution u to the parabolic problem:

$$\frac{\partial}{\partial t} u - \Delta u = 0 \text{ in } Q_V, \quad \frac{\partial}{\partial n_t} u + \langle V(t), n_t \rangle u = 0$$

$$\text{on the moving boundary } \Gamma_t, \quad u(0) = u_0. \tag{3.1}$$

This b.c. cannot be written as $\frac{\partial}{\partial \nu} u = 0$ on the lateral time-space boundary Σ.

3.1. The weak formulation

$\forall \psi \in C^1([0,\tau] \times R^n)$ with $\psi(\tau) = 0$,

$$\int_0^\tau \int_{\Omega_t} \left(-u \frac{\partial}{\partial t} \psi + \nabla u.\nabla \psi \right) dt dx = \int_{\Omega_0} \psi(0)(x)\, dx. \tag{3.2}$$

Introducing $U(t,x) = u(t) o T_t(V)(x)$, the transported solution on the cylindrical domain, we get U as solution to the parabolic boundary value problem:

$$U_t + U(\ln J)_t - J^{-1}\text{div}(U\, J\, DT^{-1}.V(t) o T_t(V))$$
$$- J^{-1}\text{div}(J\, DT^{-1}.(DT^*)^{-1}.\nabla U) = 0 \tag{3.3}$$

with the boundary condition

$$\langle DT_t(V)^{-1}.(DT_t(V)^{-1})^*.\nabla U, n \rangle$$
$$+ \langle DT_t(V)^{-1}.V(t) o T_t(V), n \rangle U = 0. \tag{3.4}$$

Given some element $U_d \in L^2(D)$ and $\sigma > 0$, we introduce the cost functional in the following form

$$j(\zeta, V) = 1/2 \int_0^\tau \int_D \zeta((u - U_d)^2 + |\nabla u|^2) dx dt$$

$$j_1(\zeta, V) = j(\zeta, V) + \sigma/2 \int_0^\tau (|V(t)|_E^2 + |\nabla \zeta|_{M^1(D,R^N)}) dt. \tag{3.5}$$

We consider the minimization problem

$$Inf\ \{j_1(\zeta, V) \ | \ (\zeta, V) \in \mathbf{T}_{\Omega_0}\}. \tag{3.6}$$

For elements (ζ, V) in $\mathbf{T}_{\Omega_0}^{lip}$ we have $\zeta = \zeta_{\Omega_0} o T_t(V)^{-1}$. The element ζ is uniquely associated to the vector field V.

Stability of weak relaxed solution: Given a tube $(\zeta, V) \in \mathbf{T}_{\Omega_0}^{lip}$, we consider the solution u to the weak parabolic problem:

$$\forall \psi \in C^1([0,\tau] \times R^n) \text{ with } \psi(\tau) = 0, \ \mathbf{H} = \nabla u$$

$$\int_0^\tau \int_D \zeta \left(-u \frac{\partial}{\partial t} \psi + \mathbf{H}.\nabla \psi \right) dt dx = \int_{\Omega_0} \psi(0)(x)\, dx. \tag{3.7}$$

For the optimal control purpose we introduce the adjoint problem: the weak relaxed dual problem is the following one:

$$\forall \psi \in C^1([0,\tau] \times R^n) \text{ with } \psi(0) = 0, \tag{3.8}$$

$$\int_0^\tau \int_D \zeta \left(p \frac{\partial}{\partial t} \psi + \mathbf{H}.\nabla\psi \right) dtdx + \int_0^\tau \int_{\Omega_t} p\,\psi\,v\,d\Gamma_t\,dt = \int_{\Omega_\tau} \psi(\tau)(x)\,dx.$$

Solution stability: In order to get $\mathbf{H}^0 = (\nabla u)^0$ in (3.7), we need the tube to be an open set or at least such property for a.e. t concerning the set Ω_t such that $\zeta(t) = \chi_{\Omega_t}$, a.e.t. A technique to that approach is to "replace" the perimeter by the "density perimeter" $P_\gamma \partial\Omega_t$. As far as the functional j_1 is concerned, we have the following stability result:

Proposition 3.1. *Assume that (ζ_v, V_n) is a sequence of smooth tubes, $\zeta_n = \chi_{Q_n}$, we say that Q_n is builds by smooth speed vector fields V_n. For each n we have a solution u_n. Assume that (V_n, ζ_n) converges to (V, ζ) in $\sigma(L^2, L^2) \times L^1$ topology, with $\zeta_n = \zeta_{\Omega_0} o T_t^{-1}(V_n)$. Moreover assume that u_n^0 (the extension by zero) weakly converges in $L^2(0, \tau, D)$ to a limit element u as well as $(\nabla u_n)^0$ to some element $\mathbf{H}$. Then $(u, \mathbf{H}) \in L^2(Q)^{N+1}$ and is solution to the problem (3.7) in Q.*

In order to prove the existence result for a functional governed by the heat equation we need open sets and Hausdorff complementary convergence, then we modify the optimal control problem as follows:

$$j_2(\zeta, V) = j(\zeta, V) + \sigma/2 \int_0^\tau \left(||V(t)||_E + ||P_\gamma(\partial\Omega_t)||_{BV(0,\tau)} \right) dt. \tag{3.9}$$

We consider the minimization problem

$$\text{Inf } \{j_2(\zeta, V) \mid (\zeta, V) \in \mathbf{T}^{\text{lip}}\}. \tag{3.10}$$

Let (V_n, ζ_n) be a minimizing sequence. The tube Q_n is smooth and the solution u_n of heat equation is classically defined. Obviously the null-extensions to the cylinder $[0, \tau] \times D$ of both u_n and the gradients ∇u_n are bounded in $L^2([0, \tau] \times D)$. We consider a weakly converging subsequence, still denoted u_n, weakly converging to u and ∇u_n weakly converging to some vector field Z. On the other hand, as the $BV(0, \tau)$ norm of $P_\gamma(\partial\Omega_t^n)$ is bounded there exists a subsequence, still denoted Ω^n, such that $P_\gamma(\partial\Omega_t^n)$ converges in $L^1(0, \tau)$ to some integrable function f. Then for almost every t, $P_\gamma(\partial\Omega_t^n) \to f(t)$. As a result a.e.t, $P_\gamma(\partial\Omega_t^n) \le M(t)$ then for almost every time t, the open set Ω_t^n converges to some open set Ω_t both in H^c and L^p topologies, $\text{meas}(\partial\Omega_t) = 0$ and $P_\gamma(\partial\Omega_t) \le \liminf_{n->\infty} P_\gamma(\partial\Omega_t^n) = f(t)$

Let $\phi \in \mathbf{D}([0, \tau] \times D)$ such that a.e.t, $\phi(t) \in \mathbf{D}(\Omega_t)$. For $n \ge N_t$ we have $\phi(t) \in \mathbf{D}(\Omega_t^n)$ so that, a.e. t, we have:

$$\int_{\Omega_t^n} \langle \nabla u_n(t), \phi(t) \rangle dx = - \int_{\Omega_t^n} u_n(t)\,\text{div}\,\phi(t)\,dx.$$

Obviously,

$$\int_D \langle \nabla u_n(t), \phi(t) \rangle \, dx = \int_{\Omega_t^n} \langle \nabla u_n(t), \phi(t) \rangle \, dx,$$

and the same concerning u, so that in the limit we get $Z = \nabla u$.

We consider the Gateaux derivative of that functional, for $(\zeta, V) \in \mathbf{T}_{\Omega_0}^{\text{lip}}$ that is for $V \in E^{lip}$ and $\zeta(t) = \chi_{\Omega_0} o T_t(V)^{-1}$.

3.2. The dual problem

$$-\frac{\partial}{\partial t} p - \Delta p = 0 \text{ in } Q_V \tag{3.11}$$

$$p(\tau) = u_\tau \tag{3.12}$$

$$\frac{\partial}{\partial n_t} p = 0 \quad \text{on the moving boundary} \, \Gamma_t. \tag{3.13}$$

The adjoint weak formulation is the following one:

$$\forall \psi \in C^1([0, \tau] \times R^n) \text{ with } \psi(0) = 0, \tag{3.14}$$

$$\int_0^\tau \int_{\Omega_t} \left(p \frac{\partial}{\partial t} \psi + \nabla u . \nabla \psi \right) dt dx + \int_0^\tau \int_{\partial \Omega_t} p \psi v \, d\Gamma_t \, dt = \int_{\Omega_\tau} \psi(\tau)(x) \, dx.$$

Setting $P = p o T_t(V)$ we get the same equation as (3.3), but with final condition at $t = \tau$, and the following boundary condition:

$$(DT_t(V)^{-1}.(DT_t(V)^{-1})^*.\nabla P, n > + \langle DT_t(V)^{-1}.V(t) o T_t(V), n \rangle P \tag{3.15}$$

$$+ \langle V(t) o T_t(V), (DT_t(V)^{-1})^*.n \rangle P = 0.$$

4. Energy stabilization in wave equation by moving domain

We consider the wave equation in a moving domain and the variation of the "acoustic" energy with respect to the dynamical boundary. We analyse passive and active controls in order to decrease the energy for given initial conditions under Dirichlet or Neumann boundary conditions.

In 1985 [34], 1987 [35] we addressed the shape stabilization issue for the wave equation in a moving domain under homogeneous Dirichlet boundary conditions. We revisit that results for Neumann homogeneous boundary conditions in view of modelling of smart actuators acting as a periodical (may be small) moving part of the boundary in the acoustic wave equation. As in the previous work the key point is that the time derivative of the energy turns out to be a cubic term with respect to the normal component of the boundary speed.

4.1. Wave equation

Let $(a, b) \in H^1(\Omega) \times L^2(\Omega)$ be the initial datum. We consider the wave equation:

$$\frac{\partial^2}{\partial t^2} y - \Delta y = 0 \text{ in } Q, \tag{4.1}$$

with initial conditions:

$$y(0) \;=\; a, \quad \frac{\partial}{\partial t}y(0) \;=\; b \text{ in } \Omega, \tag{4.2}$$

and one of the three boundary conditions

$$y \;=\; 0 \text{ on } \partial\Omega_t \tag{4.3}$$

or

$$\frac{\partial}{\partial n_t}y(0) \;=\; 0 \text{ on } \partial\Omega_t \tag{4.4}$$

or

$$\frac{\partial}{\partial n_t}y(0) \;+\; \langle V(t), n_t \rangle\, y_t \;=\; 0 \text{ on } \partial\Omega_t. \tag{4.5}$$

We refer to problem (4.1), (4.2), (4.3) as the Dirichlet problem, and to (4.1), (4.2), (4.5) as the Neumann problem which is known ([46]) to be well posed in the energy norm when the boundary is smooth enough. Note that the system (4.1), (4.2), (4.4) is not known to be well posed.

4.2. Energy functional

Let

$$E(t) = 1/2 \int_{\Omega_t} (y_t^2 + |\nabla y|^2)dx, \tag{4.6}$$

and

$$\mathbf{E}_t(V) \;=\; \int_0^t E(s)\,ds. \tag{4.7}$$

4.3. Weak formulation of the Neumann problem (4.1), (4.2), (4.5)

Let y be such that $y(0) = a$ in Ω with y and $y_t \in L^2(0, \tau, L^2(\Omega_t(V))$ together with $\nabla y \in L^2(0, \tau, L^2(\Omega_t(V)), R^N))$, and verifying, $\forall \psi \in C^1([0, \tau], H^1(R^N))$ with $\psi(\tau) = 0$:

$$\int_0^\tau \int_{\Omega_t(V)} (\langle \nabla y, \nabla \psi \rangle - y_t\, \psi_t)dt + \int_\Omega b\,\psi(0)\,dx = 0. \tag{4.8}$$

Lemma 4.1. *The following equation holds:*

$$\int_0^\tau \int_{\Omega_t(V)} z\psi_t dx dt = -\int_0^\tau \int_{\Omega_t(V)} z_t\psi dx dt - \int_0^\tau \int_{\partial\Omega_t(V)} z\psi\langle V(t), n_t\rangle d\Gamma_t dt$$

$$+ \int_{\Omega_\tau(V)} z(\tau, x)\psi(\tau, x)dx - \int_{\Omega_0(V)} z(\tau, 0)\psi(\tau, 0)dx. \tag{4.9}$$

Then from (4.8) we get the boundary condition (4.5) on the moving boundary.

338   J.-P. Zolésio

4.3.1. Hyperbolic adjoint problem.

$$\frac{\partial^2}{\partial t^2}p - \Delta p \;=\; 2\Delta y \text{ in } Q, \tag{4.10}$$

with initial conditions:

$$p(t) \;=\; 0, \quad \frac{\partial}{\partial t}p(t) \;=\; y(t) \text{ in } \Omega_t(V) \tag{4.11}$$

and one of the two boundary conditions

$$p \;=\; 0 \text{ on } \partial\Omega_t, \tag{4.12}$$

or (respectively)

$$\frac{\partial}{\partial n_t}p(t) \;+\; \langle\, V(t),\, n_t \,\rangle\, p_t \;=\; -\,2\frac{\partial y}{\partial n} \text{ on } \partial\Omega_t. \tag{4.13}$$

4.4. Derivative with respect to the vector field V

We have

$$2\mathbf{E}'_t(V,W) = \int_0^t \int_{\Omega_t} 2(\nabla y.\nabla y' + y_t y'_t)dxdt + \int_0^t \int_{\Gamma_t} ((y_t)^2 + |\nabla y|^2)\langle Z(t),\, n_t\rangle\, d\Gamma_t dt \tag{4.14}$$

where $Z(t)$ is the transverse field (see Section 4.5). Making use of Green's theorem and of Lemma 4.1 we get:

$$= \int_0^t \int_{\Omega_t} 2(-\Delta y\, y' - y_{tt}y')dxdt + \int_0^t \int_{\Gamma_t} \left(2\frac{\partial y}{\partial n_t}y' + ((y_t)^2 + |\nabla y|^2)z\right) d\Gamma_t dt$$

$$- \int_0^\tau \int_{\partial\Omega_t(V)} y_t\, y\, \langle V(t), n_t\rangle d\Gamma_t\, dt + \int_{\Omega_\tau(V)} (y_t\, y)(\tau, x)dx - \int_{\Omega_0} (y_t\, y)(0, x)dx$$

As $\Delta y = y_{tt}$ we get:

$$= -4\int_0^t \int_{\Omega_t} \Delta y y' dxdt + \int_0^t \int_{\Gamma_t} \left(2\frac{\partial}{\partial n_t}yy' + ((y_t)^2 + |\nabla y|^2)z\right) d\Gamma_t dt \tag{4.15}$$

$$- \int_0^\tau \int_{\partial\Omega_t(V)} y_t\, y\, \langle V(t), n_t\rangle d\Gamma_t\, dt + \int_{\Omega_\tau(V)} (y_t\, y)(\tau, x)dx - \int_{\Omega_0} (y_t\, y)(0, x)dx.$$

We have now to consider the two different boundary conditions:

4.4.1. Dirichlet condition 4.3.

$$2\mathbf{E}'_t(V,W) = -4\int_0^t \int_{\Omega_t} \Delta y y' dxdt + \int_0^t \int_{\Gamma_t} \left(2\frac{\partial}{\partial n_t}yy' + ((y_t)^2 + |\nabla y|^2)z\right) d\Gamma_t dt$$

$$\int_{\Omega_\tau(V)} (y_t\, y)(\tau, x)dx - \int_{\Omega_0} (y_t\, y)(0, x)dx.$$

From $y(t, x(t)) = 0$ we get

$$\frac{\partial}{\partial t}y(t, x(t)) = y_t(t, x) + \nabla y(t, x).V(t, x) = 0$$

so that $y_t = -\frac{\partial}{\partial n_t} y \, \langle V(t), n_t \rangle$, then:

$$2\mathbf{E}'_t(V, W) = -4 \int_0^t \int_{\Omega_t} \Delta y y' \, dx dt$$

$$+ \int_0^t \int_{\Gamma_t} \left(2\frac{\partial}{\partial n_t} y y' + \left(\frac{\partial}{\partial n_t} y\right)^2 (\langle V(t, n_t)^2 + 1 \rangle \langle Z(t), n_t \rangle \right) d\Gamma_t dt$$

$$+ \int_{\Omega_\tau(V)} (y_t\, y)(\tau, x) dx - \int_{\Omega_0} (y_t\, y)(0, x) dx \, .$$

4.4.2. Neumann condition (4.5).

$$2\mathbf{E}'_t(V, W) = -4 \int_0^t \int_{\Omega_t} \Delta y y' \, dx dt + \int_0^t \int_{\partial \Omega_t} ((y_t)^2 + |\nabla y|^2) \, \langle Z(t),\, n_t \rangle \ d\Gamma_t dt$$

$$+ \int_0^\tau \int_{\partial \Omega_t(V)} \frac{\partial}{\partial n_t} y \ (y \langle V(t), n_t \rangle + 2\, y') \ d\Gamma_t \, dt$$

$$+ \int_{\Omega_\tau(V)} (y_t\, y)(\tau, x) dx - \int_{\Omega_0} (y_t\, y)(0, x) dx \, .$$

4.4.3. Characterisation of y' for the Neumann condition (4.5). From (4.8) we obtain that

$$\int_0^\tau \int_{\Omega_t(V)} (\nabla y'.\nabla \psi - y'_t \, \psi_t) dx dt + \int_0^\tau \int_{\partial \Omega_t(V)} (\nabla y.\nabla \psi - y_t \psi_t) Z(t).n_t \, ds_t \, dt = 0.$$

Then we obtain y' being solution to the homogeneous wave equation:

$$- \Delta y' + \frac{\partial^2 \, y'}{\partial t^2} = 0 \, .$$

Concerning the boundary condition, we make use of the following "boundary version" of Lemma 4.1

Lemma 4.2. *It holds:*

$$\int_0^\tau \int_{\partial \Omega_t(V)} f \, \psi_t ds_t \tag{4.16}$$

$$= - \int_0^\tau \int_{\partial \Omega_t(V)} \left(f_t\, \psi \ + \left(H(t) \, f\psi + \frac{\partial f}{\partial n_t} \psi + \frac{\partial \psi}{\partial n_t} f \right) \langle V(t), n_t \rangle \ \right) ds_t \, dt.$$

Making use of Green's theorem and of Lemma 4.1 we get:

$$\int_0^\tau \int_{\Omega_t(V)} (\nabla y'.\nabla \psi - y'_t \, \psi_t) dx dt$$

$$= \int_0^t \int_{\Omega_t} (-\Delta y' \, \psi + y'_{tt}\psi) dx dt + \int_0^t \int_{\partial \Omega_t} (\frac{\partial y'}{\partial n_t} + y'_t \, \psi \, \langle V(t), n_t \rangle) \, \psi \ d\Gamma_t \, dt$$

$$- \int_{\Omega_\tau(V)} (y_t\, y)(\tau, x) dx + \int_{\Omega_0} (y_t\, y)(0, x) dx \, .$$

While from Lemma 4.2, with $f = y_t \langle Z(t), \nabla b_{\Omega_t(V)} \rangle$, we have:

$$\int_0^\tau \int_{\partial \Omega_t(V)} (\nabla y . \nabla \psi - y_t \psi_t) Z(t).n_t \, ds_t \, dt = \int_0^\tau \int_{\partial \Omega_t(V)} (\nabla y . \nabla \psi) Z(t).n_t$$

$$+ f_t \psi + \left(H(t) f \psi + \frac{\partial f}{\partial n_t} \psi + \frac{\partial \psi}{\partial n_t} f \right) \langle V(t), n_t \rangle) ds_t \, dt$$

then

$$\frac{\partial y'}{\partial n_t} + v \, y_t' - \operatorname{div}_{\partial \Omega_t(V)} (\langle Z(t), n_t \rangle y(t))$$

$$+ (f_t + H(t) f \langle V(t), n_t \rangle + \frac{\partial f}{\partial n_t} \langle V(t), n_t \rangle,$$

where $f = y_t \langle Z(t), \nabla b_t \rangle$ and

$$f_t = y_{tt} \langle Z(t), n_t \rangle + y_t \left\langle \frac{\partial}{\partial t} Z(t), n_t \right\rangle + y_t \left\langle Z(t), \nabla \left(\frac{\partial}{\partial t} b_{\Omega_t(V)} \right) \right\rangle.$$

But

$$\nabla \left(\frac{\partial}{\partial t} b_{\Omega_t(V)} \right) = - D^2 b_{\Omega_t(V)}.V(t)op_t - D^*(V(t)op_t).\nabla b_{\Omega_t(V)},$$

so that

$$(f_t)|_{\partial \Omega_t(V)} = y_{tt} \langle Z(t), n_t \rangle + y_t \langle Z'(t), n_t \rangle$$

$$- y_t \langle Z(t), D^2_{\Omega_t(V)}.V(t) + D^* V(t).n_t \rangle.$$

Concerning the term $\frac{\partial}{\partial n_t} f$ we have

$$\frac{\partial}{\partial n_t} f = - \frac{\partial}{\partial n_t} y(t) \langle Z(t), n_t \rangle - y(t) \langle DZ(t).n_t, n_t \rangle.$$

That is

$$\frac{\partial y'}{\partial n_t} + \langle V(t), n_t \rangle y_t' = \operatorname{div}_{\partial \Omega_t(V)} (\langle Z(t), n_t \rangle \nabla_{\partial \Omega_t(V)} y(t)) \qquad (4.17)$$

$$+ (y_{tt} \langle Z(t), n_t \rangle + y_t \langle Z'(t), n_t \rangle - y_t \langle Z(t), D^2 b_{\Omega_t(V)}.V(t) + D^* V(t).n_t \rangle$$

$$+ H(t) y_t \langle Z(t), n_t \rangle \langle V(t), n_t \rangle$$

$$- \left(\frac{\partial}{\partial n_t} y(t) \langle Z(t), n_t \rangle + y(t) \langle DZ(t).n_t, n_t \rangle \right) \langle V(t), n_t \rangle.$$

The gradient $\mathbf{G}(V)$ of the functional $\mathbf{E}(V)$ is such that

$$\mathbf{E}'(V; W) := \int_0^\tau \int_D \langle \mathbf{G}(t, x), W(t, x) \rangle_{R_x^N} \, dx dt,$$

where the space integral over D should be understood as distribution over D with compact support included in the moving boundary. In order to get an explicit expression of that gradient we could introduce an adjoint state p in order to "eliminate" y' from the previous expression of the derivative. That adjoint will permit to use the characterization of y' whose previous problem leads to explicit expressions in terms of normal fields $\langle V(t), n_t \rangle$ but also $\langle Z(t), n_t \rangle$ on the moving boundary. On that Neumann boundary condition (4.5) which, in the weak formulation, corresponds to "free test functions" ψ lying in the whole linear space $H := H^1(0,\tau,L^2(D)) \cap L^2(0,\tau,H^1(D))$ we shall make use of the min max Lagrangian approach that we briefly describe here:

4.4.4. Min Max approach. We have

$$\mathbf{E}_\tau(V) = \text{Min}_{\phi \in H}\ \text{Max}_{\psi \in H}\{\ \mathbf{L}(\phi,\psi)\ |\ \ \psi(\tau)=0\,,\ \ \phi(0)=a\ \},$$

where the Lagrangian is

$$\mathbf{L}(\phi,\psi) = \int_0^\tau \int_{\Omega_t(V)} [1/2((\phi_t)^2 + |\nabla\phi|^2) + \nabla\phi.\nabla\psi - \phi_t\,\psi_t\,]dxdt + \int_{\Omega_0} b\psi(0)dx.$$

Proposition 4.3. *We have*

$$\mathbf{E}'_\tau(V,W) = \int_0^\tau \int_{\partial\Omega_t(V)} [1/2((y_t)^2 + |\nabla y|^2) + \nabla y.\nabla p - y_t\,p_t\,]\langle Z(t),n_t\rangle\,ds_t\,dt,$$

$$(4.18)$$

where the *adjoint* state solves:

$$\forall\phi \in H,\ \phi(0)=0,\ \int_0^\tau \int_{\Omega_t(V)} [\phi_t\,y_t + \nabla\phi.\nabla y + \nabla\phi.\nabla p - \phi_t\,p_t\,]dxdt = 0,\ \ (4.19)$$

that is

$$\int_0^\tau \int_{\Omega_t(V)} [-\Delta p + p_{tt}]\,\phi dxdt + \int_0^\tau \int_{\partial\Omega_t(V)} \left(\frac{\partial}{\partial n_t}p - \langle V(t),n_t\rangle\,p_t\right)\phi\,d\Gamma_t\,dt$$

$$= -\int_0^\tau \int_{\Omega_t(V)} -2\Delta y\,\phi\,dxdt - \int_0^\tau \int_{\partial\Omega_t(V)} \left(\frac{\partial}{\partial n_t}y + \langle V(t),n_t\rangle\,y_t\right)\phi\,d\Gamma_t\,dt,$$

and as, from (4.5) $\langle V(t),n_t\rangle\,y_t = \frac{\partial}{\partial n_t}y$ we get:

$$\int_0^\tau \int_{\Omega_t(V)} [-\Delta p + p_{tt}]\,\phi dxdt + \int_0^\tau \int_{\partial\Omega_t(V)} \left(\frac{\partial}{\partial n_t}p - \langle V(t),n_t\rangle\,p_t\right)\phi\,d\Gamma_t\,dt$$

$$= -\int_0^\tau \int_{\Omega_t(V)} -2\Delta y\,\phi\,dxdt - \int_0^\tau \int_{\partial\Omega_t(V)} 2\frac{\partial}{\partial n_t}y\,\phi\,d\Gamma_t\,dt\,.$$

so that p solves the following *backward* wave equation:

$$p(\tau) = 0,\ -\Delta p + p_{tt}\ =\ 2\,\Delta y,$$

with the boundary condition:

$$\frac{\partial}{\partial n_t}p - \langle V(t),n_t\rangle\,p_t\ =\ -2\frac{\partial}{\partial n_t}y.\qquad(4.20)$$

Notice that this backward problem is well posed as it boundary condition (3.14) is similar to (4.5). This is not obvious, but we have to keep in mind that when reversing the time variable (i.e., taking $s = \tau - t$) then we consider the function $\tilde{p}(s) = p(\tau - s)$ which is defined in the reverse tube which is itself built from the "initial domain" $\Omega_\tau(V)$ by the speed vector field $\tilde{V}(s) := -V(\tau - s)$ so that the sign is unchanged in the difference which occurs in the boundary condition (3.14) expressed in terms of V and p or in terms of $\tilde{V}$ and $\tilde{p}$.

4.5. Transverse derivative

For two given vector fields V and W the *transverse* field Z is the solution to the Lie bracket evolution

$$H_V.Z = W, \quad \text{where} \quad H_V.Z := \frac{\partial}{\partial t}Z + [Z, V], \quad [Z, V] := DZ.V - DV.Z. \quad (4.21)$$

4.5.1. Free divergences vector fields. For the sake of simplicity we assume first that V and W are free divergence vector fields: $\mathrm{div}V = \mathrm{div}W = 0$. Then the transverse field Z is itself a free divergence field. We consider the operator
$$H_V \in \mathbf{L}(\mathbf{D}(D, R^N), \mathbf{D}(D, R^N)).$$
Its transposed $H_V^* \in \mathbf{L}(\mathbf{D}'(D, R^N), \mathbf{D}'(D, R^N))$ is given by:

$$H_V^*.\Lambda := -\frac{\partial}{\partial t}\Lambda - D\Lambda.V - D^*V.\Lambda. \quad (4.22)$$

Lemma 4.4. *We have:*

$$H_V.(\zeta\, \vec{e}) = \left(\frac{\partial}{\partial t}\zeta + \nabla\zeta.V\right)\vec{e} + \zeta\, H_V.\vec{e}, \quad (4.23)$$

$$H_V^*(\zeta\, \vec{f}) = -\left(\frac{\partial}{\partial t}\zeta + \nabla\zeta.V\right)\vec{f} - \zeta\, H_V^*.\vec{f}. \quad (4.24)$$

We see that if (ζ, V) is a tube then the two previous expressions simplify as the first term vanishes by (2.1). In particular, we have

$$H_V^*.(\zeta\, \nabla\lambda) = -\zeta\left(\frac{\partial}{\partial t}\nabla\lambda + D\nabla\lambda.V + D^*V.\nabla\lambda\right) = -\zeta\nabla\left(\frac{\partial}{\partial t}\lambda + \nabla\lambda.V\right).$$

Then if we set the "scalar" operator h_V by:

$$h_V.\psi := \frac{\partial}{\partial t}\psi + \nabla\psi.V$$

we get, (ζ, V) being a tube,

$$H_V^*(\zeta\, \nabla\lambda) = -\zeta\, \nabla(\, h_V.\lambda). \quad (4.25)$$

Notice that when $\mathrm{div}V = 0$ the adjoint operator verifies $h_V^* = -h_V$. By "reversing" the time, that operator h_V is then self-adjoint.

4.5.2. Adjoint for the transverse field Z. Assume we have a functional derivative in the form

$$J'(V, W) = \int_0^\tau \int_{\Gamma_t(V)} g(t)\, \langle Z(t), n_t \rangle\, d\Gamma_t\, dt.$$

Let us consider $\tilde{g}$ as being *any* extension of $g(t)$ to D (the most useful choice being $\tilde{g}(t) = g(t)op_t$, where $p_t := I_d - b_{\Omega_t}\nabla b_{\Omega_t}$, the projection mapping is well defined in a neighborhood of the moving boundary when it is smooth enough, see ([24]), ([1]). Then, as the fields are divergence free, we have

$$J'(V,W) = \int_0^\tau \int_{\Omega_t(V)} \operatorname{div}(\tilde{g}(t)\, Z(t)\,)dxdt = \int_0^\tau \int_D \chi_{\Omega_t(V)}\, \langle \nabla \tilde{g}(t),\, Z(t)\rangle\, dxdt.$$

Then considering the adjoint problem

$$H_V^* \Lambda = \chi_{\Omega_t(V)}\, \nabla \tilde{g}(t), \quad \Lambda(\tau) = 0, \tag{4.26}$$

we get

$$J'(V,W) = \int_0^\tau \int_D \langle H_V^*.\Lambda, Z\rangle_{R^N}\, dxdt$$

$$= \int_0^\tau \int_D \langle \Lambda,\, H_V.Z\rangle_{R^N}\, dxdt = \int_0^\tau \langle \Lambda,\, W\rangle_{R^N}\, dxdt.$$

Thus Λ is the linear mapping $W \to J'(V;W)$. Let us consider the adjoint problem (4.26): we search for a solution in the form

$$\Lambda(t,x) = \chi_{\Omega_t(V)}(x)\, \nabla_x \lambda(t,x).$$

Then, as

$$H_V^*(\chi_{\Omega_t}\, \nabla \lambda) = \chi_{\Omega_t(V)}\, \nabla(\, h_V^*.\lambda\,),$$

it is sufficient to choose λ as solution to the *scalar adjoint* backward problem:

$$-\frac{\partial}{\partial t}\lambda - \nabla\lambda.V = \tilde{g}, \quad \lambda(\tau) = 0. \tag{4.27}$$

Therefore we see that the solution Λ to the problem (3.14) is given by $\Lambda = \chi_{\Omega_t(V)}\, \nabla\lambda$ where λ is the solution to the problem (4.27), and that expression is independent of the choice of the extension $\tilde{g}$ of g outside the lateral boundary of the tube. Let us consider any extension such that $\frac{\partial}{\partial n_t}\tilde{g}(t) = 0$ on $\partial\Omega_t(V)$ (this is the case when we choose $\tilde{g}(t) = g o p_t$). Then the term $\nabla\lambda\tilde{g}(t)$ simplifies to $\nabla_{\Gamma_t}g(t)$ which is the *tangential* gradient of $g(t)$ on the surface $\Gamma_t := \partial\Omega_t(V)$. Finally, assuming the vector fields V, W to be divergence free (that is the measure of the moving domains to be preserved), we consider the tangential equation

$$\lambda_t + \nabla_{\Gamma_t(V)}\lambda.V = g(t) \ \text{ on } \Gamma_t, \quad \lambda(\tau) = 0, \ \text{ on } \Gamma_\tau \tag{4.28}$$

and

$$J'(V;W) = \int_0^\tau \int_{\Gamma_t} g(t)\langle Z(t), n_t\rangle d\Gamma_t dt = \int_0^\tau \langle \Lambda(t), W(t)\rangle dt$$

$$= \int_0^\tau \int_D \langle -\chi_{\Omega_t(V)}\, \nabla\lambda, W\rangle dxdt = \int_0^\tau \int_{\Omega_t(V)} \operatorname{div}(\lambda W)dxdt$$

$$= \int_0^\tau \int_{\Gamma_t(V)} \lambda(t)\, \langle W(t), n_t\rangle\, d\Gamma_t\, dt.$$

As a result:

Proposition 4.5. *Let V, W be given in $L^1(0, \tau, W_0^{1,\infty}(D, R^N))$, and $g \in L^1(0, \tau, L^1(\Gamma_t(V)))$. Consider the solution λ to the tangential backward problem (4.28). Then we have*

$$\int_0^\tau \int_{\Gamma_t(V)} g(t) \langle Z(t), n_t \rangle d\Gamma_t\, dt = \int_0^\tau \int_{\Gamma_t} \lambda(t) \langle W(t), n_t \rangle\, d\Gamma_t\, dt.$$

4.6. Optimal control problem for moving domain in D with given measure

We consider divergence free fields V, W, Z and we introduce

$$\lambda_t + \nabla_{\Gamma_t(V)}\lambda.V = 1/2((y_t)^2 + |\nabla y|^2) + \nabla y.\nabla p - y_t\, p_t\,, \quad \lambda(\tau) = 0.$$

Then we get:

$$\mathbf{E}_\tau'(V, W) = \int_0^\tau \int_{\partial\Omega_t(V)} \lambda \langle W(t), n_t \rangle\, ds_t\, dt. \tag{4.29}$$

In order to consider the vector field V as a control parameter we shall now "increase" the cost functional by a "small" term (which would be "the price" of the control V):

$$\sigma > 0, \quad \mathbf{E}_{\tau,\sigma}(V) = \mathbf{E}_\tau(V) + \sigma/2\,||\Delta V||^2\,. \tag{4.30}$$

The gradient of $\mathbf{E}_{\tau,\sigma}(V)$ is given by:

$$\mathbf{E}_{\tau,\sigma}'(V; W) = \int_0^\tau \int_{\partial\Omega_t(V)} \lambda(t) \langle W(t), n_t \rangle\, d\Gamma_t\, dt + \sigma \int_0^\tau \int_D \langle \Delta^2 V, W \rangle\, dx dt\,.$$

The Optimal Speed Synthesis is the following *equation*:

$$V^*(t) = (\Delta^2)^{-1}.\, \lambda(t)\nabla\chi_{\Omega_t(V^*)} + \nabla_x\Pi(t),$$

$$\lambda_t + \nabla_{\Gamma_t(V^*)}\lambda.V^* = 1/2((y_t)^2 + |\nabla y|^2) + \nabla y.\nabla p - y_t\, p_t\,, \quad \lambda(\tau) = 0,$$

$$p(\tau) = 0, \quad \frac{\partial}{\partial n_t}p - \langle V(t), n_t \rangle p_t = -2\frac{\partial}{\partial n_t}y,$$

$$y(0) = 0, \ y - t(0) = b, \ \frac{\partial}{\partial n_t}y - \langle V(t), n_t \rangle y_t = 0. \tag{4.31}$$

4.6.1. General case: V is not a divergence free vector field. We shall just notice the (small) changes in the previous analysis: the adjoint problem is

$$H_V^*\Lambda = -\frac{\partial}{\partial t}\Lambda - D\Lambda.V - D^*V.\Lambda - (\mathrm{div}V)\,\Lambda.$$

$$H_V^*(\zeta \vec{f}) = \left(-\frac{\partial}{\partial t}\zeta - \nabla\zeta.V\right) \vec{f} + \zeta\, H_V^*.\vec{f}$$

So that if (ζ, V) is a tube we simply get as before

$$H_V^*(\zeta\vec{f}) = \zeta\, H_V^*.\vec{f}.$$

Now

$$H_V^*(\nabla\lambda) = -\nabla\,\lambda_t \; - \; D(\nabla\lambda).V \; - \; D^*V.\nabla\lambda \; - \; \mathrm{div}V\,\nabla\lambda$$
$$= -\nabla\,\lambda_t \; - \; \nabla(\,\nabla\lambda.V\,) \; - \; \mathrm{div}V\,\nabla\lambda,$$

and

$$H_V^*(\chi_{\Omega_t(V)}\,\nabla\lambda) = -\chi_{\Omega_t(V)}\left[\nabla\left(\frac{\partial}{\partial t}\lambda + \nabla\lambda.V\right) \; + \; \mathrm{div}V\,\nabla\lambda\right].$$

But also:

$$H_V^*(-\lambda\,\nabla\chi) = \left(\frac{\partial}{\partial t}\lambda + \nabla\lambda.V\right)\nabla\chi + \lambda\left(\,\nabla\left(\frac{\partial}{\partial t}\chi + \nabla\chi.V\right) \; + \; \mathrm{div}V\,\nabla\chi\right)$$
$$= \left(\frac{\partial}{\partial t}\lambda + \nabla\lambda.V \; + \; \lambda\,\mathrm{div}V\right)\,\nabla\chi.$$

Hence we have:

Proposition 4.6. *Let* $\lambda \in L^1(0,\tau,L^1(\Gamma_t))$ *solve the backward problem*

$$\frac{\partial}{\partial t}\lambda + \nabla\lambda.V \; + \; \lambda\,\mathrm{div}V = g, \quad on \; \Gamma_t, \;\; \lambda(\tau) = 0.$$

Then we have:

$$\int_0^\tau \int_{\Gamma_t} g\,\langle Z(t),n_t\rangle d\Gamma_t dt = \int_0^\tau \int_{\Gamma_t} \lambda\,\langle W(t),n_t\rangle d\Gamma_t dt.$$

4.6.2. Optimality condition: the general case. We consider the minimization of the previous functional $\mathbf{E}_\sigma(V)$ without any constraint on the divergence of the vector field V. We obtain the following characterisation:

$$V^*(t) = \frac{1}{\sigma}\,(\Delta^2)^{-1}.\,\lambda(t)\nabla\chi_{\Omega_t(V^*)},$$

$$\lambda_t + \nabla_{\Gamma_t(V^*)}\lambda.V^* + \lambda\,\mathrm{div}V^* = 1/2((y_t)^2 + |\nabla y|^2) + \nabla y.\nabla p - y_t\,p_t\,, \quad \lambda(\tau) = 0,$$

$$p(\tau) = 0, \;\; \frac{\partial}{\partial n_t}p - \langle V(t),n_t\rangle\,p_t = -\,2\frac{\partial}{\partial n_t}y,$$

$$y(0) = 0, \; y - t(0) = b, \;\; \frac{\partial}{\partial n_t}y - \langle V(t),n_t\rangle\,y_t = 0. \qquad (4.32)$$

In order to simplify that system, an idea would be to search for a solution V^* proportional to the normal field $n_{\Gamma_t(V^*)}$.

A possibility is to parametrize the domain as a level set of some function $\Phi(t,x)$ say

$$\Omega_t := \{x \in D \;\mid\; \Phi(t,x) > 0 \; \}.$$

Then the vector field can be taken as

$$V(t,x) = -\frac{\partial}{\partial t}\Phi\,\frac{\nabla\Phi}{||\nabla\Phi||}$$

which indeed is proportional to $n_{\Gamma_t} = \frac{\nabla \Phi}{||\nabla \Phi||}$. In this case the equation for the scalar boundary adjoint term λ drastically simplifies to the following:

$$\frac{\partial}{\partial t}\lambda + \lambda \operatorname{div} V = g := 1/2((y_t)^2 + |\nabla y|^2) + \nabla y.\nabla p - y_t\, p_t.$$

In the last section we shall investigate the fixed point theory for solving that non-linear problem in terms of level sets.

In order to avoid the analysis of the free boundary value problem (the first order necessary optimality condition) we propose here a different analysis which may be "suboptimal" but permits to handle algorithms for the decay of the energy functional $\mathbf{E}(V)$. It is based on the so-called *cubic* derivative, that is the cubic (with respect to the normal component $v(t) = \langle V(t), n_t \rangle$) expression for the energy density time derivative $E'(t)$. We recall now these old results and extend to the previous Neumann condition (4.5) for the wave equation in a moving domain.

4.7. Time derivative of the energy: a passive shape control approach

The cubic expression for the energy derivative was established for the homogeneous Dirichlet condition. We extend it here, as a new result, to the previous "Neumann condition" $\frac{\partial}{\partial \nu^*} u = 0$ on the moving boundary:

4.7.1. Dirichlet problem.

We recall [34] the cubic expression of the derivative for the wave equation with Dirichlet boundary condition (4.3) on the moving domain.

Proposition 4.7. *Let y be the solution of (4.1), (4.2), (4.3), and $v(t) = \langle V(t), n_t \rangle$ be the normal component of the speed vector field at the moving boundary $\partial \Omega_t$. Then*

$$E'(t) = 1/2 \int_{\partial \Omega_t} \left(\frac{\partial}{\partial n_t} y\right)^2 \left(v^3(t) - v(t)\right) d\Gamma_t. \tag{4.33}$$

Proof. Assuming smooth enough solution y, we get

$$E'(t) = \int_{\Omega_t} \frac{\partial}{\partial t}y\, \frac{\partial^2}{\partial t^2}y + \nabla y.\nabla \left(\frac{\partial}{\partial t}y\right) dx + \int_{\Gamma_t} 1/2 \left(\left(\frac{\partial}{\partial t}y\right)^2 + ||\nabla y||^2\right) v\, d\Gamma_t$$

$$= \int_{\Omega_t} \frac{\partial}{\partial t}y\left(\frac{\partial^2}{\partial t^2}y - \Delta y\right) dx + \int_{\Gamma_t}\left[1/2\left(\left(\frac{\partial}{\partial t}y\right)^2 + ||\nabla y||^2\right) v + \frac{\partial}{\partial n_t}y\, \frac{\partial}{\partial t}y\right] d\Gamma_t$$

using the boundary condition we conclude the proof.

Corollary 4.8.

$$E(t) = 1/2 \int_{\Omega} (a^2 + |\nabla b|^2)dx + 1/2 \int_0^t \int_{\partial \Omega_s}\left(\frac{\partial}{\partial n_s}y(s)\right)^2 (v^3(s) - v(s))\, d\Gamma_s\, ds \tag{4.34}$$

$$\mathbf{E}_t(V) = tE(0) + 1/2 \int_0^t (t-s)\int_{\partial \Omega_s}\left(\frac{\partial}{\partial n_s}y\right)^2 (v^3(s) - v(s))\, d\Gamma_t\, ds. \tag{4.35}$$

4.8. Neumann condition (4.5)

Proposition 4.7 extends to the Neumann condition with some generalisation in the cubic expression.

Proposition 4.9. *Let y be the solution of (4.1), (4.2), (4.5) $v(t) = \langle V(t), n_t \rangle$ be the normal component of the speed vector field at the moving boundary $\partial\Omega_t$. Then*

$$E'(t) = 1/2 \int_{\partial\Omega_t} (y_t)^2 (v^3(t) - v(t)) \, d\Gamma_t + 1/2 \int_{\partial\Omega_t} |\nabla_{\Gamma_t} y|^2 v(t) d\Gamma_t. \quad (4.36)$$

In view of (4.36) the energy derivative, as in the previous Dirichlet case, is governed by a cubic in $v = \langle V(t), n_t \rangle$ at the boundary, here it takes a different form:

$$E'(t) = 1/2 \int_{\partial\Omega_t} ((y_t)^2 v^3(t) + (|\nabla_\Gamma y|^2 - (y_t)^2) v(t)) \, ds_t. \quad (4.37)$$

Corollary 4.10.

$$E(t) = E(0) + 1/2 \int_0^t \int_{\partial\Omega_s} (y_t(s))^2 (v^3(s) - v(s)) \, d\Gamma_s \, ds$$
$$+ 1/2 \int_0^t \int_{\partial\Omega_s} |\nabla_{\Gamma_s} y(s)|^2 v(s) d\Gamma_s \quad (4.38)$$

$$\mathbf{E}_t(V) = tE(0) + 1/2 \int_0^t (t-s) \int_{\partial\Omega_s} [(y_s)^2 (v^3(s) - v(s)) + |\nabla_{\Gamma_s} y(s)|^2 v(s)] \, d\Gamma_s \, ds. \quad (4.39)$$

4.9. Energy derivative in term of the normal derivative of y

Making use of (4.5) we rewrite (4.37) in term of the *normal derivative* $\frac{\partial y}{\partial n_t}$, as it was the case for the expression we got for the Dirichlet case, we obtain a much more suitable expression (4.40). What is to be observed is that in (4.40) "the minus sign of (4.37) has been changed" for a plus (as $(\frac{\partial y}{\partial n})^2 + |\nabla_\Gamma y|^2 = |\nabla y|^2$)), which will make all the difference in the next result:

$$E'(t) = 1/2 \int_{\partial\Omega_t(V)} \frac{1}{v} (|\nabla y(t)|^2 v^2 - (\frac{\partial y(t)}{\partial n_t})^2) \, ds_t. \quad (4.40)$$

4.10. The case for a damped material

Assume that the Neumann boundary condition takes the slightly different following form: let $d_0 > 0$ and

$$\frac{\partial y}{\partial n_t} + (d_0 + \langle V(t), n_t \rangle) y_t = 0 \quad \text{on } \partial\Omega_t(V). \quad (4.41)$$

Then we get:

$$E'(t) = 1/2 \int_{\partial\Omega_t} ((y_t)^2 v^3(t) + (|\nabla_\Gamma y|^2 - (y_t)^2) v(t)) \, ds_t \quad (4.42)$$

$$+ \int_{\partial\Omega_t} d_0 (y_t)^2 (v^2 + 1/2 \, d_0 \, v - 1).$$

The integrand in the additive term is negative for v^2 "small enough", that is

$$v_-^* \; < \; v \; < \; v_+^*$$

with $v_{+,-}^* = -1/4 \, d_0 \; + - \sqrt{1 + 1/16 \, d_0^2}$

$$E'(t) = 1/2 \int_{\partial \Omega_t(V)} \frac{1}{v} \left(|\nabla y(t)|^2 \, v^2 - \left(\frac{\partial y(t)}{\partial n_t} \right)^2 \right) \, ds_t. \qquad (4.43)$$

4.11. Passive boundary control

4.11.1. Dirichlet problem. In order to have a decay of $E(t)$ we can, *formally,* choose the normal speed component $v(t)$ on the moving boundary such that $E'(t) \leq 0$. For this it is sufficient to impose the following condition

$$0 \leq v(t) \leq 1 \; \text{ or } \; v(t) \leq -1. \qquad (4.44)$$

4.11.2. Neumann problem. In the same formal approach, following (4.37) the sufficient condition in order to derive $E'(t) \leq 0$ would be to "choose" the normal speed verifying the following conditions (which are in fact an equation on v):

or
$$v(t) \geq 0 \text{ and } v(t)^2 \geq \left(\frac{|\nabla_{\Gamma_t} y(t)|^2}{(y_t)^2} - 1 \right),$$
$$v(t) \leq 0 \text{ and } v(t)^2 \leq \left(\frac{|\nabla_{\Gamma_t} y(t)|^2}{(y_t)^2} - 1 \right). \qquad (4.45)$$

Obviously (4.45) is "often" an empty condition.

In view of (4.40) we shall obtain a negative derivative $E'(t) \leq 0$, for the energy if the integrand is non positive at almost every $x \in \partial \Omega_t(V)$.

Proposition 4.11. *Assume $\nabla y(t, x)$ different from zero for a.e.$(t,)$, $x \in \partial \Omega_t(V)$, then let*

$$K(t, x) = \left(\frac{(\frac{\partial y(t)}{\partial n_t})^2}{|\nabla y|^2} \right)^{1/2} \leq 1.$$

Assume that the speed field V verifies:

If $v(t, x) > 0$, then $v(t, x) \leq K(t, x)$; if $v(t, x) < 0$, then $v(t, x) \leq -K(t, x)$.
$$\qquad (4.46)$$
Then $E'(t) \leq 0$ and that inequality is strict if, on a subset of $\partial \Omega_t(V)$ of non zero measure one of the inequalities (4.46) is strict.

Notice that the conditions (4.46) is not really a feedback loop between the state $y(t)$ and the control v at the boundary, but we cannot say that, as in the Dirichlet case, the synthesis (4.46) is a passive control. It can be written in more compact form as follows:

$$a.e.t, a.e.x \in \partial \Omega_t(V), \; v(t, x) \in \mathbf{K} =] - \infty, \, - K(t, x)] \cup [0, \; K(t, x)] = \mathbf{K}^- \cup \mathbf{K}^+.$$
$$\qquad (4.47)$$

4.12. An example

In order to satisfy (4.47) choose

$$v(t, x) = 1/\sqrt{2}\ K \ \text{ or } \ v(t, x) = -\sqrt{2}K. \tag{4.48}$$

then the integrand term in the expression (4.40) of $E'(t)$ takes the following form:

$$-\frac{1}{\sqrt{2}}\left|\frac{\partial y}{\partial n}\right| |\nabla y| \quad (\text{when } v = -\sqrt{2}\ K \text{ or } v = +1/\sqrt{2}K). \tag{4.49}$$

Proposition 4.12. *Assume v satisfies (4.48), then*

$$E'(t) = -\frac{1}{2\sqrt{2}} \int_{\partial\Omega_t(V)} \left|\frac{\partial y}{\partial n_t}\right| \|\nabla y\|\, ds_t \leq -\frac{1}{2\sqrt{2}} \int_{\partial\Omega_t(V)} \left|\frac{\partial y}{\partial n_t}\right|^2 ds_t. \tag{4.50}$$

4.13. Smart material modelling

We impose a constraint on the normal component of the speed field which translates the fact that during the time, the shape evolution is periodical or "locally periodical". Assume that only small parts of the boundary centered on given points $x_1, \ldots, x_m$ of $\partial\Omega_0$ are moving with the speed of a "rigid body motion" for example.

4.13.1. One actuator. Assume a single "small part" γ_t of the boundary $\partial\Omega_t(V)$ is moving, keeping zero curvatures, and with a constant space normal speed $v(t)$. That is that $\Omega_t(V)$ is a moving domain those boundary $\partial\Omega_t(V)$ moves under a normal a speed $V(t, x).n_t = v$ that is zero out of $\bar{\gamma}_t$. Now we assume $v(t, .)$ to be constant on γ_t and γ_0 flat (without curvatures) so that in fact $\gamma_t = \gamma_0 + d(t)\vec{n}_0$ (is moved by translations). The boundary $\partial\Omega_t = (\partial\Omega_0 - \bar{\gamma}_0)\bigcup \bar{\gamma}_t \bigcup \mathbf{O}_t.$ where $\mathbf{O}_t$ is a cylindrical (variable with t) open set on which V is tangential but $v = 0$. Of course the moving boundary $\partial\Omega_t(V)$ is not of class C^1 but is Lipschitzian smooth (which is enough for justifying the previous considerations). From (4.40) we obtain:

$$E'(t) = 1/2\ \frac{1}{v(t)} \left(v(t)^2 \int_{\gamma_t} \|\nabla y(t, x)\|^2\, dx \ - \int_{\gamma_t} \left(\frac{\partial y(t, x)}{\partial n_t}\right)^2 dx \right). \tag{4.51}$$

We introduce

$$\mathbf{K}_{\gamma_0}(t) = \frac{\int_{\gamma_t} (\frac{\partial y(t, x)}{\partial n_t})^2\, dx}{\int_{\gamma_t} \|\nabla y(t, x)\|^2\, dx} \ \leq\ 1. \tag{4.52}$$

Proposition 4.13. *Assume that $\int_{\gamma_t} \|\nabla y(t, x)\|^2\, dx$ is different from zero, and $v(t)$ verifies:*

$$v(t) > 0 \ \text{ and } \ v(t) \leq \mathbf{K}_{\gamma_0}(t), \quad \text{or} \ \ v(t) < 0 \ \text{ and } \ v(t) \leq -\mathbf{K}_{\gamma_0}(t). \tag{4.53}$$

That is

$$v(t) \in\,]-\infty,\, -\mathbf{K}_{\gamma_0}(t)]\bigcup [0,\, \mathbf{K}_{\gamma_0}(t)].$$

Then $E'(t) \leq 0$.

For example chose

$$v(t) = +1/\sqrt{2}\,\mathbf{K}_{\gamma_0}(t) \ \text{ or } \ v(t) = -\sqrt{2}\,\mathbf{K}_{\gamma_0}(t).$$

Then we get:

$$E'(t) = -\frac{1}{2\sqrt{2}} \left(\int_{\gamma_t} \left(\frac{\partial y(t,x)}{\partial n_t} \right)^2 dx \right)^{1/2} \left(\int_{\gamma_t} \|\nabla y(t,x)\|^2\, dx \right)^{1/2}. \qquad (4.54)$$

This displacement is periodical by the choice of adequate time intervals.

5. The level set approach for moving domain: Asymptotic analysis

The boundary of a moving domain $\Omega_t \subset D$ can be parametrized as a *level set* of a one parameter smooth function $\Phi(t,.) \in H^1(D)$ such that, for example, $\Phi(t,.)+1 \in H^1_0(D)$ (that is to say that $\Phi(t,.) = -1, \ x \in \partial D$).

In that situation

$$\Omega_t := \{\, x \in D,\ s.t.\ \Phi(t,x) > 0 \,\} \qquad (5.1)$$

is a quasi open subset in D (it is an open subset up to *zero capacity* subset), and verifies $\bar{\Omega}_t \subset D$.

Of course for a given open tube $\ Q = \bigcup_{0<t<\tau} \{t\} \times \Omega_t \ \subset]0,\tau[\times D$, such function Φ is *not unique*.

5.1. Intrinsic geometry function

It is classical that $\Phi(t) = \Phi(t,.)$ can be decomposed as

$$\Phi(t) = \Phi(t)^* o\beta_{\Phi(t)}, \qquad (5.2)$$

where the *monotone rearrangement* $\Phi(t)^*$ is a monotone (increasing) mapping defined from the interval $[0, meas(D)]$ into R, while the *intrinsic geometry mapping* $\bar{\beta}(t) =: \beta_{\Phi(t)}$ is defined from D into the interval $[0, meas(D)]$ as:

$$\bar{\beta}(t)(x) = meas(\{\, y \in D\ s.t.\ \Phi(t)(y) < \Phi(t)(x)\ \}). \qquad (5.3)$$

Assume that

$$\partial\Omega_t = \{\, x \in D\ s.t.\ \Phi(t)(x) = 0 \,\} = \Phi(t)^{-1}(0). \qquad (5.4)$$

Then we verify that this property (5.4) just depends on the *intrinsic geometry mapping* $\bar{\beta}(t)$, that is: (5.4) holds true for any monotone increasing continuous mapping $\Phi(t)^*(s)$.

The *intrinsic* view point consists in taking $\Phi(t)^*(s) = s, \ \forall s \in [0, meas(D)]$.

In that situation we see that $\Phi(t) = \bar{\beta}(t)$, that is to say that (5.3) holds for the function $\Phi(t) = \bar{\beta}(t)$:

$$\forall x \in D,\ \bar{\beta}(t)(x) = meas(\{\, y \in D\ s.t.\ \bar{\beta}(t)(y) < \bar{\beta}(t)(x)\ \}) \qquad (5.5)$$

and

$$\partial\Omega_t = \bar{\beta}(t)^{-1}(0).$$

(5.1) can be rewritten as

$$\Omega_t = \{\, x \in D \ s.t. \ \Phi(t)(x) > 0 \,\} \ = \ \{\, x \in D \ s.t. \ \bar{\beta}(t)(x) > 0 \,\}. \tag{5.6}$$

Of course there are several functions $\bar{\beta}(t)$ verifying (5.6) and (5.5).

5.2. Speed vector field

Assuming (5.1), it is classical ([8], [24]) that a speed vector (whose flow mapping carries that moving domain) is

$$V^\phi(t,x) = -\frac{\partial}{\partial t}\phi(t,x)\ \frac{\nabla_x\phi(t,x)}{||\nabla_x\phi(t,x)||^2}. \tag{5.7}$$

Any other field building this tube is of the form $W = V^\phi + Z$ with $\langle Z(t), n_t \rangle = 0$ on Γ_t.

We understand that the presence in the denominator of the term $||\nabla_x\phi||$ will not help to define correctly the flow mapping of that vector field V^ϕ. At least we shall have to control the term $||V^\phi(t,x)||_{R^N} = |\frac{\partial\phi}{\partial t}|/||\nabla_x\phi||$ in order to use the shape differential equation technique. The way to bypass that difficulty is as follows. Consider any shape gradient descent method for minimizing a shape functional $J(\Omega)$ whose shape gradient $G(\Omega) \in \mathbf{D}'(R^N, R^N)$ is a vector distribution with compact support included in the boundary of the domain (which, in *smooth* situation, takes the form $G(\Omega) = \gamma^*_{\partial\Omega}.(g\,n)$ where g, the *shape gradient density* is a scalar distribution on the boundary with zero transverse order, n being the normal field). The Shape differential equation consists in solving the non linear problem (see [7], [8], ..., [24]):

$$\boxed{\ \forall t, \ \ 0 \le t \le \tau, \ \ V(t,.) = -A^{-1}.G(\Omega_t(V))\ } \tag{5.8}$$

leading to the decrease of the functional:

$$J(\Omega_t(V)) \le J(\Omega_0) - \alpha \int_0^t ||V(s,.)||^2_{\mathbf{D}(A^{1/2})}\,ds \ . \tag{5.9}$$

Let V^* be *a solution* to (5.8), the problem is then to find a function ϕ such that the associated vector speed V^ϕ builds the same tube $Q^* = Q_{V^*}$. The necessary and sufficient condition (under some smoothness) is that the normal components of the two vector fields are equal *on the lateral boundary* Σ_V, that is

$$-\frac{\partial}{\partial t}\phi/|\nabla_x\phi| = \langle V^*(t), n_t\rangle \ \ \text{on} \ \ \partial\Omega_t(V^*). \tag{5.10}$$

Assume that ϕ solves the equation (5.10) on the lateral boundary Σ_{V^*}, by "multiplying" that equation by the non negative term $|\nabla\phi(t)|$, we obtain that ϕ solves the problem:

$$\frac{\partial}{\partial t}\phi(t,x) \ + \langle \nabla_x\phi(t,.), V^*(t)\rangle = 0 \ \ \text{on} \ \ \partial\Omega_t(V^*). \tag{5.11}$$

An obvious way to solve that problem (5.11) is to consider the *global* convection problem:

$$\frac{\partial}{\partial t}\phi(t,x) \; + \; \langle \nabla_x\phi(t,.), \, \bar{V}^*(t) \rangle \; = \; 0 \; \text{ in } \; D, \tag{5.12}$$

where $\bar{V}^*$ is *any admissible extension* to the cylinder $(0,\tau) \times D$ of the vector field $V^*|_{\Sigma_{V^*}}$ (the restriction of V^* to the lateral boundary of the tube Q_{V^*}). A possible choice of such vector field $\bar{V}^*$ is V itself, but there are many other examples, one of them is $\bar{V}^* = V^* o p^*$ where p^* stands for the projection mapping or the local (or *narrow*) $(p^*)^h$ projection onto the boundary $\partial\Omega_t(V^*)$. In this paper we furnish existence and uniqueness results for this convection problem (5.12) when $\bar{V}^*$, div $\bar{V}^*$ are in $L^1(0,\tau, L^2(D))$ while the initial data satisfies $\phi_0 \in L^\infty(D)$.

Let V^* and ϕ be solutions to (5.8) and (5.12), respectively, then we get

$$\frac{\partial}{\partial t}\phi(t,x)/|\nabla\phi| = \langle \nabla_x\phi(t,.)/|\nabla\phi|, \bar{V}^* \rangle,$$

so that

$$\boxed{ \; \left| \frac{\partial}{\partial t}\phi(t,x)/|\nabla\phi| \; \right| \; \leq \; ||\bar{V}^*(t,x)||_{R^N} \; . \;} \tag{5.13}$$

Assume that $\bar{V}^* = V^*$, then if $V^* \in E = L^1(0,\tau, L^2(D, R^N))$ with ϕ being solution to (5.12), we get $V^\phi \in E$. Assume that div $V^\phi \in E$ then for any given $\phi_0 \in L^\infty(D)$ we get the existence of a solution to (5.12). In the classical setting (developed in [7], [8], ...) the shape gradient G of the shape functional J is bounded (in some "negative" Sobolev space of distributions over the universe D) and continuous (with respect to the *Courant metric*, see [24]), then, $\forall k, \; k \geq 1$, the shape differential equation possesses smooth solutions $V^* \in \mathbf{C}^{0,k} = C^0([0,\tau], C^k(\bar{D}, R^N)) \subset E$. Then the flow mapping $T_t(V^*)$ is classically defined and the unique solution to the convection problem 5.12, if $\bar{V}^*$ is also chosen in $\mathbf{C}^{0,k}$, is given by

$$\phi(t) = \phi_0 o T_t(\bar{V}^*)^{-1}. \tag{5.14}$$

As $t \to T_t(\bar{V}^*)^{-1}(.) \in \mathbf{C}^{1,k}$ (see [5]) assuming the initial data $\phi_0 \in C^k(\bar{D})$, we get $\phi \in \mathbf{C}^{0,k-1}(D \setminus K_\phi)$, with the compact set $K_\phi = \{x \in D \; s.t. \; \nabla\phi(x) = 0 \; \}$.

As $\phi(t) = \phi_0 o T_t^{-1}$, we get $\nabla\phi(t) = ((DT_t)^{-1}.\nabla\phi_0)o T_t^{-1}$ so that K_{ϕ_0} is void implies that, $\forall t, \; K_{\phi(t)}$ is empty. Then $V^\phi \in \mathbf{C}^{0,k-1}$ (assuming now that $k \geq 2$) and $\Omega_t(V^\phi = \Omega_t(\bar{V}^*)\Omega_t(V^*)$. (In other words, the three vector fields V^*, $\bar{V}^*$, V^ϕ build the same tube Q_V as they have the same normal speed v on the lateral boundary Σ_V.) From (5.8) we get

$$-\frac{\partial}{\partial t}\phi(t,x) \; + \; \langle \nabla_x\phi(t,.), \, A^{-1}.G(\Omega_t(V^\phi)) \rangle = 0 \tag{5.15}$$

which is a Hamilton-Jacobi like equation for the function ϕ. From 5.9, we get

$$J(\Omega_t(V)) \leq J(\Omega_0) - \alpha \int_0^t ||\frac{\partial\phi(s,.)}{\partial t} / \nabla_x\phi(s,.)||^2_{\mathbf{D}(A_k^{1/2})} ds. \tag{5.16}$$

Briefly we could say that the shape differential equation 5.8 is solved by the fixed point method (see [7], [26]) in a classical setting which does not permit the change of topology in the moving domain. We introduce here the weak setting which permits to handle that equation with possible topological changes by avoiding any homeomorphism.

5.3. Example of the operator A

Let D be a bounded domain in R^N with smooth boundary. Let $\vec{G}$ be a vector distribution with compact support in D, that is $\vec{G} \in \mathbf{E}'(D, R^N)$, of the form $\vec{G} = \gamma_\Gamma^*.(g\,\vec{n})$, where Γ is the boundary, a manifold of regularity C^1, of the domain Ω, with $\bar{\Omega} \subset D$.

The trace (or restriction) operator is $\gamma_\Gamma \in \mathbf{L}(H_0^1(D, R^N), H^{1/2}(\Gamma, R^N))$ and its adjoint operator is $\gamma_\Gamma^* \in \mathbf{L}(H^{-1/2}(\Gamma, R^N), H^{-1}(D, R^N))$; the normal field on Γ outgoing from Ω is denoted by n and g given in $L^1(\Gamma)$, is a scalar function defined on the boundary Γ.

We consider the linear operator $A \in \mathbf{L}(H_0^1(D, R^N), H^{-1}(D, R^N))$ defined by $A.F = (-\Delta F_1, \ldots, -\Delta F_N)$ for any $F_i \in H_0^1(D), 1 \leq i \leq N$.

We consider the element $U \in H_0^1(D, R^N), \ U = -A^{-1}.\vec{G}$. It solves the problem $-\Delta U_i = 0$ in $D \setminus \Gamma$, $[DU.n] = g\,\vec{n}$ in $H^{-1/2}(\Gamma)$, where $[\vec{E}]_\Gamma$ stands for the *jump* term on Γ.

5.4. Shape gradient estimate

In order to perform the fixed point approach in the non linear shape differential equation (5.8) we require the shape gradient G to verify an estimate as follows: there exist two positive constants s and M such that:

$$\boxed{\forall \Omega \subset D, \ s.t. \ \partial\Omega \ \text{is a} \ C^1 \ \text{manifold}, \ \ ||G(\Omega)||_{H^{-s}(D,R^N)} \leq M.} \qquad (5.17)$$

We can immediately give several such examples. Consider the following *distributed* functionals: E being a measurable subset in Ω, $y = y(\Omega) := (-\Delta)^{-1}.f$, f given in $L^2(D)$,

$$J(\Omega) = \int_E (y - Y_d)^{1/2} dx. \qquad (5.18)$$

The Eulerian derivative is given by:

$$dj(\Omega, V) = \langle G, V(0) \rangle$$

with

$$\langle G(\Omega), W \rangle = \int_\Omega \langle A'(W).\nabla y, \nabla p \rangle \ dx \qquad (5.19)$$

$$+ \int_\Omega (\ \langle \nabla f, W \rangle\, p - \chi_E\,(y - Y_d)\,\langle \nabla y, W \rangle\ )dx,$$

where the symmetric matrix $A'(W) := 2\epsilon(W) - \text{div} W\, I_d$ is an element of

$$C^0([0, \tau], C^0(\bar{D}, R^{N^2}) \subset L^\infty([0, \tau] \times D, R^{N^2})$$

(with $2\epsilon(W) := DW + DW^*$). The usual estimates holds:

$$||y(\Omega)||_{H_0^1(D)} \leq 1/\sqrt{\lambda_1(D)} \ ||f||_{L^2(D)},$$

so that $||p(\Omega)||_{H_0^1(D)} \le M$ and then there exist a constant $C > 0$ such that for all $W \in W_0^{1,\infty}(D, R^N)$ we have

$$|\langle G(\Omega), W\rangle| \le C \, ||W||_{W^{1,\infty}(D,R^N)}.$$

As for $s > 1 + \frac{1}{N}$ we have $H_0^s(D, R^N) \subset W^{1,\infty}(D, R^N)$, we get the boundedness of the gradient in $H^{-s}(D, R^N)$:

There exists a positive constant $C(||f||_{L^2(D)}, \lambda_1(D), ||Y_d||_{L^2(E)})$ such that

$$\forall \Omega \subset D, \ ||G(\Omega)||_{H^{-s}(D,R^N)} \le C.$$

Choose for example $s = 2$ and the operator

$$\mathbf{A} := A^2 \in \mathbf{L}(H_0^2(D, R^N), H^{-2}(D, R^N)),$$

we have

$$\langle \mathbf{A}.W, W\rangle_{H^{-2}(D,R^N) \times H_0^2(D,R^N)} = \int_D \Sigma_{i=1,\ldots,N} \, (\Delta W_i)^2 \, dx = \int_D ||\Delta W||^2 \, dx.$$

5.5. Existence of solution to the Shape Differential Equation (5.8)

The domain is bounded in R^N. For $k \ge 1$ we consider:

$$F^k = \{V \in C^0([0,1], C^k(\bar{D}, R^N) \cap H_0^1(D, R^N))\}$$

Given a domain $\Omega_0 \subset D$ we consider the family

$$\mathbf{O}_{D,\Omega_0}^k := \{\Omega \subset D, \ \partial\Omega \in C^1, \ \exists V \in F^k, \ s.t. \ \Omega = T_1(V)(\Omega_0)\},$$

The family $\mathbf{O}_{D,\Omega_0}^k$ is equipped with the courant metric d_k, it is a complete metric space. Let $\Omega \in \mathbf{O}_{D,\Omega_0}^k$, we consider $B^k(\Omega) = \{\Omega' \in \mathbf{O}_{D,\Omega_0}^k \ s.t. \ d_k(\Omega, \Omega') \le 1\}$. Then, for $k \ge 2$, $B^K(\Omega)$ is compact in $\mathbf{O}_{D,\Omega_0}^{k-1}$. Let us consider

$$\mathbf{F}_{D,\Omega_0}^k = \{(t \to \Omega_t) \in C^0([0,1], \mathbf{O}_{D,\Omega_0}^k)\}$$

and the *shape gradient mapping*

$$G \in C^0(\mathbf{O}_{D,\Omega_0}^1 \ \mathbf{E}'(D, R^N)),$$

where $\mathbf{E}'(D, R^N)$ is the linear space of vector Distributions on the open set D with compact support (and then with finite order). In practice that shape gradient mapping will range in some "negative Sobolev" space over D, say:

$$G \in C^0(\mathbf{O}_{D,\Omega_0}^1 \ H^{-s}(D, R^N) \cap \mathbf{E}'(D, R^N)).$$

We assume that $G(.)$ verifies the boundedness assumption (5.17). We apply the Leray fixed point theorem on the closed convex

$$K_M = \{V \in F^1 \ s.t. \ ||V(t)||_{C^1(\bar{D},R^N)} \le M\}$$

and the mapping

$$f: \ V \in K_M \to \Omega_t(V) \in \mathbf{K}_{D,\Omega_0}^k \to G(\Omega_t(V)) \in \mathbf{F}_{D,\Omega_0}^k$$

$$\to -\mathbf{A}^{-1}.G(\Omega_t(V)) \in F_{D,\Omega_0}^k.$$

When M is large enough, the mapping f ranges in K_M: $f(K_m) \subset K_M$.

Theorem 5.1. *The gradient mapping g being continuous and bounded (5.17), there exists a vector field $V \in F^1$ such that $f(v) = V$. In other words, there exists $V \in F^1$ solution to the shape differential equation (5.8).*

The proof follows from the equicontinuity of the family of mappings $(t \to T_t(V)\,)$, when V describes the convex set K_M and from the complete continuity (compactness) of the linear injection mapping $H^s(D) \subset H^{s-\epsilon}(D)$, $\epsilon > 0$ as D is a smooth domain (see [7] or [6] for a complete proof and several weaker results in that direction).

5.6. Asymptotic domains

Under the global continuity and boundedness assumption (5.17) we can successively apply the previous existence result on all time interval $[n, n+1]$ so that we derive the existence of a continuous solution V to equation (5.8) for any time $t > 0$. The asymptotic problem is then to characterise the situation in the limit as $t \to \infty$. From the decrease of the shape functional J whose G is the shape gradient under consideration we get, for any solution V^* of equation (5.8):

$$j(\Omega_t(V)) \leq J(\Omega_0) - \alpha \int_0^t \|V(s)\|^2 \, ds\,.$$

Thus, assuming $J(.) \geq 0$ (or more generally with a finite lower bound) we have

$$\int_0^\infty \|V^*(s)\|^2 \, ds \; < \; J(\Omega_0),$$

and hence,

$$V^* \in L^2(R_+, \; C^1(\bar{D}, R^N) \cup H_0^1(D, R^N)\,)\,.$$

From (5.8) we get

$$G^*(t) := G(\Omega_t(V^*)) \; \in \; L^2(0, \infty, H^{-1}(D, R^N))\,.$$

Thus there exist sequences $t_n \to \infty$ such that $V^*(t_n) \to 0$ and $G^*(t_n) \to 0$, in the respective topologies.

The main question is what is arriving to the domains Ω_{t_n}. From classical compactness of family of open subsets in D in the complementary Hausdorff topology, there exists an open set Ω_∞ in D such that $\Omega_{t_n} \to \Omega_\infty$ in the complementary Hausdorff topology. The main property for our purpose is the well-known following "compactivorous" stability:

$$\forall \psi \in \mathbf{D}(\Omega_\infty), \; \exists N_{K(\psi)}, \; s.t. \; n \geq N_{K(\psi)}, \; \text{implies } \psi \in \mathbf{D}(\Omega_{t_n})\,. \tag{5.20}$$

Here $K(\psi)$ designates the compact support of the function ψ. Let $\zeta_n = \chi_{\Omega_{t_n}}$ be the characteristic function of the set Ω_{t_n}, we have the weak-* convergence of ζ_n to some function $\lambda \in L^\infty(D)$, $0 \leq \lambda \leq 1$.

5.6.1. The asymptotic analysis for the Dirichlet problem.
Let us consider that the governing state equation of the shape functional J is the solution $y = y(\Omega) \in H_0^1(\Omega) \subset H_0^1(D)$, solution to $y = (-\Delta)^{-1}.F|_\Omega$, for given data $F \in L^2(D)$ (here we denote by $F|_\Omega$ the restriction of F to the open set Ω).

In that case we do have, with $y_n := y^0(\Omega_{t_n})$ (the extension by zero, element of $H^1_0(D)$):

$$(1 - \zeta_n)\, y_n = 0,$$

so that in the limit (as $\|y_n\|_{H^1_0(D)} \leq 1/\sqrt{\lambda_1(D)}\, \|F\|_{L^2(D)}$ we assume, after extraction of a new subsequence, that y_n weakly converges in $H^1_0(D)$, then strongly in $L^2(D)$, to some element $y_\infty \in H^1_0(D)$):

$$(1 - \lambda(x))\, y_\infty(x) = 0 \ a.e.x \in D.$$

Now, as $y_\infty \in H^1_0(D)$, we consider

$$\tilde{\Omega}_\infty := \{x \in D \ s.t. \ y_\infty(x) > 0\},$$

is a quasi open subset in D (open up to a zero capacity subset) and then we get

$$\lambda \geq \chi_{\tilde{\Omega}_\infty}. \tag{5.21}$$

From the definition of y_n we get:

$$\forall \phi \in \mathbf{D}(\Omega_{t_n}), \quad \int_D \left(\langle \nabla y_n, \nabla \psi \rangle - F\,\psi \right) dx = 0.$$

(This integral should be taken over the smooth open set Ω_{t_n}, but as ψ is compactly supported in Ω_{t_n} we can write it over the larger domain D.) From (5.20) this equality can be extended to any element $\psi \in \mathbf{D}(\Omega_\infty)$ as soon as $N \geq N_{K(\psi)}$. Then, in the limit we derive:

$$\forall \phi \in \mathbf{D}(\Omega_\infty), \quad \int_D \left(\langle \nabla y_\infty, \nabla \psi \rangle - F\,\psi \right) dx = 0. \tag{5.22}$$

Now we must pay attention to the fact that the open sets Ω_∞ and $\tilde{\Omega}_\infty$ are not known to have smooth boundaries (not even with N-dimensional zero measures) so we have to distinguish between the two main definitions for the Sobolev space H^1_0. Let us consider

$$H^1_0(\Omega_\infty) := \{\phi \in H^1_0(D) \ s.t. \ \phi(x) = 0 \ q.e.x \in D \setminus \Omega_\infty\}$$

and the smaller linear subspace:

$$\mathbf{H}^1_0(\Omega_\infty) = \text{closure of } \mathbf{D}(\Omega_\infty)$$

$$\text{in } H^1_0(D) = cl_{\{H^1_0(D)\}}(\mathbf{D}(\Omega_\infty)) \subset H^1_0(\Omega_\infty) \subset H^1_0(D).$$

A priori we do have $y_\infty \in H^1_0(\Omega_\infty)$ verifying the weak equation (5.22) for any test functions ψ in the *smaller* space $\mathbf{H}^1_0(\Omega_\infty)$. A sufficient condition in order to derive the equality of these two subspaces of $H^1_0(D)$ is a Wiener condition on the local capacity on the complementary of Ω_∞ see [41], [42], [43]. In dimension 2 these conditions are fulfilled, see [45] and the original 2D result in [44]. The idea is that the initial domain Ω_0 being smooth is such that its complementary in D has a finite number of connected components. (which we denote by $\#(D \setminus \Omega_0)$). Now, this number is lower semi continuous for the Hausdorff complementary convergence of open subset of D, that is $\#(D \setminus \Omega_\infty) \leq \liminf \#(D \setminus \Omega_{t_n}) = \#(D \setminus \Omega_0)$. In 2D the non connectivity of two closed subsets implies that there exist a linear segment reaching these sets and give a lower bound on the local relative capacity which enables us to conclude and we get the shape gradient asymptotic stability: As we have the following convergence in $H_0^1(D)$:

$$y_n \to y(\Omega_\infty)$$

$$p_n \to p(\Omega_\infty)$$

$$\langle G(\Omega_{t_n}), W \rangle = \int_{\Omega_{t_n}} \langle A'(W).\nabla y_n, \nabla p_n \rangle \; dx$$

$$+ \int_{\Omega_{t_n}} (\; \langle \nabla f, W \rangle \, p_n \; - \; \chi_E \,(y_n - Y_d)\, \langle \nabla y_n, W \rangle \;)dx \tag{5.23}$$

$$= \int_D \langle A'(W).\nabla y_n, \nabla p_n \rangle \; dx$$

$$+ \int_D (\; \langle \nabla f, W \rangle \, p_n \; - \; \chi_E \,(y_n - Y_d)\, \langle \nabla y_n, W \rangle \;)dx \tag{5.24}$$

$$\longrightarrow \int_D \langle A'(W).\nabla y_\infty, \nabla p_\infty \rangle \; dx$$

$$+ \int_D (\; \langle \nabla f, W \rangle \, p_\infty \; - \; \chi_E \,(y_\infty - Y_d)\, \langle \nabla y_\infty, W \rangle \;)dx. \tag{5.25}$$

As the elements are zero a.e. in $D \setminus \Omega_\infty$ we get:

Proposition 5.2. *The shape gradient $G(\Omega_{t_n})$ defined by (13.1) (weakly) converges as element of $(F^1)'$ to a distribution $G(\Omega_\infty) \in W_0^{1,\infty}(D, R^N)'$ characterised by:*

$$\forall W \in W_0^{1,\infty}(D, R^N), \; \langle G(\Omega_\infty), W \rangle = \int_{\Omega_\infty} \langle A'(W).\nabla y_\infty, \nabla p_\infty \rangle \; dx$$

$$+ \int_{\Omega_\infty} (\; \langle \nabla f, W \rangle \, p_\infty \; - \; \chi_E \,(y_\infty - Y_d)\, \langle \nabla y_\infty, W \rangle \;)dx. \tag{5.26}$$

5.7. Topological change in finite time

The previous asymptotic analysis when $t \to \infty$ can be brought back at time $s = 1$ by the following change of ("time" variable):

Let $s = 1 - (t+1)^{-1}$ so that $t = \frac{s}{1-s}$ and $\frac{\partial t}{\partial s} = \frac{1}{(1-s)^2}$. When t describes the line $[0 + \infty[$, the variable s describes the interval $[0, 1]$. We set

$$\tilde{V}(s) := V(\frac{s}{1-s}), \; \tilde{\Omega}_s := \Omega_{\frac{s}{1-s}}, \; \tilde{y}_s := y(\Omega_{\frac{s}{1-s}}), \ldots.$$

Let $\Phi(t, x)$ be a solution to

$$\forall t, \; t > 0 \;, \; \frac{\partial}{\partial t}\Phi + \langle \nabla_x \Phi, V \rangle = 0, \; \Phi(0,.) = \Phi_0(.).$$

We also set $\tilde{\Phi}(s, x) = \Phi(\frac{s}{1-s}, x)$ and this function solves the problem

$$\forall s \in [0, 1[, \; \frac{\partial}{\partial s}\tilde{\Phi}(s, x) + \langle \nabla_x \tilde{\Phi}(s, x), \frac{1}{(1-s)^2} \tilde{V}(s, x) \rangle = 0, \; \tilde{\Phi}(0,.) = \Phi_0(.).$$

358   J.-P. Zolésio

Consider

$$\mathbf{V}(s, x) := \frac{1}{(1-s)^2}\, \tilde{V}(s, x).$$

Then $\tilde{\Phi}$ solves the **V**-convection problem

$$\forall s \in [0, 1[, \quad \frac{\partial}{\partial s}\tilde{\Phi} + \langle \nabla_x \tilde{\Phi}(s, x), \mathbf{V} \rangle = 0, \quad \tilde{\Phi}(0, .) = \Phi_0(.). \tag{5.27}$$

If **V** was smooth enough on the finite interval $s \in [0, 1]$ we would have, as previously for V the convected unique solution to (5.27): $\tilde{\Phi}(s) = \Phi_0 o T_s(\mathbf{V})^{-1}$. But now the field **V** is non-smooth on the closed interval $[0, 1]$. Effectively from (5.17) we see that $||V(t)||$ is bounded so that

$$||\mathbf{V}(s, .)||_{H^2(D, R^N) \cap H_0^1(D, R^N)} = \mathbf{O}(\frac{1}{(1-s)^2}), \quad s \to 1.$$

Moreover we get:

$$\tilde{V}(s)/(1-s) \in L^2(0, 1, H^2(D, R^N) \cap H_0^1(D, R^N)),$$

so that we cannot conclude for $\mathbf{V}(s, .)$ to be in $L^p(0, 1, H^2(D, R^N) \cap H_0^1(D, R^N))$, for some $p \geq 1$, which would be necessary in order to derive existence and also uniqueness to the solution of the Hamilton-Jacobi-like equation (5.15) using the results derived in [33] (based on the use by L. Ambrosio [22] of Alberti rank one theorem). One way to get an existence result is to modify the choice of the speed vector field in the previous fixed point construction as follows:

5.7.1. Existence result for the Hamilton-Jacobi equation with possible topological changes. Let us consider a solution V^{**} to the following fixed point problem:

$$t \leq 0, \quad V^{**}(t) := -(t+1)^{-2}\, A^{-1}.G(\Omega_t(V^{**})).$$

We get

$$\int_0^\infty (t+1)^2\, ||V^{**}(t)||^2\, dt \leq \frac{1}{\alpha}\, J(\Omega_0).$$

Then by change of variable:

$$\int_0^1 ||V^{**}(\frac{s}{1-s})||^2\, \frac{1}{(1-s)^4}\, ds \leq \frac{1}{\alpha}\, J(\Omega_0).$$

We set

$$\bar{\mathbf{V}}(s, x) := \frac{1}{(1-s)^2}\, V^{**}\left(\frac{s}{1-s}\right) \in L^2(0, 1, L^2(0, 1, H^2(D, R^N) \cap H_0^1(D, R^N))).$$

Again with

$$\Phi(t, .) := \Phi_0 o T_t(V)^{-1},$$

Φ solves the convection problem in strong form over $[0, \infty[$

$$\Phi(0, .) = \Phi_0, \quad \frac{\partial}{\partial t}\Phi + \nabla\Phi.V(t) = 0.$$

Let

$$\Phi^*(s, x) := \Phi(\frac{s}{1-s}, x).$$

Then it also solves the convection problem in strong form for $0 \le s < 1$:

$$\Phi(0,.) = \Phi_0, \quad \frac{\partial}{\partial s}\Phi + \nabla\Phi.\bar{V}(s) = 0.$$

With the vector field $\bar{V} \in L^2(0,\tau,H)$ we get existence and uniqueness of solution on any interval $(0,\tau)$ hence this solution can be extended for larger values of s so that we obtain solution through the possible topological change.

6. Tubes associated to BV vector fields

Let $V \in \mathbf{E}^p$. Equipped with the graph norm $\mathbf{E}^p$ is a Banach space.

Proposition 6.1. *Let $1 \le p < \infty$, then $H_0^1(D)$ is a dense subspace in $\mathbf{E}^p(D)$.*

The proof is done in three steps.

Lemma 6.2. *For any $V \in \mathbf{E}^p(D)$ let V^0 designate the extension by zero outside of D. Then we have $\operatorname{div}V^0 = (\operatorname{div}V)^0$ and $V^0 \in \mathbf{E}^p := \{ V \in L^p(R^N, R^N), \ \operatorname{div}V \in L^p(R^N) \}$.*

Let b designate the oriented distance function to the bounded domain D. For given $h > 0$, a small enough parameter, consider the "cut oriented distance function" with support in the narrow band h, $b_h := \rho o b$ where the cutting function ρ is smooth, positive, with support in the interval $[-2h, +2h]$ and taking the value $\rho = 1$ on the subinterval $[-h, +h]$. We introduce the flow mapping $T_h := T_h(\nabla b_h)$ and consider the following mapping

$$\mathbf{T}_h : V \in \mathbf{E}^p(D) \longrightarrow V_h := \det(DT_h)\, DT_h\, V^0 oT_h.$$

Lemma 6.3. $\operatorname{div}(V_h) = \det DT_h\, \operatorname{div}(V^0)oT_h.$

It follows that $\operatorname{div} V_h = \det DT_h(\, (\operatorname{div} V)^0)oT_h \in L^2(R^N)$. Moreover, as $h > 0$, we have $V_h.n = 0$, so that

$$V_h \in \mathbf{E}^p(D).$$

We consider a mollifier $r_h \to \delta_0$ as $h \to 0$ with support of $r_h \subset B(0, h/2)$ so that the support verifies (for $h > 0$) $\operatorname{supt}\{\, r_h * (\, \mathbf{T}_h.V\,)\,\} \subset D$. Thus

$$V^h := r_h * (\, \mathbf{T}_h.V\,) \ \in \mathbf{D}(D, R^N) \subset H_0^1(D, R^N).$$

It is now immediate to verify that $V^h \to V$ strongly in $\mathbf{E}^p(D)$, which establishes the density result without any geometrical assumption on the domain D. Then we may restrict to the bounded domain D the results from [22]:

Theorem 6.4. *Let $V \in \mathbf{E}^{1,1} \cap L^1(0,\tau, BV(D, R^N))$. Assume that $(\operatorname{div} V)^+$ (resp. $(\operatorname{div} V)^-$) is in $L^1(0,\tau, L^\infty(D, R^N))$, then problem (7.5) (resp. (7.2)) has a unique solution ζ such that $(\zeta, V) \in \mathbf{T}_{\Omega_0}^1$.*

Then we see that some regularity on V implies that, for given V, the characteristic solution is unique. We denote it by ζ_V. The converse is false: for given ζ the set of V such that $(\zeta, V) \in \mathbf{T}_\Omega^{p,q}$ is a closed convex set that we denote by $\mathbf{K}(\zeta)$. For $1 < p, q < \infty$, In that convex set we can find a *minimal* element V_ζ (which minimizes the associated norms in $\mathbf{K}(\zeta)$.

6.1. Tube energy – *variational problem*

For any given positive constants $a > 0$, $\sigma \geq 0$, $\mu \geq 0$ and $\nu \geq 0$ we shall consider the minimization associated to the following functionals:

$$J^a(V) = 1/2 \int_I \int_D (a + \xi_V) \left(|V(t,x)|^2 + (\operatorname{div} V(t,x))^2 \right) dx dt, \qquad (6.1)$$

$$J^a_{\sigma,\mu,\nu}(V) = J^a(V) + \sigma \int_0^\tau \|\nabla(\xi_V(t))\|_{M^1(D)} dt$$

$$+ \mu\Theta(V, \Omega_0) + \nu \int_0^\tau \int_D DV..DV \, dx dt. \qquad (6.2)$$

We shall consider the three situations associated with $\sigma + \mu + \nu > 0$ and $\sigma\mu\nu = 0$. When ν is zero the terms σ and μ will play a surface tension role at the dynamical interface while the case $\sigma + \mu = 0$ should be considered as a mathematical regularisation, as in the non usual variational interpretation developed in the previous section $\nu > 0$ does not lead to the usual viscosity term (i.e., does not lead to the Navier-Stokes equations)

Theorem 6.5. *Assuming $V \in \mathbf{E}^2$, $a > 0$, $\sigma.\eta.\nu > 0$, there exists $V \in \mathbf{E}^2$ such that, $\forall W \in \mathbf{E}^2$:*

$$J^a_{\sigma,\eta,\nu}(V) \leq J^a_{\sigma,\eta,\nu}(W).$$

7. Saddle points

Let us consider the following more general situation:

$$\phi(0) = \phi_0, \quad \frac{\partial}{\partial t}\phi + \nabla\phi.V = f, \qquad (7.1)$$

$$\psi(0) = \psi_0, \quad \frac{\partial}{\partial t}\psi + \operatorname{div}(\psi V) = g. \qquad (7.2)$$

Let two real numbers (p, q) be given, with $1 < p \leq \infty$, $1 \leq q < \infty$, and the linear space for speed vector fields:

$$E^{p,q} = \{ V \in L^p(0, \tau, L^q(D, R^N)) \text{ s.t. } \operatorname{div} V \in L^p(0, \tau, L^q(D)) \},$$

$$\mathbf{E}^{p,q} = \{ V \in E^{p,q}, \text{ s.t. } V.n = 0 \text{ in } W^{-1,1}(\partial D) \}. \qquad (7.3)$$

The null condition on the normal component of the vector field at the boundary can be weakly written as

$$\forall\phi \in L^2(I, C^1(\bar{D})), \quad \int_I \int_D (\operatorname{div} V \, \phi + \nabla\phi.V) \, dt dx = 0. \qquad (7.4)$$

Proposition 7.1. ([13]) *Assume* $V \in \mathbf{E}^{2,2}$. *If* $(\operatorname{div} V)^+ \in L^1(0, \tau, L^\infty(D))$, *problem* (7.1) *has solutions*

$$\phi \in L^\infty(0, \tau, L^2(D)) \cap H^1(0, \tau, H^{-1}(D)) \subset C^0([0, \tau], H^{-1/2}(D)).$$

If $(\operatorname{div} V)^- \in L^1(0, \tau, L^\infty(D))$ *problem* (7.2) *has solutions*

$$\psi \in L^\infty(0, \tau, L^2(D)) \cap H^1(0, \tau, H^{-1}(D)) \subset C^0([0, \tau], H^{-1/2}(D)).$$

The first idea would be to consider $\operatorname{div} V \in L^1(0, \tau, L^\infty(D))$. Then both problems have solutions. They are, formally, adjoint problems of each other, and we could be tempted to conclude uniqueness to both problems. That argument does not apply as one of the two solutions ϕ or ψ should be smooth in order to be "put in duality". Then under previous poor regularity on V we will not get existence nor uniqueness for shape convection problem (7.5):

$$\zeta(0) = \zeta_{\Omega_0}, \quad \frac{\partial}{\partial t} \zeta + \nabla \zeta . V = 0, \quad \zeta = \zeta^2 . \tag{7.5}$$

Functional setting: To give sense to the product $\nabla \zeta . V$ in (7.5), we write it as

$$\nabla \zeta . V = \operatorname{div}(\zeta V) - \zeta \operatorname{div} V.$$

Then, as soon as $\zeta \in L^\infty((0, \tau) \times D)$, that term makes sense in $L^1(0, \tau, W^{-1,1}(D))$ when V and its divergence $\operatorname{div} V$ are in $L^1(0, \tau, L^1(D))$. we consider a vector fields V in $\mathbf{E}^{2,2}$ and any function $G \in L^\infty(D)$ verifying $G \geq \alpha > 0$. We consider the Lagrangian expression for the functional

$$g(V) = \operatorname{Inf}_{\zeta \in \mathbf{U}_V} \operatorname{Sup}_{\phi \in \mathbf{H}_V} \mathbf{L}_V(\zeta, \phi),$$

with

$$\mathbf{L}_V(\zeta, \phi) = \int_0^\tau \int_D \left\{ 1/2 \zeta^2 \, G \; + \zeta \left(\frac{\partial}{\partial t} \phi + \operatorname{div}(\phi V) \right) \right\} dx dt - \int_{\Omega_0} \phi(0) dx ,$$

$$\mathbf{H}_V = \left\{ \phi \in L^2(0, \tau, L^2(D)) \; s.t. \; \frac{\partial}{\partial t} \phi + \operatorname{div}(\phi V) \in L^2(I \times D) , \; \phi(\tau) = 0 \right\} ,$$

$$\mathbf{U}_V = \left\{ \zeta \in L^2(I \times D), \; s.t. \; \frac{\partial}{\partial t} \zeta + \nabla \zeta V) \in L^2(I \times D) , \; \zeta(0) = \chi_{\Omega_0} \right\} .$$

Notice that the elements of $\mathbf{H}_V$ are continuous ($\phi \in C^0([0, \tau], W^{-1/2,1}(D))$ so that $\phi(\tau)$ makes sense.

The Lagrangian $\mathbf{L}_V$ is concave-convex on $\mathbf{U}_V \times \mathbf{H}_V$. $L^2(I \times D) \times \mathbf{E}^{2,2}$.

Saddle points (ξ, λ) are solution to the system composed of equation (7.1) (with $\Phi_0 = \chi_{\Omega_0}$, $f = 0$) and the following backward "adjoint equation"

$$\frac{\partial}{\partial t} \lambda + \operatorname{div}(\lambda V) = -\xi_V \, G , \quad \lambda(\tau) = 0. \tag{7.6}$$

The converse is true when we have an extra density condition on V and $\operatorname{div} V$:

$$\text{Assumption on V}: \quad \{ \phi \in C^\infty(I \times D) \cap \mathbf{E}^{2,2} \} \text{ is dense in } \mathbf{H}_V. \tag{7.7}$$

The weakly coupled system (7.1), (7.6) possesses solutions when

$$(\operatorname{div} V)^+ \in L^1(I, L^\infty(D)).$$

We derive the following uniqueness results for the convection problem (7.1):

Proposition 7.2. *Assume* $V \in \mathbf{E}^{2,2}$, *verifying* (7.7) *and* $(\operatorname{div} V)^+ \in L^1(I, L^\infty(D))$.
Then, with $f = 0$ *and* $\Phi_0 = \chi_{\Omega_0}$ *(convection problem), or more generally* $\Phi_0 \in L^\infty(D)$, *the problem* (7.1) *possesses a unique solution* ζ_V *verifying*

$$0 \leq \zeta_V \leq 1 \quad a.e.(t,x) \in I \times D$$

or (in the more general setting)

$$\text{Infess } \Phi_0 \leq \zeta_V \leq \text{Supess } \Phi_0.$$

We have the monotonicity: $\Omega_0^1 \subset \Omega_0^2$ *(or, in the more general setting* $\Phi_0^1 \leq \Phi_0^2$*) implies* $\zeta_V^1 \leq \zeta_V^2$.

Proof. From the strong assumption (7.7) the set of saddle points is not empty and is completely characterized by the system (7.1)–(7.6). Let us denote by $\mathbf{S}_V$ the set of saddle points. We know that it can be written as $\mathbf{S}_V = A_V \times B_V$, which means that if (ζ^i, λ^i), $i = 1,2$ are saddle points then ζ^1, λ^2 and ζ^2, λ^1 are also saddle points. We infer that equation (7.6) with right-hand side $G\,\zeta^i$ has solutions and we derive uniqueness of ζ_V (from the fact that $G > 0$), being the single element in A_V. From uniqueness we know that $0 \leq \zeta \leq 1$.

7.1. Derivative with respect to the speed field V

Functionals J in form of min max have a well-known Gateaux derivative with respect to V. Now, it is important to notice that in the present saddle point formulation the linear vector spaces depend on the parameter V.

With $\operatorname{div} \in L^1(I, L^\infty(D))$ the Ambrosio results apply as we have $H^1(D) \subset W^{1,1}(D) \subset BV(D)$.

Applying the results from [2] and [18] for differentiation under uniqueness of the saddle point, and assuming

$$V \in B := L^2(0, \tau, H_0^1(D, R^N)), \quad \mathbf{L}_V^\nu := \mathbf{L}_V + \nu/2 \int_0^\tau \int_D DV..DV\,dx\,dt,$$

we have

$$J'(V, W) = \langle (a + \zeta)V - \nabla(\zeta \operatorname{div} V) - \nu \Delta V - \lambda \nabla \zeta, W \rangle_{F' \times F}.$$

We shall give now a sense to the term $\lambda \nabla \zeta$, using the concept of a *transverse field* Z which has been introduced in [12] and further developed in [11], [13], [25], [5], [17], [26]. Indeed we have

$$\lambda \nabla \zeta = \nabla(\zeta \lambda) - \zeta \nabla \lambda.$$

We give a sense to that last term $\zeta \nabla \lambda$ through the *transverse field* problem and its adjoint.

7.2. Transverse derivative

For two given vector fields V and W the *transverse* field Z is the solution to the Lie bracket evolution

$$H_V.Z = W, \ H_V.Z := \frac{\partial}{\partial t}Z + [Z, V], \quad [Z, V] := DZ.V - DV.Z. \tag{7.8}$$

For *avoiding some technicalities* we assume now (as it will be the case in the first example associated with modelling of arteris) that V and W are free divergence vector fields: $\operatorname{div} V = \operatorname{div} W = 0$. The adjoint operator is then:

$$H_V^*.\Lambda := -\frac{\partial}{\partial t}\Lambda - D\Lambda.V - D^*V.\Lambda. \tag{7.9}$$

Lemma 7.3.

$$H_V.(\zeta\,\vec{e}) = \left(\frac{\partial}{\partial t}\zeta + \nabla\zeta.V\right)\vec{e} + \zeta\,H_V.\vec{e} \tag{7.10}$$

$$H_V^*(\zeta\,\vec{f}) = -\left(\frac{\partial}{\partial t}\zeta + \nabla\zeta.V\right)\vec{f} - \zeta\,H_V^*.\vec{f}. \tag{7.11}$$

We see that if (ζ, V) is a tube then the two previous expressions simplifies as the first term vanishes from 7.5. Specifically we have

$$H_V^*.(\zeta\,\nabla\lambda) = -\zeta\left(\frac{\partial}{\partial t}\nabla\lambda + D\nabla\lambda.V + D^*V.\nabla\lambda\right) = -\zeta\nabla\left(\frac{\partial}{\partial t}\lambda + \nabla\lambda.V\right).$$

Then, if we set the "scalar" operator h_V by:

$$h_V.\psi := \frac{\partial}{\partial t}\psi + \nabla\psi.V$$

we get, (ζ, V) being any tube:

$$H_V^*(\zeta\,\nabla\lambda) = -\zeta\,\nabla(\,h_V.\lambda). \tag{7.12}$$

Notice that here, as $\operatorname{div} V = 0$, the adjoint operator verifies $h_V^* = -h_V$. By "reversing" the time, that operator h_V is then self-adjoint. Nevertheless, it is *not* maximally defined in the admissible setting so that it is not possible to use same classical abstract setting of evolution linear system in order to solve (7.1) and/or (7.2).

8. Shape tube metric for smooth sets

Assuming the domain Ω_0 to be smooth, i.e., its boundary being a C^k manifold, for $k \geq 1$, we consider the subfamily

$$\mathbf{O}_{\Omega_0}^k = \{\Omega \in \mathbf{O}_{\Omega_0}, \Omega = \zeta(1), \ \zeta \text{ having lateral boundary } \Sigma \text{ piecewise } C^k\}.$$

Let Ω_1, Ω_2 be given in $\mathbf{O}_{\Omega_0}^k$, we consider

$$T_k(\Omega_1, \Omega_2) = \{\ \zeta \in T(\Omega_1, \Omega_2) \ s.t. \Sigma \text{ is piecewise } C^k\ \}$$

and introduce the metric

$$\delta(\Omega_1, \Omega_2) := \operatorname{Inf}_{\{\zeta \in T_k(\Omega_1, \Omega_2)\}} \ ||v||_{L^1(I, L^1(\Gamma_t))}.$$

In general the infimum is not a minimum. Notice also that, as for any $F \in C^0(\bar{D})$:

$$\int_\Sigma f \, d\Sigma = \int_0^1 dt \int_{\Gamma_t} f(t,x) \sqrt{1+v^2} \, d\Gamma_t(x),$$

and we have

$$\delta(\Omega_1, \Omega_2) = \mathrm{Inf}_{\{\zeta \in T_k(\Omega_1,\Omega_2)\}} \int_\Sigma \frac{|v|}{\sqrt{1+v^2}} d\Sigma$$

$$\leq \mathrm{Min} \left\{ \mathrm{Inf}_{\{\zeta \in T_k(\Omega_1,\Omega_2)\}} \int_\Sigma |v| \, d\Sigma \ , \quad \mathrm{Inf}_{\{\zeta \in T_k(\Omega_1,\Omega_2)\}} \int_\Sigma d\Sigma \right\}.$$

It can be verified that the two majorant terms to be a metric. The first one does not verify the triangle axiom and the second does not verify the first metric axiom (as it cannot be zero with $\tau = 1$).

Proposition 8.1. *Let $k \geq 1$, then, equipped with δ, $\mathbf{O}^k_{\Omega_0}$ is a metric space.*

Obviously δ is non-negative. Assume that $\delta(\Omega_1, \Omega_2) = 0$, then $\forall t \in I$, $v(t,x) = 0$ on Γ_t; then the time space normal vector field satisfies

$$\nu(t,x) = \frac{1}{\sqrt{1+v(t,x)^2}} (-v(t,x), \vec{n}_t(t,x)) = (0, \vec{n}_t(x)).$$

When the tube is a cylinder and the domain Ω_t does not depend on t, then $\Omega_1 = \Omega_2$. The symmetry of δ is immediate by taking the backward tube: $\zeta'(t,.) := \zeta(1-t,.)$.

Concerning the triangle inequality, assume three domains Ω_i in D and tubes ζ^1 connecting Ω_1, Ω_2, ζ^2 connecting Ω_2, Ω_3 with both of them realizing the infimum in the δ definition up to some given $\epsilon > 0$.

Let us consider the following new piecewise C^1-tube defined (through its characteristic function ζ) as follows:

$$\zeta(t,x) = \zeta^1(2t, x), \qquad 0 \leq t \leq 1/2$$
$$\zeta(t,x) = \zeta^2(2t - 1, x), \qquad 1/2 \leq t \leq 1.$$

It can be easily verified that its normal speed v is given by

$$v(t,x) = 2v^1(2t, x), \ 0 \leq t \leq 1/2$$
$$v(t,x) = 2v^2(2t - 1, x), \ 1/2 \leq t \leq 1.$$

Now by construction ζ connects Ω_1 to Ω_3, $\in \bar{\mathbf{T}}_k(\Omega_1, \Omega_3)$, hence we get

$$\delta(\Omega_1, \Omega_3) \leq \int_0^1 \| v(t) \|_{L^1(\Gamma_t)} \, dt = \int_0^{1/2} \| v(t) \|_{L^1(\Gamma_t)} \, dt + \int_{1/2}^1 \| v(t) \|_{L^1(\Gamma_t)} \, dt$$

$$= \int_0^{1/2} \| 2v^1(2t) \|_{L^1(\Gamma^1_{2t})} \, dt + \int_{1/2}^1 \| 2v^2(2t - 1) \|_{L^1(\Gamma_{2t-1})} \, dt$$

$$= \int_0^1 2 \| v^1(r) \|_{L^1(\Gamma^1_r)} 1/2 \, dr + \int_0^1 2 \| v^2(u) \|_{L^1(\Gamma_u)} 1/2 \, du,$$

as v^1 (resp. v^2) realizes the infimum (in the definition of $\delta(\Omega_1, \Omega_2)$ (resp. $\delta(\Omega_2, \Omega_3)$) up to $\epsilon > 0$. Then $\forall \epsilon > 0$ we get

$$\delta(\Omega_1, \Omega_3) \leq \delta(\Omega_1, \Omega_2) + \delta(\Omega_2, \Omega_3) + 2\,\epsilon.$$

9. Geodesic for the metric $\bar{\delta}$

9.1. Derivative of the metric terms

9.1.1. Admissible variations for the geodesic: transverse fields $\mathbf{Z}$ preserving the extremities of the tube.
Let us assume that a tube $(\zeta, V) \in \mathbf{C}_{\Omega_1, \Omega_2}$ is a minimizer for $\delta_2(\Omega_1, \Omega_2)$. Let $\mathbf{Z}(s; t, x) \in R^{N+1}$ be a "horizontal vector field" $\mathbf{Z}(s; t, x) = (0, Z(s; t, x))$, where $Z(s; t, x) \in R^N$. The variable s will be the (tube) perturbation parameter while the time t is an independent parameter in the horizontal (transverse) flow mapping $T_s(Z(t))$ (here $Z(t) = Z(s; t, x) := Z(t)(s, x)$). The R^{N+1} flow mapping is

$$T_s(\mathbf{Z})(t, x) = (t, x) + \left(0, \int_0^s Z(\sigma; t, T_\sigma(Z(t))(x)\,d\sigma\right),$$

that is $T_s(\mathbf{Z})(t, x) = (t, T_s(Z(t))(x))$. We are looking for necessary optimality conditions solved by the vector field V (or at least by its normal component $v(t, .) = \langle V(t, .), n_t(.)\rangle$ on the lateral boundary. Consider any Z such that $Z(s; 0, x) = Z(s; 1, x) = 0$. Then if Q is a tube which connect the two sets Ω_1 and Ω_2, as $T_s(, Z(0)) = T_s(Z(1)) = 0$, the tubes

$$Q^s := T_s(\mathbf{Z})(Q) \quad \text{connects the two sets} \quad \Omega_1, \ \Omega_2.$$

Let v^s be the normal speed of Σ^s. As usual

$$v^s o T_s(\mathbf{Z}) = ||\operatorname{cof}(D_{t,x} T_s(\mathbf{Z}).\nu||_{R^{N+1}}^{-1} \operatorname{cof}(D_{t,x} T_s(\mathbf{Z}).\nu,$$

where the cofactor is given by $cof A := det A\,(A^*)^{-1}$. From that expression we get the explicit representation for

$$v^s o T_s = \frac{1}{\sqrt{1 + (v^s o T_s)^2}}\,(-v o T_s,\ \vec{n}_t^s).$$

We will be interested in the expression for $v^s o T_s$ and its (material) derivative

$$\dot{v}(t, s) := \frac{d}{ds}(v^s o T_s(\mathbf{Z}))|_{s=0} \quad \text{on } \Gamma_t.$$

Then the *shape boundary derivative* is implicitly given through Z as:

$$v(t)'_{\Gamma_t} := \dot{v} - \nabla_{\Gamma_t} v(t).Z(t)_{\Gamma_t}$$

(see the books [9], [24], [10], [26]). We shall prove the

Theorem 9.1.

$$v(t)\,(v(t)^s)'_{\Gamma_t} = v(t)\,\frac{\partial}{\partial t}(\langle Z(0, t, .), n_t(.)\rangle\,). \tag{9.1}$$

9.2. Optimality condition for the vector field V

Let $p \geq 1$, and consider that the tube Q_V minimizes the term

$$J := \int_0^1 \left(\int_{\Gamma_t(V)} |\langle V(t), n_t \rangle| \, d\Gamma_t \right)^p dt\,.$$

Then, considering a "transverse horizontal" field $\mathbf{Z} = (0, Z(s; t, x))$, we get $j(0) \leq j(s)$, with

$$j(s) := \int_0^1 \left(\int_{T_s(\mathbf{Z})(\Gamma_t(V))} |v^s| \, d\Gamma_t^s \right)^p dt, \quad J = j(0),$$

where of course $\Gamma_t^s = T_s(\mathbf{Z})(\Gamma_t)$, $\Sigma^s = T_s(\mathbf{Z})(\Sigma) = \cup_{0<t<1} \{t\} \times \Gamma_t^s$ and v^s is the normal field on Σ^s in the form $v^s = \frac{1}{\sqrt{1+(v^s)^2}}(-v^s, n_t^s)$ where n_t^s is the N-dimensional (horizontal) normal field to Γ_t^s. With

$$a(t) := p \left(\int_{\Gamma_t(V)} |\langle V(t), n_t \rangle| \, d\Gamma_t \right)^{p-1},$$

we have:

$$j'_+(0) = \int_0^1 a(t) \left\{ \int_{\Gamma_t} \operatorname{sgn}(v)\, v(t)'_{\Gamma_t}(Z) \, d\Gamma_t \right\} dt$$

$$+ \int_0^1 a(t) \left\{ \int_{\Gamma_t \cap \{v(t)^{-1}(0)\}} |v(t)'_{\Gamma_t}(Z)| \, d\Gamma_t \right\} dt.$$

Hence from Theorem 9.1 we obtain:

Theorem 9.2. *Let ζ be a minimizer of $d(\Omega_1, \Omega_2)$. Then for any smooth vector field Z such that $Z(0,.) = Z(1,.) = 0$, we have*

$$\int_0^1 a(t) \left\{ \int_{\Gamma_t} \operatorname{sgn} v(t) \, \frac{\partial}{\partial t}(\langle Z(t,.), n_t(.) \rangle) \, d\Gamma_t \right.$$

$$\left. + \int_{\Gamma_t \cap \{v(t)^{-1}(0)\}} |v(t)'_{\Gamma_t}(Z)| \, d\Gamma_t \right\} dt \geq 0\,.$$

Formal calculus (for $p = 1$) would lead to

$$j'(0) = -\int_0^1 \left\{ \int_{\Gamma_t} [\, \frac{\partial}{\partial t} \operatorname{sgn} v(t)\, (\langle Z(0,t,.), n_t(.) \rangle) \right.$$

$$+ H(t) \operatorname{sgn} v(t)\, v(t)\, \langle Z(0,t,.), n_t \rangle\,] d\Gamma_t \left. \right\} dt$$

$$+ \int_0^1 \left\{ \int_{\Gamma_t \cap \{v(t)^{-1}(0)\}} |v(t)'_{\Gamma_t}(Z)| \, d\Gamma_t \right\} dt$$

(as we already assumed that Z is chosen such that $\frac{\partial}{\partial n_t}(\langle Z(t), n_t \rangle) = 0$ on Γ_t (as well as the same for $v(t) = \langle V(t), n_t \rangle$). As $\operatorname{sgn} v(t)\, v(t) = |v(t)|$, the necessary condition leads to the following geodesic condition:

Proposition 9.3. *Let $(\zeta, V) \in \mathbf{T}(\Omega_1, \Omega_2)$ be a smooth minimizer of $d(\Omega_1, \Omega_2)$. Then the time-space normal field ν verifies*

$$\exists\, c(t), \ s.t. \ \forall t, \ \ 0 < t < 1, \ \ -\frac{\partial}{\partial t}(\operatorname{sgn} v(t)op_t)$$

$$+ (H(t)\,|v(t)|\,)op_t = c(t)\,H(t)op_t \ \ on \ \Gamma_t. \tag{9.2}$$

Of course the derivative $\frac{\partial}{\partial t}(\operatorname{sgn} v(t)op_t)$ is a problem, even assuming the boundaries to be smooth. The idea is to generalize definition of the metric, the easiest case being to replace the L^1 norm by the L^2 norm such that

$$d_2(\Omega_1, \Omega_2) = \inf \int_0^1 \int_{\Gamma_t} 1/2 \, v(t, x)^2 \, d\Gamma_t(x) \, dt.$$

Then the same calculus would lead to a nice necessary condition (where $\operatorname{sgn} v$ is replaced by $v(t)$). But now the theoretical basement of the metric partially vanishes; d_2 remains a metric but is not a complete one.

9.3. Explicit expression for $v(t)'_{\Gamma_t}$ (proof of Theorem 9.1)

The direct calculation of that representation is rather complicate due to the term $\operatorname{cof} D_{t,x} T_s(\mathbf{Z})$ (mainly because of the time derivatives in $D_{t,x} \ldots$). We proceed in a more intrinsic way: making the calculus of the derivative

$$\frac{d}{ds} \int_{\Sigma^s} F(s, t, x) \, d\Sigma(t, x)$$

by two different techniques: the $N+1$ boundary integral derivative and the N-dimensional one. From the Proposition 9.6 and Theorem 9.5 below, we have:

$$\frac{d}{ds}\left(\int_{\Sigma^s} F(s, t, x)\, d\Sigma(t, x)\right)_{s=0} = \int_0^1 \int_{\Gamma_t} \left\{ \frac{1}{\sqrt{1+v^2}}(-v\partial_t F + \partial_{n_t} F) \right. \tag{9.3}$$

$$+ \left[-\frac{1}{\sqrt{1+v^2}^3}(\langle \partial_t V(t), n_t\rangle - \langle \nabla_{\Gamma_t} v(t), V_{\Gamma_t}\rangle) + \frac{H(t)}{\sqrt{1+v^2}} \right] F \left. \right\} \langle Z(t), n_t\rangle d\Gamma_t \, dt.$$

On the other hand we may consider

$$\frac{d}{ds}\left(\int_{\Sigma^s} F(s, t, x)\, d\Sigma(t, x)\right)_{s=0}$$

$$= \int_0^1 \left(\frac{d}{ds} \int_{T_s(Z(t))(\Gamma_t)} F(s, t, x)\sqrt{1+(v^s)^2}\, d\Gamma_t^s\right)_{s=0} dt$$

$$= \int_0^1 \int_{\Gamma_t} \left\{ (F(s, t, x)\sqrt{1+(v^s)^2}\,)'_{\Gamma_t} \right.$$

$$\left. + F(0, t, x)\sqrt{1+v(t, x)^2}\,t)\,H(t)\,\langle Z(t), n_t\rangle \right\} d\Gamma_t(x) dt,$$

where $H(t, x)$ is the mean curvature of the boundary Γ_t, $H(t) = \operatorname{div}_{\Gamma_t}(n_t)$. Also we have:

$$(F(s, t, x)\sqrt{1+(v^s)^2}\,)'_{\Gamma_t} = F(s, t, x)'_{\Gamma_t}\sqrt{1+v^2} + F(s, t, x)\,(\sqrt{1+(v^s)^2}\,)'_{\Gamma_t}.$$

Let us consider specific functions F such that F is independent of the perturbation parameter s and $F(t,x) = F(t, op_{\Gamma_t}(x))$ in a neighborhood of the unperturbed lateral boundary Σ, where p_t is the horizontal N-dimensional projection mapping onto the boundary Γ_t. Then we have $\frac{\partial}{\partial n_t} F = 0$ and

$$F(t)'_{\Gamma_t} = \frac{\partial}{\partial s} F(t) + \frac{\partial}{\partial n_t} F(t) \langle Z(t), n_t \rangle = \frac{\partial}{\partial s} F(t) = 0.$$

In these expressions $V(t,x) \in R^N$ is any vector field building the tube Q. We choose

$$V(t,x) = v(t) op_{\Gamma_t} \nabla b_{\Omega_t},$$

so that on Γ_t we have

$$\left\langle \frac{\partial}{\partial t} V(t), n_t \right\rangle \frac{\partial}{\partial t} (v(t) op_{\Gamma_t}) = 0, \text{ and } V(t)_{\Gamma_t} = 0.$$

Then

$$\left(\frac{d}{ds} \int_{\Sigma^s} F(s) d\Sigma \right)_{s=0}$$

$$= \int_0^1 \int_{\Gamma_t} \left\{ -\frac{v(t)}{\sqrt{1+v(t)^2}} \frac{\partial}{\partial t} F(t) \left[-\frac{1}{\sqrt{1+v^2}^3} \partial_t(vop) + \frac{H(t)}{\sqrt{1+v^2}} \right] F(0,t) \right\}$$

$$\times \langle Z(t), n_t \rangle d\Gamma_t dt$$

$$= \int_0^1 \int_{\Gamma_t} \left\{ F(0,t)(\sqrt{1+(v(t)^s)^2})'_{\Gamma_t} \right.$$

$$\left. + F(0,t)\sqrt{1+v(t)^2}) H(t) \langle Z(t), n_t \rangle \right\} d\Gamma_t(x) dt.$$

We have

$$(\sqrt{1+(v(t)^s)^2})'_{\Gamma_t} = \frac{v(t)}{\sqrt{1+v(t)^2}} v(t)'_{\Gamma_t}.$$

For general choices of F we have:

$$\int_0^1 \int_{\Gamma_t} F(0,t) (\sqrt{1+(v(t)^s)^2})'_{\Gamma_t} d\Gamma_t$$

$$= \int_0^1 \int_{\Gamma_t} \left\{ -\left(\frac{v(t)}{\sqrt{1+v(t)^2}} \langle z(t), n_t \rangle \right) \frac{\partial}{\partial t} F(t) \right.$$

$$\left. - F(0,t)\sqrt{1+v(t)^2}) H(t) \langle Z(t), n_t \rangle \right\} d\Gamma_t(x) dt$$

$$+ \int_0^1 \int_{\Gamma_t} \left[-\frac{1}{\sqrt{1+v^2}^3} \partial_t(vop) + \frac{H(t)}{\sqrt{1+v^2}} \right] F(0,t) \langle Z(t), n_t \rangle d\Gamma_t dt.$$

Now we have (F is independent on the perturbation variable s but not on t)

$$F'_{\Gamma_t} = \frac{\partial}{\partial t} F + \frac{\partial}{\partial n_t} F \langle V(t), n_t \rangle,$$

so that we get the following formula (integration by parts on Σ):

$$\int_0^1 \int_{\Gamma_t} \partial_t F\, G\, d\Gamma_t\, dt = -\int_0^1 \int_{\Gamma_t} F\, \partial_t G\, d\Gamma_t dt - \int_0^1 \int_{\Gamma_t} H(t) F(t) G(t) \langle V(t), n_t \rangle\, d\Gamma_t dt$$

$$+ \int_{\Gamma_1} (FG)(1) d\Gamma_1 - \int_{\Gamma_0} (FG)(0) d\Gamma_0$$

$$+ \int_0^1 \int_{\Gamma_t} F(t) \frac{\partial}{\partial n_t} G(t) \langle V(t), n_t \rangle d\Gamma_t dt\,.$$

So that, with $G = -\frac{v}{\sqrt{1+v^2}} \langle Z(t), n_t \rangle$:

$$\int_0^1 \int_{\Gamma_t} G(t) \frac{\partial}{\partial t} F(t)\, d\Gamma_t dt = -\int_0^1 \int_{\Gamma_t} \frac{\partial}{\partial t} \left\{ \sqrt{1+v(t)^2}^{\,-1} v(t)\, \langle Z(t), n_t \rangle \right\} F(t) d\Gamma_t dt$$

$$- \int_0^1 \int_{\Gamma_t} H(t)\, \sqrt{1+v(t)^2}\, v(t)\, F(t)\, \langle Z(t), n_t \rangle d\Gamma_t dt\,.$$

Moreover

$$\int_0^1 \int_{\Gamma_t} F(0,t)\, (\sqrt{1+(v(t)^s)^2}\,)'_{\Gamma_t}\, d\Gamma_t$$

$$= -\int_0^1 \int_{\Gamma_t} F(0,t) \sqrt{1+v(t)^2}\,)\, H(t)\, \langle Z(t), n_t \rangle \}\, d\Gamma_t(x) dt$$

$$+ \int_0^1 \int_{\Gamma_t} \frac{\partial}{\partial t} \left\{ \frac{v(t)}{\sqrt{1+v(t)^2}}\, \langle Z(t), n_t \rangle \right\} F(t) d\Gamma_t dt$$

$$+ \int_0^1 \int_{\Gamma_t} H(t)\, \frac{v(t)^2}{\sqrt{1+v(t)^2}}\, F(t)\, \langle Z(t), n_t \rangle d\Gamma_t dt$$

$$+ \int_0^1 \int_{\Gamma_t} \left[-\frac{1}{\sqrt{1+v^2}^{\,3}} \partial_t(vop) + \frac{H(t)}{\sqrt{1+v^2}} \right] F(0,t) \langle Z(t), n_t \rangle d\Gamma_t dt$$

$$+ \int_0^1 \int_{\Gamma_t} \frac{\partial}{\partial n_t} \left(\frac{v(t)}{\sqrt{1+v(t)^2}} \langle Z(t), n_t \rangle \right)\, v(t)\, F(0,t) d\Gamma_t dt\,.$$

The last term is zero as we have chosen $\frac{\partial}{\partial n_t} v(t) = \frac{\partial}{\partial n_t} z(t) = 0$. That is

$$(\sqrt{1+(v(t)^s)^2}\,)'_{\Gamma_t}$$

$$= \left\{ -\sqrt{1+v(t)^2}\,)\, H(t) + H(t)\, \frac{v(t)^2}{\sqrt{1+v(t)^2}} \right.$$

$$+ \left. \left[-\frac{1}{\sqrt{1+v^2}^{\,3}} \partial_t(vop) + \frac{H(t)}{\sqrt{1+v^2}} \right] \right\} \langle Z(t), n_t \rangle + \frac{\partial}{\partial t} \left\{ \frac{v(t)}{\sqrt{1+v(t)^2}} \langle Z(t), n_t \rangle \right\}$$

$$= -\frac{1}{\sqrt{1+v^2}^{\,3}} \partial_t(vop) \langle Z(t), n_t \rangle + \frac{\partial}{\partial t} \left\{ \frac{v(t)}{\sqrt{1+v(t)^2}} \langle Z(t), n_t \rangle \right\}\,.$$

We obtain

$$\left(\sqrt{1 + (v(t)^s)^2}\right)'_{\Gamma_t} = \frac{v(t)}{\sqrt{1 + v(t)^2}} \; \frac{\partial}{\partial t}\left(\langle Z(t), n_t \rangle\right). \tag{9.4}$$

Hence Theorem 9.1 is proved.

We turn now to the proof of Proposition 9.6 below that we used at 9.3, in the beginning of that section: we have to compute in Theorem 9.5 the time-space mean curvature $\mathbf{H}$ of the lateral surface Σ in R^{N+1} and then the $N+1$-dimensional boundary derivative concept f'_Σ.

9.4. Mean curvature H of the lateral time-space boundary

Assuming the moving domain "smooth enough", we consider the normal speed v chosen as $v = \langle V(t), \nabla b_{\Omega_t(V)} \rangle$ and

$$\frac{\partial}{\partial t}\left(\frac{v}{\sqrt{1 + v^2}}\right) = \frac{1}{(\sqrt{1 + v^2})^3} \; \frac{\partial}{\partial t} v.$$

But

$$\frac{\partial}{\partial t} v = \left\langle \frac{\partial}{\partial t} V, \nabla b \right\rangle + \left\langle \frac{\partial}{\partial t} \nabla b, \; V \right\rangle.$$

Now, we have

$$\frac{\partial}{\partial t} b_{\Omega_t(V)} = -\langle V(t), n_t \rangle o p_t,$$

where p_t is the projection onto the boundary $\Gamma_t(V) = \partial\Omega_t(V)$. Moreover,

$$\frac{\partial}{\partial t} \nabla b_{\Omega_t(V)} = -(\nabla_{\Gamma_t} \langle V(t), n_t \rangle) o p_t$$

and hence,

$$\frac{\partial}{\partial t} v = \left\langle \frac{\partial}{\partial t} V(t), n_t \right\rangle - \langle (\nabla_{\Gamma_t} \langle V(t), n_t \rangle) o p_t, V_{\Gamma_t} \rangle.$$

we obtain

Proposition 9.4.

$$\frac{\partial}{\partial t}\left(\frac{v}{\sqrt{1 + v^2}}\right) = \frac{1}{(\sqrt{1 + v^2})^3} \left\langle \frac{\partial}{\partial t} V(t), n_t \right\rangle - \langle (\nabla_{\Gamma_t} \langle V(t), n_t \rangle) o p_t, V_{\Gamma_t} \rangle. \tag{9.5}$$

On the other hand we have:

$$\mathrm{div}_{\Gamma_t}\left(\frac{1}{(\sqrt{1 + v^2})} n\right) = -\left\langle \nabla_{\Gamma_t} \frac{1}{\sqrt{1 + v^2}}, n \right\rangle + \frac{1}{(\sqrt{1 + v^2})^3} \; \mathrm{div}_{\Gamma_t} n.$$

so that we get

$$\mathrm{div}_{\Gamma_t}\left(\frac{1}{\sqrt{1 + v^2}} n\right) = -\frac{1}{(\sqrt{1 + v^2})^3} \langle \epsilon(V).n_t, n_t \rangle + \frac{H_t}{\sqrt{1 + v^2}},$$

where $\epsilon(V) = 1/2\,(DV + DV^*)$ is the deformation tensor. We consider the situation in which the field V verifies the following property:

$$V(t) = V(t) o p_t \;\; \text{in a neighbourhood of } \; \Gamma_t, \tag{9.6}$$

where p_t is the R^N projection mapping onto Γ_t ("horizontal" projection). Then we get:

$$p_t = I_d - b_{\Omega_t(V)} \nabla b_{\Omega_t(V)},$$

and

$$\frac{\partial}{\partial t} p_t = -\frac{\partial}{\partial t} b_{\Omega_t(V)} \nabla b_{\Omega_t(V)} - b_{\Omega_t(V)} \nabla (\frac{\partial}{\partial t} b_{\Omega_t(V)}).$$

The restriction to the boundary Γ_t leads to the distance $b_{\Omega_t(V)} = 0$ so the expressions simplify as follows (also we shall now denote by b_t that distance function):

$$\frac{\partial}{\partial t} p_t|_{\Gamma_t} = \langle V(t), n_t \rangle n_t,$$

and on the boundary $\Gamma_t(V)$ we get $DV(t).n_t = 0$.

9.4.1. Time-space mean curvature of the lateral boundary Σ.

Theorem 9.5. *Assume that the field V verifies for each t: $V(t) = V(t)op_t$. Then, on the boundary $\Gamma_t(V)$ the mean curvature $\mathbf{H} := \mathbf{Div}_\Sigma \nu$ is given by:*

$$\mathbf{H} = -\frac{1}{(\sqrt{1+v^2})^3} \left(\left\langle \frac{\partial}{\partial t} V, n_t \right\rangle - \langle \nabla_{\Gamma_t}(\langle V(t), n_t \rangle), V(t)_{\Gamma_t} \rangle \right) + \frac{1}{\sqrt{1+v^2}} H_t.$$

9.5. Lateral boundary derivative

9.5.1. Transverse horizontal vector field.
The normal component of any horizontal field $\tilde{Z} = (0, Z(t, x))$ is given by:

$$\langle \tilde{Z}, \nu \rangle = \frac{1}{\sqrt{1+v^2}} \langle Z, n_t \rangle.$$

If $f(\Sigma)$ is the restriction to the lateral boundary Σ of a function $F(t, x)$ defined over R^{N+1}, we get the (lateral) shape $N + 1$-dimensional boundary derivative $f'_\Sigma(\tilde{Z})$ in the direction of the horizontal field $\tilde{Z}$ as follows: $f'_\Sigma(\tilde{Z}) = \frac{\partial}{\partial \nu} F$

9.5.2. Lateral shape derivative f'_Σ.
We recall that (see [9], [24])

$$f'_\Sigma(\tilde{Z}) = (\frac{d}{ds}(f(\Sigma_s)o\mathbf{T}_s(\tilde{Z}))_{s=0} - \langle \nabla_\Sigma f(\Sigma), \tilde{Z}_\Sigma \rangle.$$

Notice that the operator ∇_Σ, as a tangential differential operator of the space time surface Σ, is itself a time-space manifold, and we get

$$f'_\Sigma(\tilde{Z}) = \dot{f}(\Sigma, \tilde{Z}) - \frac{vz}{1+v^2} \frac{\partial}{\partial t} f - \langle Z - \frac{z}{1+v^2} n_t, \nabla f \rangle.$$

Consider a given function $F \in C^1([0, \tau] \times \bar{D})$. In a first step we assume that F is zero in the neighborhood of $t = \tau$ so that the following derivative of the lateral boundary integral could be considered as derivative of the integral on the total boundary of the tube (as it will generate no term on the top $t = \tau$ of the tube).

Then the usual derivative expressions apply: we consider the derivative of the lateral integral.

$$\Sigma^s = \{\, (t, T_t(V + sW)(x)) \mid x \in \partial\Omega_0 \,\},$$

$$\frac{\partial}{\partial s}\Big|_{s=0}\left(\int_{\Sigma^s} F\, d\Sigma^s\right) = \int_{\Sigma}\left(\frac{\partial}{\partial \nu}F + \mathbf{H}_\Sigma F\right)\langle \mathbf{Z}, \nu\rangle_{R^{N+1}}\,)\, d\Sigma,$$

where $\mathbf{H}_\Sigma$ is the mean curvature of the lateral boundary of the tube. At each point $(t, x) \in \Sigma$ we have:

$$\langle \mathbf{Z}(t, x), \nu(t, x)\rangle_{R^{N+1}} = \frac{1}{\sqrt{1 + \langle V(t), n_t\rangle^2}}\,\langle Z(t), n_t\rangle.$$

Moreover,

$$\frac{\partial}{\partial \nu}F = \frac{1}{\sqrt{1 + (\langle V(t), n_t\rangle)^2}}\left(-\langle V(t), n_t\rangle \frac{\partial}{\partial t}F + \frac{\partial}{\partial n_t}F\right).$$

Then,

$$\frac{\partial}{\partial s}\Big|_{s=0}\left(\int_{\Sigma^s} F\, d\Sigma^s\right) = \int_{\Sigma}\left[\frac{1}{\sqrt{1 + v^2}}\left(-v\frac{\partial}{\partial t}F + \frac{\partial}{\partial n_t}F\right)\right. \tag{9.7}$$

$$\left(-\frac{1}{(\sqrt{1 + v^2})^3}\left(\left\langle \frac{\partial}{\partial t}V, n_t\right\rangle - \langle \nabla_{\Gamma_t} v, V(t)_{\Gamma_t}\rangle\right)\right.$$

$$\left.\left. +\frac{1}{\sqrt{1 + v^2}}H_t\right)F\right]\frac{1}{\sqrt{1 + v^2}}\langle Z, n\rangle_{R^N}\, d\Sigma.$$

Proposition 9.6. *Assume the vector field V in the canonical form $V(t) = V(t)\circ p_t$ in a neighborhood of the lateral boundary Σ and let $v = \langle V(t), n_t\rangle$ on Γ_t, then we have:*

$$\frac{\partial}{\partial s}\Big|_{s=0}\left(\int_{\Sigma^s} F\, d\Sigma^s\right) = \int_0^\tau \int_{\Gamma_t}\left[\frac{1}{\sqrt{1 + v^2}}\left(-v\frac{\partial}{\partial t}F + \frac{\partial}{\partial n_t}F\right)\right. \tag{9.8}$$

$$+ F\left(-\frac{1}{(\sqrt{1 + v^2})^3}\left(\left\langle \frac{\partial}{\partial t}V, n_t\right\rangle - \langle \nabla_{\Gamma_t} v, V(t)_{\Gamma_t}\rangle\right)\right.$$

$$\left.\left. +\frac{1}{\sqrt{1 + v^2}}H_t\right)\right]\langle Z, n\rangle_{R^N}\, d\Gamma_t\, dt.$$

9.5.3. Tube with minimal lateral boundary. In the specific case where $F = 1$ all the derivatives of F cancel and we have the derivative of the lateral surface of the tube:

$$\frac{\partial}{\partial s}\Big|_{s=0}\left(\int_{\Sigma^s} d\Sigma^s\right) = \int_{\Sigma}\left[-\frac{1}{(1 + v^2)^2}\left(\left\langle \frac{\partial}{\partial t}V, n_t\right\rangle - \langle \nabla_{\Gamma_t} v, V(t)_{\Gamma_t}\rangle\right)\right. \tag{9.9}$$

$$\left. +\frac{1}{1 + v^2}H_t\right]\langle Z, n\rangle_{R^N}\, d\Sigma.$$

The optimality condition for a minimal surface tube is easily obtained via the adjoint problem solution λ as

$$\frac{\partial}{\partial s}\bigg|_{s=0}\left(\int_{\Sigma^s} d\Sigma^s\right) = \int_{\Sigma} \lambda \,\langle W, n_t\rangle \,d\Sigma, \tag{9.10}$$

where λ solves:

$$\lambda(\tau) = 0, \tag{9.11}$$

$$-\frac{\partial}{\partial t}\lambda - \mathrm{div}(\lambda\, V) = -\frac{1}{(1+v^2)^2}\left(\left\langle\frac{\partial}{\partial t}V, n_t\right\rangle - \langle\nabla_{\Gamma_t} v,\, V(t)_{\Gamma_t}\rangle\right) + \frac{1}{1+v^2}\,H_t.$$

The optimality condition for a tube with minimal lateral surface is be

$$\boxed{-\frac{1}{(1+v^2)}\left(\left\langle\frac{\partial}{\partial t}V, n_t\right\rangle - \langle\nabla_{\Gamma_t} v,\, V(t)_{\Gamma_t}\rangle\right) + H_t = 0\,.} \tag{9.12}$$

10. Non-smooth tubes analysis

We denote by $\mathbf{H}^k$ the family of tubes ζ having a lateral boundary Σ piecewise C^k (in the precise sense of the previous section), with $\zeta = \chi_Q$. Let $\zeta \in \mathbf{H}^k$, we consider the $N+1$-dimensional perimeter

$$P_{I\times D}(Q) = \|\nabla_{t,x}\zeta\|_{M^1(I\times D)} = \int_0^1\int_{\Gamma_t}\sqrt{1+v^2}\,d\Gamma_t\,dt$$

$$\leq \int_0^1 P_D(\Omega_t)\,dt + \int_0^1\int_{\Gamma_t}|v(t)|\,d\Gamma_t\,dt.$$

Consider also the fact that

$$\left\langle\frac{\partial}{\partial t}\zeta,\, g\right\rangle_{\mathbf{M}(I\times D)\times C^0_{\mathrm{comp}}(I\times D)} = \int_0^1\int_{\Gamma_t}v\,g\,d\Gamma_t\,dt = \int_0^1\int_{\Omega_t}\mathrm{div}(g\,V)\,dx\,dt,$$

(where V is any smooth extension to $I\times D$ of v).

As

$$\int_0^1\int_{\Gamma_t}|v|\,d\Gamma_t\,dt = \left\|\frac{\partial}{\partial t}\zeta\right\|_{M^1(I\times D)}, \quad P_D(\Omega_t) = \|\nabla_x\zeta(t)\|_{M^1(D,R^N)}$$

we have:

$$\|\nabla_{t,x}\zeta\|_{M^1(I\times D)} \leq \left\|\frac{\partial}{\partial t}\zeta\right\|_{M^1(I\times D)} + \int_0^1\|\nabla_x\zeta(t)\|_{M^1(D,R^N)}\,dt. \tag{10.1}$$

We shall consider the weak closure of such smooth tubes ζ and verify that the estimate (10.1) still holds true on the closure:

Proposition 10.1. *let $\zeta_n \in \mathbf{H}^k$ be a sequence of tubes such that*

$$\left\|\frac{\partial}{\partial t}\zeta_n\right\|_{M^1(I\times D)} + \int_0^1\|\nabla_x\zeta_n(t)\|_{M^1(D,R^N)}\,dt \leq M\,. \tag{10.2}$$

Then there exists a subsequence (still denoted ζ_n) and ζ such that $\zeta_n \to \zeta$ strongly in $L^1(I \times D)$ (so that $\zeta = \zeta^2$) and:

$$||\nabla_{t,x}\zeta||_{M^1(I\times D)} \leq \liminf ||\frac{\partial}{\partial t}\zeta_n||_{M^1(I\times D)} + \int_0^1 ||\nabla_x\zeta_n(t)||_{M^1(D,R^N)} \, dt. \quad (10.3)$$

Proof. From (10.1) we get: $||\zeta_n||_{BV(I\times D)} \leq M + \text{meas}(D),$ so that the classical compact embedding of BV in L^1 leads to

$$||\nabla_{t,x}\zeta||_{M^1(I\times D,R^{N+1})} \leq \liminf \ ||\nabla_{t,x}\zeta_n||_{M^1(I\times D,R^{N+1})},$$

$$||\frac{\partial}{\partial t}\zeta||_{M^1(I\times D)} \leq \liminf ||\frac{\partial}{\partial t}\zeta_n||_{M^1(I\times D)},$$

$$\int_0^1 ||\nabla_x\zeta(t)||_{M^1(D,R^N)} \, dt \leq \liminf \int_0^1 ||\nabla_x\zeta_n(t)||_{M^1(D,R^N)} \, dt.$$

We introduce the weak closure $\mathbf{H}^*$ of $\mathbf{H}^k$ in $BV(I \times D)$:

$$\mathbf{H}^* = \{\zeta = \zeta^2 \in L^1(I \times D), \ s.t. \ \exists \zeta_n \in \mathbf{H}^k, \ \zeta_n \to \zeta \text{ in } L^1(I \times D) \ ,$$

$$\nabla_{t,x}\zeta_n \to \nabla_{t,x}\zeta \, (\text{wealky in }) M^1(I \times D) \ \}.$$

In order to extend the metric δ to that setting we would like to define the families $\mathbf{O}^*_{\Omega_0}$ (resp. $T^*(\Omega_1,\Omega_2)$) similar to $\mathbf{O}_{\Omega_0}$ (resp. $T(\Omega_1,\Omega_2)$). But the difficulty is that elements ζ in the closure $\mathbf{H}^*$ are not continuous so that the connection property cannot be defined. In order to recover that continuity on the closure we propose two directions following the two next Propositions 11.1 and 11.2.

11. Compactness result

Proposition 11.1. *Consider ζ_n bounded in $L^1(I, BV(D))$ together with $\frac{\partial}{\partial t}\zeta_n$ bounded in $L^p(I, M^1(D))$ for some $p > 1$. Then there exists a subsequence and an element*

$$\zeta \in L^1(I, BV(D)) \cap W^{1,1}(I, M^1(D)) \subset C^0(I, M^1(D))$$

such that ζ_n strongly converges to ζ in $L^1(I, L^1(D))$ with $\nabla\zeta \in L^p(I, M^1(D, R^N))$ verifying

$$||\zeta||_{L^1(I,BV(D))} \leq \liminf \ ||\zeta_n||_{L^1(I,BV(D))}$$

and

$$||\frac{\partial}{\partial t}\zeta||_{L^p(I,M^1(D))} \leq \liminf \ ||\frac{\partial}{\partial t}\zeta_n||_{L^p(I,M^1(D))}.$$

Proposition 11.2. *Consider ζ_n bounded in $L^1(I, BV(D))$ together with*

$$a.e.t \quad ||\frac{\partial}{\partial t}\zeta_n(t)||_{M^1(D)} \leq \theta(t) \text{ for some } \theta \in L^1_{\text{loc}}(0,1).$$

Then there exists a subsequence and an element

$$\zeta \in L^1(I, BV(D)) \cap W^{1,1}(I, M^1(D)) \subset C^0(I, M^1(D))$$

such that ζ_n strongly converges to ζ in $L^1(I, L^1(D))$ with $\nabla\zeta \in L^p(I, M^1(D, R^N))$ verifying

$$\|\zeta\|_{L^1(I, BV(D))} \le \liminf \, \|\zeta_n\|_{L^1(I, BV(D))}$$

and

$$\left\|\frac{\partial}{\partial t}\zeta\right\|_{L^1(I, M^1(D))} \le \liminf \, \left\|\frac{\partial}{\partial t}\zeta_n\right\|_{L^1(I, M^1(D))}.$$

Obviously, if the sequence verifies $(\zeta_n)^2 = \zeta_n$, then in the strong limit, we get $\zeta^2 = \zeta$.

At that point notice that, in both Propositions 11.1 and 11.2,

$$\zeta \in W^{1,1}(I, M^1(D)) \text{ implies } \zeta \in C^0(I, L^1(D)),$$

more precisely:

Proposition 11.3. *Let $\zeta(t,x) = \zeta^2(t,x)$, $a.e.(t,x) \in I \times D$; $\zeta \in W^{1,1}(I, M^1(D))$ then $\zeta \in C^0(I, L^1(D))$ and the mapping:*

$$t \in \bar{I} \to p(t) := \|\nabla_x\zeta(t)\|_{M^1(D,R^N)} \text{ is lower semi continuous.} \tag{11.1}$$

Proof. Let $t_k \to t$ as $k \to \infty$, as $W^{1,1}(I, M^1(D)) \subset C^0(I, M^1(D))$ we get $\zeta(t_k) \to \zeta(t)$ so that:

$$\forall g \in C^0_{\text{comp}}(D), \int_D (\zeta_k(t,x) - \zeta(t,x)) \, g(x) \, dx \to 0.$$

But there exists a subsequence (still denoted t_k) with $\zeta(t_k)$ weakly $L^2(D)$ convergent to some $\mu, 0 \le \mu \le 1$ over D. From the previous convergence it turns out that $\mu = \zeta(t) = \zeta^2 = \mu^2$. Then that convergence is also strong in $L^2(D)$, hence in $L^1(D)$ (as $\zeta^2 = \zeta$). Being defined as supremum of continuous terms:

$$p(t) = \sup_{\{\, g \in C^1_{\text{comp}}(D,R^N), \, \|g(x)\|_{R^N} \le 1 \,\}} \int_D \zeta(t,x) \, \text{div}_x \, g \, dx$$

then p is l.s.c. We introduce the weak closures $\mathbf{H}_p^{c,*}$ and $\mathbf{H}_\theta^{c,*}$ of $\mathbf{H}^k$:

$$\mathbf{H}_p^{c,*} = \{\zeta = \zeta^2 \in \mathbf{H}^c \cap \mathbf{H}^*, \, s.t. \, \exists \zeta_n \in \mathbf{H}^k, \, \zeta_n \to \zeta \text{ in } L^1(I \times D) \, ,$$

$$\nabla_{t,x}\zeta_n \to \nabla_{t,x}\zeta \, , \, \sigma M^1(I \times D),$$

$$\text{with } \frac{\partial}{\partial t}(\zeta_n - \zeta) \to 0, \, \sigma L^P(I, M^1(D)) \, \}$$

$$\mathbf{H}_\theta^{c,*} = \{\zeta = \zeta^2 \in \mathbf{H}^c \cap \mathbf{H}^*, \, s.t. \, \exists \zeta_n \in \mathbf{H}^k, \, \zeta_n \to \zeta \text{ in } L^1(I \times D) \, ,$$

$$\nabla_{t,x}\zeta_n \to \nabla_{t,x}\zeta \text{ (weakly in) } M^1(I \times D),$$

$$\text{with } a.e.t \in I, \, \|\frac{\partial}{\partial t}\zeta_n(t)\|_{M^1(D)} \le \theta(t) \, \}.$$

12. Fully Eulerian metric spaces

As soon as the speed vector field V verifies the assumption of Theorem 6.4, there is a unique tube associated to V. Thus then we have an application $V \to \zeta_V$, and with such regularity on V we can revisit the complete metric d: the non-differentiable perimeter and curvature terms that we were obliged to introduce in order to apply the compactness theorems are not any more necessary. From the previous tube analysis we consider several interesting choices for the spatial regularity of the speed vector field (together with its divergence field). Let E be a closed subspace in $BV(D) \cap \mathbf{E}^{1,1}$ such that any element $V \in E$ verifies the assumptions of Theorem 6.4. A first example is, when working with prescribed volume for the moving domain,

$$E_0 = \{\, V \in BV(D, R^N) \cap \mathbf{E}^{1,1}, \; s.t. \; \operatorname{div} V = 0 \; a.e. \; (t,x) \in I \times D \;\}$$

V be a free divergence vector field with $\operatorname{div} V = 0$, $V \in L^1(I, E_0))$, where $E = BV(D, R^N)$ or any closed subspace (for example $E = \{ V \in H_0^1(D, R^N)$, $s.t.\operatorname{div} V = 0\}$). An *obvious* metric is to consider the set

$$\mathbf{V}(\Omega_1, \Omega_2) = \{ V \in \mathbf{E}^{1,1} \; s.t. \, V, \; \operatorname{div} V \in L^p(I, E_0), \; s.t. \; \zeta_0 = \chi_{\Omega_1}, \; \zeta(1) = \chi_{\Omega_2} \}$$

$$\delta_{E_0}(\Omega_1, \Omega_2) = \operatorname{Inf}_{V \in \mathbf{V}(\Omega_1, \Omega_2)} \int_0^1 \|V(t)\|_{E_0} \, dt. \tag{12.1}$$

As V is divergence free the previous boundedness assumption on the divergence are verified and to each V a tube ζ_V is associated trough the convection. Then following the same proof we get the

Proposition 12.1. *Let E be any subspace of $BV(D, R^N) \cap \mathbf{E}^{1,1}$, such that any element V satisfies to assumptions of theorem 6.4. Then equipped with δ_E the family $\mathbf{O}_{\Omega_0}^E$ is a metric space.*

$$d_{E_0}(\Omega_1, \Omega_2) = \operatorname{Inf}_{V \in \mathbf{V}(\Omega_1, \Omega_2)} \quad \|V\|_{L^1(I, E_0)} \; + \; \left\| \frac{\partial}{\partial t} V \right\|_{L^1(I, M^1(D, R^N))}. \tag{12.2}$$

Theorem 12.2. *Let E be any subspace of $BV(D, R^N) \cap \mathbf{E}^{1,1}$, such that any element V satisfies to assumptions of Theorem 6.4. Then equipped with d_E the family $\mathbf{O}_{\Omega_0}^E$ is a complete metric space.*

12.1. Geodesic

The previous transverse tube perturbation will apply. In that setting we are concerned with vector fields $Z(s, t, x) \in R^N$ such that $Z(s, 0, x) = Z(s, 1, x) = 0$, so that the extremities of the perturbed tube are preserved. The previous study for the transverse field implies that for given such a vector field Z, with $\operatorname{div}_x Z(s, t, x) = 0$ we get the admissible perturbation of the field V in the following form $V + sW(s, t, x)$ with

$$W(s, t, x) = (\partial/\partial t)Z(s, t, x) + [Z, V]$$

more precisely define the Lipschitz-continuous connecting set

$$\mathbf{V}^{1,\infty}(\Omega_1, \Omega_2) = \{\, V \in L^1(I, W^{1,\infty}(D, R^N)) \cap \mathbf{E}^{1,1}, \; s.t. \; \zeta_V \in \bar{\mathbf{T}}(\Omega_1, \Omega_2) \;\}.$$

And the set of smooth transverse vector fields:

$$\mathbf{Z} = \{\ Z(t,x) \in C^{\infty}_{\mathrm{comp}}(I \times D, R^N)\ \}.$$

(Notice that such Z verifies $Z(0,.) = Z(1,.) = 0$ on D.)

Proposition 12.3. *Let $V \in \mathbf{V}(\Omega_1,\Omega_2)$ and $Z(t,x) \in \mathbf{Z}$. The transformation $\mathbf{T} = T_s(Z)oT_t(V)$ maps $\Omega_t(V)$ onto $\Omega_t^s := T_s(Z)(\Omega_t(V))$ so that*

$$\forall s,\ \forall Z,\ V^s(t,x)$$

$$= \frac{\partial}{\partial t}\mathbf{T}o\,\mathbf{T}^{-1} = \left(\frac{\partial}{\partial t}T_s(Z(t))\right) + DT_s(Z(t)).V(t)\ )oT_s(Z(t))^{-1} \in \mathbf{V}^{1,\infty}(\Omega_1,\Omega_2).$$

We obtain:

Lemma 12.4.

$$\frac{\partial}{\partial s}V^s(t,x)|_{s=0} = \frac{\partial}{\partial t}Z(t) + [Z(t),\,V(t)]. \tag{12.3}$$

Corollary 12.5. *Consider a functional $\mathbf{J}(V) = j(\zeta_V)$ and let $\bar V$ be a minimizing element of $\mathbf{J}$ on $\mathbf{V}(\Omega_1,\Omega_2)$ then we have*

$$\forall Z \in \mathbf{Z},\ \frac{\partial}{\partial s}\mathbf{J}(\bar V^s)_{s=0} = J'\left(\bar V;\ \left(\frac{\partial}{\partial s}V^s\right)_{s=0}\right)$$

$$= \mathbf{J}'\left(\bar V;\ \frac{\partial}{\partial t}Z(t) + [Z(t),\,V(t)]\right)\ \geq\ 0. \tag{12.4}$$

That variational principle extends to vector field $V \in E$ for which the flow mapping $T_t(V)$ is poorly defined. The element $\zeta_V \in \mathbf{H}^c$ is uniquely defined. For any $Z \in \mathbf{Z}$ we have $\zeta_V^s := \zeta_V oT_s(Z)^{-1} \in \bar{\mathbf{T}}(\Omega_1,\Omega_2)$. Moreover we have

Proposition 12.6. $\zeta_V^s = \zeta_{V^s}$ *with*

$$V^s(t,.) := -DT_s^{-1}(-Z(t)).(\,V(t)oT_s(Z(t))^{-1}) - \frac{\partial}{\partial t}T_s(-Z(t))\,).$$

In other words:

$$\frac{\partial}{\partial t}\zeta + \nabla\zeta.V = 0\ \ implies\ \ \frac{\partial}{\partial t}(\zeta oT_s(Z(t))^{-1}) + \nabla(\zeta oT_s(Z(t))^{-1}).V^s = 0.$$

It can also be verified that the expression (12.3) for the derivative of the field still holds true so that the variational principle (12.4) is valid for any functional $\mathbf{J}$ minimized over the Lipschitzian connecting family $\mathbf{V}^{1,\infty}(\Omega_1,\Omega_2)$.

And more generally, without assuming V in E we have:

Proposition 12.7. *Let $(\zeta,V) \in \mathbf{T}^{p,q}(\Omega_1,\Omega_2)$, then for all $s > 0$ and $Z \in \mathbf{Z}$ we have:*

$$(\zeta oT_s(Z)^{-1},\,V^s) \in \mathbf{T}^{p,q}(\Omega_1,\Omega_2).$$

Notice that, in order to get a differentiable metric we could consider

$$\tilde d(\Omega_1,\Omega_2) = \mathrm{Inf}_{V \in \mathbf{V}(\Omega_1,\Omega_2)}\ \int_0^1 \left(\ \|V(t)\|_{H_0^1 \cap E_0} + \|\frac{\partial}{\partial t}V\|_{L^2(D)}\right)\,dt.$$

Equipped with $\tilde{d}$, $\mathbf{O}_{\Omega_0}$ would be a complete metric space but $\tilde{d}$ fails to be a metric because of the triangle axiom.

The advantage is that now the associated functional is differentiable with respect to V and we can apply the previous variational principle with transverse vector field Z.

Let $\bar{V}$ be a minimizer in $\mathbf{V}(\Omega_1, \Omega_2)$ for $\tilde{d}(\Omega_1, \Omega_2)$. Then $\forall Z \in \mathbf{Z}$ we have

$$\int_0^1 \{ \|V(t)\|^{-1} \langle V(t), Z_t + [Z,V] \rangle + |V'(t)|^{-1} ((V'(t)(Z_t + Z, V)')) \} dt = 0,$$

where $\langle, \rangle$ is the $H_0^1(D, R^N)$ inner product while $((,))$ is the $L^2(D, R^N)$. In order to recover a differentiable complete metric we introduce again the constraint on the perimeter as in the beginning and set

$$p \geq 1, \quad \delta_{H^1, p}(\Omega_1, \Omega_2) = \operatorname{Inf}_{V \in \mathbf{V}(\Omega_1, \Omega_2)} \int_0^1 \|V(t)\|^p_{H_0^1 \cap E_0} dt. \tag{12.5}$$

The optimality condition is:

$$\forall Z \in \mathbf{Z} \; s.t. \operatorname{div} Z = 0 \text{ and } \int_0^1 \int_{\Gamma_t} H(t) \langle Z(t), n_t \rangle d\Gamma_t \, dt = 0,$$

$$p \int_0^1 \|V(t)\|^{p-2} \langle V(t), Z_t + [Z,V] \rangle dt = 0.$$

From (7.9), that last condition can be rewritten as

$$\langle \|V(t)\|^{p-2} V(t), \; H_V.Z \rangle = 0.$$

The adjoint operator (for free divergence vector field V) is given by:

$$H_V^*.\Lambda := -\frac{\partial}{\partial t}\Lambda - D\Lambda.V - D^*V.\Lambda. \tag{12.6}$$

Finally, the second condition turns to be:

$$\langle H_V^*.(\|V(t)\|^{p-2} V(t)), \; Z \rangle = 0$$

$$\exists c(t), P \quad s.t. \quad \frac{\partial}{\partial t}(\|V(t)\|^{p-2} V(t)) + \|V(t)\|^{p-2} (DV(t).V + D^*V.V(t))$$

$$= \nabla P + c \, \chi_{\Gamma_t} \operatorname{div}_{\Gamma_t}(n_t) \, n_t.$$

That is,

$$(p-2)\|V\|^{p-4}((V, \frac{\partial}{\partial t}V))V + \|V(t)\|^{p-2} (\frac{\partial}{\partial t}V + DV(t).V + D^*V.V(t))$$

$$= c \, \chi_{\Gamma_t} \operatorname{div}_{\Gamma_t}(n_t) \, n_t, \tag{12.7}$$

which can be written as (with the notations $\bar{V} = \|V\|^{-1} V$, $\Pi = P - 1/2|V|^2$):

$$\operatorname{div} V = 0,$$

$$\boxed{\frac{\partial}{\partial t}V + (p-2)\left(\left(\frac{\partial}{\partial t}V, \bar{V}\right)\right) \bar{V} + DV.V = \nabla\Pi + c(t)\|V\|^{2-p} \chi_{\Gamma_t} \operatorname{div}_{\Gamma_t}(n_t) \, n_t.}$$

$$\tag{12.8}$$

13. Level set formulation for the tube shape metric

13.1. Shape Gradient approximation

From the shape derivative structure theorem, we know that any shape gradient takes the following form

$$\int_{\partial\Omega} g \, \langle V(0), n \rangle \, d\Gamma, \tag{13.1}$$

where g is the so-called *density gradient*, a measure on the boundary, and $v = \langle V(0), n \rangle$ is the normal component of the vector field. In the level set setting, assume that $\Omega = \{ x \in D \mid \Phi(t, x) > 0 \}$ then $V = -\frac{\partial}{\partial t}\Phi \, \frac{\nabla_x \Phi}{||\nabla_x \Phi||}$, so that obviously, $v = -\frac{\partial}{\partial t}\Phi / ||\nabla_x \Phi||$.

From Federer measure decomposition theorem we have:

$$\int_{\mathbf{U}_h(\Gamma)} F(x) \, dx = \int_{-h}^{+h} \left(\int_{\Phi^{-1}(z)} \frac{F}{||\nabla_x \Phi||} \, d\Gamma \right) dz,$$

where

$$\mathbf{U}_h^{\Phi}(\Gamma) = \{ x \in D \mid |\Phi(x)| < h \, \}.$$

Assuming the mapping $z \in (-h, +h) \rightarrow (\int_{\Phi^{-1}(z)} \frac{F}{||\nabla_x \Phi||} \, d\Gamma \,)$ to be continuous we obtain

$$\int_{\Gamma} \frac{F(x)}{||\nabla_x \Phi(x)||} \, d\Gamma(x) = \frac{1}{2h} \int_{\mathbf{U}_h(\Gamma)} F(x) \, dx \, + \, o(1), \quad h \to 0.$$

Applying that approximation in the previous shape derivative we obtain, for any smooth enough extension $\tilde{g}$ of g to the neighborhood $\mathbf{U}_h^{\Phi}(\Gamma)$:

$$\int_{\partial\Omega} g \, \langle V(0), n \rangle \, d\Gamma = - \frac{1}{2h} \int_{\{ x \in D, \, |\Phi(x)| < h \, \}} \tilde{g}(x) \, | \frac{\partial}{\partial t}\Phi(t, x)| \, dx \, + \, o(1), \quad h \to 0.$$

The point being that the denominator $||\nabla_x \Phi(t)||$ has been eliminated.

13.2. h-scale metric d_h

Two open subsets $\Omega_i \subset D$ being given, for $i = 1, 2$, (resp. quasi open subsets) we associate two continuous functions (resp. elements of $H^1(D)$) $\phi_i \in C^o(\bar{D})$ such that $x \in \Omega_i$ iff $\phi_i(x) > 0$, $x \in \partial\Omega_i$ iff $\phi(x) = 0$ (then $x \in D \setminus \bar{\Omega}$ iff $\phi(x) < 0$). We consider the following closed convex set

$$\mathbf{K}(\Omega_1, \Omega_2) := \left\{ \psi(t, x) \in L^2(0, 1, H^1(D)), \, \int_0^1 \psi(t, x) \, dt = \phi_1(x) - \phi_2(x) \right\}. \tag{13.2}$$

Then for any element $\psi \in \mathbf{K}(\Omega_1, \Omega_2)$ we consider the level set function

$$\Phi(t, x) := \phi_1(x) \, + \, \int_0^t \psi(s, x) \, ds,$$

and the moving domain

$$\Omega_t := \{ \, x \in D, \quad \Phi(t, x) > 0 \, \} \in \bar{\mathbf{T}}(\Omega_1, \Omega_2) \,.$$

Then we set

$$d_h(\Omega_1,\Omega_2):=\mathrm{Inf}_{\{\psi\in\mathbf{K}(\Omega_1,\Omega_2)\}}\int_0^1\left[-\frac{1}{2h}\int_{\{|\Phi(t,x)|<h\}}|\psi(t,x)|\,dx+\|\psi(t)\|_{H^1(D)}\right]dt.\tag{13.3}$$

That metric turns to be numerically tractable and several experiments are performed at INRIA with J. Picard and L. Blanchard. The choice of h has to be tuned to the pixels density.

References

[1] M.C. Delfour and J.P. Zolésio. *Oriented distance function and its evolution equation for initial sets with thin boundary.* SIAM J. Control Optim. 42 (2004), no. 6, 2286–2304

[2] M. Cuer and J.P. Zolésio. *Control of singular problem via differentiation of a min-max.* Systems Control Lett. 11 (1988), no. 2, 151–158.

[3] D. Bucur and J.P. Zolésio. *Free Boundary Problems and Density Perimeter.* J. Differential. Equations 126(1996), 224–243.

[4] D. Bucur and J.P. Zolésio. *Boundary Optimization under Pseudo Curvature Constraint.* Annali della Scuola Normale Superiore di Pisa, IV, XXIII (4), 681–699, 1996.L

[5] R. Dziri and J.P. Zolésio. *Dynamical Shape Control in Non-cylindrical Navier-Stokes Equations.* J. convex analysis, vol. 6, 2, 293–318, 1999.

[6] N. Gomez and J.P. Zolésio. *Shape sensitivity and large deformation of the domain for norton-hoff flow.* Volume 133 of Int. Series of Num. Math., pages 167–176, 1999.

[7] J.P. Zolésio. *Identification de domaine par déformations. Thèse de doctorat d'état,* Université de Nice, 1979.

[8] J.P. Zolésio. In Optimization of Distributed Parameter structures, vol. II, (E. Haug and J. Céa eds.), Adv. Study Inst. Ser. E: Appl. Sci., 50, Sijthoff and Nordhoff, Alphen aan den Rijn, 1981.
  i) *The speed method for Shape Optimization.* 1089–1151.
  ii) *Domain Variational Formulation for Free Boundary Problems,* 1152–1194.
  iii) *Semiderivative of repeated eigenvalues,* 1457–1473.

[9] J. Sokolowski and J.P. Zolésio. *Introduction to shape optimization,* sci, 16, Springer Verlag, Heidelberg, N.Y., 1991

[10] B. Kawohl, O. Pironneau, L. Tartar, J.P. Zolésio. *Optimal Shape Design,* (yellow) Lecture Notes in Mathematics, 1740, Springer Verlag, Heidelberg, N.Y., 1998.

[11] J.P. Zolésio. *Variational Principle in the Euler Flow.* In G. Leugering, editor, *Proceedings of the IFIP-WG7.2 conference, Chemnitz,* volume 133 of *Int. Series of Num. Math.,* 1999.

[12] J.P. Zolésio. *Shape Differential with Non Smooth Field.* In Computational Methods for Optimal Design and Control. J. Borggard, J. Burns, E. Cliff and S. Schreck eds., volume 24 of *Progress in Systems and Control Theory,* pp. 426–460, Birkhäuser, 1998.

[13] J.P. Zolésio. *Weak set evolution and variational applications* in *Shape optimization and optimal design*, lecture notes in pure and applied mathematics, vol. 216, pp. 415–442, Marcel Dekker, N.Y., 2001.

[14] M.C. Delfour and J.P. Zolésio. *Structure of shape derivatives for non smooth domains*, Journal of Functional Analysis, 1992, 104.

[15] P. Cannarsa, G. Da Prato and J.P. Zolésio. *The damped wave equation in a moving domain*, Journal of Differential Equations, 1990, 85, 1–16.

[16] M.C. Delfour and J.P. Zolésio. *Shape analysis via oriented distance functions*, J. Funct. Anal., 1994, 123, 1–56.

[17] R. Dziri and J.P. Zolésio. *Dynamical shape control in non-cylindrical hydrodynamics*, Inverse Problem, 1999, 15, 1, 113–122.

[18] M.C. Delfour and J.P. Zolésio. *Shape sensitivity analysis via min max differentiability*, SIAM J. Control Optim., 1988, 26, 4, 834–862.

[19] G. Da Prato and J.P. Zolésio. *Dynamical Programming for non Cylindrical Parabolic Equation*, Sys. Control Lett., 11, 1988.

[20] G. Da Prato and J.P. Zolésio. *Existence and Control for wave equation in moving domain*, L.N.C.I.S Springer Verlag, 144, 1988.

[21] P. Acquistapace. *Boundary control for non-autonomous parabolic equations in non-cylindrical domains* in *Boundary control and variation* (Sophia Antipolis, 1992), 1–12, L. N. P. A. Math., 163, Dekker, New York, 1994.

[22] L. Ambrosio. *Lecture notes on optimal transport problems. Mathematical aspects of evolving interfaces* (Funchal, 2000), 1–52, Lecture Notes in Math., 1812, Springer, Berlin, 2003. 49Q20 (49-02)

[23] K. Burdzy, Z. Chen and J. Sylvester. *The heat equation and reflected Brownian motion in time-dependent domains.* Ann. Probab. 32 (2004), no. 1B, 775–804.

[24] M. Delfour and J.P. Zolésio. *Shape and Geometry* Advances in Design and Control, 04, SIAM, 2001.

[25] J.P. Zolésio. *Set Weak Evolution and Transverse Field, Variational Applications and Shape Differential Equation* INRIA report RR-464, 2002. (http://www-sop.inria.fr/rapports/sophia/RR-464)

[26] M. Moubachir and J.P. Zolésio. *Moving Shape Analysis and Control: application to fluid structure interaction.* Pure and Applied Mathematics series, CRC, 2006.

[27] M.C. Delfour and J.P. Zolésio. *Structure of shape derivatives for non smooth domains. Journal of Functional Analysis*, 104, 1992.

[28] M.C. Delfour and J.P. Zolésio. *Shape analysis via oriented distance functions. Journal of Functional Analysis*, 123, 1994.

[29] F.R. Desaint and J.P. Zolésio. *Manifold derivative in the laplace-beltrami equation. Journal of Functionnal Analysis*, 151(1): 234, 269, 1997.

[30] J.P. Zolésio. *Introduction to shape optimization and free boundary problems.* In Michel C. Delfour, editor, *Shape Optimization and Free Boundaries*, volume 380 of *NATO ASI, Series C: Mathematical and Physical Sciences*, pages 397, 457, 1992.

[31] J.P. Zolésio. *Shape Topology by Tube Geodesic.* In Information Processing: Recent Mathematical Advances in Optimization and Control. Presses de l'Ecole des Mines de Paris, pages 185–204, 2004.

[32] Raja Dziri, J.P. Zolésio *Tube Derivative of Non-Cylindrical Shape Functionals and Variational Formulations* in proc. ifip7.2 conf. Houston, Dec. 2004, R. Glowinski, J.P. Zolésio eds., CRC press book, 2006.

[33] J.P. Zolésio. *Tubes Analysis* in proc. ifip7.2 conf. Houston, Dec. 2004, R. Glowinski, J.P. Zolésio eds., CRC press book, 2006.

[34] J.P. Zolésio *Shape Stabilization of Flexible Structures.* Lecture notes in Control and Information Sciences, vol. 75 *Distributed Parameter Systems* ed. F. Kappel, K. Kunish, W. Schapacher, proc. of the 2nd Inter. conf. in Vorau (Austria), Springer-Verlag, Berlin-Heidelberg-N.Y., 1985.

[35] C. Truchi, J.P. Zolésio *Shape stabilization of wave equation* (read) Lecture Notes in Control and Information Sciences, vol. 100 *Boundary Control and Boundary Variations*M, ed. J.P. Zolésio, proc. ifip conf. June 86 Nice (France), Springer-Verlag, Berlin-Heidelberg-N.Y., 1987.

[36] J.P. Zolésio. *Shape Formulation for Free Boundary Problem with Non Linearized Bernoulli Condition* (read) Lecture Notes in Control and Information Sciences, vol 178 *Boundary Control and Boundary Variations*M, ed. J.P. Zolésio, proc. ifip conf. June 1991, Sophia Antipolis (France), Springer-Verlag, Berlin-Heidelberg-N.Y., pp. 362–392, 1992.

[37] J.P. Zolésio. *Weak shape formulation of free boundary problems. Ann. Scuola Norm. Sup. Pisa Cl. Sci.* (4), 21(1):11–44, 1994.

[38] J.P. Zolésio. *Numerical algorithms and existence result for a Bernoulli-like steady free boundary problem. Large Scale Systems*, 6(3):263–278, 1984.

[39] J.P. Zolésio. *Some results concerning free boundary problems solved by domain (or shape) optimization of energy.* In *Modelling and inverse problems of control for distributed parameter systems (Laxenburg, 1989)*, volume 154 of *Lecture Notes in Control and Inform. Sci.*, pages 161–170. Springer, Berlin, 1991.

[40] J.P. Zolésio. *Shape differential equation with a non-smooth field.* In *Computational methods for optimal design and control (Arlington, VA, 1997)*, volume 24 of *Progr. Systems Control Theory*, pages 427–460. Birkhäuser Boston, Boston, MA, 1998.

[41] Dorin Bucur and Jean-Paul Zolésio. *Flat cone condition and shape analysis.* In *Control of partial differential equations (Trento, 1993)*, volume 165 of *Lecture Notes in Pure and Appl. Math.*, pages 37–49. Dekker, New York, 1994.

[42] Dorin Bucur and Jean-Paul Zolésio. *Optimisation de forme sous contrainte capacitaire. C. R. Acad. Sci. Paris Sér. I Math.*, 318(9):795–800, 1994.

[43] Dorin Bucur and Jean-Paul Zolésio. *N-dimensional shape optimization under capacitary constraint. J. Differential Equations*, 123(2):504–522, 1995.

[44] V. Sverak. *On optimal shape design.* J. Math. Pures Appl. (9) 72 (1993), no. 6, 537–551.

[45] D. Bucur, INLN preprint, 1994, Sophia Antipolis, France.

[46] J. Cooper and W.A. Strauss. *Energy boundedness and decay of waves reflecting off a moving obstacle.* Indiana Univ. Math. J. 25, 671–690 (1976).

Jean-Paul Zolésio
CNRS/ INRIA, 2004 Route des Lucioles
F-06902 Sophia Antipolis, France
e-mail: `jean-paul.zolésio@sophia.inria.fr`

International Series of Numerical Mathematics (ISNM)

International Series of Numerical Mathematics (ISNM)

Edited by

Karl-Heinz Hoffmann, TU München, Germany

Hans D. Mittelmann, Arizona State University, Tempe, USA

ISNM is open to all aspects of numerical mathematics. Some of the topics of particular interest include free boundary value problems for differential equations, phase transitions, problems of optimal control and optimization, other nonlinear phenomena in analysis, nonlinear partial differential equations, efficient solution methods, bifurcation problems and approximation theory. When possible, the topic of each volume is discussed from three different angles, namely those of mathematical modeling, mathematical analysis, and numerical case studies.

ISNM 136: Steinbach, J., A Variational Inequality Approach to Free Boundary Problems with Applications in Mould Filling (2002). ISBN 978-3-7643-6582-0

ISNM 135: Hillermeier, C., Nonlinear Multiobjective Optimization. A Generalized Homotopy Approach (2001). ISBN 978-3-7643-6498-4

ISNM 134: Hoffmann, K.-H. / Tang, Q., Ginzburg-Landau Phase Transition Theory and Superconductivity (2001). ISBN 978-3-7643-6486-1

ISNM 133: Hoffmann, K.-H. / Leugering, G. / Tröltzsch, F. (Eds.), Optimal Control of Partial Differential Equations (1999). ISBN 978-3-7643-6151-8

ISNM 132: Müller, M.W. / Buhmann, M.D. / Mache, D.H. / Felten, M. (Eds.), New Developments in Approximation Theory (1999). ISBN 978-3-7643-6143-3

ISNM 131: Gautschi, W. / Golub, G.H. / Opfer, G. (Eds.), Applications and Computation of Orthogonal Polynomials (1999). ISBN 978-3-7643-6137-2

ISNM 130: Fey, M. / Jeltsch, R. (Eds.), Hyperbolic Problems: Theory, Numerics, Applications. Volume II (1999). ISBN 978-3-7643-6087-0

ISNM 129: Fey, M. / Jeltsch, R. (Eds.), Hyperbolic Problems: Theory, Numerics, Applications. Volume I (1999). ISBN 978-3-7643-6080-1

ISNM 129/130: (Set) ISBN 978-3-7643-6123-5

ISNM 128: Meyer-Spasche, R., Pattern Formation in Viscous Flows. The Taylor-Couette Problem and Rayleigh-Bénard Convection (1999). ISBN 978-3-7643-6047-4

ISNM 127: Gerke, H.H. / Hornung, U. / Kelanemer, Y. / Slodicka, M., Optimal Control of Soil Venting: Mathematical Modeling and Applications (1999). ISBN 978-3-7643-6041-2

ISNM 126: Desch, W. / Kappel, F. / Kunisch, K. (Eds.), Control and Estimation of Distributed Parameter Systems (1998). ISBN 978-3-7643-5835-8

ISNM 125: Nürnberger, G. / Schmidt, J.W. / Walz, G., Multivariate Approximation and Splines (1997). ISBN 978-3-7643-5654-5

ISNM 124: Schmidt, W. / Heier, K. / Bittner, L. / Bulirsch, R. (Eds.), Variational Calculus, Optimal Control and Applications (1998). ISBN 978-3-7643-5906-5

ISNM 123: Bandle, C. / Everitt, W.N. / Losonczi, L. / Walter, W. (Eds.), General Inequalities 7 (1997). ISBN 978-3-7643-5722-1

ISNM 122: Khludnev, A.M. / Sokolowski, J., Modelling and Control in Solid Mechanics (1997). ISBN 978-3-7643-5238-7

ISNM 121: Jeltsch, R. / Mansour, M. (Eds.), Stability Theory. Hurwitz Centenary Conference (1996). ISBN 978-3-7643-5474-9

ISNM 120: Hackbusch, W., Integral Equations. Theory and Numerical Treatment (1995). ISBN 978-3-7643-2871-9

ISNM 119: Zahar, R. (Eds.), Approximation and Computation. A Festschrift in Honor of Walter Gautschi (1995). ISBN 978-0-8176-3753-8

ISNM 118: Desch, W. / Kappel, F. / Kunisch, K. (Eds.), Control and Estimation of Distributed Parameter Systems: Nonlinear Phenomena (1994). ISBN 978-3-7643-5098-7

ISNM 117: Bank, R.E. / Bulirsch, R. / Gajewski, H. / Merten, K. (Eds.), Mathematical Modelling and Simulation of Electrical Circuits and Semiconductor Devices (1994). ISBN 978-3-7643-5053-6

ISNM 116: Hemker, P.W. / Wesseling, P. (Eds.), Multigrid Methods IV (1994). ISBN 978-3-7643-5030-7

ISNM 115: Bulirsch, R. / Kraft, D., Computational Optimal Control (1994). ISBN 978-3-7643-5015-4

ISNM 114: Douglas, J. / Hornung, U. (Eds.), Flow in Porous Media (1993). ISBN 978-3-7643-2949-5

ISNM 113: Quartapelle, L., Numerical Solution of the Incompressible Navier-Stokes Equations (1993). ISBN 978-3-7643-2935-8

ISNM 112: Brass, H. / Hämmerlin, G. (Eds.), Numerical Integration IV (1993). ISBN 978-3-7643-2922-8

ISNM 111: Bulirsch, R. / Miele, A. / Stoer, J. / Well, K., Optimal Control. Calculus of Variations, Optimal Control Theory and Numerical Methods (1993). ISBN 978-3-7643-2887-0

ISNM 110: Hörnlein, H.R. / Schittkowski, K., Software Systems for Structural Optimization (1993). ISBN 978-3-7643-2836-8